Lecture Notes on Data Engineering and Communications Technologies

Volume 107

Series Editor

Fatos Xhafa, Technical University of Catalonia, Barcelona, Spain

The aim of the book series is to present cutting edge engineering approaches to data technologies and communications. It will publish latest advances on the engineering task of building and deploying distributed, scalable and reliable data infrastructures and communication systems.

The series will have a prominent applied focus on data technologies and communications with aim to promote the bridging from fundamental research on data science and networking to data engineering and communications that lead to industry products, business knowledge and standardisation.

Indexed by SCOPUS, INSPEC, EI Compendex.

All books published in the series are submitted for consideration in Web of Science.

More information about this series at https://link.springer.com/bookseries/15362

Zhengbing Hu · Sergey Petoukhov ·
Matthew He
Editors

Advances in Artificial Systems for Medicine and Education V

Springer

Editors
Zhengbing Hu
International Center of Informatics
and Computer Science, Faculty
of Applied Mathematics
National Technical University of Ukraine
“Kyiv Polytechnic Institute”
Kyiv, Ukraine

Sergey Petoukhov
Laboratory of Biomechanical Systems
Mechanical Engineering Research Institute
of the Russian Academy of Sciences
Moscow, Russia

Matthew He
Halmos College of Arts and Sciences
Nova Southeastern University
Ft. Lauderdale, FL, USA

ISSN 2367-4512 ISSN 2367-4520 (electronic)
Lecture Notes on Data Engineering and Communications Technologies
ISBN 978-3-030-92536-9 ISBN 978-3-030-92537-6 (eBook)
https://doi.org/10.1007/978-3-030-92537-6

This Springer imprint is published by the registered company Springer Nature Switzerland AG
The registered company address is: Gewerbestrasse 11, 6330 Cham, Switzerland

Preface

The creation of artificial intelligence systems and their applications in a variety of industries is one of modern science and technology's most pressing challenges. One of these fields is medical engineering, where its application aims to improve the effectiveness of identifying and treating various ailments. The rapid advancement of artificial intelligence systems necessitates an intensified training program for an increasing number of relevant individuals. Simultaneously, artificial intelligence systems have tremendous potential for application inside education technologies itself to improve the quality of specialized training by taking into account the specialists' personal qualities and the advent of new computing gadgets.

Scientists are attempting to replicate the hereditary intellectual capacities of humans and other biological organisms using digital systems of artificial intelligence. A thorough examination of genetic systems and inherited biological processes may expose living nature's bio-information patents and provide new avenues for developing increasingly powerful ways of artificial intelligence. As a result, intense development of bio-mathematical research of genetic and sensory motor systems is necessary, based on current achievements in mathematics, computer and quantum informatics, biology and physics, among others. In other words, the research of genetic systems and the development of artificial intelligence approaches should occur concurrently to complement one another.

The Fifth International Conference of Artificial Intelligence, Medical Engineering, Education (October 1–3, 2021, Moscow, Russia) has its purpose to present new thematic approaches, methods and achievements of mathematicians, biologists, physicians and technologists and also to attract additional interest of different specialists to this perspective theme. Its proceedings additionally includes articles on specific tasks in various fields, where artificial intelligence systems can be applied in the future with great benefit.

The conference is organized jointly by Mechanical Engineering Research Institute of the Russian Academy of Sciences (IMASH RAN) and the International Research Association of Modern Education and Computer Science (RAMECS).

The staging of such conferences exemplifies the rising Russian–Chinese cooperation in various spheres of science and education.

The best contributions to the conference were selected by the program committee for inclusion in this book out of all submissions.

Zhengbing Hu
Sergey Petoukhov
Matthew He

Contents

Advances in Medical and Technological Approaches

Advances in Mathematics and Bio-mathematics

Modeling and Application of Ferry Crossing Channel Frequency Based on AIS Data

Shengwei Li[1], Zhi Huang[2](✉), Tian Chai[2], and Zhiping Ren[3]

[1] School of Navigation, Wuhan University of Technology, Wuhan 430063, China
[2] Navigation College, Jimei University, Xiamen 361021, China
[3] Xiamen Navigation Aids Office, East China Sea Navigation Support Center, Xiamen 361012, China

Abstract. Based on the existing ship traffic flow AIS data, the frequency model of ship crossing channel (FMSCC) is established. In addition, the main channel of the eastern crossing channel is taken as an example, for the passenger ship crossing the main channel of the eastern crossing channel on the route from Songyu wharf in Xiamen port to Neicuo'ao wharf in Gulangyu Island. We used the mathematical model to carry out systematic and effective statistics and analysis on the ship traffic flow AIS data of the eastern crossing channel. Whether the close demonstration can shorten the departure interval of passenger ships, reveal the distribution law of departure time window, to provide theoretical basis for the adjustment of departure arrangement of ferry passenger ships, provide theoretical decision support for relevant departments of the port, optimize the marine traffic environment, and improve the efficiency of traffic organization.

Keywords: AIS data · Crossing · Channel · Frequency

1 Introduction

Due to the ferry always need cross the main channel, and the main channel within the various types of ships along the waterway, the parameters such as speed, the ship's length, the time of arriving the gate line, will affect the safety of the ferry crossing. On the other hand, the ferry itself, such as sailing speed and the width of the main channel is also the important factors that affect its safety crossing. Domestic and foreign scholars generally use queuing theory and simulation methods to study the arrival law and passage ability of ships [1].

Jiri Cejka and Josef Šedivý focused on the use of operational research methods in logistics technologies, in particular to stationary traffic design. They solved the presented problem through the application of Queuing Theory. Moreover, they use this method to determine the occupancy of a car park at a pre-selected location [2]. Li Hongxiang built a statistical system of ship traffic flow composed of AIS on board, AIS on shore side and related application software, to count the ship traffic flow automatically in a cross-section of the channel. Then, they adopted the perceptual neural

Z. Hu et al. (Eds.): AIMEE 2021, LNDECT 107, pp. 3–16, 2022.
https://doi.org/10.1007/978-3-030-92537-6_1

network verification method, to carry out automatic statistics and verification experiments on the ship traffic flow, through the cross-section of Wuhan Yangtze River Bridge [3]. Huang Xianxin combined with the characteristics of γ distribution function, established the time-distance distribution model between ships through mathematical statistical analysis, and discussed its special cases. They found that the γ distribution function has strong adaptability to the variation of time distance between ships [4]. Jia Li presented an analytical queuing system to obtain the bounds of its performance, without studying the details of individual queues. The analysis developed a new technology based on the closed solution of the generalized queuing model with a single bottleneck. They used its variation structure, to derive the upper and lower bounds of the total queue length in the series bottleneck system, and discuss its significance to the corresponding motion wave [5].

Although it is convenient to obtain AIS information, because of the large number of ships in the port and the large amount of AIS information, the data processing work is relatively complex [6]. This paper establishes the frequency model of ship crossing channel (FMSCC), through the processing and analysis of ship AIS data. In addition, it takes the passenger ship crossing the main channel of Dongdu Channel between Songyu wharf and Gulangyu Neicuo'ao wharf as an example [7]. We use the mathematical model to make systematic and effective statistics and analysis on ship traffic flow AIS data of Dongdu Channel. And we strictly demonstrate whether the passenger ship departure interval can be shortened, and to reveal the distribution law of departure time window, so as to provide theoretical basis for the adjustment of passenger ship departure arrangement and theoretical decision support for relevant coastal departments, and improve the efficiency of traffic organization, and optimize the marine traffic environment [4, 8, 9].

2 Modeling of Frequency of Ship Crossing the Channel

2.1 Problem Analysis and Mathematical Modeling

With the increase of passenger flow of passenger ferry, especially the passenger flow brought by holidays, to a certain extent, the waiting time of tourists will be too long, and even lead to the phenomenon of passenger detention. Therefore, based on the traffic flow data, we use the mathematical (modeling) method to demonstrate whether the ship departure interval which can be shortened, to effectively alleviate this situation [10, 11].

1) Flow distribution (in unit hour), ship length distribution and ship speed distribution.
   Assume that the interval length and interval speed is Δl (m) and Δv (kn), and then count the number of ships in the interval of $[l_0, l_0 + \Delta l]$, $[l_0, l_0, \Delta l + 2\Delta l]$,..., $([v_0, v_0 + \Delta v], [v_0 + \Delta v, v_0 + 2\Delta v], \ldots)$

2) Ship arrival rule and ship time distance distribution.

(1) Ship arrival rule:

Suppose one day, there were n ships arrived in the gate line, the unit time interval is Δt_0 [12]. The time that they arrive at the gate line is set as a random variable as $X_1, X_2, \ldots X_n$, then the probability of successive k ships arriving in time of Δt_0 is:

$$p(k) = p\{X_k - X_1 \leq \Delta t_0\} + P\{X_{k+2} - X_s \leq \Delta t_0\} + \ldots p\{X_n - X_{n-k+1} \leq \Delta t_0\} \\ (k = 1, 2, \ldots n) \tag{1}$$

It is clearly that $p(1) + p(2) + \ldots + p(n) = 1$, and $X_1, X_2, \ldots X_n$ is independent, and we can deem it as normal distribution approximately. Moreover, every term of the right sum of formula and every term of the neutralization formula in the $p(i)$ and $p(j)$ are equal, so the probability can be calculated by frequency [13, 14].

And now, suppose that the time when the ships arrive at the gate line in turn is $t_1, t_2, t_3, \ldots t_n (t_1 < t_2 < \ldots t_n)$, and unit time interval is Δt. And suppose that there are n intervals between $[t_i, t_i + \Delta t]$, $(i = 1, 2, \ldots n)$, the number of intervals where k ships arrive is q_k. So that $q_1 + q_2 + \ldots q_n = n$. According to the law of large numbers, we can use the frequency as an approximation of probability under certain conditions. When the time interval is Δt, so the probability of reaching the ship is approximately:

$$p(k) = \frac{q_k}{q_1 + q_2 + \ldots q_n} = \frac{q_k}{n} \tag{2}$$

(2) Ship time interval distribution

Suppose that the time when the ships arrive at the gate line in turn is $t_1, t_2, t_3, \ldots t_n (t_1 < t_2 < \ldots t_n)$, and interval time is Δt_1. Time interval between two successive ships passing through the gate line is $Intl_i = t_{i+1} - t_i$, assume $w_k = \{i | (k-1)\Delta t_1 < Intl_i \leq k\Delta t_1\}$, then the probability of successive two ships arriving at the gate line between $(k-1)\Delta t_1$ and $k\Delta t_1$ is about $\frac{size(w_k)}{n}$, therefore, the time interval for $size(w_k)$ ships to reach the gate line is between $(k-1)\Delta t_1$ and $k\Delta t_1$, where $size(w_k)$ is the number of elements in the aggregate w_k.

2.2 Statistical Simulation of the Number of Traversable Times

Model assumption (1): The observation point and the gate line are on the same straight line, and the time when the ship's arrival is referring to the time when the bow reaches the gate line.

It is assumed that the ship arrives at the gate line in turn at the time $t_1, t_2, t_3, \ldots t_n (t_1 < t_2 < \ldots t_n)$ (including all ships in the main channel in two directions), the length of the ship is $l_1, l_2, l_3, \ldots l_n$ in turn, and the speed of ships is $v_1(t), v_2(t), \ldots v_n(t)$ in turn, and the time required for the vessel to cross the main channel is Δt_2 (adjustable parameter) [15]. According to navigation practice, the unsafe distance ahead of the ship is: $S_0 = 7 \times l_n + v_n \times \Delta t_2$.

Where 7 is the empirical multiple (adjustable parameter, or set to 5 times of the ship length).

In order to ensure the ship be able to cross the Dongdu Channel safely, the crossing ship must be beyond the unsafe distance of any ship, otherwise there will be the risk of collision. For a given ship, suppose the ship's number is i, and suppose the moment when the bow of the ship reaches an unsafe distance is a_i, and the moment when the stern leaves the gate line is b_i, there would be:

$$\begin{cases} \int_{a_i}^{t_i} v(t)dt = 7 \cdot l_i + v(a_i) \cdot \Delta t_2 \\ \int_{t_i}^{b_i} v(t)dt = l_i \end{cases} \tag{3}$$

In the above model, when the data is sufficient, $v(t)$ can be obtained by polynomial fitting, and the integral of the polynomial is known. Since it is difficult to determine the ship speed when observing ships in VTS center, especially for small ships, we can assume the model that all ships move in a straight line with uniform speed in the main channel, i.e.

Model assumption (2): the ship moves in a straight line at a constant speed before and after reaching the gate line.

Based on hypothesis (2), we can simplify the model (3) as follows:

$$\begin{cases} v_i(t_i - a_i) = 7 \cdot l_i + v \cdot \Delta t_2 \\ v_i(b_i - t_i) = l_i \end{cases} \tag{4}$$

Solving the equation set, we get the result as follows:

$$\begin{cases} a_i = t_i - \frac{7 \times l_i}{v_i} - \Delta t_2 \\ b_i = t_i + \frac{l_i}{v_i} \end{cases}$$

Therefore, the interval $[a_i, b_i], (i = 1, 2, \ldots, n)$ is the unsafe period for the ship to cross the Dongdu Channel, that is, during these time periods, the ship cannot cross the Dongdu Channel. In other periods, if the length is longer than Δt_2 which of the ship crossing the Dongdu Channel from Songyu, we can launch the ship. Based on this idea, the pseudo code of the algorithm is as follows:

Step 1: Calculate the period for each ship to occupy the crossing water area (gate line), the period for the $i(i = 1, 2, \cdots, n)$ ship to occupy the crossing channel is as follows:

$$[a_i, b_i] = \left[t_i - \frac{7 \times l_i}{v_i} - \Delta t_2, t_i + \frac{l_i}{v_i}\right] \tag{5}$$

Step 2: We can use the bubble method to sort the n periods occupying the cross channel according to the starting time. Assume that the period after sorting is: $[T_i, T_i^{'}], (i = 1, 2, \ldots, n)$, and $T_1 < T_2 < \ldots T_n$.

Step 3: For any $i(1 \le i < n)$, if $T_{i+1}{'} - T_i \ge \Delta t_2$, the ship can safely cross the east crossing channel in period $\left[T_i, T_{i+1}{'}\right]$.

3 Application of FMSCC Model in Dongdu Channel of Xiamen Port

3.1 Data Acquisition and Processing

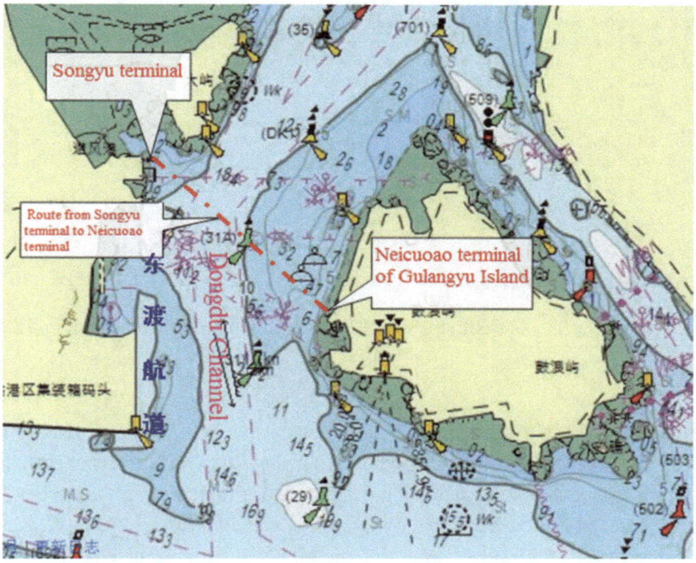

Fig. 1. Route from Songyu terminal to Neicuo'ao terminal

With the increasing number of tourists coming to Xiamen, the carrying capacity of the existing passenger ship route between Songyu Terminal in Haicang and Gulangyu (as shown in Fig. 1) has gradually been unable to meet the passenger demand. The current departure time of passenger ship from Songyu Terminal to Neicuao Terminal in Gulangyu Island is about 30 min on average. Passenger ships crossing the main channel of Dongdu Port between Songyu Ferry Terminal and Gulangyu Island, and there are various types of ships passing through the main channel, so the parameters such as ship speed, ship length, arrival time and so on will affect the departure arrangement of ferry ships. From October 18, 2016 to October 22, 2016, we decoded and analyzed the traffic flow AIS information of ships in Dongdu Channel to obtain the AIS data. As shown in Fig. 2.

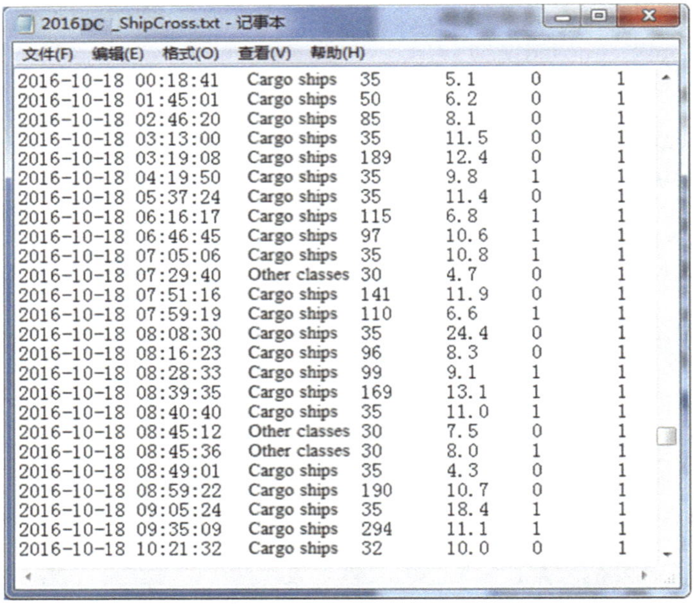
2016DC _ShipCross.txt - 记事本

文件(F) 编辑(E) 格式(O) 查看(V) 帮助(H)

2016-10-18 00:18:41	Cargo ships	35	5.1	0	1
2016-10-18 01:45:01	Cargo ships	50	6.2	0	1
2016-10-18 02:46:20	Cargo ships	85	8.1	0	1
2016-10-18 03:13:00	Cargo ships	35	11.5	0	1
2016-10-18 03:19:08	Cargo ships	189	12.4	0	1
2016-10-18 04:19:50	Cargo ships	35	9.8	1	1
2016-10-18 05:37:24	Cargo ships	35	11.4	0	1
2016-10-18 06:16:17	Cargo ships	115	6.8	1	1
2016-10-18 06:46:45	Cargo ships	97	10.6	1	1
2016-10-18 07:05:06	Cargo ships	35	10.8	1	1
2016-10-18 07:29:40	Other classes	30	4.7	0	1
2016-10-18 07:51:16	Cargo ships	141	11.9	0	1
2016-10-18 07:59:19	Cargo ships	110	6.6	1	1
2016-10-18 08:08:30	Cargo ships	35	24.4	0	1
2016-10-18 08:16:23	Cargo ships	96	8.3	0	1
2016-10-18 08:28:33	Cargo ships	99	9.1	1	1
2016-10-18 08:39:35	Cargo ships	169	13.1	1	1
2016-10-18 08:40:40	Cargo ships	35	11.0	1	1
2016-10-18 08:45:12	Other classes	30	7.5	0	1
2016-10-18 08:45:36	Other classes	30	8.0	1	1
2016-10-18 08:49:01	Cargo ships	35	4.3	0	1
2016-10-18 08:59:22	Cargo ships	190	10.7	0	1
2016-10-18 09:05:24	Cargo ships	35	18.4	1	1
2016-10-18 09:35:09	Cargo ships	294	11.1	1	1
2016-10-18 10:21:32	Cargo ships	32	10.0	0	1

Fig. 2. Decoding of the AIS data of ships in the Dongdu Channel from October 18th to October 22nd, 2016

1) Classification of ship types
   We divide all ships into two categories: Cargo ships (including cargo ships and oil tankers) and other classes (all other than cargo ships and oil tankers).
2) Processing null and invalid values
   (1) Null value of ship type: We divide all ships into cargo ship class or other classes according to the ship's length, and the classification standard is as follows: the distribution law of ship's length of existing cargo ship class and the distribution law of ship's length of the existing other classes.
   (2) Invalid or null values for time, ship's speed and length: such key data missing or invalid will lead to the failure of the subsequent modeling calculation process. We can eliminate it directly by algorithm or modify it manually.

3.2 Traffic Flow Distribution, Ship's Length Distribution and Ship's Speed Distribution

Use the endpoints of the interval as the abscissa and the number as the ordinate to make a histogram. The traffic flow distribution is similar, and the time interval is 1 h. Taking 120 h' data from 00:00 on October 18, 2016 to 23:59 on October 22, 2016 as an example, the total 309 ships sailed in the Dongdu Channel.

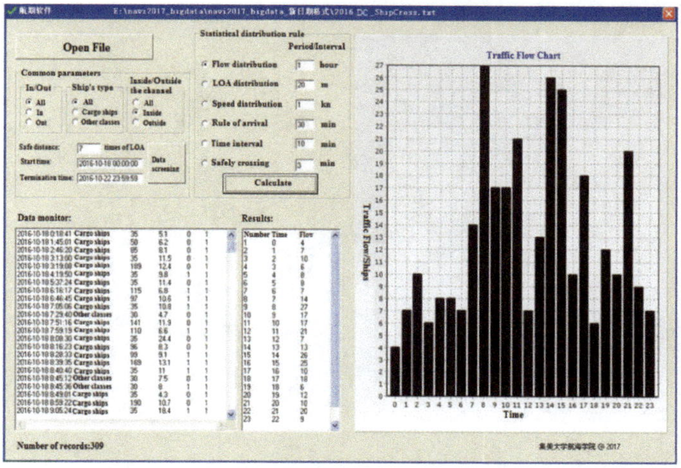

Fig. 3. Traffic flow data in the Dongdu Channel – hourly distribution of traffic flow

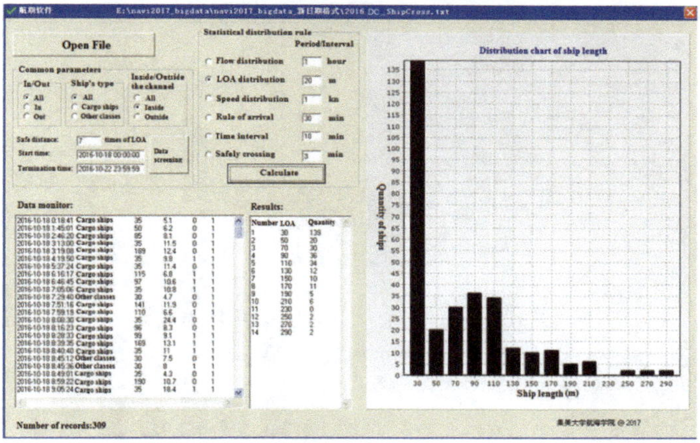

Fig. 4. Traffic flow data-ship length distribution in the Dongdu Channel

According to the traffic flow distribution (Fig. 3), the maximum and minimum traffic flow per hour in the Dongdu Channel was 27 and 4, and the number of ships was relatively small in the period from 00:00 to 07:00, and from 22:00 to 24:00.

According to the ship length distribution (Fig. 4), there were 139 ships with a length of 0–30 m and 20 ships with a length of 30–50 m in the Dongdu Channel.

The ship speed distribution (Fig. 5) is close to the normal distribution, and there were 55 ships with an expected value of about 8 kn and a speed of 8 kn in the eastbound channel.

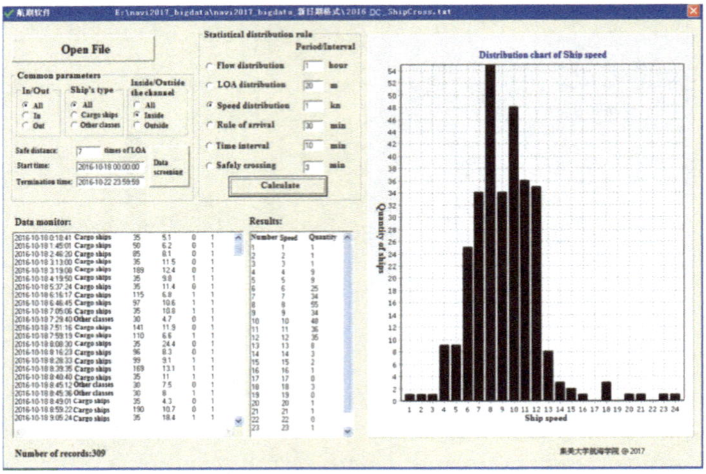

Fig. 5. Traffic flow data-ship speed distribution in the Dongdu Channel

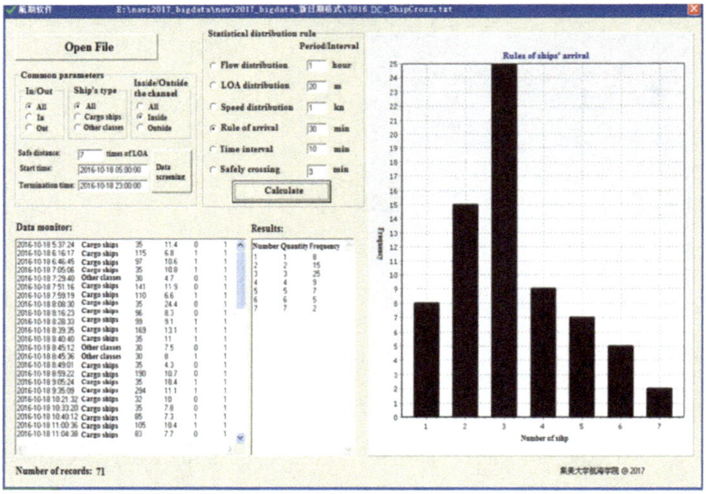

Fig. 6. Regular pattern of ship arrival in Dongdu Channel (within 30 min)

3.3 Regular Pattern of Ship Arrival

According to the time data from 05:00 to 23:00 on October 18, 2016 (within the waterway), the result is shown in Fig. 6, where $\Delta t = 30$ min.

It can be seen from the Fig. 6 that there are 8 times to reach one vessel within 30 min, 15 times to reach two vessels within 30 min, 25 times to reach three vessels within 30 min, and so on.

3.4 Distribution of Ship Time Interval

Take the time data from 05:00 to 23:00 on October 18, 2016 (within the waterway) for statistics, and take $\Delta t_1 = 2$ min. The Fig. 7 shows the statistical results.

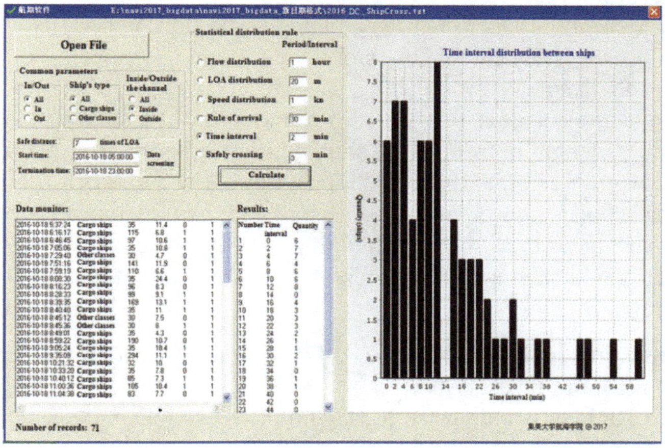

Fig. 7. Time interval distribution between ships in Dongdu Channel ($\Delta t_1 = 2$ min)

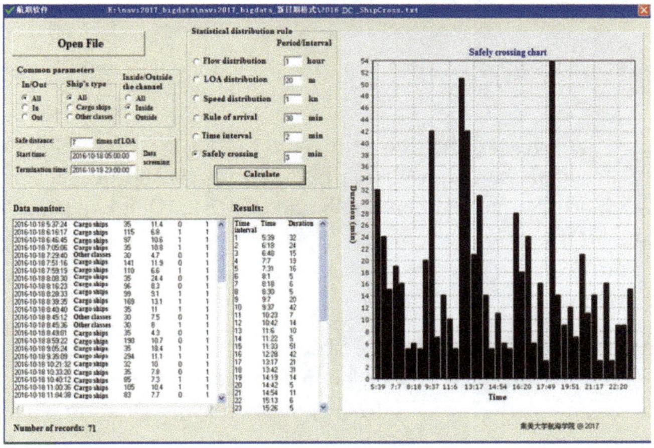

Fig. 8. Time interval distribution of ships in the Dongdu Channel (safe distance of 7 times the ship's length and safe crossing time of 3 min)

In the Dongdu Channel, there are six consecutive two ships with an interval of 0 to 2 min, seven ships with an interval of 2 to 4 min, and seven ships with an interval of 4 to 6 min.

3.5 Statistical Simulation of Traversable Times

1) Simulation statistics of safe distance seven times ship length and safe crossing time 3 min.

According to the time data from 05:00 to 19:00 on October 18, 2016 (within the channel), $\Delta t_2 = 3$ min is taken, that is, as long as the time interval is greater than or equal to 3 min, the ship can safely cross the Dongdu Channel. We can find the calculation results in Fig. 8.

According to the standard that the ship cannot cross the area which within seven times of the ship's length in front of the ship, during the period from 05:00 to 19:00 on October 18, 2016, we can find that 32 min between 05:39 and 06:18 is the safe period, 24 min between 06:18 and 06:48 is the safe period, and 15 min between 06:18 and 07:07 is the safe period, so the crossing ship can depart at any time. There are 31 traversable periods in such a day from 05:00 to 19:00, with a total length of 560 min. The details are as follows:

(1) There are 6 time periods (32′, 42′, 51′, 42′, 31′ and 54′) when crossing the Dongdu Channel for more than 30 min;
(2) There are 10 time periods when the length of crossing Dongdu Channel is greater than 20 min;
(3) There are 19 time periods when the length of crossing Dongdu Channel is greater than 10 min;
(4) The average time of the ships' crossing Dongdu Channel is 18.1 min.

2) Simulation statistics of safe distance 5 times of ship length and safe crossing time 3 min.

According to the standard that the ship cannot cross the area within 5 times of the ship's length in front of the ship, 33 min between 05:39 and 06:18, 25 min between 06:18 and 06:48, and 15 min between 06:48 and 07:07 on October 18, 2016, which are the safe period, and the crossing ship can depart at any time. In such a day from 05:00 to 19:00, there are 33 traversable periods, with a total length of 585 min. The specific situation is as follows:

(1) There are 5 time periods (33', 42', 51', 43', 31' and 55', respectively) when the length of crossing the East Ferry Channel is greater than 30 min;
(2) There are 11 time periods when the length of crossing Dongdu Channel is greater than 20 min;
(3) There are 20 time periods when the length of crossing Dongdu Channel is greater than 10 min;
(4) The average crossing time of the eastern channel is 17.7 min.

3) Simulation statistics of safety distance 5 times ship length and safety crossing time 2 min

Reduce the safe crossing time from 3 min to 2 min. Also, according to the standard that the ship cannot cross the area within 5 times of the ship's length in front of the ship, 34 min between 05:39 and 06:18, 26 min between 06:18 and 06:48 and 16 min between 06:48 and 07:07 on October 18, 2016 are the safe time, the crossing ship can be dispatched at any time. In such a day (0500–1900), there are 39 traversable periods, with a total length of 632 min. The details are as follows:

(1) There are 6 time periods (34', 43', 52', 44', 32' and 56', respectively) when the time of crossing the East Ferry Channel is longer than 30 min;
(2) There are 11 time periods when the length of crossing Dongdu Channel is greater than 20 min;
(3) There are 20 time periods when the length of crossing Dongdu Channel is greater than 10 min;
(4) The average crossing time of the eastern channel is 16.2 min.

3.6 Analysis of Measured Traffic Flow Crossing Model Results

According to the above traffic flow measurement and modeling analysis, we can find that ships can cross the channel from 05:00 to 19:00 in the 5 days from October 18 to 22, 2016, as follows:

1) Considering that the safe distance is 7 times the length of the ship and the safe crossing time is 3 min, as shown in Table 1, the total time for ships to cross the Dongdu Channel during the five-day period is 3,095 min, and the average total time for ships to cross every day is 619 min. According to the ideal situation, ships can cross the Dongdu Channel in 3 min, and the number of times they can cross the Dongdu Channel every day is 206.
2) Similarly, considering that the safe distance is 5 times the length of the ship and the safe crossing time is 3 min, the total time for ships to cross the Dongdu Channel during the five-day period is 3,168 min, and the average total time for ships to cross the Dongdu Channel every day is 634 min. According to the ideal situation that ships can cross the Dongdu Channel in 3 min, the number of times they can cross the Dongdu Channel is 211 times per day.
3) Considering that the safe distance is 5 times the length of the ship and the safe crossing time is 2 min, the total time for ships to traverse the Dongdu Channel in the five-day period is 3,381 min, and the average total time for ships to traverse every day is 676 min. According to the ideal situation, ships can cross the Dongdu Channel in 2 min, and the number of times they can cross the Dongdu Channel is 225 times per day.

Table 1. Time distribution of time interval between two adjacent ships in Dongdu Channel is more than 3 min (Safe distance is 7 times of ship length)

Serial number	Oct. 18	Oct. 19	Oct. 20	Oct. 21	Oct. 22
	The time interval (min)	The time interval (min)	The time interval (min)	The time interval (min)	The time interval (min)
1	33	98	50	7	25
2	25	19	97	36	56
3	15	53	58	7	30
4	19	3	5	27	5
5	17	9	30	18	29
6	5	3	35	14	24
7	3	36	6	29	10
8	7	7	44	3	14
9	5	17	7	35	14
10	4	15	23	12	74
11	22	12	57	14	7
12	42	4	38	6	77
13	8	28	11	4	21
14	15	55	7	4	28
15	11	11	5	11	11
16	6	6	20	3	15
17	51	32	39	3	5
18	43	21	4	16	16
19	21	3	24	28	3
20	31	20	18	18	5
21	14	6	69	6	12
22	5	8	–	4	6
23	12	9	–	6	38
24	7	29	–	5	23
25	6	10	–	4	24
26	28	3	–	33	40
27	18	28	–	45	12
28	24	35	–	16	–
29	8	39	–	3	–
30	17	–	–	–	–
31	3	–	–	–	–
32	55	–	–	–	–
33	5	–	–	–	–
Total	**585**	**619**	**647**	**417**	**624**

4 Conclusion

Based on the processing and analysis of the AIS data of the ship, this study established the FMSCC model, and took the ferry crossing the Dongdu Channel of Xiamen as the research object to verify the model, including:

1) The actual traffic flow is analyzed, and the traffic flow data is processed;
2) Traffic flow modeling is realized;
3) Statistical simulation is carried out for the traversable times of ships traversing the Dongdu Channel at Xiamen Port;
4) Obtain the analysis conclusion of the measured traffic flow-crossing model. The analysis results show that we can reduce the departure interval and the departure frequency appropriately, to alleviate the passenger detention.

The FMSCC model established in this paper provides a new idea and method for studying the regular pattern of small boats crossing the channel. However, due to the lack of statistical data, the FMSCC model does not include the ship's collision avoidance behavior. Moreover, the deceleration and direction change which caused by the ship's collision avoidance behavior, will affect the accuracy of the actual model. In addition, when modeling, we did not consider factors such as sea conditions and tides, and so on. Therefore, in order to make the constructed model more practical, we need to further optimize and perfect it in the next research.

References

1. Li, X., Xiang, L., Tian, C.: Modeling and application for traffic capacity of Xiamen Port. Navig. China **42**(02), 32–36 (2019). (in Chinese)
2. Cejka, J., Šedivý, J.: Discussion of operational transport analysis methods and the practical application of queuing theory to stationary traffic. Transp. Res. Procedia **53**, 196–203 (2021)
3. Li, H., Fang, X.: Statistical method research of vessel traffic flow based on AIS. J. Wuhan Univ. Technol. (Traff. Sci. Eng. Ed.) **35**(04), 853–857 (2011). (in Chinese)
4. Huang, X., Shao, Z., Ji, X.: Probability distribution of time-interval between ships in main channel of Xiamen Port. Navig. China **37**(04), 74–78 (2014). (in Chinese)
5. Li, J., Michael Zhang, H.: Bounding queuing system performance with variational theory. Transp. Res. Procedia **07**, 519–535 (2015)
6. Chi, X.: Study on the navigation safety of ships in the southwest sea area of Gulangyu Island of Xiamen Port. Pearl River Water Transp. **07**, 34–37 (2018). (in Chinese)
7. Li, Y., Xiao, Y., Zhou, W.: Transfer efficiency analysis of Baoshan pilotage area based on queuing model. J. Shanghai Marit. Univ. **33**, 12–17+30 (2012). (in Chinese)
8. Tong, S., Liu, J., Zhao, L.: Impact on water traffic environment based on ISM model. In: The 5th International Conference on Transportation Information and Safety in 2019 (ICTIS), pp. 1188–1191. IEEE, Liverpool (2019)
9. Das, N., Das, L., Rautaray, S., Pandey, M.: Big data analytics for medical applications. Int. J. Modern Educ. Comput. Sci. **10**(2), 35–42 (2018). https://doi.org/10.5815/ijmecs.2018.02.04

10. Zhengbing, H., Legeza, V.P., Dychka, I.A., Legeza, D.V.: Mathematical modeling of the process of vibration protection in a system with two-mass damper pendulum. Int. J. Intell. Syst. Appl. **9**(3), 18–25 (2017). https://doi.org/10.5815/ijisa.2017.03.03
11. Syed, F., UsmanDilawar, M.: Mathematical modeling and analysis of network service failure in DataCentre. Int. J. Modern Educ. Comput. Sci. **6**(6), 30–36 (2014). https://doi.org/10.5815/ijmecs.2014.06.04
12. Hilstad, K.V.: Ship traffic scheduling and disruption management for the Kiel Canal: a simulation-optimization approach. In: Norwegian University of Science and Technology (Norwegian) (2018)
13. Liu, L., Liu, J., Tan, Z.: Examination and analysis of vessel traffic flow distribution in Jingzhou bridge area. J. Dalian Maritime Univ. **36**, 8–10 (2010). (in Chinese)
14. Shamsul Alam, S.M., Saha, A.: Performance analysis of FBMC over OFDM for high data rate MIMO configurations. Int. J. Wirel. Microwave Technol. **11**(4), 20–33 (2021). https://doi.org/10.5815/ijwmt.2021.04.03
15. Pan, J., Shao, Z., Jiang, Q.: Application of data mining technology in analysis of marine traffic characteristics. Navig. China **33**, 60–62+732010. (in Chinese)

Optimization of Intelligent Warehouse Layout Based on SLP Method–Take SuNing Logistics CaiNiao Warehouse as an Example

Lu Cao(✉), Qiaoling Li, and Meng Wang

School of Logistics, Wuhan Technology and Business University, Wuhan, Hubei, China

Abstract. Under the normalization of the epidemic, the number of express delivery in China has been on a rising trend, which brings many opportunities to many logistics enterprises. If the logistics enterprise wants to maintain its long-term steady development, the operation ability of its warehouse plays a decisive role. Meanwhile, If the logistics enterprise wants to provide satisfactory services for consumers, it must improve the operation efficiency and ensure the timeliness of goods delivery. In this paper, we take SuNing logistics CaiNiao warehouse as the research object to optimize its warehouse layout. Through the relevant data analysis, the warehouse layout problems are optimized, and the SLP method is used to obtain the storage center layout plan. At the same time, the factor analysis method is used to evaluate the optimized warehouse layout, scheme and determine the final optimization layout, which can improve the operation efficiency of the logistics warehouse and reduce the enterprise costs.

Keywords: Warehousing logistics · SLP method · Layout optimization

1 Instruction

In recent years, the development of logistics industry has tended to mature stage, and more and more logistics enterprises pay more attention to the improvement of logistics speed. In the logistics supply chain, the warehouse distribution system is an important link that cannot be ignored. Therefore, warehouse distribution systematic optimization and upgrading should be given priority in improving the overall efficiency of enterprises [1]. The reasonable layout of the warehouse will effectively reduce the expenditure of management expenses and storage expenses, and increase the logistics flow rate. Mao Chao thought that reasonable layout planning is the premise of efficient warehouse work [2]. Wu Peiki believed that with the booming development of the logistics industry, the warehouse layout optimization and design is becoming more and more important for an enterprise. The reasonable layout of the warehouse will improve the operation efficiency of enterprises, save logistics costs, and then greatly enhance the competitiveness of enterprises [3]. Yang Wenxin considered that in the planning of e-commerce storage layout, SLP method is used to plan its storage layout, and then the obtained scheme adopts the hierarchical analysis method for evaluation, so as to improve the operation efficiency of logistics enterprises and reduce the operation cost

Z. Hu et al. (Eds.): AIMEE 2021, LNDECT 107, pp. 17–29, 2022.
https://doi.org/10.1007/978-3-030-92537-6_2

of enterprises [4]. Deng and Lin believed that the SLP method is applicable to system layout planning and design [5].

But since the epidemic, various industries have been impacted to varying degrees, so with the logistics industries. This year, SuNing Logistics in order to reduce costs, ready to close some of the CaiNiao front warehouse, these CaiNiao front warehouse closure will increase the operation volume of SuNing Logistics CaiNiao warehouse. However, due to the early construction of its CaiNiao warehouse, it is difficult to achieve the expected operation efficiency in the later business expansion. Therefore, improving the CaiNiao warehouse layout, and improving the operation efficiency can not wait.

2 Analysis the Current Situation of SuNing Logistics CaiNiao Warehouse

2.1 Overview of the Functional Areas of SuNing Logistics CaiNiao Warehouse

At present, the main regional divisions of Wuhan SuNing Logistics CaiNiao Warehouse are divided into purchase area, cargo inspection area, freight area, finished product area, semi-finished product area, circulation and processing area, sorting area, shipping area, equipment area and auxiliary area. The warehouse area is about 9000 m^2, of which the purchase area covers an area of 400 m^2, together with cargo inspection area 800 m^2, freight area 700 m^2, finished product area 3900 m^2, semi-finished product area 800 m^2, circulation and processing area 200 m^2, sorting area 300 m^2, shipping area 300 m^2, equipment area 1200 m^2 and auxiliary area 400 m^2. The Layout Plan of SuNing Logistics CaiNiao warehouse is shown in Fig. 1.

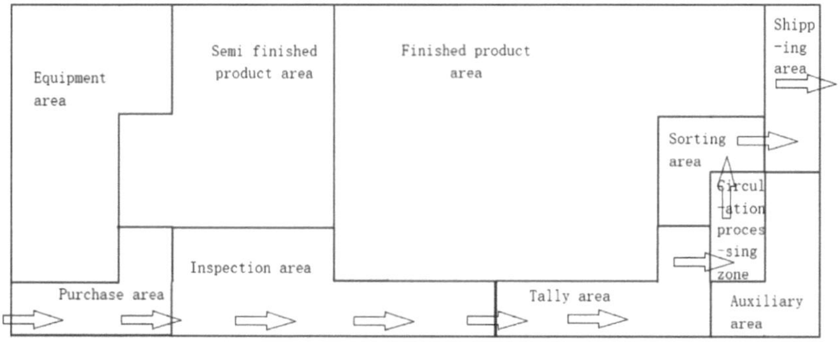

Fig. 1. The layout plan of SuNing logistics CaiNiao warehouse

2.2 Problems in the Warehouse Layout of SuNing Logistics CaiNiao Warehouse

Through Fig. 1, we can find that the warehouse layout of SuNing Logistics CaiNiao warehouse mainly has the following problems:

2.2.1 Poor Position Flexibility of Operation Unit Position

The connection between outbound operation units is close, handling activities are concentrated, especially circulation processing area, sorting area, shipping area. The location between them is too dense, making the handling work flexibility poor, and easy to cause operation congestion [6]. Congestion during electricity promotion will reduce operation efficiency, prone to explosive warehouse.

2.2.2 Operation Route Setting is Unscientific

The location relationship between some operation units is unreasonable, and there are problems such as increasing handling route length, increasing handling number, and detour route. First, the equipment area and the auxiliary area are far apart. When the auxiliary area assists each logistics link, the relevant equipment is generally obtained in the equipment area for assistance, but the distance between the equipment area and the auxiliary area will reduce the operation efficiency of the operators. Secondly, the positions between the cargo area, sorting area, circulation processing area and shipping area are too dense. These four areas are relatively dense operators, and the location relationship between them should be on a line, which is conducive to reducing the circulation distance of the goods. And the location between them is type "S", which will cause the problem of unscientific handling route, circuitous route, increased handling distance, and more handling times in the circulation of goods.

3 An Analysis of the Warehouse Layout Optimization Based on the SLP Method

System Layout Planning Method (SLP) is proposed by Richard Muse to plan the operation unit logistics and non-logistics factor analysis through the structured and structured design program mode [7]. This paper predicts the operation quantity of SuNing logistics CaiNiao warehouse to determine the logistics intensity between each functional area, and then divides the logistics relationship between the functional units to further determine the layout of the warehouse. The logistics intensity is determined through the material flow of the warehouse, and finally the logistics relationship between the functional interval and the non-logistics relationship is divided, so as to determine the warehouse layout optimization diagram.

3.1 Operation Quantity Forecast on SuNing Logistics CaiNiao Warehouse

Due to the epidemic, the actual data of CaiNiao warehouse in 2020 year is lack of reference, therefore, forecast the production and sales of CaiNiao Warehouse in 2020 and 2021 year through the production and sales of CaiNiao Warehouse in year of 2017, 2018, 2019. Which to determine the total sales volume and throughput of CaiNiao warehouse, and determine the material flow of the warehouse through the difference between them. As shown in Table 1.

Table 1. Growth table of production volume

Year	2017	2018	2019
Growing production volume (Ten thousand boxes)	480	570	660
Growing sales volume (Ten thousand boxes)	244	325	415

Determine the planned total throughput and total flow of CaiNiao Warehouse by moving averaging to predict the production and sales volume of growth in 2021.By the formula (1):

$$F_t = (A_{t-1} + A_{t-2} + A_{t-3} + \cdots + A_{t-n})/n \tag{1}$$

Computing production growth for 2020 was 570 ten thousand boxes and sales growth was 328 ten thousand boxes. Based on the calculation results in a mobile average calculation, we obtain the production growth of 600 ten thousand boxes and 356 ten thousand boxes in 2021. Therefore, the planned total throughput of the CaiNiao warehouse is 6 million boxes per year, and the total flow volume is about 3.56 million boxes per year. Thus, the total material flow of the warehouse is 2.44 million boxes a year, and the cargo acceptance volume is about 6,667 boxes per day, so as to determine the logistics intensity between each operating area.

3.2 Correlation Analysis of CaiNiao Warehouse Functional Area of SuNing Logistics

3.2.1 Correlation Analysis of the Logistics Relations of Each Functional Area

Through the SLP method, the logistics strength relationship between different functional areas is obtained. For the material flow factor, the logistics strength is divided into A (absolutely important), E (is particularly important), I (important), O (general), U (undesirable) grade, as shown in Table 2.

Table 2. Logistics strength level table

Logistics strength level	Proportion of logistics lines (%)	Material flow rate ratio (%)
A	10	40
E	20	30
I	30	20
O	40	10
U		

Through the data information of SuNing logistics system, we select the operation amount of each functional interval on one day to get the workload of each functional interval. Working relationship and daily workload between the functional zones are shown in Fig. 2.

According to the daily throughput flow between each functional units in Fig. 2, we can replace the material flow between each functional units, and calculate the percentage of each operating unit to determine the logistics intensity of each functional unit to determine the logistics intensity level between each operating area.

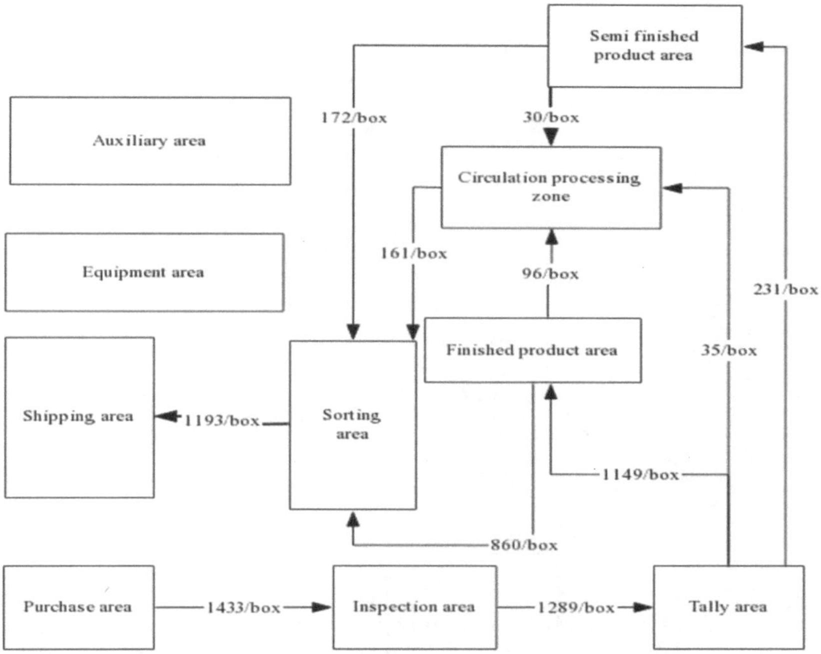

Fig. 2. Working relationship and daily workload between the functional zones

Table 3. Logistics strength grade table of each functional interval

Order number	Function Relational pairing	Material flow rate (box)	Logistics strength (%)	Logistics strength level
1	1–2	1433	21.49	A
2	2–3	1289	19.34	E
3	7–8	1193	17.89	E
4	3–4	1149	17.24	I
5	4–7	860	12.90	I
6	3–5	231	3.47	O
7	5–7	172	2.58	O
8	6–7	161	2.41	O
9	4–6	96	1.44	O
10	3–6	35	0.52	O
11	5–6	30	0.45	O

Number the purchase area, inspection area, cargo area, finished product area, semi-finished product area, circulation and processing area, sorting area, shipping area, equipment area and auxiliary area from 1 to 10 respectively, we can get the logistics strength rating table between them. As shown in Table 3.

Therefore, we obtain the logistics related diagram of each functional area, as shown in Fig. 3.

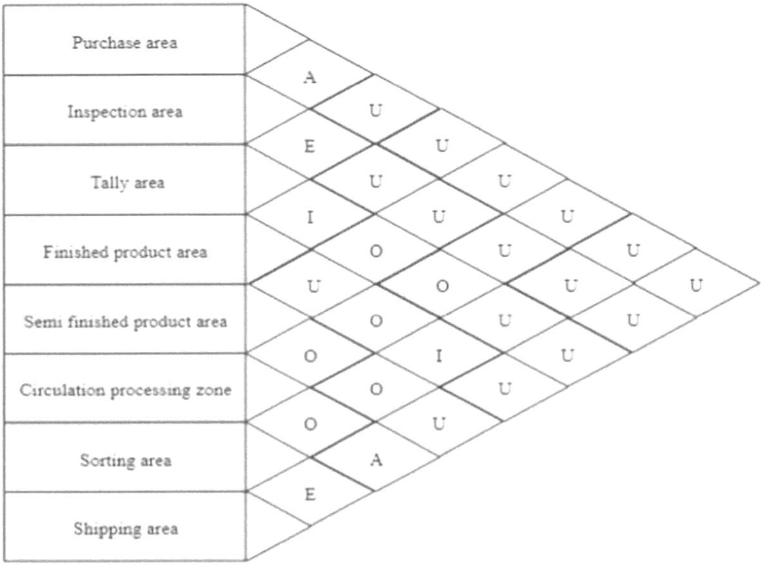

Fig. 3. Relevant diagram of logistics intensity

3.2.2 Analysis of Non-logistics Relations in Each Functional Area

The mutual relationship between the operation units is the interaction between the two, mainly called the non-logistics relationship. The correlation analysis of non-logistics activity considers the impact of other factors except for logistics accidents on the warehouse operation areas, including the process continuity of handling operation, the emergency of operation, the convenience of management, environmental hygiene and other factors [8]. Through qualitative analysis of different functional areas of SuNing Logistics, the factors affecting non-logistics relationship include six aspects.

When dividing the closeness of non-logistics relationship, common A, E, I, O, U, X are used to represent. They represent important, particularly important, important, general, unimportant, not close. Through the qualitative analysis of the functional area, we draw the non-logistics relationship between the functional area, the letters indicate the correlation between the two functional areas, and the number represents the type of relationship between the two regions, as shown in Fig. 4.

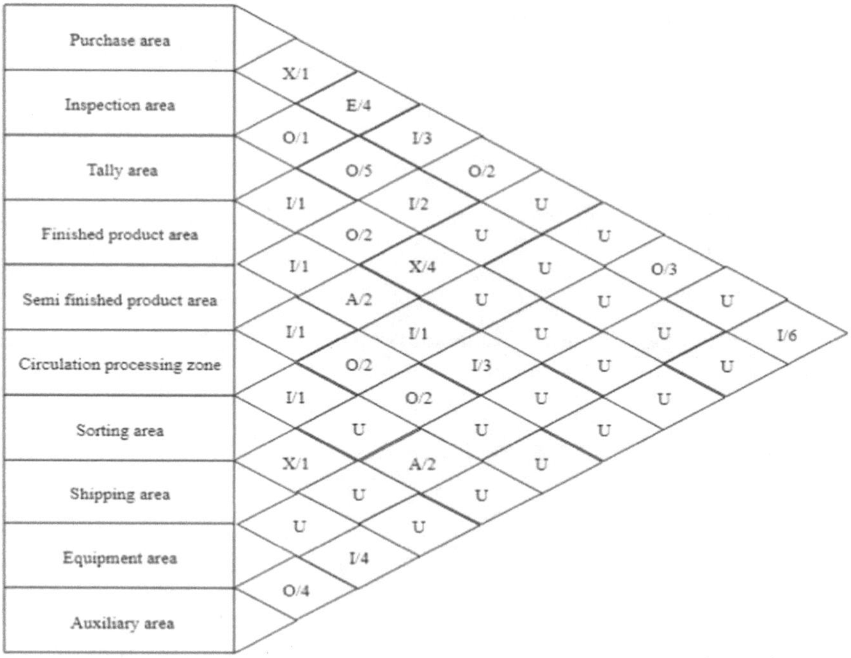

Fig. 4. Relevant diagram of non-logistics relationship of each functional area

3.2.3 Comprehensive Correlation Diagram Analysis

The comprehensive strength correlation of the warehouse can be further analyzed through the logistics and non-logistics intensity correlation diagram above. Comprehensive analysis requires the weight of logistics and non-logistics relations [9]. According to the actual situation of the warehouse, the logistics relationship dominates in the warehouse operation, and the non-logistics relationship only occupies the auxiliary position, so the weighted value proportion of the relationship between logistics and non-logistics is determined by m:n = 3: 1. A, E, I, O, U, X is 4, 3, 2, 1, 0, −1, respectively. Then we quantify the logistics and non-logistics relationship between any two functional areas i, j by MR_{ij}, non-logistics relationship between any two functional areas i, j by NR_{ij}. CR_{ij} represents the comprehensive relationship between i, j functional zones, and the calculation formula of i = j is: $CR_{ij} = m * MR_{ij} + n * NR_{ij}$. We determine the above classification calculations of comprehensive close relationship degree as shown in Table 4.

Table 4. Classification of comprehensive close level

Total points	Relation level	Work Unit logarithm	Ratio (%)
8–12	A	2	4.4
6–8	E	1	2.2
3–5	I	10	22.2
1–2	O	8	17.8
0	U	21	46.7
−1	X	3	6.7

Comprehensive logistics related diagram is obtained through the above analysis as shown in Fig. 5.

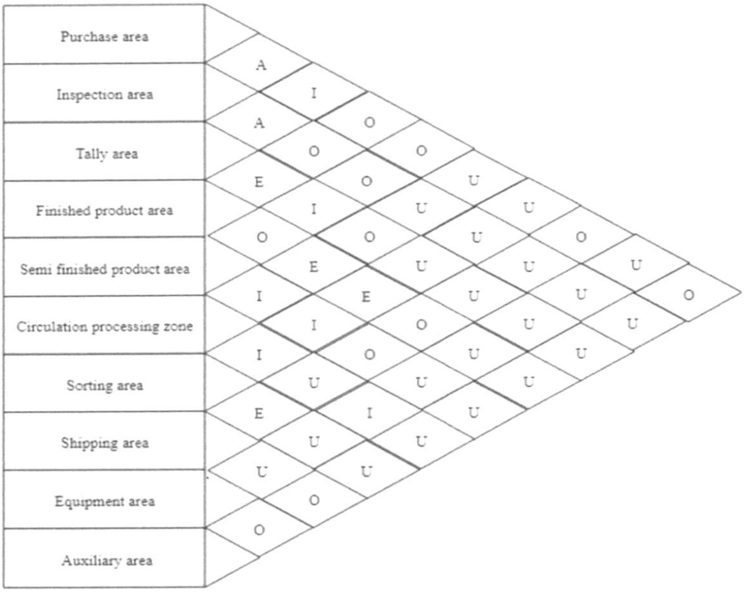

Fig. 5. Relevant drawing of comprehensive logistics

According to the comprehensive relationship between the functional zones in Fig. 5, the operating units are arranged according to their relationship level. It can be found that the relationship level between the purchase area and the inspection area is absolutely important, the relationship between the inspection area and the inventory area is absolutely important, the relationship between the inventory area and the finished product warehouse is particularly important, the relationship between the sorting area and the shipping area is particularly important, the relationship between the finished product warehouse and the semi-finished product warehouse is particularly important, and the relationship between the finished product warehouse and the circulation processing area is particularly important. It can be concluded that area 1, 2 and 3 must be connected together, area 3 and 4 together, area 7 and 8 together, and area 4 with 6 and 7. Then, the job relationships of each functional interval can be analyzed to determine their relative relationships.

In the plane layout, 100 m^2 is set as the area of a single unit, then the warehouse center plane plan is divided into 6 * 15 units, with the area required for different functional zones, as shown in Table 5. Two warehouse optimization layout schemes with an area are obtained respectively, as shown in Fig. 6 and 7.

Table 5. Required area table for each functional area

Order number	Functional area name	Required area	Number of units required
1	Purchase area	400	4
2	Inspection area	800	8
3	Freight area	700	7
4	Finished product area	3900	39
5	Semi finished product area	800	8
6	Circulation and processing area	200	2
7	Sorting area	300	3
8	Shipping area	300	3
9	Equipment area	1200	12
10	Auxiliary area	400	4

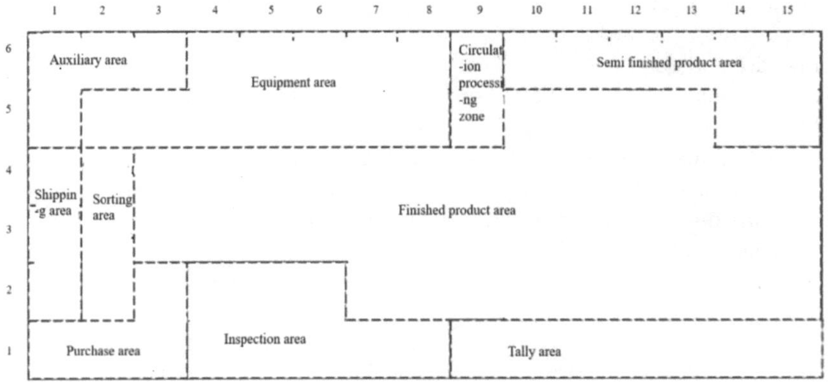

Fig. 6. Warehouse optimization layout scheme I

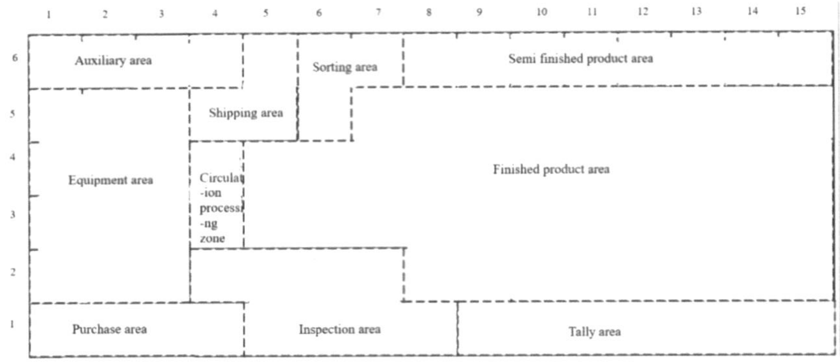

Fig. 7. Warehouse optimization layout scheme II

4 Evaluation of the Warehouse Layout Optimization Scheme

The Factor Analysis method is a quantitative and qualitative analysis method, meanwhile a common analysis method in scheme decision. It provides an important basis for decision making through data analysis and application of algorithm [10]. Through the previous analysis, we determine the final layout optimization diagram of CaiNiao warehouse.

4.1 Identify the Objective Measure

The proportion of subjective and objective factors is mainly determined by visiting and investigating the relevant personnel of CaiNiao warehouse. The 50 people used the questionnaire and collected 50 relevant personnel. Planning the warehouse, the survey results found that the subjective factor is as important as the objective factor, that is, 0.5: 0.5.

First determine the objective measure, two existing schemes can find an objective value OM_i, which is affected by the cost of the warehouse. Through the company's original warehouse, the main sources of the two schemes are economic benefits and safety management [11], as shown in Table 6.

Table 6. Cost factors table (ten thousand yuan)

Cost factors	Scheme	
	Scheme I	Scheme II
Administration cost	151	209
Producing cost	122	92
Safety costs	84	67

By the formula (2):

$$OM_i = \left[C_i \sum_{i=1}^{N} \frac{1}{C_i}\right]^{-1} \tag{2}$$

There: Ci-Total Cost of i, N-Total number of site selection schemes, we can get the results as follow:

$$(C_1)^{-1} = (357)^{-1} = 0.0028(C_2)^{-1} = (368)^{-1} = 0.0027.$$
$$OM_1 = (C_1)^{-1} \div \left((C_1)^{-1} + (C_2)^{-1}\right) = 0.0028 \div (0.0028 + 0.0027) = 0.51.$$
$$OM_2 = (C_2)^{-1} \div \left((C_1)^{-1} + (C_2)^{-1}\right) = 0.0027 \div (0.0028 + 0.0027) = 0.49.$$

By the above calculation, objective measures of Scheme I and Scheme II are: $OM_1 = 0.51$, $OM_2 = 0.49$.

4.2 Determining the Subjective Evaluation Values

By comparing the actual situation of CaiNiao warehouse, which is better in terms of the environment and higher in terms of safety, the proportion of these two non-economic factors is 0.4: 0.6. According to the formula (3):

$$S_{ik} = \frac{W_{ik}}{\sum_{i=1}^{N} W_{ik}} \tag{3}$$

There, Sik-Subjective evaluation value of K factors in i site selection scheme, Wik-Specific portion of a factor in i location scheme, Subjective evaluation of environmental and safety factors are shown in Table 7.

Table 7. Factor Summary

Factors	Scheme		
	Scheme I	Scheme II	Weight
Environmental factor	1	0	0.4
Safe factor	0	1	0.6

4.3 Identify the Subjective Measures Value

According to the formula (4):

$$SM_i = \sum_{K=1}^{M} (I_k \times S_{ik}) \tag{4}$$

There, Sik–Subjective evaluation value of I site selection scheme for A factors, M–The number of subjective factors, so: SM1 = 1 × 0.4 + 0 × 0.6 = 0.4, SM2 = 0 × 0.4 + 1 × 0.6 = 0.6.

4.4 Determine the Position Measure

Both subjective and non-subjective factors can be determined through data collection, so $X = 0.5$. According to the formula (5):

$$LM_i = X \times SM_i + (1 - X) \times OM_i \tag{5}$$

There, X-Subjective specific gravity value, (1−X)-Objective specific gravity value, then:

$$LM_1 = X * SM_1 + (1 - X) \times OM_1 = 0. * 0.4 + 0.5 * 0.51 = 0.357.$$
$$LM_2 = X * SM_2 + (1 - X) \times OM_2 = 0.5 * 0.6 + 0.5 * 0.49 = 0.392.$$

Through the above analysis, Scheme II is much better.

5 Conclusion

Advantages after the optimization of layout: First, the position of the operation unit is highly flexible, so that the position of the functional areas with large operation volume in the circulation processing area, sorting area and shipping area are relatively scattered to improve the flexibility of handling work. Second, the handling route is more scientific, so that the equipment area and the auxiliary area are relatively close, improve the operation efficiency of the operators, reduce the repeatability of the handling lines, and avoid crowding during the e-commerce promotion period.

This paper not only solves the problems existing in the current layout of SuNing Logistics CaiNiao Warehouse, but also provides a solution for improving the operation efficiency of the same type of e-commerce warehouse and reducing the operation cost. When making system layout planning, enterprises should consider cost, efficiency, flexibility, risk and other aspects, should have a long-term vision, can not blindly pursue low cost, so that the enterprise can go further in the future. Logistics enterprises have been in the stage of continuous exploration in the optimization of warehouse layout and design. If enterprises want to provide satisfactory services for customers, they must improve operation efficiency and reduce operation costs. With the continuous optimization of storage layout, they can effectively solve these problems.

Acknowledgment. This research is supported by the Project of Science and Technology Innovation Team for Outstanding Young and Middle-aged Students in Hubei Province (T201938).

References

1. Chen, C., Lin, Q., Li, Q., Pan, X.: Storage layout in steel logistics park based on EIQ-ABC-SLP method. Jimei Univ. J. (Nat. Sci. Ed.) **4**, 279–286 (2020). (in Chinese)
2. Zhang, M.T.Y.: Warehouse layout optimization research based on SLP Method -takes S warehouse as an example. Mass Invest. Guide **21**, 187–188 (2020). (in Chinese)

3. Wu, P., Zheng, Y.: Arrangement optimization design of storage center of SGCS. Logist. Eng. Manag. **41**(11), 78–80+98 (2019). (in Chinese)
4. Deng, B., Lin, G.C.: Workshop facility layout optimization by using improved slp and genetic algorithm. Modular Mach. Tool Autom. Manuf. Tech. **13**, 34–36 (2017)
5. Chen, H., Qi, Y.: Application of secondary analysis in the selection of logistics financial operation mode. Inf. Rec. Mater. **19**(03), 207–209 (2018). (in Chinese)
6. Yang, W., Chen, L., Zi, Q.K., Cheng, M.: Research on layout optimization of E-commerce warehouse of company b based on SLP and AHP. Enterp. Reform Manag. **17**, 65–67 (2019). (in Chinese)
7. Chen, L., Yang, W., Cheng, M., Zi, Q.K.: Research on layout optimization of E-commerce warehouse based on SLP. Enterp. Reform Manag. **15**, 66–67 (2019). (in Chinese)
8. Zhang, Z., Hu, M.: Research on warehouse layout optimization based on SLP method–take warehouse management of H E-Company. Times Finan. **23**, 161–163 (2020). (in Chinese)
9. Mohamed, M.: Smart Warehouse management using hybrid architecture of neural network with barcode reader 1D/2D vision technology. Int. J. Intell. Syst. Appl. **11**(11), 16–24 (2019)
10. Anitha, P., Malini, M.P.: A review on data analytics for supply chain management: a case study. Int. J. Inf. Eng. Electron. Bus. **9**(8), 30–39 (2018)
11. Zine, B., Ghalem, B., Abdelkader, N.: A cost measurement system of logistics process. Int. J. Inf. Eng. Electron. Bus. **9**(8), 23–29 (2018)

Rural Governance Based on the Coordination of Multiple Subjects Intelligent Research

Yanwei Xu[1], Yan Jiang[1](✉), Wan Zhang[1], and Wenliang Liu[2]

[1] Tianjin Agricultural University, Tianjin, China
[2] Minzu University of China, Beijing, China

Abstract. Rural governance is the center of social governance, rural governance modernization is an important embodiment of the modernization of national governance system and governance capacity, with the development of modern scientific information technology, intelligent governance has become an important development trend of rural governance. At present, china's intelligent development of rural governance has achieved some results, but it still have some problems, such as the concept of intelligence is not clear, insufficient data infrastructure, low participation of villagers, institutional mechanism is not perfect and so on, we need to actively change the concept of governance, and constantly improve the rural information infrastructure, improve the ability of villagers to govern, optimize intelligent development of the institutional mechanism, in order to make the intelligent development of rural governance better.

Keywords: Rural governance · Intelligent · Collaborative governance

1 Instruction

The report of the 19th National Congress of the Communist Party of China proposes to improve the intelligent level of social governance grass-roots governance is the focus of social governance, promote the intelligent development of rural governance at the grass-roots level in China, is to speed up the transformation of rural governance, achieve "good governance" of social governance, and promote the realization of rural revitalization of a powerful initiative.

Today's era is the era of information intelligence, modern science and technology, especially information technology, communication technology development, has brought a profound impact on the way of social governance, with the development and progress of the times, the traditional social governance has been difficult to adapt to and meet the needs of scientific and technological innovation and intelligent social construction, the development trend of intelligent social governance cannot resist [1].

Z. Hu et al. (Eds.): AIMEE 2021, LNDECT 107, pp. 30–40, 2022.
https://doi.org/10.1007/978-3-030-92537-6_3

2 Cooperative Governance of Multiple Subjects and Intelligentization of Rural Governance

2.1 Cooperative Governance of Multiple Subjects and Intelligentization of Rural Governance

The intelligent rural governance is to rely on modern scientific information technology, the ability of information collection and analysis of big data, combine governance with data, and make rural social governance more standardized, more efficient operation, more intelligent service, and more harmonious and stable social relations in rural areas [2].

2.2 Multi-subject Collaborative Governance

Collaborative governance refers to the governance activities and processes, in which differentiated governance entities form an open and orderly collaboration system, follow common governance goals, integrate their respective functional advantages, and maximize the common interests of the community through collaborative governance [3]. The participating subjects of rural governance have shown a diversified trend. The government, society and villagers are all subjects of rural governance.

2.3 The Meaning of the Intelligent Rural Governance

2.3.1 Improve the Rural Governance System, Promote the Modernization

As an important step of modern social governance, the intelligent rural governance has been studied and explored more and more widely, and the reform storm it brings is affecting everyone under the national governance system. Applying massive, dynamic, diverse, detailed, interrelated, low-cost data sets to the practice of rural governance and connecting and utilizing data sets in different ways provides the possibility for research to move from data scarcity to richness, static to dynamic, simple assumptions and models to more complex simulations and theories [3]. Thus, with data as the guide to form a new, associated with the traditional rural governance, the new theory of rural governance, enrich the connotation of rural governance, further improve the "autonomy, moral governance, rule of law" combined with the theory of rural governance system, make big data, Internet of Things, cloud computing, artificial intelligence information technology into the whole process of rural governance, can pay full attention to the analysis of rural community governance related issues and effectively resolve, can promote the realization of rural residents on the growing need for a better life, improve people's sense of gain, happiness and security. It is of great significance to promote the harmony and stability of rural society and the modernization of rural social governance [4].

2.3.2 Improve the Accuracy of Rural Governance, Realize the Link Between Supply and Demand and Maximize Service Efficiency

In traditional rural governance, the government often makes "experience-based" or "perceived" inaccurate decision-making. Many mistakes in decision-making, unreasonable allocation of governance resources, and low utilization efficiency are common consequences. The technical reason for these consequences lies in the lack of an integrated information governance interface under the existing specialized division of labor and layering system, which in turn leads to insufficient information and lack of a platform for communication and negotiation between subjects [5]. The intelligent governance platform based on information technology and data technology, through data statistical analysis, pattern recognition, expert online evaluation, and panoramic monitoring, maximizes the efficiency of government services. The application of intelligent governance platform in rural governance is necessary. This means that local governments can obtain sufficient information and integrate information resources. It means that local governments can use communication channels and coordination platforms to aggregate decision-making demands between cross-departmental, cross-level, and cross-border subjects. Intelligent platforms that use digital technologies such as big data and artificial intelligence can improve the level of scientific, refined, and intelligent rural governance decision-making, thereby realizing the link between supply and demand and maximizing service efficiency [6].

2.3.3 Improve the Modernization of the National Rural Governance System and Capacity, Building a New Pattern of Social Governance that is Shared by the Common Governance

Social governance is an important aspect of national governance, rural governance is an important basis for social governance, good rural governance is the premise and guarantee of social harmony and stability, and farmers live and work in peace. For the first time, the Third Plenary Session of the 18th Central Committee put forward the concept of "promoting the modernization of the national governance system and governance capacity" in the central document, and put forward the concept of "village governance" in the form of an official document, and clearly put forward the overall goal of "promote the modernization of the national governance system and governance capacity" as the goal of comprehensively deepening reform [7]. Intelligence provides an important condition for the transformation of governance thinking, with the development of modern information technology, the spatial pattern of people's participation in governance is broken, and based on data and intelligent analysis, change the traditional mode of thinking and operation of governance, opening a new path for the grass-roots government to enhance its governance capacity and improve the effectiveness of governance decision-making. Such as helping network information technology, the complex affairs in rural governance into the intelligent network, to achieve the organic integration of governance and network service management, the formation of multi-party, multi-factor linkage of governance pattern, efficient intelligent governance system has given the rural social development of a new vision [8]. To build an information integration platform, the grass-roots governments can share information, and villagers can participate in all aspects of rural governance to build a new pattern of rural co-construction and sharing.

3 The Status Quo of Intelligent Rural Governance in China

3.1 The Intelligent Policy System of Rural Governance is Gradually Perfected

At present, China's agricultural rural areas are rapidly moving towards the era of digital intelligent rural areas, agricultural rural economic life and agricultural production and management presents the characteristics of network, information, data, platform and intelligence, rural governance intelligence to promote the construction of China's digital villages to make an important contribution. In order to adapt to the changes brought about by information technology, improve the ability of rural governance, and speed up the construction of digital villages, the Chinese government has issued a series of policies to support the development of intelligent digitalization of rural governance (Table 1).

Table 1. The document to support the intelligence of rural governance

Times	Policy document	The main content
October 2017	Report of the 19th National Congress of the Party	Improve the level of intelligent social governance
November 2018	Opinions of the Central Committee of the Communist Party of China and the State Council on the Implementation of the Strategy for Rural Revitalization	Puts forward the implementation of digital village strategy to improve the level of intelligent rural governance
January 2019	No 1 Central Document	Clearly requires further promotion of the development of "Internet plus agriculture"
January 2019	Regulations on the Work of Rural Grassroots Organizations of the Communist Party of China	It is pointed out that we should pay attention to the use of modern information technology to improve the level of intelligent rural governance
July 2019	Notice of the General Office of the Ministry of Agriculture and Rural Affairs on the comprehensive promotion of information into the village into the household project	In-depth implementation of information into the village into the household project
January 2019	Fourth Plenary Session of the 19th Central Committee of the Communist Party of China	It is pointed out that scientific and technological support is an important guarantee point to improve the intelligence of governance
May 2020	Notice on the issuance of the > of the key points of digital rural development work < 2020	The target for digital rural development in 2020 was defined

3.2 The Effectiveness of Local Governments in Promoting the Intelligence of Rural Governance is Obvious

Starting from the second half of 2018, the reform programs of provincial local government agencies have been approved and announced to the public, several provinces in order to promote the "Internet plus government", "government network management" and "digital society" construction, according to their actual situation set up a big data bureau or related regulatory agencies. The Ministry of Industry and Information Technology and the Ministry of Finance organized the implementation of 6 batches of universal telecommunications service pilot projects, supported the construction of 130,000 administrative villages with optical fibers and 50,000 4G base stations, and gave priority to support the "three regions and three states" and other deeply impoverished areas to accelerate network coverage And the popularization and application, the ratio of both optical fiber and 4G access in administrative villages across the country has exceeded 98%, ensuring the Internet access needs of rural people. The central government and basic telecommunications companies have invested more than 50 billion Yuan in total, supporting the construction of optical fibers in 130,000 administrative villages and the construction of 36,000 base stations across the country. As of the end of 2020, the total number of rural broadband users nationwide has reached 142 million, with a net increase of 7.12 million throughout the year, an increase of 5.3% over the end of the previous year. The proportion of optical fiber and 4G in administrative villages across the country both exceeded 98%. The average download speed in the pilot areas for universal telecommunications services exceeded 70M, and the rural and urban areas achieved "same network and same speed" (Fig. 1).

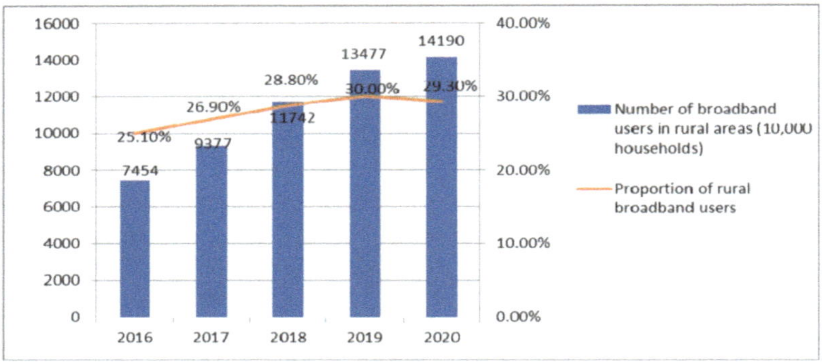

Fig. 1. 2016–2020 China's rural broadband access users and their proportions

3.3 Digital Agriculture Has Consolidated the Material Foundation of Rural Governance

The modernization of rural governance based on the intelligent perspective requires the support of capital and technology, as well as the guarantee of software and hardware facilities. The intelligent construction of rural governance belongs to the category of superstructures and cannot be separated from the solid material foundation created by the economic foundation. In recent years, the development of digital agriculture in my country has gradually improved. The digitalization of agriculture has broken the traditional production and processing methods. It has comprehensively collected and in-depth data on farmland, crops, inspection, packaging, warehousing, logistics and other links, and used satellites and remote sensing. Such technologies integrate and analyze these data, and then realize the Pareto improvement of agricultural resource allocation, promote high-quality and high-yield agricultural products, centralized transportation, and precise distribution, and give agriculture a "multiplier effect" for reducing costs and increasing efficiency. From the perspective of the integration of agriculture and the secondary and tertiary industries, the digitalization of agriculture, which is widely used in all aspects of agricultural production and management, including cloud computing, big data, Internet of Things, and artificial intelligence, and other new-generation information technologies, will promote the intelligent embedding of information into the planting industry, The development of animal husbandry, fishery, etc. has spawned a new type of business that integrates primary, secondary and tertiary industries, giving a "multiplier effect" to the efficiency and value-added of the entire rural industry chain. Regardless of whether the agricultural industry under agricultural digitalization or the reduction in costs and incomes of the entire rural industry and the increase in efficiency and value-added, it will promote the enhancement of rural endogenous economic development capabilities, thereby consolidating the material foundation for the modernization of the rural governance system [9] (Fig. 2).

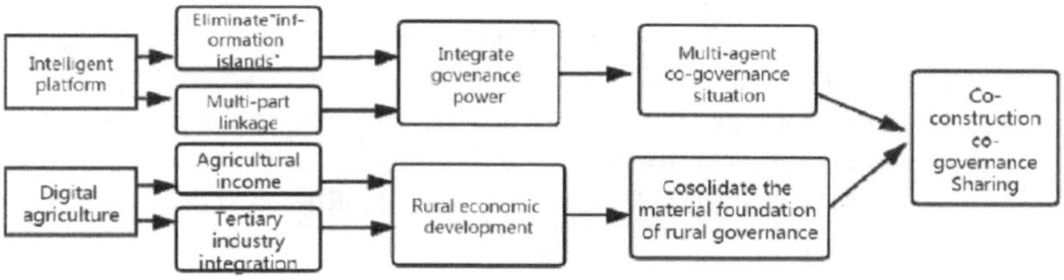

Fig. 2. The empowerment of intelligent technology rural governance

4 The Problem of the Intelligent Governance of Rural Areas in Our Country

4.1 The Government's Philosophy is Unclear and Lacks Reasonable Planning and Guidance

Collaborative and intelligent governance in rural areas is not only a product of adapting to the development of modern information technology and changes in actual social needs, but also the result of the country's initiative to delegate power. Over time, my country's rural areas have gradually moved from a society of low complexity and uncertainty to a society of high complexity and uncertainty. Increasingly complex public affairs and increasingly diverse social needs have brought unprecedented challenges to national governance capabilities [10]. Intelligent Rural Governance under

Multi-agent Cooperation is a linkage system engineering that requires the cooperation of the various governance subjects, to complete this project, it needs the overall design of the system, but due to the unclear concept, resulting in the lack of unified planning and deployment in the process of promoting the intelligent rural governance in our country, so far has not formed a set of mature and complete system design planning and standards, each subject is, lack of cooperation and coordination, and in the implementation process, "did not pay attention to the special situation in different regions, This makes it difficult to promote the intelligent construction of rural governance in our country, with only partial results.

4.2 Multi-agent Communication is not Smooth and "Information Gap" Appears

At present, the management system of the Chinese government is a "pyramid-shaped" bureaucracy. Regardless of whether it is at the central or local level, the administrative unit at the upper level has more information sources, wider channels, and more complete information. As a result, the internal information of the government management system is relatively closed, the information is asymmetry between the upper and lower levels, the information is asymmetry between departments, and the information is asymmetry between government departments and society [10]. From a horizontal perspective, the information exchange between urban and rural areas, between regions, and between farmers and citizens is not smooth, the situation of.

horizontal digital construction is uneven, and there are a lot of "data islands and data department symmetry". These are all obstacles on the road to the intelligent modernization of rural governance. From a vertical perspective, information sharing between local governments at all levels is insufficient, and an integrated rural information platform has not yet been formed, and the fragmentation of rural digital services has become prominent. All of these have aggravated the differentiation in the process of rural intelligent governance.

4.3 Villagers' Awareness of Participation in Rural Governance is Weak and Their Ability is Insufficient

The rural residents do not know enough about the content and field of intelligent participation in rural governance, which makes them lack the ability and learning motivation to participate in the intelligent participation of rural governance [11]. The sense of self-learning of rural residents is out of touch with the promotion of intelligence, and even many residents have never heard of the idea of intelligent rural governance [11]. Rural population structure and villagers' information literacy cannot adapt to the needs of intelligent construction of rural governance, in rural areas, the loss of young and middle-aged labor force, rural social subject hollowing, old and weak, at present, many elderly people in rural areas due to their own quality is not enough, backward concept, learning ability is relatively weak, for high-tech understanding is very little, and it is difficult to learn, will not use smartphones, will not access the Internet, affecting the process of rural governance intelligence.

4.4 The Intelligent Governance System Mechanism Under Multi-agent Coordination is not Sound

An innovative activity cannot be sustained without the support of the system. To promote the intelligent development of rural governance, the construction of institutional mechanism is an important guarantee and prerequisite. In recent years, China has issued a number of policy documents to standardize the intelligent development of governance, to a certain extent, improve the rural governance intelligent standardization, institutionalization level, but, on the whole, the intelligent institutional mechanism of rural governance is still not sound enough, the development of relative lag, mainly manifested in the rural governance intelligent responsibility implementation mechanism is not in place, the division of labor between departments is not clear, for example, after the problem, the departments have shirking responsibility [12]. The use of intelligent means of rural governance informatization is immature, the integration of "governance plus intelligence" is low, the degree of scientific decision-making, management refinement and service precision needs to be further improved, the experience extension mechanism of governance is not perfect, and the summary refinement and promotion of effective practices and successful experiences formed by practical exploration in various regions and departments need to be further strengthened.

5 Conclusion and Suggestions

The rural governance system is the micro foundation of the national governance system and an important guarantee for rural residents to build a better life. Intelligent technology is the key to promoting the modernization of the rural governance system. Therefore, it is necessary to strengthen the top-level design and transform the governance concept to promote multi-agent co-governance; it is necessary to encourage social forces to participate in governance and promote the development of rural economy to consolidate the material basis of governance; it is necessary to optimize the system and mechanism and strengthen the protection of intelligent development of social governance.

5.1 Strengthen the Top-Level Design and Change the Concept of Grass-Roots Governance

The government should constantly strengthen the concept of intelligent rural governance, break through the limitations of traditional governance thinking, establish the concept of data thinking, actively cultivate intelligent cadres and talents, and cultivate a new type of talents who have both intelligent ideas and are good at mastering the laws of governance [12]. In the overall planning and design at the same time, to take into account local particularities, for the countryside, intelligent governance is an unprecedented new thing, almost no reference experience, and high technical requirements, implementation is very difficult, can not blindly encourage all villages to carry out intelligent construction, to estimate the objective reality of different places, adhere to demand-oriented, according to local conditions, can not pursue unity.

5.2 Strengthen the Construction of Information Platform and Promote the Coordination of Governance Entities

The establishment of an information governance platform is the basis for realizing the intelligentization of rural governance, and it is also an important carrier for realizing multi-agent collaborative governance. It is necessary to strengthen the construction of local rural informatization infrastructure, upgrade and upgrade rural network optical fiber and popularize the network [13]. The government must take the initiative to innovate, decentralize, actively disclose information, and dare to accept social supervision. The society must invest capital and technology, and villagers actively participate in expressing their needs. As shown in Fig. 3, the government promotion, the development of information technology and the participation of society and villagers are placed in a three-dimensional coordinate graph. Only the higher the level of each party, the larger the corresponding "volume", and the level of collaborative and intelligent governance in rural areas the higher it is.

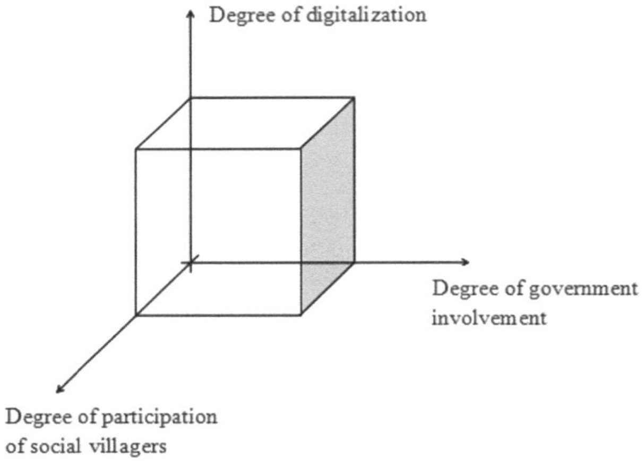

Fig. 3. Three-dimensional model of collaborative and intelligent governance in rural areas

5.3 Improve the Villagers' Awareness of Participation and Cultivate Their Ability to Govern

It is also very important to optimize the system and mechanism and strengthen the intelligent development guarantee of social governance [16]. The improvement of villagers' awareness of participation and management ability needs to be realized through training [14, 15]. Starting with the basic education of modern villagers, we consciously cultivate the villagers' consciousness of participating in governance, and actively mobilize the enthusiasm of villagers to participate in governance through classroom teaching, organizing collective learning, daily publicity, and other means, so that villagers have a sense of subjectivity. Actively carry out the application training of rural cadres on intelligent data platform, to further strengthen their awareness of intelligent governance and improve the ability of rural cadres to use modern information technology for governance. In rural areas, villagers are regularly organized to learn to use the Internet and various data application platforms, improve their own cultural literacy, and master the use of the Internet to participate in governance.

5.4 Optimize the Institutional Mechanism and Strengthen the Intelligent Development Guarantee of Social Governance

Firstly, we should strengthen organizational protection, build an intelligent responsibility implementation mechanism for rural governance, establish a reward and punishment system, and incorporate the level of intelligent development of rural governance and its effectiveness into the assessment objectives of individual departments [14]. Secondly, we should integrate the service platform and build an intelligent process optimization mechanism for rural governance. Thirdly, we should carry out pilot demonstration and construct the mechanism of exploring the intelligent practice of rural governance. Pay attention to the difficulties in the process of intelligent development of rural governance, select representative areas with good basic conditions as pilot areas, the government, and enterprises to provide financial and technical support for the pilot areas, encourage the pilot areas to explore boldly, form a typical development model of intelligent rural governance, other regions to learn from, and comprehensively promote the intelligent development of rural social governance (Fig. 4).

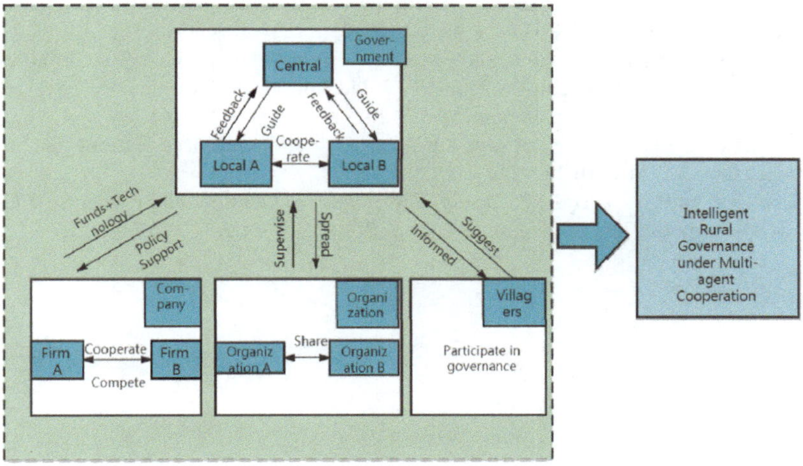

Fig. 4. The intelligent model of rural governance with multi-agent participation

References

1. Zhang, G.: The real dilemma and implementation path of intelligent development of Social Governance. Leading Science **06**, 32–34 (2020). (in Chinese)

2. Chen, Z.: Digital and intelligent transformation of governance. Natl. Gover. **03**, 20–24 (2020). (in Chinese)
3. Liu, X., Nie, J.: How to build a new pattern of rural governance in the new eras. People's Forum **02**, 70–71 (2020). (in Chinese)
4. Zhou, W.: Digital government and modernization of national governance. Administrative Management Reform **02**, 4–10 (2020). (in Chinese)
5. Zhou, X., Li, L.: Promoting the overall change of rural governance system with "Three Governances." Exploration **04**, 156–163 (2019). (in Chinese)
6. Gao, Q.: Promoting the goal of "good governance" in rural areas with intelligent governance. Natl. Govern. **19**, 29–33 (2019). (in Chinese)
7. Xu, X.: The outstanding problems and governance path of the current grass-roots governance. Natl. Govern. **26**, 9–12 (2020). (in Chinese)
8. Zhang, X.: Village "Hollow" governance under the strategy of rural revitalization-based on the investigation and analysis of rural areas in northern Henan. Int. J. Social Sci. Educ. Res. **4**(4), 320–326 (2021)
9. Ágoston, B.: The languages of village governments in the eastern stretches of dualist hungary: rights and practices. Slavonic East Eur. Rev. **99**(1), 1–30 (2021)
10. Jiang, W., Hu, M., Li, L.: Research on digital technology promoting the modernization of rural governance system. Electron. Gov. (07), 72–79 (2021). (in Chinese)
11. Lai, J.: Study on the dilemma and countermeasures of urban communitygovernance based on multiple coordination. Heihe Acad. J. **02**, 8–11 (2019). (in Chinese)
12. Listyaningsih, V., Utami, E.: Decision support system performance-based evaluation of village government using AHP and TOPSIS methods: Secang sub-district of magelang regency as a case study. Int. J. Intell. Syst. Appl. (IJISA) **10**(4), 18–28 (2018)
13. Anand, R., Saxena, S., Saxena, S.: E-Learning and its impact on ru-ral areas. Int. J. Mod. Educ. Comput. Sci. (IJMECS) **4**(5), 46–52 (2012)
14. Hu, X.: Sustainable renovation and low-energy retrofitting of houses in rural areas of southern China. In: Proceedings of the 2017 2nd International Conference on Machinery, Electronics and Control Simulation (MECS 2017), pp. 37–44 (2016)
15. De Matas, S., Keegan, B.P.: A case study on adult and workplace learning. Int. J. Educ. Manag. Eng. (IJEME) **10**(1), 11–19 (2020)
16. Bharaditya, W.P., Sukarsa, M., Buana, P.W.: Internal Control Improvement for Creating Good Governance. Int. J. Inf. Eng. Electron. Bus. (IJIEEB) **9**(3), 9–17 (2017)

Swin Transformer and Mask R-CNN Based Person Detection Model for Firefighting Aid System

Shuo Gao[1], Wei Ren[1], and Kan Hu[2(✉)]

[1] Changsha Fire Rescue Detachment, Changsha 410003, China
[2] Wuhan ONEW Technology Co., Ltd., Wuhan 430223, China

Abstract. In the complex and dynamic fire environment, firefighters in rescue missions are susceptible to visual interference, which could reduce the firefighters' efficiency in detecting people at the scene. Moreover, low visibility usually has the firefighters exposed to great threats. Deep Learning techniques have become a powerful method of learning feature representations directly from data and have led to significant breakthroughs in the field of object detection. In order to tackle the problem of low visibility in the firefighting scene, we apply the Deep learning techniques in object detection to aid the firefighters to improve their visual perception. In this paper, a Swin Transformer and Mask R-CNN based framework for person detection at firefighting scenarios is applied to a Firefighting Aid System. In this model, Swin Transformer module is considered as the backbone of Mask R-CNN model. Compared with baseline model, the experimental results demonstrate that our model is capable of accurate person detection in the complex fire environments and performs better than the baseline model.

Keywords: Firefighting · Deep learning · Object detection · Mask R-CNN · Swin transformer

1 Introduction

Fire poses a great threat to the safety of public life and property, so it is important to conduct fire rescue missions effectively and safely when a fire occurs. During the firefighting mission, the fire environment with low visibility affects firefighters' judgment, which may result in wrong decisions. In addition, firefighting is an inherently dangerous activity, as firefighters are exposed to dangers such as lack of oxygen, which may due to dense smoke, chemical leaks, and secondary explosions in the fire environment. Because the intelligent information platform can eliminate space-time limitation, maximize the utility of multi-party collaboration in fire rescue activities and expand the function to assist in recognizing the fire scenes quickly, it's crucial to improve the intelligence level of firefighting, which will contribute to improving the efficiency and coping with potential hazards in firefighting activities.

Z. Hu et al. (Eds.): AIMEE 2021, LNDECT 107, pp. 41–50, 2022.
https://doi.org/10.1007/978-3-030-92537-6_4

With the rapid development of deep learning (DL), increasing researchers in various fields tend to utilize different DL algorithms to gain new inspiration for problem solving. DL allows computational models to learn fantastically complex, subtle, and abstract representations, driving significant progress in a broad range of problems including object detection [1]. DL based object detection algorithms aim to locate object of interest from images, accurately determine the category of each object, and output the bounding box. According to the presence or absence of Region Proposal stage, the mainstream DL object detection algorithms are divided into two categories: one-stage object detection algorithms and two-stage object detection algorithms. The one-stage object detection algorithms, such as YOLO, SSD act on the image to generate detection results directly, which is fast but performs moderate in detection accuracy [2, 3]. The two-stage object detection algorithms extract the region proposal from the image and then makes a correction again based on the region proposal to get the detection results, which has a higher detection accuracy. R-CNN, Fast R-CNN, Faster R-CNN are widely used two-stage object detection algorithms [4–6]. Mask R-CNN is extended from Faster R-CNN and has higher accuracy in detection a large number of small objects [7].

This paper introduces a model based on Mask R-CNN with Swin Transformer as the backbone for object detection in firefighting scenarios, being capable of detecting the wounded trapped and firefighters simultaneously in the fire scene [7, 8]. Thus, the algorithm can be utilized in the Firefighting Aid System, which is able to provide real-time information or knowledge to off-site control centers. Such information or knowledge can assist in live operation command and enabling them to make judgments quickly and take effective actions by conveying information about object of interest to firefighters.

2 Related Works

There are many researchers applying various technologies in firefighting. Joshua Shaffer et al. propose high-level mission planner and controller to manage unmanned aerial vehicles, designed for various firefighting scenarios [9]. GPS and Path-Planning algorithm are employed respectively in path planning for firefighting robot [10, 11]. Rory M. S. Clifford et al. utilize Virtual Reality to simulate a high-fidelity and stressful firefighting scene for firefighting training [12]. There have also been several studies in assisting firefighters to perceive the environment better. With the widely application of DL models in the field of computer vision, some researchers have applied this in firefighting. Manish Bhattarai, et al. uses a trained Deep Convolutional Neural Network system to classify and identify objects of interest, informing firefighters real-time information about the environment [13]. Saponara et al. creates a smart and real-time video-surveillance system for fire/smoke detection by YOLOv2 Convolutional Neural Network [14].

As DL algorithms have emerged as powerful methods for learning feature representations automatically from data, research related to Computer Vision based on DL has increased rapidly in recent years [1]. Mask R-CNN is a generic object instance segmentation framework that was proposed in 2017 to perform object detection and instance segmentation, and has been widely used in personnel detection, autonomous driving, vehicle-damage-detection, geospatial object detection etc. [15–18]. ResNet has been adopted for feature extraction in the Mask R-CNN models in some research [15, 18]. ResNet is widely applied because of the advantages such as being able to solve the problems like deep network degeneration, as well as the accuracy of the classification increases as its network becoming deeper [19]. As a state-of-the-art vision transformer, Swin Transformer can be applied as a generic backbone for Computer Vision, which has shown excellent performance in object detection on Common Objects in Context (COCO) datasets, significantly outperforming previous state-of-the-arts, so it has powerful potential in Computer Vision. However, no research until now applies Swin Transformer framework to person identification for fire rescue activities.

3 Firefighting Aid System

The Firefighting Aid System proposed in this paper is a decision support system for firefighting activities, the operation platform of which has the functions of basic information display, event setting, event analysis, operation deployment, real-time monitoring, etc., as is demonstrated in Fig. 1. The system provides information about the firefighting action updated live, and it offers planning and decision support for the firefighters. This system can be applied in the fire rescue and firefighting exercises. In the monitor module, DL algorithms are utilized to dispose the captured images to quickly identify and highlight the objects of interest such as people. In the following sections, we focus on the object detection models for person detection.

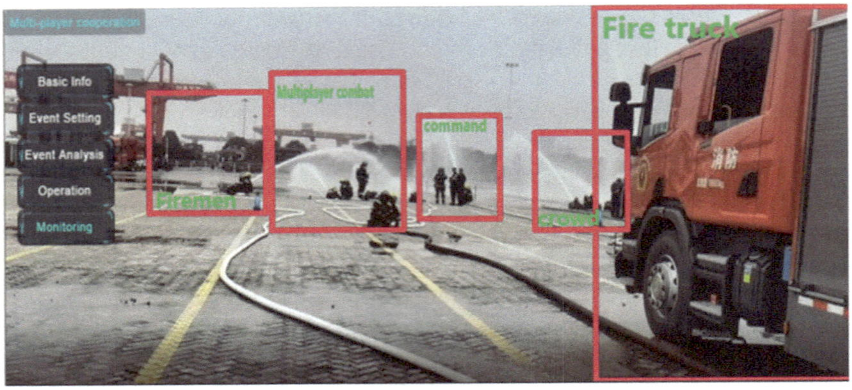

(a)

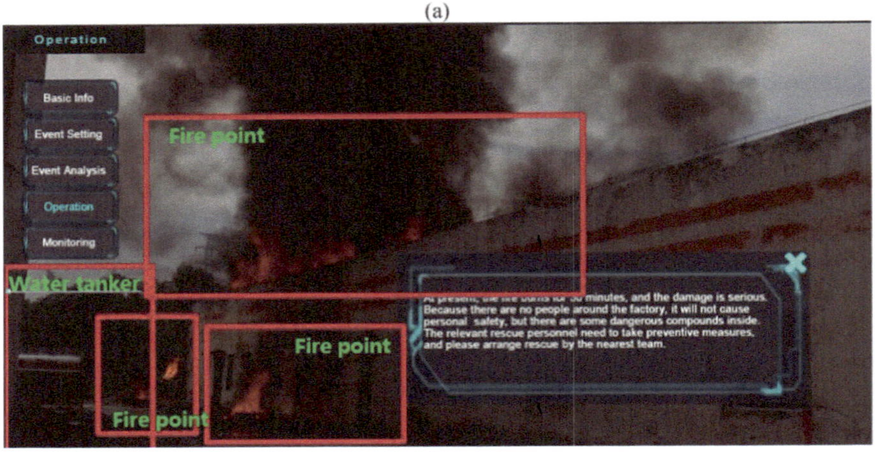

(b)

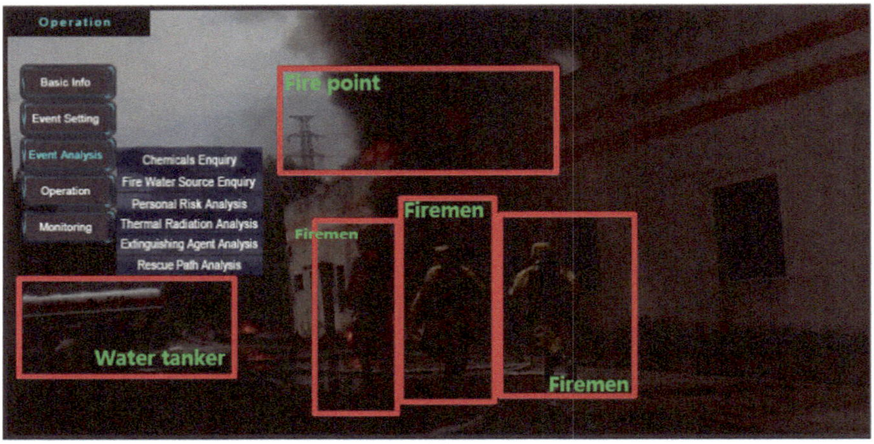

(c)

Fig. 1. Fire rescue detection aid system interface (a) Multi -player cooperation-monitoring; (b) Rescue operation -operation; (c) Rescue operation -event analysis

4 Deep Learning Based Object Detection in Firefighting Scenario

The model designed in this paper is the object detection model based on Swin Transformer + Mask R-CNN, and the trained model can detect and recognize people in the fire scene and output the bounding box of human body in the input images. (The framework of the model is shown as Fig. 2). The video of fire rescue training exercise is input and preprocessed by frame extraction at 5 frames per second and finally get 1380 images, then use LabelImg for person labeling for 1000 of the images.

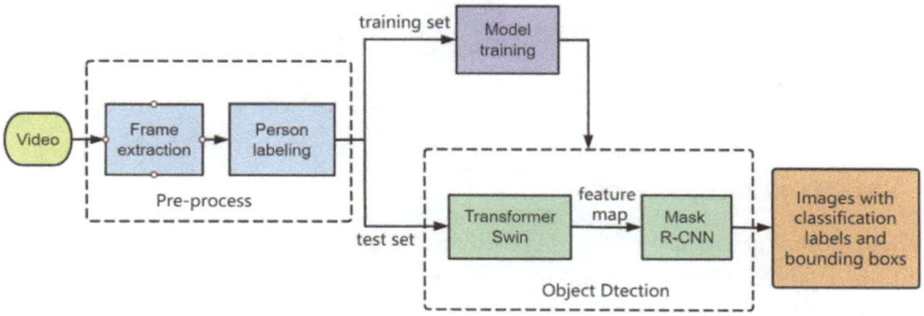

Fig. 2. The framework of the model

The model bases on Mask R-CNN with Swin Transformer series as the backbone (shown in Fig. 3.). Swin Transformer series refers to Swin Transformer's base model (Swin base) and it is variants: Swin tiny and Swin small, which are versions of about 0.25× and 2× the computational complexity of the base model. In this paper, Mask R-CNN framework is used for classification and bounding box output, and the model is trained by the images extracted from the video of fire rescue training exercise. The multitask loss function for each sample in training is defined as:

$$L = L_{cls} + L_{box} \tag{1}$$

L_{cls} is Classification loss and L_{box} represents Bounding-Box regression, both using the Fully Convolutional Network as the input.

4.1 Mask R-CNN Framework

Mask R-CNN has been widely used as an object instance segmentation framework, which was proposed by Facebook AI Research. It is extended from Fast R-CNN and creates RoIAlign layer to replace RoI Pooling layer, and adds mask output. Generally, Convolutional Neural Networks are used in a Mask R-CNN model for feature extraction, while in this paper, Swin Transformer series are used instead to output the feature map and then input it to the Region Proposal Network (RPN) to get Region of Interest (RoI). The RoI is processed in the RoIAlign layer where the fixed size feature map is produced, and finally performs classification and box regression. RoIAlign layer uses a computing method of bilinear interpolation, addressing the problem of misalignments between the RoI and the extracted features generated by two steps of quantization in RoI Pooling which is adopted by Faster R-CNN.

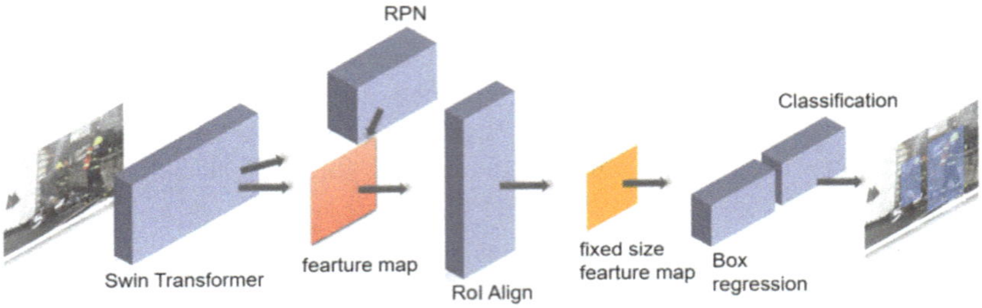

Fig. 3. The framework for Mask R-CNN

4.2 Swin Transformer Network

The Swin Transformer has a four-stage hierarchy as shown in Fig. 4. The first stage is to split the input RGB image into non-overlapping patches embedded in the embedding block through the patch splitting module, and then pass them into the swin transformer block for feature transformation. Stages 2 to 4 each firstly enter patches into Patch Merging procedure, merging features of adjacent patches, and output lower dimensional features, then pass them into the Swin transformer block. These stages enable the hierarchical representation to output the same resolution feature map as a typical Convolutional Network.

The Swin Transformer block consists of a shifted window based on multi -head self-attention (MSA), a 2-layer MLP, and a layer norm in front of each module, where the shifted window scheme is an important innovation of this new framework. The shifted window replaces the traditional transformer's slide window, restricting the self-attention computation to a non-overlapping local window, the computational complexity of a global MSA module and a window MSA module are computed as (2) and (3) respectively, where suppose the image consist of $h \times w$ patches and each window contains $M \times M$ patches. C represents an arbitrary dimension. We can find that the computational complexity of the window MSA is smaller compared with global MSA for a large hw. Besides, the MSA module allows cross-window connections, thus the model can dispose huge amounts of data while maintaining high efficiency.

$$\omega(MSA) = 4hwC^2 + 2(hw)^2C, \tag{2}$$

$$\omega(W - MSA) = 4hwC^2 + 2M^2hwC, \tag{3}$$

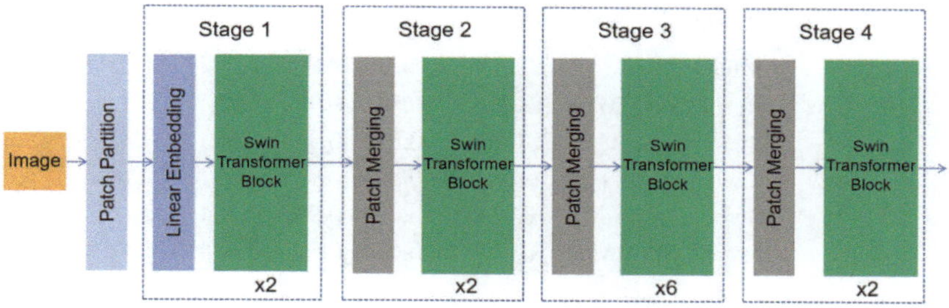

Fig. 4. The architecture of Swin transformer

5 Experiments and Analysis

The models run on a server with 4 TITAN RTX GPUs, by CUDA 10.2, using Python for programming under the PyTorch DL framework.

5.1 Evaluation Indicators

These experiments test the models including Mask R-CNN model with Swin base, Swin tiny, and Swin tiny3 (Swin tiny dilation = 3) as backbone respectively, using ResNet50 + Mask RCNN as the baseline model. To measure the object detection accuracy of the studied models, the output parameters including mAP (mean Average Precision), AP50 (The estimated average precision when the IoU threshold is 0.5.) and AP75 (The estimated average precision when the IoU threshold is 0.7) were counted as evaluation criteria map is an important measurement of object detection accuracy and is calculated as follows. Firstly, precision (p) and recall (r) is computed as (4) and (5) (TP, FP, TN, and PN indicate true positive, false positive, true negative, and false negative, respectively, where positive or negative are determined by the binary classifier).

$$p = \frac{TP}{TP + FP} \tag{4}$$

$$r = \frac{TP}{TP + FN} \tag{5}$$

The PR curve is formed by the sets of (p, r) under a range of confidence thresholds. AP equals to the area under the line of the PR curve, which can evaluate the classifier, and usually the higher the AP, the higher the accuracy of the classifier. mAP (mean average precision) is the mean value of AP of multiple categories, indicating the comprehensive performance of the object detection models.

5.2 Experimental Results and Analysis

The test set containing 300 images was input to the Mask R-CNN models combined with different backbone networks, yielding the experimental results in Table 1 (Fig. 5).

Table 1. Person detection results of the models

Model	mAP	AP50	AP75
ResNet50 + Mask R-CNN	0.542	0.904	0.65
Swin tiny + Mask R-CNN	0.555	0.938	0.633
Swin tiny3 ×+Mask R-CNN	0.543	0.933	0.631
Swin small + Mask R-CNN	0.603	0.936	0.754
Swin base + Mask R-CNN	0.677	0.938	0.834

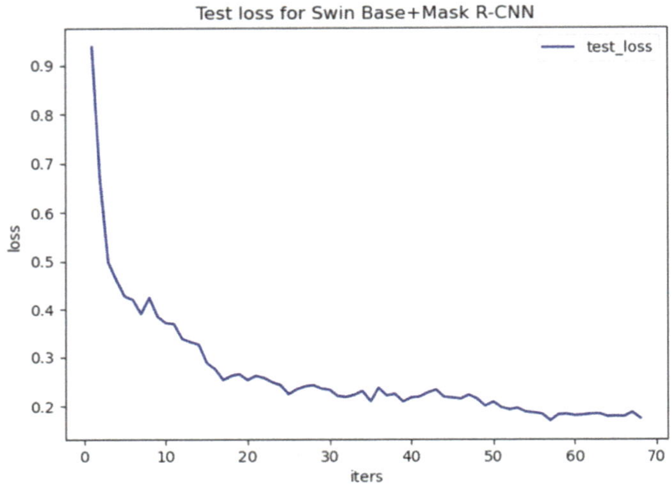

Fig. 5. Test loss for Swin Base + Mask R-CNN

As can be seen from Table 1, the Swin Transformer series have better performance than ResNet50. Among which, the Swin base is the best, with mAP, AP50, and AP75 values improves 0.135, 0.034, and 0.769 respectively compared with the baseline model, which is a significant improvement. Figure 6 displays the test loss curve of Swin Base + Mask R-CNN, indicating that the model is a good fit. A processed image is displayed in Fig. 6. The model can detect people and output the bounding box, so that people in view can be quickly discovered.

Fig. 6. Person detection results on the picture using Mask R-CNN with Swin Transformer

6 Conclusion

The DL model generally performs well in a new environment and can take care of scenarios that slightly differ but have the same feature with what it's being trained on [20]. We propose the DL based Mask R-CNN model to meet the powerful visual perception needs of firefighters during firefighting operations, which could improve the safety and efficiency of firefighting activities. The model utilizes the latest general-purpose Computer Vision backbone, Swin Transformer, which has been shown to outperform all the state-of-the-art. This study verifies that Mask R-CNN with Swin Transformer series framework has significantly higher object detection accuracy than Mask R-CNN with ResNet50 as the baseline model. Since the model accurately identifies people on pictures of fire rescue scenes, and can highlight the identified people with labels and bounding box, it can be embedded in the Firefighting Aid System, supporting the fire monitoring and fire rescue activities.

There remain a lot of challenges to output high quality results of object detection via streaming media in the Firefighting Aid System. When the scene is poorly illuminated, it is difficult to extract features of the images, and sometimes cameras and data transmission equipment can't provide robust, high-resolution images. The problems above both cause negative impact on the accuracy of object detection, so our future work will focus on solving the problems.

References

1. Liu, L., Ouyang, W., Wang, X., et al.: Deep learning for generic object detection: a survey. Int. J. Comput. Vision **128**, 261–318 (2019)
2. Redmon, J., Farhadi, A.: YOLO9000: better, faster, stronger. In: 2017 IEEE Conference on Computer Vision and Pattern Recognition, pp. 6517–6525 (2017)

3. Liu, W., et al.: SSD: single shot multibox detector. In: Leibe, B., Matas, J., Sebe, N., Welling, M. (eds.) Computer Vision – ECCV 2016: 14th European Conference, Amsterdam, The Netherlands, October 11–14, 2016, Proceedings, Part I, pp. 21–37. Springer International Publishing, Cham (2016). https://doi.org/10.1007/978-3-319-46448-0_2
4. Xie, Y., Shen, J., Wu, C.: Affine geometrical region CNN for object tracking. IEEE Access **8**, 68638–68648 (2020)
5. Girshick, R.: Fast R-CNN. In: 2015 IEEE International Conference on Computer Vision (2015)
6. Ren, S., He, K., Girshick, R., et al.: Faster R-CNN: towards real-time object detection with region proposal networks. IEEE Trans. Pattern Anal. Mach. Intell. **39**(6), 1137–1149 (2017)
7. He, K., Gkioxari, G., Dollár, P., et al.: Mask R-CNN. IEEE Trans. Pattern Anal. Mach. Intell. **42**(2), 386–397 (2020)
8. Nguyen, H.D., Vu, X.S., Le, D.T.: Modular graph transformer networks for multi-label image classification. In: Proceedings of the AAAI Conference on Artificial Intelligence, vol. 35, no. 10, pp. 9092–9100 (2021)
9. Shaffer, J., Carrillo, E., Huan, X.: Hierarchal application of receding horizon synthesis and dynamic allocation for UAVs fighting fires. IEEE Access **6**, 78868–78880 (2018)
10. Zhang, Z.: Path planning of a firefighting robot prototype using GPS navigation. In: 2020 3rd International Conference on Robot Systems and Applications (2020)
11. Ramasubramanian, S., Muthukumaraswamy, S.A.: On the enhancement of firefighting robots using path-planning algorithms. SN Comput. Sci. **2** (2021)
12. Clifford, R.M.S., Jung, S., Hoerrnann, S., et al.: Creating a stressful decision making environment for aerial firefighter training in virtual reality. In: 2019 IEEE Conference on Virtual Reality and 3D User Interfaces, pp. 181–189 (2019)
13. Bhattarai, M., Martínez-Ramón, M.: A deep learning framework for detection of targets in thermal images to improve firefighting. IEEE Access **8**, 88308–88321 (2020)
14. Saponara, S., Elhanashi, A., Gagliardi, A.: Real-time video fire/smoke detection based on CNN in antifire surveillance systems. J. Real-Time Image Proc. **18**, 889–900 (2021)
15. Chen, T., Jiang, Y., Jian, W., et al.: Maintenance personnel detection and analysis using mask-RCNN optimization on power grid monitoring video. Neural Process. Lett. **51**, 1599–1610 (2020)
16. Tahir, N.M., Bature, U.I., Baba, M.A., et al.: Image recognition based autonomous driving: a deep learning approach. Int. J. Eng. Manuf. **10**(6), 11–19 (2020)
17. Zhang, Q., Chang, X., Bian, S.B.: Vehicle-damage-detection segmentation algorithm based on improved mask RCNN. IEEE Access **8**, 6997–7004 (2020)
18. Al-Alimi, D., Shao, Y., Alalimi, A., et al.: Mask R-CNN for geospatial object detection. Int. J. Inf. Technol. Comput. Sci. **12**(5), 63–72 (2020)
19. He, K., Zhang, X., Ren, S., et al.: Deep residual learning for image recognition. In: 2016 IEEE Conference on Computer Vision and Pattern Recognition, pp. 770–778 (2016)
20. Aamir, M., Rahman, Z., Abro, W.A., et al.: An optimized architecture of image classification using convolutional neural network. Int. J. Image Graph. Signal Process. **11**(10), 30–39 (2019)

Investigation of the Dynamics of the Formation of Collective Immunity in a Pandemic Using Multi-agent Modeling and Selective Neural Network

Mikhail Mazurov(✉)

Russian University of Economics Moscow, Moscow, Russia

Abstract. The review of the known methods of mathematical modeling of the pandemic, based on the use of ordinary differential equations, probabilistic methods and the use of agent-based modeling, is given. It is shown that the probabilistic method allows us to estimate the percentage of collective immunity necessary to stop the epidemic process. However, it does not give an idea of the dynamics of the formation of collective immunity. A mathematical model has been developed for the study of the pandemic and the formation of collective immunity, based on the use of a multi-agent method in combination with selective neural networks. A nonlinear function reflecting changes in the state of agents is defined. The used mathematical model of the pandemic and the formation of collective immunity allowed us to qualitatively investigate the following tasks: the impact of quarantine measures on the dynamics of the pandemic; the impact of vaccination on the dynamics of the pandemic. Training with an electoral neural network is reduced to the formation of clusters of an electoral neural network. The developed software allows us to establish the main properties of the dynamics of the formation of a coronavirus infection.

Keywords: Pandemic · Quarantine measures · Vaccination · Collective immunity · Hybrid multi-agent model · Selective neural networks

1 Introduction

The formation of collective immunity (CI) is not only a medical problem, but to a large extent a social and economic problem that has a significant impact on the state of economic development. Quarantine measures and vaccination are a radical means of combating the pandemic. All quarantine measures provide only a temporary weakening of the development of the pandemic. As soon as a sick person appears among the remaining healthy people, the epidemic will break out with a new force.

People who are immune to the disease act as a barrier to the spread of the disease, slowing down or preventing the transmission of the disease to others. Immunity can be acquired after recovery from an illness or after vaccination. When a critical part of the population acquires immunity, called the threshold of collective immunity, the disease may no longer persist in the population and cease to be endemic. The formation of collective immunity is possible only in the case of vaccination of a significant part of

Z. Hu et al. (Eds.): AIMEE 2021, LNDECT 107, pp. 51–66, 2022.
https://doi.org/10.1007/978-3-030-92537-6_5

the population in combination with quarantine measures, there is no other way. The development of collective immunity, which significantly depends on the social behavior of the population, comes to the fore. Mathematical modeling is an effective means of predicting and evaluating the effectiveness of such measures.

1.1 Literature Review

The classical approach for studying the dynamics of the pandemic is mathematical models based on differential equations. The first model was proposed in 1927 by Kermak and McKendrick in the form of a system of differential equations and described the change in the number of susceptible (S), infected (I) and recovered individuals (R), called the SIR model. The theory developed by Kermak and McKendrick marked the beginning of a new direction in the study of infectious diseases – mathematical epidemiology. The SIR model was further developed, as a result of which the following similar more advanced models appeared [1–5]. The limitations of this approach are the assumption of the homogeneity of each of the groups of individuals under consideration, the instantaneous mixing of the entire population, the inability to assess the influence of individual subjects.

One of the ways to overcome these limitations is the agent-based modeling method [6–17]. The agent-based modeling method is based on the representation of the system under study as a set of interacting and differing agents. The properties of agents can change over time as a result of the interaction of agents with each other and with the environment. Agent-based modeling is an intuitive method for studying complex and poorly formalized systems, for which much is known about the properties of their constituent agents and little about the behavior of the system as a whole. The first agent models appeared in the social sciences in order to study how the properties of individual individuals (micro-properties), which can lead to local social shifts, lead to the reaction of society as a whole (macro-property). In agent models of epidemic processes, the parameters are the duration and proximity of contacts between infectious and sensitive agents. These parameters are observable, they can be measured and they can be controlled, for example, through quarantine measures, vaccination. The methodology for constructing agent models includes the following components: For example, in epidemiological models, a simplified description of the environment makes it difficult to analyze the effectiveness of various quarantine measures. Too detailed description of the environment is time-consuming, complicates the software implementation of the model and increases the computational costs of modeling. From a practical point of view, the level of detail of the description of the environment and agents is determined by the goals of modeling and the available resources. The work [18] contains the latest information on the dynamics of the development of the pandemic. The monograph [19] contains quite detailed information about the properties of agent-oriented models.

1.2 Probabilistic Method for Calculating Collective Immunity

Termination of the epidemic process is possible when collective immunity is achieved. One of the ways to study collective immunity is a method based on probabilistic estimates. According to the known data on the development of a pandemic, when implementing vaccination, collective immunity occurs in the following cases: 1) All objects of the population have been ill, and they have immunity; 2) Most of the objects of the population have been vaccinated; 3) Some of the objects have been ill, and some of the objects of the population have been vaccinated. The required percentage of vaccination to stop the pandemic P is equal to

$$P = (1 - \frac{1}{R}) \cdot 100$$

Where, R is the number of healthy people infected by one patient [18]. Graph of the function P = f (R) shown in Fig. 1.

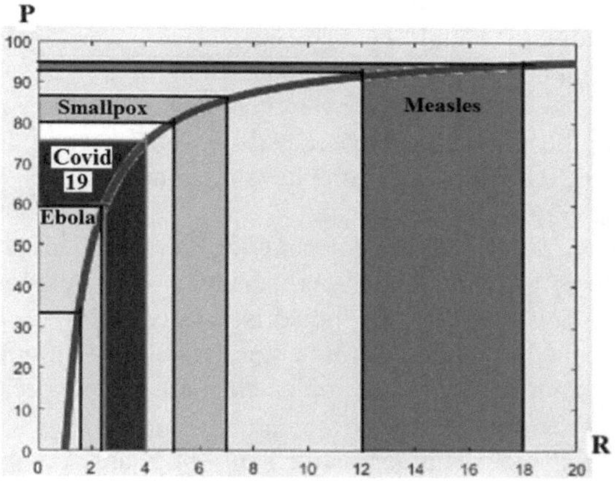

Fig. 1. The graph of the function P = f (R)

Note that the horizontal axis shows the intervals that characterize the number of healthy people infected by one patient for various pandemics. The vertical axis shows the intervals that characterize the percentage of collective immunity necessary to stop the epidemic.

It is assumed that infection occurs when a certain amount of infection is exceeded for one patient. According to experts, the spread rate of the COVID-19 coronavirus in the absence of deterrent measures is 2.5–4. Thus, if the spread coefficient is 2.5–4, then, according to this formula, in order to develop collective immunity, (60–75)% of the population must be ill.

The above formula allows us to estimate in general terms the necessary percentage of collective immunity necessary to stop the epidemic process. However, it does not give an idea of the dynamics of the formation of collective immunity, does not allow us to find the parameters of the epidemic process that can effectively slow down and stop the epidemic.

1.3 Multifactorial Mature of the Dynamics of the Pandemic and Vaccination

The development of the dynamics of the pandemic and vaccination, as pointed out by the President of the Russian Academy of Sciences, Sergeev A.M., is influenced by a huge number of factors and it is impossible to take them all into account. However, in the future, it is assumed that a large number of influencing social factors will be taken into account and a more complete picture of the development of the pandemic and the formation of CI will be formed. President of the Russian Academy of Sciences (RAS) Sergeev A.M., speaking at an online press conference at the International Information Agency (MIA) "Russia Today", said that more empirical data is needed to model the spread of COVID-19, which is not yet available. He recalled that in the spring, during the first wave of coronavirus, scientists in different countries proposed several models for the development of the pandemic, but none gave a forecast that was confirmed in the future. "It should be a lesson and an understanding of the fact that there are so many unknown numerical characteristics in COVID. That whatever the best, most powerful model you have for the most powerful supercomputer, it will not work. Because empirical numerical characteristics are inserted into any model to analyze, to look at the dynamics. We don't have them and no one else does," the head of the Russian Academy of Sciences said.

An effective mathematical model for studying the dynamics of the formation of collective immunity is a hybrid multi-agent model in combination with a selective neural network. The advantage of this model is the possibility of a visual qualitative study of a large number of modes of formation of collective immunity, as well as a simple selection of parameters for the implementation of the specified modes of formation of CI.

The purpose of this work is a qualitative study of a large number of modes of CI formation, the selection of parameters for the implementation of the specified modes and the use of the information obtained for effective management of CI formation. It is planned to develop a qualitative and visual picture of the dynamics of the formation of a pandemic, collective immunity.

Reducing the number of agents in the total sample allows us to accelerate the dynamics of the pandemic process and observe the details of this process. Therefore, the number of agents in the sample was taken to be equal to 10000. On a regular personal computer, it is possible to work with the number of agents equal to 106, which corresponds to the number of residents of Moscow, but the visibility of observing the phases of the development of the pandemic is lost. The task of creating accurate methods for accurately predicting the dynamics of the pandemic and the formation of CI is not set. The method under consideration allows us to make a qualitative prediction of the dynamics of the formation of a pandemic, collective immunity.

2 Mathematical Model of the Pandemic Based on a Multi-agent Model in Combination with Selective Neural Networks

We will give a description of the hybrid mathematical model used: 1) The model consists of agents with specified properties and rules for changing states; 2) Rules and methods of interaction between agents; 3) The environment in which agents exist and the rules for interaction of agents with the environment; 4) The model includes training of a selective neural network with selective agent neurons; 5) The formation of random initial data. The block diagram of the multi-agent model of the epidemic process is shown in Fig. 2.

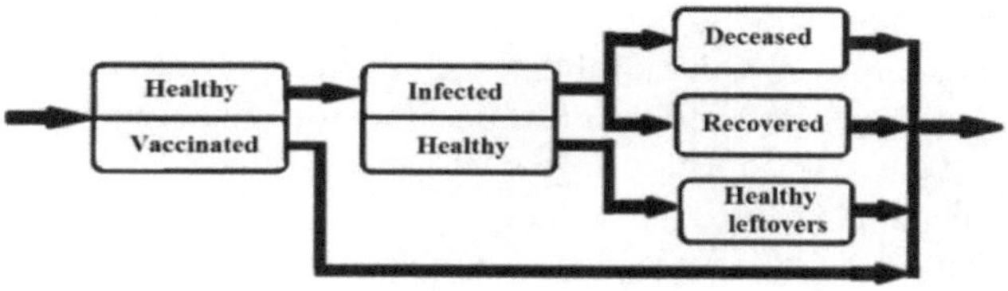

Fig. 2. Block diagram of a multi-agent model of the epidemic process

The mathematical model performs the following functions: 1) describes the process of vaccination (dark gray pixels are transformed into light gray); 2) describes the infection of healthy people (dark gray pixels are transformed into dark dark gray); 3) describes the process of improving the health of patients (dark dark gray pixels are transformed into slightly gray (almost white); 4) describes the death of patients (dark dark gray pixels are transformed into black). We will give a more detailed description of the hybrid multi-agent mathematical model.

Each agent belongs to one of five types: a) healthy; b) sick; c) recovered; d) deceased; e) vaccinated. Each agent, depending on its state, is modeled by a pixel of a certain color. The coloring of pixels is illustrated in Table 1.

Table 1. The coloring of pixels modeling the state of agents

Agents number	Agent status	Pixel color	Pixel color
1	Healthy		Dark gray
2	Ill		Dark dark gray
3	Recovered		Slightly gray (almost white)
4	Deceased		Black
5	Vaccinated		Light gray

The pixel colors of certain types of agents are characterized by the following gray tones: a healthy agent is dark gray, a sick agent is dark dark gray, a recovered agent is slightly gray (almost white), a dead agent is black; a vaccinated agent is light gray, The states of agents can change as follows: a healthy person turns into a patient (dark gray pixels transform into dark dark gray), a patient turns into a recovered person (dark dark gray pixels transform into slightly gray (almost white), a patient turns into a deceased person (dark dark gray pixels transform into black), a healthy person turns into a vaccinated person (dark gray pixels transform into light gray).

Each agent has the properties of a selective neuron that receives information from surrounding agents. It is useful to use the properties of a selective neuron to simulate the infection of a healthy agent from a patient due to the peculiarities of the infection procedure.

In selective neurons and selective neural networks, selectivity is achieved not by selecting weighting coefficients, as in well-known neural networks based on the use of McCulloch-Pitts neurons and due to the selective clustering of communication channels in neurons according to the information properties of the input signals, that is, clustering is provided by adjusting to the properties of the input signals. Selective properties can be implemented in neural networks operating both on binary input signals and on pulse, spike input signals. The selective neuron proposed in [20–28] is shown in Fig. 3 on the right, for comparison, Fig. 3 on the left shows the well-known model of the McCulloch-Pitts neuron.

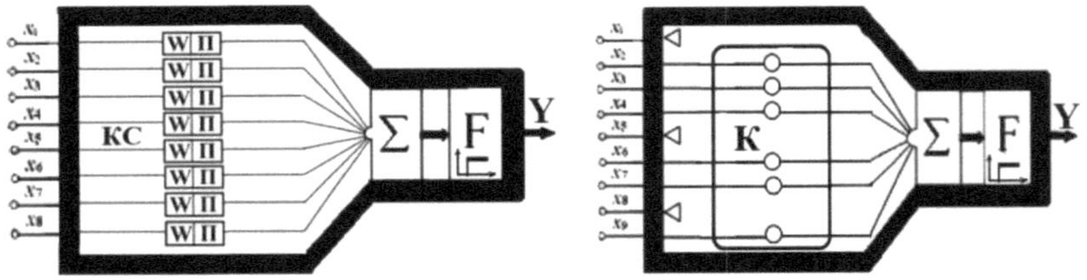

Fig. 3. Structure of mathematical models of neurons

Note. On the left – the McCulloch-Pitts neuron, on the right-the selective neuron. The following designations are used in the figure: $\boldsymbol{x} = (x_1, \ldots, x_n)$, W - weight functions; П - multipliers of input signals by weight coefficients; KC-communication channels; $\sum$ - adder, F-nonlinear threshold function, K-cluster of communication channels in a selective neuron. Some input signals are blocked during the formation of communication channels.

The properties of selective neurons are due to the fact that internal communication channels are tuned to specific components of input signals. The selective neurons do not respond to the other components.

In the multi-agent model, each agent-selective neuron is exposed to neurons from certain clusters of surrounding neurons. Therefore, it is advisable not to calculate the weight coefficients using complex procedures, but to set them based on the required properties of the system. For this purpose, a mathematical model is used, in the form of a two-dimensional selective neural network with selective agent neurons. A more detailed description of selective neurons and neural networks is contained in [20–28].

Let's determine the connections of each agent-neuron with others. We will consider the agents of the nearest environment around the selected one, which is indicated in a rectangular area. In the future, it is planned to use areas that take into account the specific forms of the epidemic area. For a rectangular area, each element is surrounded by eight elements, as shown in Fig. 4 on the left.

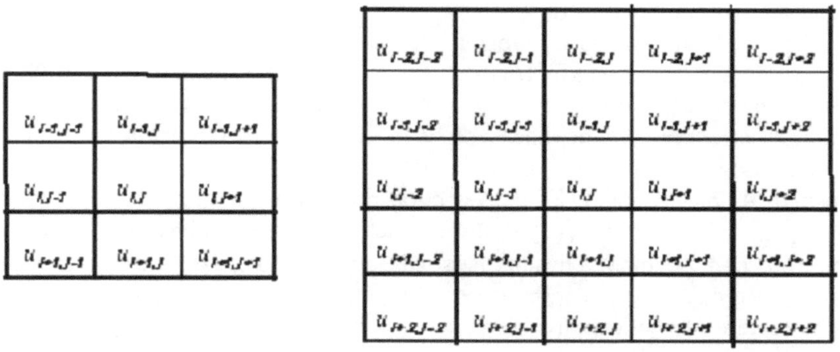

Fig. 4. The environment of the central element modeling a specific agent

In this paper, we will use the environment of the agent-neuron, shown in Fig. 4 on the left. In the future, in subsequent works, it is assumed to consider the two-layer surrounding region shown in Fig. 4 on the right. Let's establish the connections of each cell with the surrounding elements. We will consider the cells of the nearest environment around the selected element, which is indicated $\boldsymbol{u}_{i,j}$.

The selective neural network character of the proposed mathematical model is used in modeling the infection of healthy agents from patients. At the same time, the green pixels describing healthy agents are repainted into red ones describing patients. Infection occurs from patients in the cluster surrounding a healthy agent, provided that the sum of patients in the cluster with the specified weight coefficients exceeds the specified threshold value U_P, that is, if.

$$S_{(ij)(kl)} = \sum_{ij \in R} \sum_{K_{(ij)(kl)}} \lambda_{(ij)(kl)} u_{(ij)(kl)} \geq U_P, \qquad (1)$$

where $\lambda_{(ij)(kl)}$ is the specified value of the weight coefficients characterizing the contagion, $K_{(ij)(kl)}$ - the cluster surrounding the element, u_{ij}, U_P - the specified threshold value.

As a cluster in this work, a single-layer set of elements surrounding the element is used u_{ij}, shown in Fig. 4 on the left. In the following works, it is assumed to use a cluster of a two-layer set of elements, shown in Fig. 4 on the right. Let's consider the formation of weight coefficients when an element u_{ij} is connected to the nearest environment. All weight coefficients can be represented as a vector $\boldsymbol{\lambda}_{(ij)(kl)}$ with components.

$$\boldsymbol{\lambda}_{(ij)(kl)} = (\lambda_{(ij)(i-1,j+1)}, \lambda_{(ij)(i,j+1)}, \lambda_{(ij)(i+1,j+1)}, \ldots, \lambda_{(ij)(i-1,j-1)}, \lambda_{(ij)(i,j-1)}, \lambda_{(ij)(i+1,j-1)}).$$

Let's list the possible values of the weighting coefficients: one of the coefficients is equal to 1, the rest are equal to 0; two coefficients are equal to 0.5, the rest are equal to 0; three coefficients are equal to 0.34, the rest are equal to 0; four coefficients are equal to 0.25, the rest are equal to 0.

In the cases under consideration, for a given number of coefficients m < 8, the weight coefficients are equal to 1/m, the remaining coefficients are equal to 0.

We assume that in the initial stage, in the absence of protective measures, one patient infects a person who does not have immunity, through direct contact. The method of infection is drip, contact, inhalation of infected air and others are not considered. At the beginning of the infection process, we assume that any weighting factor $\lambda_{(ij)(kl)} = 1$. This means that any contact between a healthy person and a sick person leads to infection of a healthy person. This is a natural assumption, since contact in the absence of alertness can be very tight.

In the future, we believe, assuming the appearance of quarantine measures, that the value in the area $\lambda_{(ij)(kl)}$ of the immediate environment decreases, $(k,l) \in R$. We assume that after a few days or several steps in the program $\lambda_{(ij)(kl)} = 0.5$, with a further increase in time, there is a further decrease in the weights of the links, for example, a decrease in the weight coefficients to the values $\lambda_{(ij)(kl)} = 0.34$, and so on.

We define nonlinear functions that reflect changes in the state of agents in the proposed multi-agent model. These functions are specific to specific transformations of agents. We introduce a vector function $\boldsymbol{\Phi}(\Phi_1, \Phi_2, \ldots \Phi_n)$ for converting the color of a pixel that characterizes an agent of a multi-agent system. $\Phi_1(S)$-the function converts a healthy agent into a sick one or formally a dark gray pixel into a dark dark gray. We will write this function in the form

$$\Phi_1(S_{(ij)(kl)(green)}) = \begin{cases} red & \Big| S_{(ij)(kl)(green)} \geq U_P \\ green & \Big| S_{(ij)(kl)(green)} < U_P \end{cases} \tag{2}$$

$\Phi_2(green) = blue$ - a function describing vaccination. It is accepted that vaccination is performed for each P- next healthy agent. The value of P-determines the resulting percentage of vaccination;

$\Phi_3(green) = green$- function of the identical transformation of dark gray color;

$\Phi_4(blue) = blue$- function of the identical transformation of light gray color;

$\Phi_5(red) = red$ - function of the identical transformation of dark-dark gray color;

$\Phi_6(black) = black$ - function of the identical transformation of black color;

$\Phi_7(red) = yellow$ - function that converts a dark-dark gray color into a slightly gray (almost white) characterizes the recovery of the patient. It is assumed that the patient's recovery occurs on average q days after the onset of the disease.

$\Phi_8(red) = black$ - function that characterizes the death of a patient.

Consider the spread of a pandemic and the formation of collective immunity in a two-dimensional rectangular area of size N×N, where N is the length of the side of the area. In the future, it is assumed to use a two-dimensional region that accurately reflects the shape of the area under study. For example, when studying the pandemic and the formation of CI in Moscow, this region will be the area where Moscow is located, for the Moscow region it will be the area where it is located.

The considered collaboration of all participants of the epidemic is a multi-agent system. In this case, it is advisable to endow agents with the properties of selective neurons, and to consider the connections between agent–neurons as connections in a selective neural network. Each agent-neuron receives a signal in the form of a weighted sum from the surrounding agent – neurons at a subsequent time, characterized by an upper index. In the two-dimensional case

$$S^{(n+1)}_{(ij)(kl)(color)} = \sum_{ij\in R}\sum_{kl\in R} \lambda^{(n)}_{(ij)(kl)} u^{(n)}_{ij(color)} \tag{3}$$

where u_{ij} is the state of the neuron equal to 0 or 1; $\lambda_{(ij)(kl)}$ - connections described as weighting coefficients, we believe that the neuron in the cell u_{ij} is connected with agents - neurons, $u_{kl} \in R_i$, R - the area of really influencing connections.

The relation (3) determines the state of the agent at the time characterized by the upper index (n + 1), through the values $\lambda^{(n)}_{(ij)(kl)}$, $u^{(n)}_{ij(color)}$ at the previous time n.

The appearance of the portrait of a multi-agent system, representing a beautiful five-color image, changes over time, which is expressed in a change in the color palette at each iteration. Each iteration corresponds to the next moment in time. If the iterations change every other day, then the n-th iteration describes the state of the system for the n-th day. An iterative procedure describing the time dynamics can be represented by a difference equation.

$$\Phi^{n+1} = \Phi^n + \Phi(\Phi^n) \tag{4}$$

Φ^n is the state of the system or its color portrait for the nth iteration. A special case of an iterative process for calculating transformations of healthy agents into patients can be described by a difference equation.

$$\Phi_1^{n+1} = \Phi_1^n + \Phi(S^n_{(ij)(kl)(green)})$$

A sufficiently adequate model of the pandemic will be a model that takes into account: 1) the strength of connections or the influence of contagion, taking into account the strength of the connection in the interval [0,1]; 2) the presence of the threshold of infection under the influence of the sum of connections; 3) the selective nature of connections or the possibility of infection only from a certain group (cluster) of patients; 4) taking into account the presence of non – infected – vaccinated or having innate immunity; 5) the dynamics of infection-recovery or death of individuals.

Training of a hybrid multi-agent model with a selective neural network is reduced to adjusting the weight coefficients of the selective neural network to adequately reflect the development of the pandemic process, quarantine measures, the process of infection of healthy people, recovery and death of patients, the dynamics of the vaccination process. In practice, training is reduced to the selection of parameters in formulas (1)–(3). The selection of parameters is carried out in order to obtain a qualitative adequacy of the model to the dynamics of the pandemic and the formation of the dynamics of vaccination. In general, the training of a hybrid multi-agent model with a selective neural network is illustrated in Fig. 5.

The work provides for the creation of an initial distribution of sick, healthy people and people who died from the pandemic. In the future, we will call them agents according to the terminology adopted in social networks. The initial data is generated using a two-dimensional matrix of random numbers, with a uniform distribution in the interval [0, 1]. Additionally, in the interval [0, 1], 4 intervals were set to create three groups of agents: healthy (green), sick (red), with innate immunity (blue).

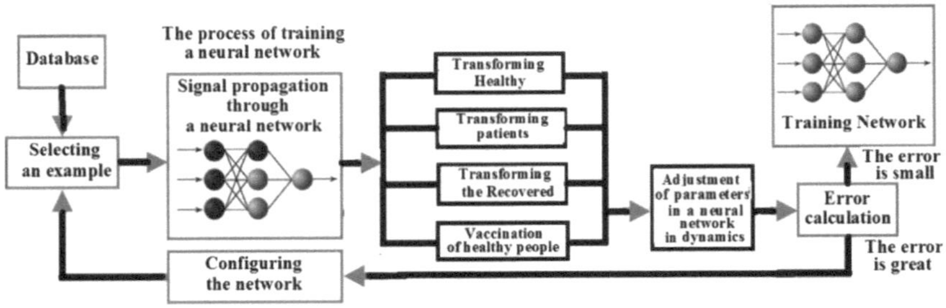

Fig. 5. Training of a hybrid multi-agent model with a selective neural network

3 Description of the Software

The software for calculating the dynamics of the pandemic, the dynamics of the formation of collective immunity was implemented in Matlab 7. The software allowed us to qualitatively study the dynamics of the development of pandemic regimes and the dynamics of the formation of collective immunity.

A software package has been developed for the formation of the dynamics of the development of collective immunity in the coronavirus pandemic on the basis of a hybrid multi-agent model with selective neural networks.

The developed software product allows us to establish the main properties of the dynamics of the formation of coronavirus infection, namely the influence of the qualitative characteristics of quarantine measures, the influence of the percentage of vaccination on the formation of coronavirus infection, to establish the value of the main parameters on the formation of coronavirus infection. The product is a software in the Matlab 7 language, designed in the form of a procedure that is convenient for practical use.

The input parameters are: 1) Parameters of the random distribution of healthy people, patients and those with innate immunity; 2) Parameters of the beginning and end of vaccination; 3) The value of the percentage of vaccination; 4) The number of frames or days of the course of the disease; 5) The percentage value of the parameter characterizing the severity of quarantine measures; 6) The average recovery time of patients.

The program has parameters that characterize the severity of quarantine measures. The stiffness is adjusted by changing the weight coefficients, as well as the parameters of the time intervals during which the adjustment is carried out. In the future, it is planned to introduce parameters that characterize the disease in a mild form.

The calculation is carried out in a rectangular area. When modifying the program, it is assumed that the coronavirus infection will be implemented in an area of arbitrary shape and the heterogeneous nature of the area itself. First of all, the calculation of coronavirus infection in the regions: Moscow; Moscow region; the territory of the entire Russian Federation.

The developed software package for solving the problem of studying the dynamics of the development of collective immunity for any pandemic and specifically for the COVID-19 coronavirus pandemic allows us to study and identify not only the optimal vaccination parameters, but also the degree of rigidity of the required quarantine measures. The rigidity of quarantine measures was taken into account by adjusting the weight coefficients in the main formula (1) given above. A decrease in weight coefficients characterizes an increase in the rigidity of quarantine measures. Zero weight coefficients or close to zero characterize the most stringent quarantine measures or lockdown.

4 Results

We present the practical results of the study of the dynamics of the formation of collective immunity with different percentages of vaccination. Let's consider the dynamics of the development of the epidemic process in the absence of vaccination. In this case, the dynamics of the epidemic development shows that with insufficiently strict quarantine measures, the disease occurs in all participants of the population. There are no healthy participants – there are no green pixels left. The time is counted in conditional days. After a certain time, equal to the average period of the disease, some patients recover, the corresponding pixels are colored yellow, some patients die and such pixels are colored black. We present the results of the development of the pandemic in the absence of quarantine measures and vaccination, shown in Fig. 6.

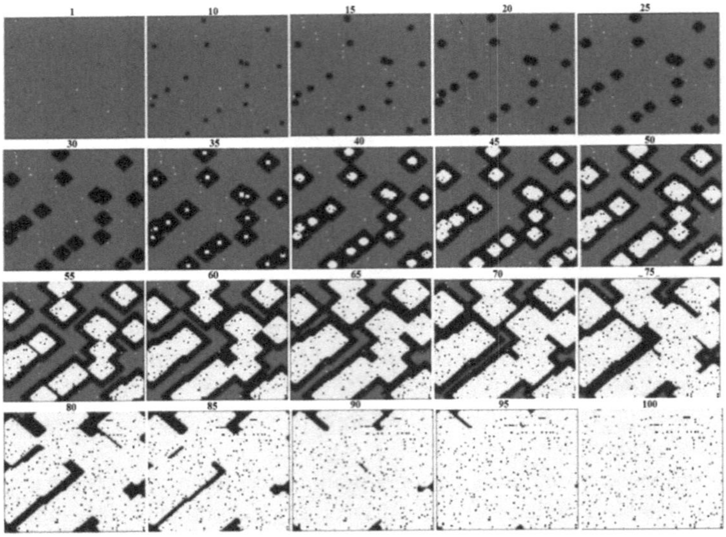

Fig. 6. Implementation of the dynamics of the epidemic process in the absence of quarantine measures and vaccination

The obtained results of the computational experiment show that in the absence of quarantine measures and vaccination, the pandemic covers the entire contingent of healthy people. The spread of the pandemic has a rapid explosive character.

Let's consider the impact of vaccination on the dynamics of the pandemic. The vaccination itself can proceed at different rates, depending on many reasons. These reasons may be of a social, economic, or political nature, which we are currently observing. The economic reasons may be due to the insufficient quantity of the vaccine, the competitive relationship between the vaccine manufacturers, the social reasons are due to the distrust of the population, the political reasons are due to the relationship between individual countries and can be very diverse. If there are favorable conditions, vaccination can be carried out quickly or be stretched over time. In our work, in a computational experiment, the implementation of the vaccination process in a given specific time interval is adopted, which is one of the options for its implementation.

The study of the dynamics of the development of collective immunity with a vaccination rate of 30% shows that collective immunity is not achieved, the number of sick people is constantly increasing. An illustration of the computational experiment is shown in Fig. 7.

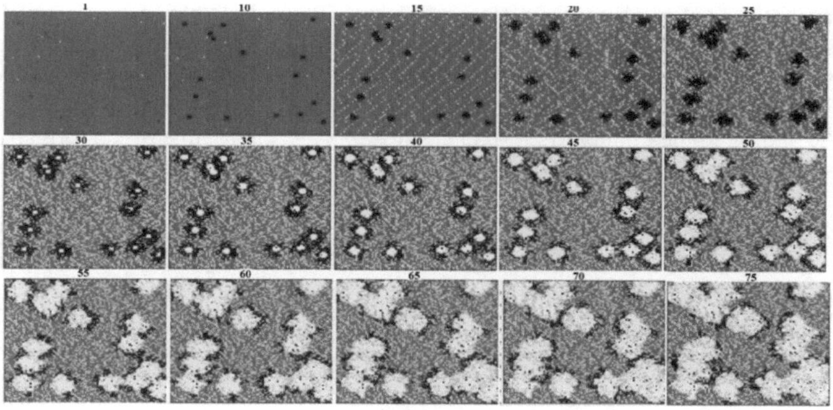

Fig. 7. Implementation of the dynamics of the development of the epidemic process with vaccination of 30% carried out in the interval from 15 to 30 days

The study of the dynamics of the development of collective immunity at a vaccination rate of 40% shows that there is an approximation to collective immunity. However, the number of sick people is quite large. Let's consider the dynamics of the development of the epidemic process during vaccination of 65% and vaccination from 15 to 25 days. The implementation of the epidemic process with vaccination of 65% is shown in Fig. 8.

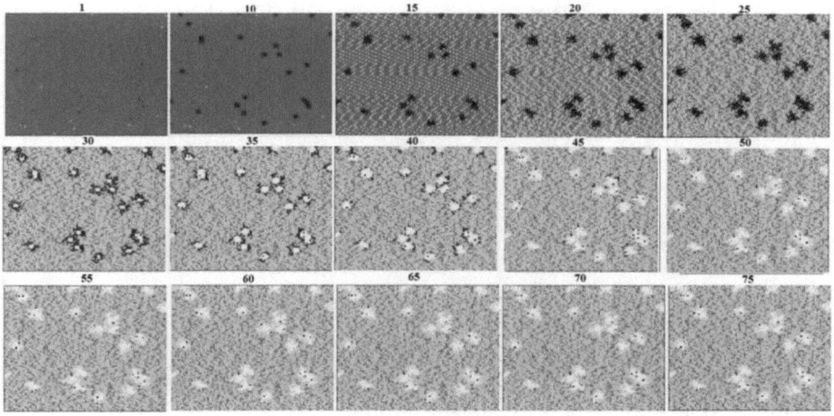

Fig. 8. Implementation of the dynamics of the epidemic process during vaccination 65%

Consideration of the dynamics of the development of collective immunity with a vaccination rate of 65% shows that collective immunity is achieved at a minimum cost, and the number of sick people is less than 10%. This is the level of cases that was before the vaccination, the study of dynamics also shows the presence of healthy unvaccinated people. They are protected from the disease by a network of people who have been vaccinated and recovered from the coronavirus disease. A more detailed study shows that for values of the vaccination rate greater than 65%, the pandemic ends almost at the same time as for the vaccination rate of 65%.

According to the totality of all statistical data for the entire time of the formation of collective immunity for samples with different percentages of vaccination, dependencies are constructed (Fig. 9): 1) the number of sick people from the percentage of vaccination graph on the left (red); 2) the number of healthy people who did not get sick from the percentage of vaccination graph in the center (green); 3) the number of deaths with different percentages of vaccination graph on the right (black).

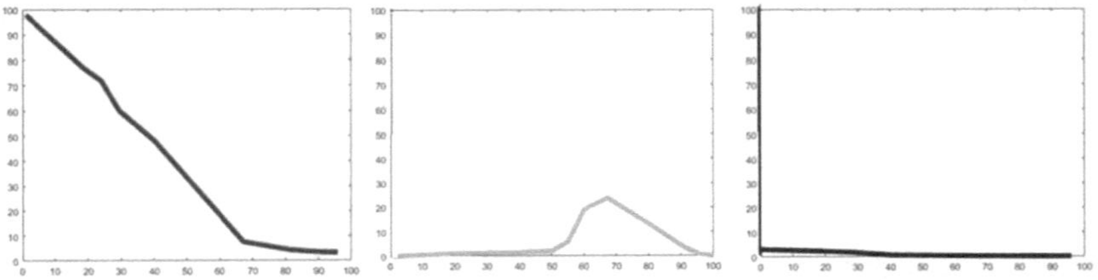

Fig. 9. Integral dependencies

Note. Integral dependencies: the number of cases from the percentage of vaccination graph on the left (red); the number of healthy people who did not get sick from the percentage of vaccination graph in the center (green); the number of deaths with different percentages of vaccination graph on the right (black).

The obtained integral characteristics show that effective collective immunity is achieved for the COVID-19 pandemic with vaccination of about 65%. There is a certain number of protected, unvaccinated healthy people with collective immunity. Protection is due to the presence of a network of vaccinated people who prevent the spread of infection.

5 Conclusion

It is established that neural networks are an effective mechanism for studying and predicting processes in complex dynamic systems [29–32]. In this paper, based on a hybrid multi-agent model in combination with selective neural networks, modeling of the dynamics of the formation of collective immunity and the impact of quarantine measures in the coronavirus pandemic is considered in the two-dimensional domain. Agents are endowed with the properties of selective neurons, and the connections between agent-neurons are considered as connections in the form of weighting coefficients with selective clusters. Each agent-neuron receives a signal from the surrounding agent – neurons in the form of a weighted sum. A nonlinear function is defined that reflects changes in the state of agents, namely, transformations of the pixel color that characterizes the agent, depending on the degree of connection with the surrounding agents, which is specific for specific transformations of agents. The used mathematical model of the pandemic and the formation of collective immunity allowed us to qualitatively investigate the following tasks: the impact of quarantine measures on the dynamics of the pandemic; the impact of vaccination on the dynamics of the pandemic; the joint impact of quarantine measures and vaccination on the dynamics of the pandemic; the influence of the percentage of vaccination on the achievement of collective immunity.

The proposed hybrid multi-agent model based on selective neural networks can be considered as a cellular automaton with a large possible set of states. In particular, such a cellular automaton is the well-known game "Life" by Conway. Such a cellular automaton allows for natural improvement by increasing the number of possible states. At the same time, it is essential that the recalculation of the properties of the model as a whole when adding new states is either not required, or is necessary to a minimum extent. As new additional conditions, we can consider: a mild form of the agent's disease, modifications of the disease, for example, for COVID-19, these are more contagious forms: British, African, Indian, Russian and others. Thus, there are opportunities to improve the proposed model of the pandemic, an increase in the number of its significant parameters, the need for such improvement was indicated by the President of the Russian Academy of Sciences, Academician A. M. Sergeev.

References

1. Nucci, M.C., Leach, P.G.L.: An integrable SIS model. Math. Anal. Appl. **290**(2), 506–518 (2004)
2. Li, M.Y., Graef, J.R., Wang, L., Karsai, J.: Global dynamics of a SEIR model with varying total population size. Math. Biosci. **160**(2), 191–213 (1999)
3. Fontanet, A., Cauchemez, S.: COVID-19 herd immunity: where are we? Nat. Rev. Immunol. **20**(10), 583–584 (2020)
4. Randolph, H.E., Barreiro, L.B.: Herd immunity: understanding COVID-19. Immunity **52**(5), 737–741 (2020)
5. Hethcote, H.W.: The mathematics of infectious diseases. SIAM Rev. **42**(4), 599–653 (2000)
6. Billari F.C., Prskawetz A.: Agent-based computational demography: Using simulation to improve our understanding of demographic behaviour. Springer, Heideleberg (2012)
7. Perez, L., Dragicevic, S.: An agent-based approach for modeling dynamics of contagious disease spread. Int. J. Health Geograph. **8**(1), 50 (2009)
8. El-Sayed, A.M., Scarborough, P., Seemann, L., Galea, S.: Social network analysis and agent-based modeling in social epidemiology. Epidemiol. Perspect. Innov. **9**(1), 1 (2012)
9. Janssen, M.A., Ostrom, E.: Empirically based, agent-based models. Ecol. Soc. **11**(2) (2006)
10. Wang, Z., Butner, J.D., Kerketta, R., Cristini, V., Deisboeck, T.S.: Simulating cancer growth with multiscale agent-based modeling. In: Seminars in Cancer Biology, vol. 30, p. 70. Academic Press (2015)
11. Bonabeau, E.: From classical models of morphogenesis to agent-based models of pattern formation. Artif. Life **3**(3), 191–211 (1997)
12. Bauer, A.L., Beauchemin, C.A., Perelson, A.S.: Agent-based modeling of host-pathogen systems: the successes and challenges. Inf. Sci. **179**(10), 1379–1389 (2009)
13. Frias-Martinez, E., Williamson, G., Frias-Martinez, V.: An agent-based model of epidemic spread using human mobility and social network information. In: IEEE Conference on Social Computing (2011)

14. Hunter, E., Namee, B.M., Kelleher, J.: An open-data-driven agent-based model to simulate infectious disease outbreaks. PLOS One. **13**(12), e0208775 (2018)
15. Merler, S., et al.: Spatio-temporal spread of the Ebola 2014 outbreak in Liberia and the effectiveness of non-pharmaceutical interventions: a computational modelling analysis. Lancet. Infect. Dis **15**(2), 204–211 (2015)
16. Kim, Y., Ryu, H., Lee, S.: Agent-based modeling for super-spreading events: a case study of MERS-CoV transmission dynamics in the Republic of Korea. Int. J. Environ. Res. Public Health **15**(11), 2369 (2018)
17. Hackl, J., Dubernet, T.: Epidemic spreading in urban areas using agent-based transportation models. Future Internet. **11**(4), 92 (2019)
18. Vlad, A.I., Sannikova, T.E., Romaniukha, A.A.: Modeling of the spread of respiratory viral infections in the city: a multi-agent approach. Math. Biol. Bioinf. **15**(2), 338–356 (2020)
19. Makarov, V.L., Bakhtizin, A.R.: Social Modeling – a Scientific Computer Breakthrough (Agent-Oriented Models), p. 295. Ekonomika, Moscow (2013)
20. Kartvelishvili, V.M., Mazurov, M.E., Petrov, L.F.: Applied System-Dynamic Models. Plekhanov Russian University of Economics, Moscow (2018)
21. Mazurov, M.E.: Identification of Mathematical Models of Nonlinear Dynamical Systems, p. 284. Lenand (2019)
22. Mazurov, M.E.: An impulse neuron close to the real one. Patent for invention No. 2598298. Accessed 09 Feb 2015
23. Mazurov, M.E.: A neuron modeling the properties of a real neuron. Patent for invention No. 2597495. Accessed 07 Nov 2014
24. Mazurov, M.E.: Single-layer perceptron based on selective neurons. Patent for invention No. 2597497. Accessed 13 Jan 2015
25. Mazurov, M.E.: A single-layer perceptron modeling the properties of a real perceptron. Patent for the invention No. 2597496. Accessed 24 Feb 2015
26. Mazurov, M.E.: Modeling of the dynamics of collective immunity of the coronavirus pandemic on the basis of a multi-agent structure with neuropodic elements. In: XIX All-Russian Scientific Conference "Neurocomputers and their Application" Plenary Report (2021)
27. Mazurov, M., Mikryukov, A.: Prediction of the rating of the university using hybrid cognitive maps and selective dendritic networks of neurons. In: Proceedings of the 10th International Scientific and Practical Conference named after A. I. Kitov "Information Technologies and Mathematical Methods in Economics and Management, Moscow, pp. 360–373 (2020)
28. Mikryukov, A.A., Mazurov, M.E.: Solving the classification problem based on the Committee of electoral neural networks. In: Collection of Scientific Papers Part 2 XX1 International Scientific and Technical Conference, Neuroinformatics-2019, pp. 186–194 (2019)
29. Abdulhabib, S., Alzaeemi, S., Sathasivam, S., Velavan, M.: Agent-based modeling in doing logic programming in fuzzy hopfield neural network. Int. J. Mod. Educ. Comput. Sci. (IJMECS) **13**(2), 23–32 (2021)
30. Hu, Z., Ivashchenko, M., Lyushenko, L., Klyushnyk, D.: Artificial neural network training criterion formulation using error continuous domain. Int. J. Mod. Educ. Comput. Sci. (IJMECS) **13**(3), 13–22 (2021)
31. Gustisyaf, A.I., Sinaga, A.: Implementation of convolutional neural network to classification gender based on fingerprint. Int. J. Mod. Educ. Comput. Sci. (IJMECS) **13**(4), 55–67 (2021)
32. Ahmed, R., Robin, T.I., Shafin, A.A.: Automatic environmental sound recognition (AESR) using convolutional neural network. Int. J. Mod. Educ. Comput. Sci. (IJMECS) **12**(5), 41–54 (2020)

Managing Self-organized Criticality in Neural Networks on the Edge of Stability and Chaos

Mikhail Mazurov(✉)

Russian University of Economics Moscow, Moscow, Russia

Abstract. The features of the modes and their physical properties in the region close to the bifurcation points in systems with self-organized criticality are considered. It is shown that a mode is possible when a system with self-organized criticality operates in the instability region without leaving this region. This mode is called working on the edge of stability or chaos. It is shown that these modes near the bifurcation point can in some cases be considered as a threshold synchronization of nonlinear relaxation self-oscillating systems. It is shown that for this it is necessary that as the amplitude of the unstable mode increases, nonlinear negative feedback occurs, which will stabilize the unstable mode and turn it into a stable one. This situation occurs in most biological systems and living organisms. Examples of the work of physiological sensory systems in which the generation of powerful pulse flows occurs from very weak receptor potentials are considered. Examples of reactions in the implementation of unconditional and conditional reflexes are considered. To study the modes near the bifurcation point, a mathematical model of synchronization of relaxation self-oscillations based on the modified Wiener-Rosenblut axiomatic model and the properties of uniform almost-periodic functions is used.

Keywords: Bifurcation point · System with self-organized criticality · Stability · Chaotic dynamics · Positive feedback

1 Introduction

An important and poorly studied question is the possible regimes of systems with self-organized criticality (SSC) near the bifurcation point [1–3]. This work is not devoted to the description of the modes after the critical point and subsequent events, but to the study of possible stable modes when working in the immediate vicinity of the critical point before its passage. The stability of unstable modes is achieved due to nonlinear negative feedback. It is shown that in the considered unstable region, it is possible to exist regimes with remarkable properties that are widely used in all living things, physics, and technical devices invented by man. These are the following properties: 1) a significant (up to millions of times) amplification of input signals; 2) a significant increase in selectivity to external influences; 3) a significant increase in development opportunities.

Z. Hu et al. (Eds.): AIMEE 2021, LNDECT 107, pp. 67–75, 2022.
https://doi.org/10.1007/978-3-030-92537-6_6

An illustration of these remarkable capabilities is realized based on the formation of impulse flows going to the central nervous system (CNS) under the influence of a small receptor potential in physiological sensory systems, neural networks that generate reflexes: unconditional, conditional, complex conditional, imprinting.

2 Critical Points are Bifurcation Points of Systems with Self-organized Criticality

Self-organized criticality is a property of dynamical systems that have bifurcation points. The behavior in the vicinity of this point is characterized by the fact that with a small perturbation, the system can pass the bifurcation point, thereby completely changing its behavior model. Classical examples of self-organized criticality are the phase transition or the sand heap model [1–3]. During the passage of the bifurcation point, two stable states may appear, which is illustrated in Fig. 1.

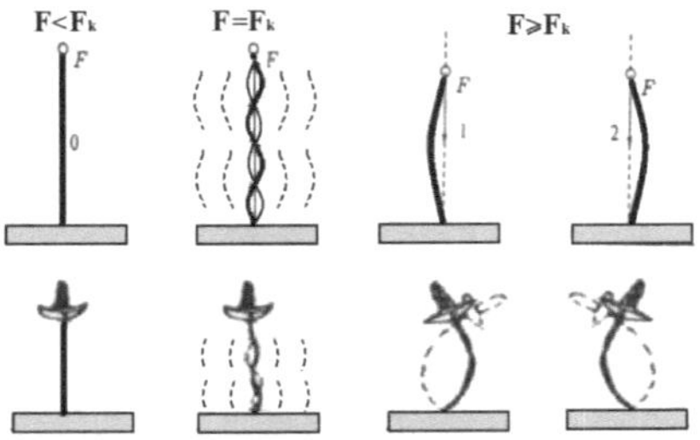

Fig. 1. Bifurcation points on the example of a rod (top) and a sword (bottom)

At the critical point, the system becomes unstable and can move to one of two stable states. In general, the system can move from a critical point to n stable states.

A system with self-organized criticality (SSC) is a complex system characterized by the inclusion of subsystems that have the following properties: the presence of internal energy, the presence of positive feedback, the presence of nonlinear negative feedback, the presence of internal organization, the presence of synchronization of internal relaxation self-oscillations, and others. The structure of the SSC is shown in Fig. 2 on the left.

The question arises how the SSC develops after the parameter passes the critical point. The list of the main problems that arise is shown in Fig. 2 on the right. Forecasting of these regimes, including catastrophic ones, is carried out using specific signs called disaster flags [1–3]. In this paper, disaster flags and their characteristics are not considered.

The question arises how the SSC develops after passing the critical point. The list of the main emerging modes is shown in Fig. 2 on the right. The prediction of these modes, including catastrophic ones, is made with the help of specific signs called disaster flags. In this paper, disaster flags and their characteristics are not considered.

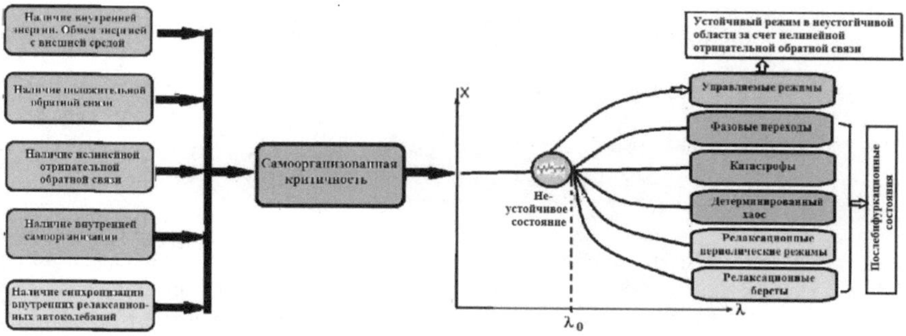

Fig. 2. The structure of systems with self-organized criticality (left) and possible modes after the transition of the bifurcation point (right). A possible mode is shown when working in an unstable region located before passing through the bifurcation point in an unstable region (shown by a blue circle)

Another interesting mode is possible, when the SSC operates in the instability region shown in figure with a blue circle, without touching this region. This mode will be called working on the edge of stability or chaos. Is it possible to work steadily in an unstable area, shown in Fig. 2 by the blue circle. It turns out that it is possible. But for this, it is necessary that as the amplitude of the unstable mode increases, nonlinear negative feedback occurs, which will stabilize the unstable mode and turn it into a stable one. This situation occurs in most biological systems and living organisms. Two examples of many possible ones will be considered below: 1) physiological sensory systems in which the generation of powerful impulse flows occurs from very weak receptor potentials; 2) reactions during the implementation of unconditional and conditional reflexes when exposed to weak but dangerous external influences.

3 Mathematical Model for the Study of Modes in Systems with Self-organized Criticality Near the Bifurcation Point

Let us briefly define the property of systems with self-organized criticality. In the nonlinear dynamics of systems, an example of a SC is an autogenerator with positive feedback, shown in Fig. 3.

The block diagram of autogenerators with positive feedback and negative nonlinear feedback is shown in Fig. 3 below. Near the self-excitation threshold, the autogenerator is an SSC. The smallest input signal is enough to trigger an auto-generator response with a significant amplitude. From a physical point of view, the autogenerator is self-excited. The expression for the output signal contains a linear link having a transmission coefficient k equal to, has the form $y = \frac{kx}{1-k}$, where x, y, are the input and output signals.

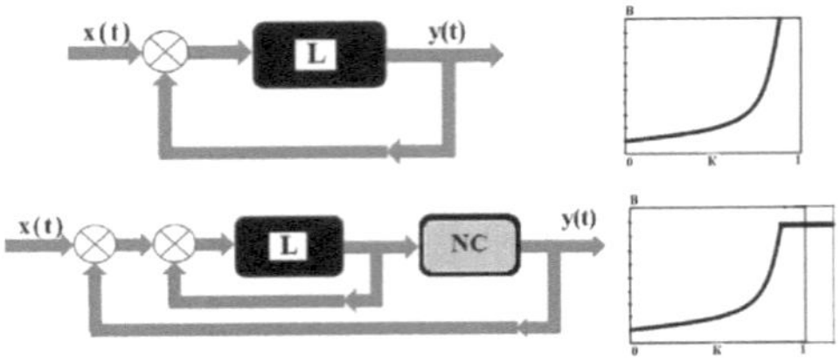

Fig. 3. Block diagram of an autogenerator with positive feedback from above. The unlimited amplitude characteristic is shown on the right.. Structural diagram of an autogenerator with positive and negative feedback (from below). Its limited amplitude characteristic is shown on the right

The critical gain level $k = 1$, while the signal y in the absence of nonlinearity tends to ∞. This behavior of the SSC is illustrated in Fig. 3 from the top right. With a smooth approach k to 1, the system becomes unstable and cannot be implemented in practice. To use the useful properties of systems with self-organized criticality at the threshold of self-excitation or in the bifurcation mode, negative nonlinear feedback can be used. It allows you to significantly increase the sensitivity of the conversion, and at the same time will prevent the occurrence of instability. This allows you to significantly increase the quality of the system. A system with self-organized criticality in the form of an autogenerator with positive feedback supplemented with negative nonlinear feedback is shown in Fig. 3 below.

In some cases, the modes of operation of the SSC near the bifurcation point can be considered essentially as a threshold synchronization of nonlinear relaxation self-oscillating systems. Currently, two methods are known for the synchronization of relaxation self-oscillatory systems: the method of ordinary differential equations with small parameters before derivatives [4] and the method using the modified Wiener-Rosenblut axiomatic method in combination with the properties of uniform almost-periodic functions [5–9]. According to the first method, systems of differential equations with a small parameter for derivatives are used to study the synchronization of relaxation self-oscillations.

$$\varepsilon\frac{d\mathbf{x}}{dt} = \mathbf{f}(\mathbf{x}, \mathbf{y}, \boldsymbol{\beta}_1(\mathrm{t})), \ \frac{d\mathbf{y}}{dt} = \mathbf{g}(\mathbf{x}, \mathbf{y}, \beta_2(\mathrm{t})),$$

Where, $\mathbf{x} = (x_1, \ldots, x_k)$, $\mathbf{y} = (y_1, \ldots, y_l)$, $\mathbf{f} = (f_1, \ldots, f_k)$, $\mathbf{g} = (g_1, \ldots, g_l)$, $\boldsymbol{\beta}_1(\mathrm{t})$, $\boldsymbol{\beta}_2(\mathrm{t})$ - the synchronizing effects. The exact analytical solutions of the above equations are unknown, and there are also no approximate analytical methods for solving these equations [4].

In [5–9], a method for obtaining accurate analytical solutions to the synchronization problem is proposed, the equations of the system are given in a modified axiomatic form. The simplest form of the relaxation system was proposed by N. Wiener and A.

Rosenblut in 1946 [5]. The geometric interpretation of the form of relaxation oscillations in the proposed model and the Wiener-Rosenblut model are shown in Fig. 4 on the left and right.

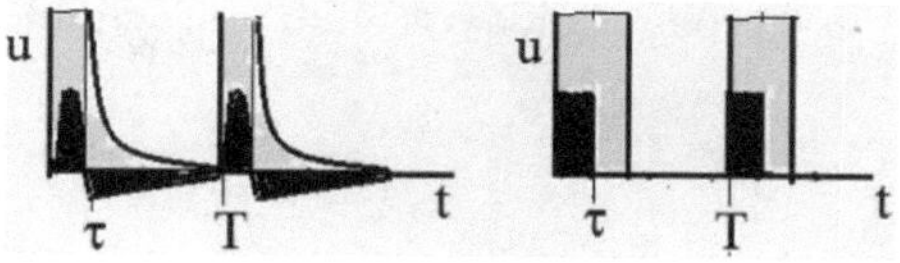

Fig. 4. Geometric interpretation of the form of relaxation oscillations shown in black in the figure in the proposed model and the Wiener-Rosenblut model are shown on the left and right in the figure. The refractory area is shown in light gray

A modified axiomatic model of the relaxation system is given, in combination with the properties of uniform almost-periodic functions, is proposed in [6–8]. It has the form

$$f(t) = \begin{cases} f_1(t) & 0 \le t \le t_1 \\ f_2(t) & t_1 < t \le T \end{cases}; \quad f_\partial(t) = \begin{cases} \infty & 0 \le t \le t_1 \\ f_3(t) & t_1 < t \le T \end{cases}; \tag{1}$$

$$f(R) + U_c \ge f_\partial(R); \quad R = t - \mathrm{mod}(t, T_c); \tag{2}$$

$$-\varepsilon < (mT_c - nT) < 0; \quad \varepsilon = f_\partial^{-1}(U_c), \tag{3}$$

where: $f(t)$ - a function that characterizes the form of relaxation self - oscillation; $f_1(t)$, $f_2(t)$ - functions in the interval of "fast" and "slow" changes in relaxation self-oscillation; $f_\partial(t)$ - the dynamic excitation threshold; $f_3(t)$ a function that characterizes the dynamic excitation threshold for the "slow" phase; R - the function of the remainder of the division of t by T_c $R = t - \mathrm{mod}(t, T_c)$; $\mathrm{mod}(t, T_c)$ - the designation of the function-the quotient of the division of t by T_c; (3) - the Kronecker inequality [9], T_c, T - the periods of the synchronizing signal and the relaxation oscillator; $\varepsilon = f_\partial^{-1}(U_c)$; f_∂^{-1} - function is the inverse of the function f_∂; U_c - the amplitude of the excitation pulse.

4 Control of Systems with Self-organized Criticality in the Generation of Pulse Flows in Sensor Systems

The generator potential in sensory neurons is excited from the continuously changing potential coming from the receptor. The place where the receptor potential is converted into single pulses or pulse flows in the form of spikes or periodic pulse flows are special places of neurons called the axon hole. The occurrence of impulse responses in the neuron of the abdomen of the river cancer and the human visual analyzer is illustrated in Fig. 5.

The receptor is located in a specialized muscle bundle lying on the dorsal surface of the abdominal muscles. When the abdomen is flexed, these muscles stretch and the receptor is activated. On the left are recordings of the electrical responses of the receptor to weak and strong stimulation. On the right, the functions of various parts of the neuron are indicated. Graduated receptor currents arising in the stretch-sensitive membrane of dendrites in the pulse initiation zone turn into electrical signals such as pulse sequences or spikes.

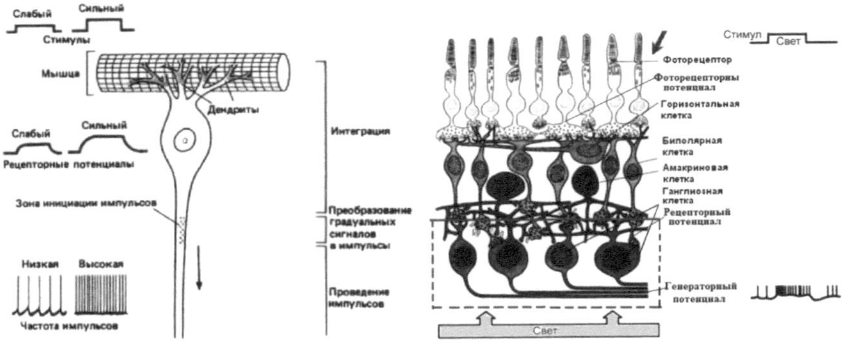

Fig. 5. The stretch receptor from the abdomen of the river crayfish, consisting of a sensory neuron with stretch-sensitive dendrites and a pulse flow generator (left). The structure of the human visual analyzer - (right) [10]

The source of pulse flows in the river cancer receptor is the shaded area shown in Fig. 5 from the bottom left, and in the human visual analyzer are the ganglion cells shown in Fig. 5 from the bottom right [10]. The receptor potential is usually very small, on the order of microvolts, while the generator potential has the order of tens of millivolts, that is, 100 and 1000 times more. It is significant that the receptor potential can change in magnitude within significant limits by tens or hundreds of times. In this case, the pulse repetition frequency of the generator neuron should increase in proportion to the intensity of the stimulus, which follows from the experimental data.

How are such remarkable properties of generator neurons achieved? Ordinary neurons with a threshold of about 10 mv are not capable of such feats. All these unusual and seemingly incompatible properties are achieved in biological systems that have adopted the principle of self-organized criticality during evolution. When using a neuron in the near-threshold mode, synchronization is possible, which implements the generation of pulse flows with a very low excitation threshold. The reduction of the excitation threshold due to positive feedback is illustrated in Fig. 6 on the left.

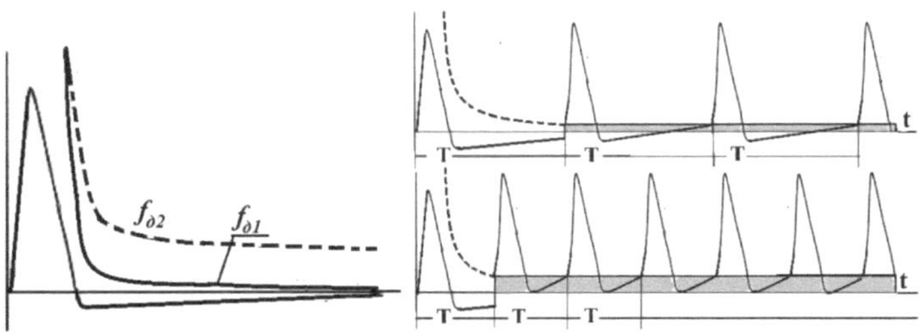

Fig. 6. The excitation threshold $f_{\partial 1}(t)$ in the presence of positive feedback is shown on the left by a solid line

The threshold of excitation $f_{\partial 2}(t)$ in the absence of positive feedback is shown on the left by the dotted line. On the right, the generation of a pulse sequence with different amplitudes of excitation is shown; from below, the excitation with an amplitude approximately 2 times greater than from above is shown.

Figure 6 on the left shows the values of the excitation thresholds in the usual case $f_{\partial 2}(t)$ and with increased feedback $f_{\partial 1}(t)$. A decrease in the excitation threshold makes it possible to realize an effective increase in the excitation frequency with an increase in the amplitude of the input pulse at a linearly increasing dynamic threshold $f_{\partial 1}(t)$. This case is illustrated in Fig. 6 on the right. The figure shows that as the excitation amplitude increases, the period of the generated pulse sequence decreases, and the frequency increases accordingly.

It is possible to generate neural spikes, that is, a series of pulses of any predetermined structure. The model allows us to implement responses with a pulse repetition frequency proportional to the intensity of the input receptor potential.

In the numerical implementation, the allocation of ε-almost-periods of an equal-dimensional almost-periodic function obtained as a result of synchronization is carried out by allocating the quotient of the current time of the process by the period of the synchronizing oscillation using the built-in procedure $\mathrm{mod}(t, T_c)$ of the Matlab7 language and then allocating the remainder $R = t - \mathrm{mod}(t, T_c)$.

An essential property of the conversion of the receptor potential into the generator potential is the property of the linear dependence of the pulse repetition frequency on the intensity of irritation. The results obtained in a computational experiment using the mathematical model (1)–(3) show that the pulse frequency increases with an increase in the intensity of irritation.

5 Management of Self-organized Criticality in the Implementation of Reflexes in Neural Networks

An essential area of implementation of self-organized criticality management is systems that generate reflexes: unconditional, conditional, complex conditional, imprinting [11]. The controlled SSC system, which implements the unconditional knee reflex, is shown in Fig. 7.

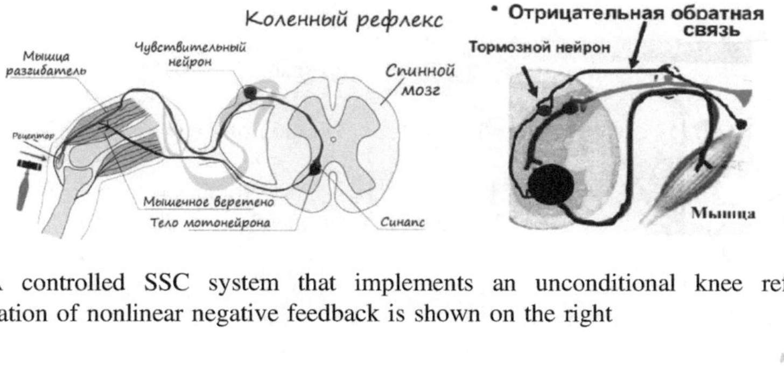

Fig. 7. A controlled SSC system that implements an unconditional knee reflex. The implementation of nonlinear negative feedback is shown on the right

The organization of conditioned reflexes is carried out by means of control in the CNS, which includes an additional channel of positive feedback created in the central nervous system due to the combined supply of stimuli for triggering unconditional and conditioned reflexes. The basic structure for the implementation of the conditioned reflex salivation, the I. P. Pavlov reflex, is shown in Fig. 8 [11].

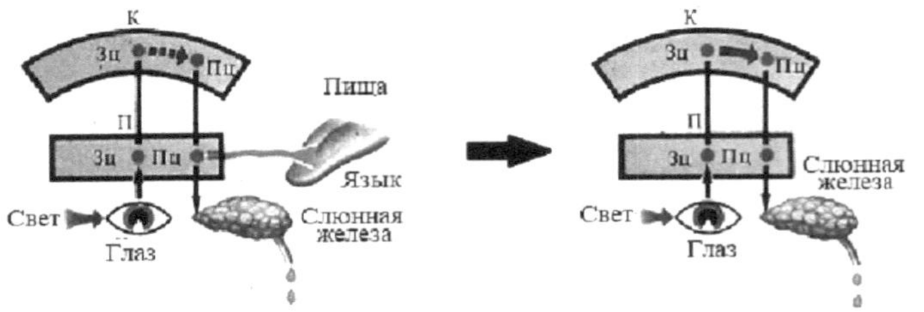

Fig. 8. A controlled SSC system that implements a conditional food reflex of saliva release to light. The right shows the formation of a stable connection in the brain, which is used to control the salivary gland

6 Conclusions

It is established that neural networks are an effective mechanism for studying and predicting processes in complex dynamic systems [12–15]. The existence of remarkable properties of the SSC modes near the critical point-the bifurcation point – is revealed: the possibility of a significant increase in the sensitivity and selectivity of the system, the possibility of accelerated development.

A mathematical method for studying modes near a critical point-the bifurcation point – has been developed. These are the modes of threshold synchronization of the relaxation self-oscillating system. As examples, the work of the SSC near the bifurcation point during the generation of pulse flows by generator neurons in physiological sensory systems and during the implementation of unconditional and conditional reflexes is studied.

References

1. Bak, P.: How nature works: the theory of self-organized criticality. Translated from the English, p. 276. URSS, Moscow (2013)
2. Kelso, J.A.S.: Dynamic Patterns: The Self-Organization of Brain and Behavior. MIT Press, Cambridge (1995)
3. Malinetsky, G.G., Potapov, A.B., Podlazov, A.V.: Nonlinear Dynamics. Approaches, Results, Hopes URSS Series: Synergetics: from the Past to the Future, p. 280. URSS, Moscow (2016)
4. Mishchenko, U.F., Rozov, N.H.: Differential Equations with a Small Parameter and Relaxation Oscillations, p. 247. Nauka, Moscow (1975)

5. Wiener, N., Rosenblueth, A.: The mathematical formulation of th problem of conduction of impulses in network of connected exitable elements, specificaly in cardiac muscle. Arch. Inst. Cardiologia de Mexico. **205**(16), 3–4 (1946)
6. Mazurov, M.E.: Solution of diophantine inequalities in problems of synchronization of relaxation oscillations. ZHVM MF USSR Acad. Sci. **31**(11), 1619–1636 (1991)
7. Mazurov, M.E.: Synchronization of relaxation self-oscillating systems, synchronization in neural networks. IZV RAS Phys. Ser. **82**(1), 83–87 (2018)
8. Mazurov, M.E.: Nonlinear dynamics and synchronization of neural ensembles in the formation of attention. IZV RAS Ser Phys. **84**(3), 451–456 (2020)
9. Levitan, B.M.: Almost-periodic functions, p. 396. GITTL (1953)
10. Mazurov, M.E.: Modeling of Intellect with the use of complex conditional reflexes and selective neural network technologies. In: International Conference of Artificial Intelligence, Medical Engineering, Education, Moscow, Russia, pp. 365–376 (2018)
11. Ostrovsky, M.A.: Current trends in modern brain science. Bull. Russ. Acad. Sci. **80**(5–6), 402–415 (2010)
12. Abdulhabib, S., Alzaeemi, S., Sathasivam, S., Velavan, M.: Agent-based modeling in doing logic programming in fuzzy hopfield neural network. Int. J. Mod. Educ. Comput. Sci. (IJMECS) **13**(2), 23–32 (2021)
13. Hu, Z., Ivashchenko, M., Lyushenko, L., Klyushnyk, D.: Artificial neural network training criterion formulation using error continuous domain. Int. J. Mod. Educ. Comput. Sci. (IJMECS) **13**(3), 13–22 (2021)
14. Gustisyaf, A.I., Sinaga, A.: Implementation of convolutional neural network to classification gender based on fingerprint. Int. J. Mod. Educ. Comput. Sci. (IJMECS) **13**(4), 55–67 (2021)
15. Ahmed, R., Robin, T.I., Shafin, A.A.: Automatic environmental sound recognition (aesr) using convolutional neural network. Int. J. Mod. Educ. Comput. Sci. (IJMECS) **12**(5), 41–54 (2020)

Method for Rapid Calculation of Urban Public Transport Passenger Flows in Residential Areas of Large Cities

Andrey Zatonskiy[1], Julia Volodina[1], Kalmira Ismailova[2], and Ruslan Bazhenov[3(✉)]

[1] Perm National Research Polytechnic University, Perm 614990, Russia
[2] Osh State University, Osh 723500, Kyrgyz Republic
[3] Sholom-Aleichem Priamursky State University, Birobidzhan 679015, Russia

Abstract. Passenger traffic is usually measured at multiple stops and in multiple vehicles. A lot of personnel are necessary to carry out measurements. The article suggests a way to reduce the number of measurements while maintaining accuracy. The method is based on a model of the emergence of transport needs. The model is based on open access information about some apartments in buildings. Zones of attraction of transport stops are determined by the coordinate method with an optimal step founded by authors. A queue GPSS simulation model of individual properties of passengers is used to determine some parameters of the main model. The stop importance is determined using a conservative passenger traffic model. The number of jobs, the number of stores, and other attractors are taken into account along with the apartment count. The authors propose to measure passenger traffic only at a few typical stops. By induction results can be extended to the remaining stops by original technique developed and verified in a city with a population of about 150 thousand people. It is shown an error of technique not more 20%, it seems to be good considering the error of the initial measurements.

Keywords: Urban public transport · Modeling · Measurement · Control

1 Instruction

1.1 Formulation of the Problem

Urban public transport management has great importance to both the economy and society as a whole [1, 2]. Changes in the city, changes in passenger preferences, and other factors require periodic reorganization of the transportation network. At the same time, measuring passenger traffic at stops and in transport is a problem because simultaneous and continuous measurements are required. We cannot measure only in transport as described in reference 3 and many other issues because the population chooses transport taking into account an existing network of routes [3]. Therefore, the measurement results in transport cannot be used for changing the transportation network, only for a schedule. Many authors note problems with measurements at stops: laboriousness and low accuracy [4, 5].

Z. Hu et al. (Eds.): AIMEE 2021, LNDECT 107, pp. 76–84, 2022.
https://doi.org/10.1007/978-3-030-92537-6_7

Here are many simulation tools for transport research: VISSIM [6], VISUM [7], Merakas/PIKAS [8] and others. PTV VISUM has wide capabilities for analyzing traffic flows in the city as a whole [9] and public transport as a subset [10]. For example, the software allows us to calculate passenger flows at stops, isochrones of reachability from the starting point to the target, taking into account pedestrian sections, etc. It also offers tools for marketing, service planning, calculating the benefits and costs of transport as a decision support system. The idea of 'knots of attraction' proposed in VISUM/VISSIM successfully used after modification in this paper later.

1.2 Disadvantages of Exist Software and Problem Clarification

However, the VISUM developers do not disclose the algorithmic and model foundations of their software. It is impossible to state the clarity of this issue despite the efforts of the researchers [11, 12]. A traditional settlement and graphical scheduling methods ensure the fulfillment of a known volume of passenger traffic along routes, taking into account the available number of rolling stock but it is poorly suited for imitation and optimization [13]. Most traffic simulation systems like PIKAS and many others are designed to plan traffic within a network of stops, not to change it. It is not the only obstacle to the use of such systems as well as systems with an unknown model part. Anyway, the quality of the initial data is the very first problem.

Software for solving particular problems of scheduling for a ready-made route network uses supposedly known passenger flows along routes determined by measurements at stops and in transport. However, the provision of each route with transport units has a formative influence on passenger traffic, which is rarely taken into account. If the passenger knows in advance how often public transport runs on a particular route, he will take this circumstance into account when choosing a route or stop. Therefore, it is impossible to optimize the route network or the number of transport units on the route, relying on the passenger traffic measured under the conditions of some actual provision of routes with transport. It seems practically important and relevant to be able to calculate the distribution of passengers based on the minimum and relatively easily obtained information, on widespread information, when initial assessments of the possibility of improving the urban transport is being produced. In this sense, it seems to us an interesting idea [14] about the integration of flows in discrete cells. But the authors of this work consider only the cells along the transport routes, but not in the spatial residential part of the city. Therefore, such a model cannot be applied directly to simulate urban public transport.

We will find and present the original measurement technique in this article. Reducing the number of measurements is possible by using new models: a model of passengers' behavior [15], a model of an emergence of a necessity for transportation, a model of distribution of passengers' traffic.

2 Materials and Methods

2.1 Analysis of City Population Distribution

The main problem of a model of transportation needs creating is the inaccessibility of the information on the spatial distribution of city population. The residents of houses create primary passenger traffic, which is then repeatedly 'reflected' from shops, workplaces, and other attractors. Information about citizens is protected by the 'Law on the protection of personal data' in Russia, and the national concept of 'registration on a place of residence' does not apply in other countries. The area of apartments in buildings is used for roughly determinate the number of residents in a stop neighborhood. Documentation about residential buildings is open access, so is the total number of city residents (about 150 thousand residents in our case). A special software was created, and data about all city houses (1479 houses in our case) was entered in it. Then, we could get the number of residents in each square of the map as a proportion between the sum of the area of apartments in the entire city and the square. An example of city sampling and stops of various types are in Fig. 1.

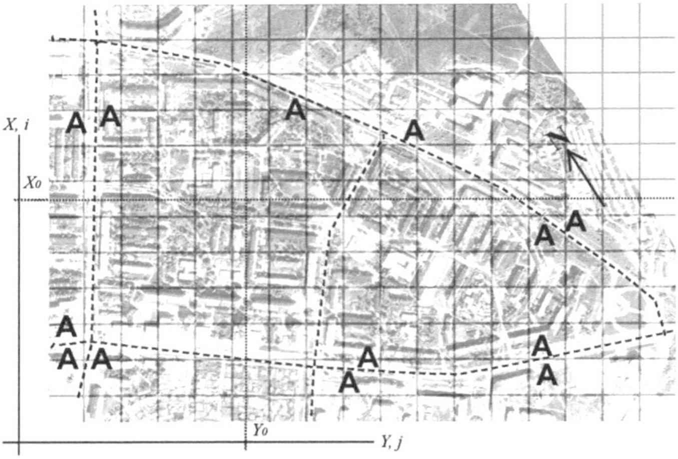

Fig. 1. An example of city sampling and stops of various types

An optimal sampling step of the map (the size of the square) was founded, at which the abrupt change in the average median population density per square stopped. It was 260 m on the ground and corresponded to the distance of 70–300 m recommended in Russia.

2.2 Basics of Passenger Conservation Model

Let the passenger feels the need to move from the square *{i, j}* to the square *{k, m}*, and there are no stops in the source and destination squares (Fig. 2). This approach is used successfully in [16].

The passenger will choose the stops that fit their expectations best of all. The selection criterion is the travel time if the fare is the same in any transport. Such criteria are used in many kinds of research [17, 18], are convincing examples.

An algorithm has been formalized for choosing a passenger's travel path, taking into account psychological characteristics [19]. This algorithm is not the only possible one, but it represents the basic properties of the object. It is in the following:

a) The passenger makes the set of all initial stops available in a reasonable time from the initial point (*ij* lines on Fig. 2).

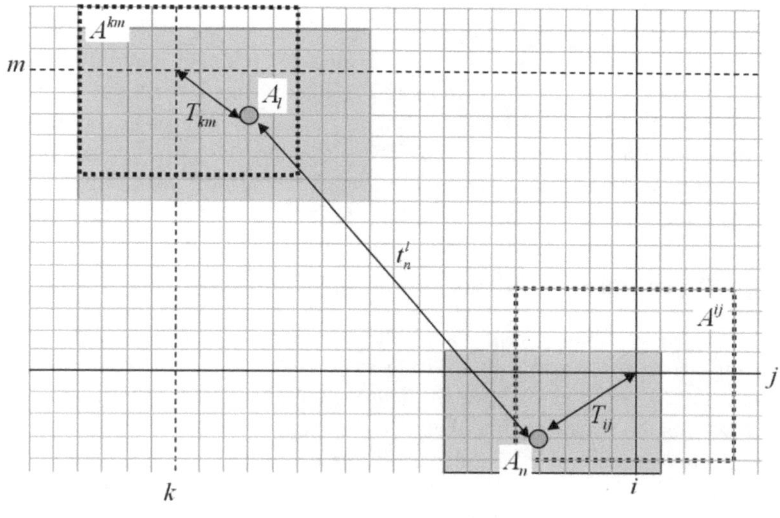

Fig. 2. An illustration for travel time computation: zones of attraction of stops are shaded

b) The passenger ranks the stops of the set according to the access time, roughly estimated by the passenger, we get a list of stops $A^{ij} = \{A_1^{ij}, A_2^{ij}, \ldots\}$: $T\left(ij, A_q^{ij}\right) \le T\left(ij, A_{q+1}^{ij}\right)$, where $T\left(ij, A_q^{ij}\right)$ is the time of movement from the *ij*-point to the stop A_q^{ij}, subjectively estimated by the passenger.

c) A set of target stops A^{km} is determined, in the zone of attraction of which the target *km*-th element is located.

d) The stops of the set are ranked according to the travel time from the stop to the *km*-th element, we get a list of stops.

$$A^{km} = \{A_1^{km}, A_2^{km}, \ldots\} : T\left(A_p^{km}, km\right) \le T\left(A_{p+1}^{km}, km\right)$$

e) For each stop $A_q^{ij} \in A^{ij}, q = 1, 2, \ldots$, a list of transport routes is compiled on which you can get to one of the $A_p^{km} \in A^{km}, p = 1, 2, \ldots$.

f) If the list is not empty, the routes in it are ranked according to the subjectively estimated sum of the waiting times at the initial stop and the travel time to the target stop, the passenger chooses the first route (with the shortest time), thus choosing for himself the initial stop $A_n = A_q^{ij}$ and the target stop $A_l = A_p^{km}$. The algorithm is finished.

g) If the list of urban public transport routes is empty, the passenger considers the next stop with $q: = q + 1$, then step e).

h) If the whole set of stops A^{ij} has been exhausted, and all the lists of public transport routes are empty, the problem is unsolvable. The passenger refuses to use public transport. The algorithm is finished.

The passenger will most likely refuse the trip or take a taxi if the waiting time will be too long. The limitation time was determined by polls. A conditional *impatient* passenger will not agree to wait for more. A linear function of the probability of denying a trip was built. The stop-flow queue model, taking into account the transportation needs, is implemented in GPSS. If the transport goes frequently enough, the effect of denying is almost nonexistent. However, here is a nonlinear dependence of the number of passengers on the time at the stop, if we increase the interval of movement.

2.3 Stops Types and Its Influence to Passenger Flows

Three basic types of stops were identified: an end stop with a large number of attractors, an intermediate nodal stop with a large number of attractors, and a common intermediate stop. Final stops in residential areas do not differ from intermediate stops. Primary flows for them can be calculated in the same way as for intermediate ones.

A conservative model of passenger traffic based on the formula:

$$I_1 = \iint_{x,y} \int_t^{t+\Delta t} \zeta(x, y, t) dx\, dy\, dt \approx \Delta t \sum_i \sum_j \zeta_{ij}(t), \quad (1)$$

where *{x, y}* are the coordinates of the centre of the square, *{i, j}* are the indices of the square, $\zeta(x, y, t) = \zeta_{ij}(t)$ is the intensity of passengers' arrival in the square for the period $[t, t + \Delta t]$. We have taken into account enterprises, which are characterized by short intervals of transport loading [20]. As a result, a trend of the intensity is alike as shown in Fig. 3.

On the other hand, the flow of departing passengers is defined as.

$$I_2 = \left(\psi^*(t) \cdot M + \sum_i \sum_j \varphi_{ij}(t)\right) \cdot \Delta t, \quad (2)$$

where ψ_{ij}^{*} is a specific need for the movement of residents of the square *{i, j}*, *M* is the total passenger quantity, φ_{ij} is the secondary passenger flows in squares, determined by attractors in the influence zone. It is obvious that $\varphi_{ij}(t) = \zeta_{ij}(t - \Delta T_{ij})$, where ΔT_{ij} is a characteristic time of passengers' stay in the attractors.

This intensity will be different for different types of stops predefined above. The first type is the final or close to the final stop, which is surrounded by civil objects with high traffic. For example, this is a square located in a densely populated area, around which there are many shops. Type 2 is a similar stop in the middle of a city, hence in the middle of any route. And type 3 is a stop in a residential part of the city, where no significant attractors for passengers.

For each stop and all stops together, a *passenger conservation equation I1* = I_2 gives us a conservative model that is valid also for the city as a whole

$$\sum_{i}\sum_{j}\left(\psi^{*}(t)\cdot M_{ij}+\varphi_{ij}(t)\right)=\sum_{k}\sum_{m}\zeta_{km}\left(t+T_{ij}^{km}\right) \quad (3)$$

where T_{ij}^{km} is the time of the passenger's movement from the square *{i, j}* to the square *{k, m}* taking into account the walking segment.

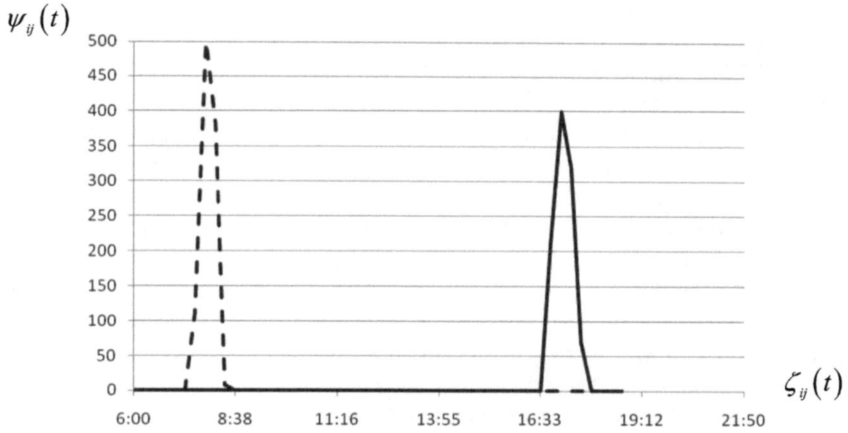

Fig. 3. An example of arrival and departure intensity on some stop

3 Results and Discussion

We found the dependence of passenger traffic on the value of a stop and the population in the vicinity by measurements and calculations. Measurement errors and experts' mistakes were compensated by synthesis an approximating polynomial $S_{\text{in,out}} = f\cdot\delta_f\cdot K_f + r\cdot\delta_r + p\cdot\delta_p\cdot K_p = a_0 + \sum_{i=0}^{I}\sum_{j=0}^{I} a_{ij}r^i f^j$, where *r* is the number of places of work, *f* is the capacity of attractors near the stop, K_f and K_p are unknown normalized coefficients for equalization the value of the attractor and the population with the value of the workplace, δ_f, δ_r and δ_p – signs of whether jobs, attractors and the population participate in the formation of flows (outcoming or incoming) in a given time interval; $\delta_i = \delta_i(\Delta t)$ and $K_i = K_i(\Delta t)$, Δt is the time interval for measuring passenger traffic (7–9 h, 12–14 h, 17–19 h).

Thus, we have obtained a method for calculating the value of any stop in the city if we know normalized coefficients for the whole city. The formula is valid for both the outcoming value (passengers departing) and the incoming value (passengers arriving) with appropriate coefficients.

Incoming/outcoming passenger traffic ratios K_{in}/K_{out} was calculated for each stop to find the normalized coefficients of the model, and

$$\langle K_f(\Delta t), K_p(\Delta t)\rangle : \begin{cases} \left(K^A_{\text{out}}(\Delta t) - K^B_{\text{out}}(\Delta t)\right)^2 + \left(K^A_{\text{out}}(\Delta t) - K^C_{\text{out}}(\Delta t)\right)^2 \rightarrow \min \\ \left(K^A_{\text{in}}(\Delta t) - K^B_{\text{in}}(\Delta t)\right)^2 + \left(K^A_{\text{in}}(\Delta t) - K^C_{\text{in}}(\Delta t)\right)^2 \rightarrow \min \end{cases} \tag{4a}$$

subject to the obvious conditions:

$$\forall \Delta t : \begin{cases} K_f(\Delta t) \geq 0 \\ K_p(\Delta t) \geq 1 \end{cases}, \tag{4b}$$

where < A, B, C > are the typical stops listed above. As a result we found the follow values of coefficients on time intervals < 7–9, 12–14, 17–19 > hours: $K_f =$ < 7.36, 152.13, 122.72 >, $K_p =$ < 1, 1.26, 1 >. The obtained coefficients do not contradict common sense. The dimensionless coefficient K_p of the transition from jobs to population is close to 1 (the excess in the middle of the day indicates that jobs almost do not form passenger flows). The dimensional coefficient K_f of the transition from the number of attractors to the number is minimal in the morning (for example, they do not go to shops and cultural institutions much in the morning) and approximately the same the rest of the time.

To estimate the error of the method (including the fatal measurement error), the values of incoming and outgoing passenger flows for all stops and the average values of $\overline{K}_{\text{out}}(\Delta t)$ and $\overline{K}_{\text{in}}(\Delta t)$, as well as the relative calculation errors. The results of our calculations are shown in Table 1.

The obtained results show us the following. The measured data can be expanded to all stops in the city with an error of ± 22.3% basing on the maximum average error of the approximated coefficients. The value of the error is not very small, but it does not exclude the hypothesis above.

Similar calculations for the summer weekend variant of the schedule show that the average values of the coefficients are correct with an error of ± 19.4. At the same time, the importance of jobs for the passenger traffic formation here is 3.7 times less, which does not contradict common sense (not all enterprises work on weekends). The same result can be obtained for the winter schedule.

Table 1. The results of calculations

Stop or value	Time interval (hours)		
	7–9	12–14	17–19
Input passenger flow (K^i_{in})			
Type 1	0.29540	0.01510	0.03231
Type 2	0.32887	0.02439	0.04112
Type 3	0.30672	0.02026	0.04599
Output passenger flow (K^i_{out})			
Type 1	0.11730	0.01046	0.01729
Type 2	0.11730	0.01046	0.01729
Type 3	0.08607	0.02274	0.03027
Average mean for all the stops			
$\overline{K}_{\text{in}}(\Delta t)$	0.31033	0.01992	0.03981
$\overline{K}_{\text{out}}(\Delta t)$	0.13835	0.01961	0.03058
Errors estimation			
$\sum_i \left(\frac{\overline{K}_{\text{in}}(\Delta t) - K^i_{\text{in}}(\Delta t)}{\overline{K}_{\text{in}}(\Delta t)}\right)^2$	0.00602	0.10936	0.06070
$\sum_i \left(\frac{\overline{K}_{\text{out}}(\Delta t) - K^i_{\text{out}}(\Delta t)}{\overline{K}_{\text{out}}(\Delta t)}\right)^2$	0.44673	0.33736	0.38633
$\overline{\varepsilon}_{\text{in}}(\Delta t) = \frac{1}{3}\sum_{i=1}^{3} \varepsilon^i_{\text{in}}(\Delta t)$	2.6%	11.0%	8.2%
$\overline{\varepsilon}_{\text{out}}(\Delta t) = \frac{1}{3}\sum_{i=1}^{3} \varepsilon^i_{\text{out}}(\Delta t)$	22.3%	19.4%	20.7%
$\max_i \varepsilon^i_{\text{in}}(\Delta t)$	6.0%	24.2%	18.8%
$\max_i \varepsilon^i_{\text{out}}(\Delta t)$	53.0%	46.7%	44.4%
$\min_i \varepsilon^i_{\text{in}}(\Delta t)$	1.2%	1.7%	3.3%
$\min_i \varepsilon^i_{\text{out}}(\Delta t)$	15.2%	15.9%	1.0%

4 Conclusion

As a result, a comprehensive methodology for determining passenger traffic at stops, for example, $N^i_{\text{source}}(\Delta t) = \overline{K}_{\text{source}}(\Delta t) \cdot S^i_{\text{source}}(\Delta t)$, is created. The technique differs from others both in the small number of required measurements and in the acceptable accuracy about 20%. Since the technique is based on conservative equations, there is no doubt about the possibility of its application. We applied the methodology for a city with the population of about 150 thousand inhabitants and received acceptable results. Next, we can use the calculated passenger flows to schedule traffic or reorganize the route network.

References

1. Banerjee, N., Morton, A., Akartunal, K.: Passenger demands forecasting in scheduled transportation. Eur. J. Oper. Res. **286**(3), 797–810 (2020)

2. Chang, Y.-L., Yuan, C.-H., Sun, C., Zhang, P.: Calculation method for actual impedance of urban network based on improved cell transmission model. J. Jilin Univ. (Eng. Technol. Edn.) **50**(1), 132–139 (2020)
3. Zeroual, A., Messai, N., Kechida, S., Hamdi, F.: A piecewise switched linear approach for traffic flow modeling. Int. J. Autom. Comput. **14**(6), 729–741 (2017)
4. Kaszczyszyn, P., Sypion-Dutkowska, N.: Walking access to public transportation stops for city residents: a comparison of methods. Sustainability **11**(14), 3758 (2019)
5. Perzyński, T.: Measurement and analysis of the noise level in selected zones of bus stops in the city of radom. Commun. Sci. Lett. Univ. Zilina **23**(2), D26–D33 (2021)
6. Lownes, N.E., Machemehl, R.B.: VISSIM: a multi-parameter sensitivity analysis. In: Proceedings of the 2006 Winter Simulation Conference, pp. 1406–1413 (2006)
7. Crisan, H.G., Filip, N.: Traffic modeling aspects using visum software and effects on the traffic optimization. In: Proceedings of the European Automotive Congress EAEC-ESFA, pp. 495–506 (2015)
8. Chernova, A., Semchugova, E., Negrov, N.: The analysis of passengers travel time to the distant from a central area transportation districts in the city of rostov-on-don. MATEC Web Conf. EDP Sci. **132**, 4006 (2017)
9. Baskan, O., Ozan, C.: Combined solution of capacity expansion and signal setting problems for signalized road networks. Transp. Res. Proc. **10**, 61–70 (2015)
10. Gulhan, G., Ceylan, H., Özuysal, M., Ceylan, H.: Impact of utility-based accessibility measures on urban public transportation planning: a case study of Denizli, Turkey. Cities **32**, 102–112 (2013)
11. Vijay, N., Aruna, T.: Free open source software in electronics engineering education: a survey. Int. J. Mod. Educ. Comput. Sci. **5**, 15–25 (2014)
12. Song, Z., Wang, H., Sun, J., Tian, Y.: Experimental findings with vissim and transmodeler for evaluating environmental and safety impacts using micro-simulations. Transp. Res. Rec. **2674**(8), 566–580 (2020)
13. Youan, W., Xumei, C., Yanhui, H., Shuxia, G.: Forecast of passenger and freight traffic volume based on elasticity coefficient method and grey model. Proc. Soc. Behav. Sci. **96**, 136–147 (2013)
14. Öztürk, M.M.: Tuning stacked auto-encoders for energy consumption prediction: a case study. Int. J. Inf. Technol. Comput. Sci. **11**(2), 1–8 (2019)
15. Cevallos-Torres, L., Botto-Tobar, M.: Case study: logistical behavior in the use of urban transport using the Monte Carlo simulation method. Stud. Comput. Intell. **824**, 97–110 (2019)
16. Zatonsky, A.V.: Balance model of conversation of passengers flows in residential part of city. World Appl. Sci. J. **30**(12), 1772–1775 (2014)
17. Avila-Torres, P., et al.: The urban transport planning with uncertainty in demand and travel time: a comparison of two defuzzification methods. J. Ambient. Intell. Humaniz. Comput. **9** (3), 843–856 (2018)
18. Nur, K.D.: Review of vehicle surveillance using IoT in the smart transportation concept. Int. J. Eng. Manuf. **11**(1), 29–36 (2021)
19. Park, P.Y., Sahaji, R.: Safety network screening for municipalities with incomplete traffic volume data. Accid. Anal. Prev. **50**, 1062–1072 (2013)
20. Hu, X.-J., Wang, W., Sheng, H.: Urban traffic flow prediction with variable cell transmission model. J. Transp. Syst. Eng. Inf. Technol. **10**(4), 73–78 (2010)

Genetic Informatics and the Bit-reversible Holography - The Problem of Inherited Holographic Features of Organisms

Sergey V. Petoukhov[1,2(✉)], Elena S. Petukhova[1], and Vladimir V. Verevkin[1]

[1] Mechanical Engineering Research Institute, Russian Academy of Sciences, M. Kharitonievsky pereulok, 4, Moscow, Russia
[2] Moscow State Tchaikovsky Conservatory, Bolshaya Nikitskaya, 13/6, Moscow, Russia

Abstract. The article is devoted to algebraic interrelations between the system of structured DNA alphabets and methods of algebraic holography known in technologies of storing and processing digital informatics. Matrix representations of DNA alphabets of 4 nucleotides, 16 doublets, 64 triplets, 256 tetraplets, etc., which are constructed based on binary-oppositional traits of nucleotides, are used. These matrices have phenomenological mosaics and demonstrate unexpected properties under their analysis by a known method of the bit-reversible holography. The presented author's results show previously unknown abilities and directions for developing genetic informatics on holographic principles in parallels with holographic approaches actively developed in modern theoretical physics and science on brain functions. In addition, an unexpected connection of these matrices with split-quaternions and the Poincare disk model of Lobachevski hyperbolic geometry is revealed. This connection can explain known facts of the relationship of physiological phenomena with hyperbolic geometry. In general, the presented results of the author's studies of the structural rules of genetic informatics give pieces of evidence in favor of the efficiency of a model approach to living organisms as quantum-informational algebraic-harmonic essences.

Keywords: DNA alphabets · Holography · Bit-reversible permutation · Binary oppositions · Matrices · Tensor product · Split-quaternions · Poincare disk model

1 Introduction

There are many ways to try to create artificial intelligence that should reproduce the properties of natural intelligence, whose physiological foundations are genetically transmitted from generation to generation. When choosing ways to create artificial intelligence, it is advisable not to lose sight of the genetic roots of the ability for intellectual activity. As it turns out, the structures of genetic informatics are associated with holographic principles, especially popular in physiology and cognitive research, as well as in modern physics with its concepts of the holographic universe. As it is known,

Z. Hu et al. (Eds.): AIMEE 2021, LNDECT 107, pp. 85–94, 2022.
https://doi.org/10.1007/978-3-030-92537-6_8

holographic methods are widely used in storing and processing digital information as extremely effective. In particular, they have great perspectives on the creation of artificial intelligence devices.

It has long been known that living organisms have properties that resemble those of holography with its non-local informatics. For example, in embryology, even 100 years ago, G. Driesch showed that by separating the blastomeres of sea urchin eggs from each other, it is possible to grow from one embryonic cell (blastomere) normal (albeit reduced) larvae with all their organs [1]. Such "holographic" phenomena of distributed information have been repeatedly confirmed in many taxonomic groups of multicellular organisms - from sponges to mammals.

The book listed in the second references on the holographic principles of the brain emphasizes that holographic description is unmatched for explaining perceptual problems, especially imaging problems and fantastic recognition ability [2]. There is a wealth of physiological data to support the holographic concept. For example, K. Lashley showed in 1929 that no matter which part of the rat's brain we remove, we couldn't destroy the memories of how to perform the complex actions that the rat learned before the operation [3]. Many similar phenomena are described, for example, in the book [4]. But all morphogenetic, brain, and other physiological structures are genetically inherited, and therefore it is natural to look for the original holographic properties in a richly structured system of genetic coding.

Recall that structures with holographic properties are realized not only when using both laser beams and photosensitive plates, then also in algebraic models without the use of these devices. It is the methods of algebraic holography that are being developed in theoretical physics and digital informatics. The main goal of the present study is to identify possible algebraic relationships between the structural features of DNA informatics and the methods of algebraic holography.

2 Research Methods

In this study, we use methods of algebraic holography known in digital informatics to model structural features of DNA informatics. These methods include:

1) The bit-reversible holography (or bit-reversible permutations);
2) Holographic coding by Walsh-Hadamard transformations with randomization and permutation of data;
3) Logical holography based on Walsh functions.

We found that all these three methods are consistent with the structures of DNA informatics (details are given, for example, in the preprint [5]). This article only describes consistency with bit-reversible holography. Bit reversion means reading binary numbers in the opposite direction: for example, 001 becomes 100 (which correspond to decimal numbers 1 and 5).

Figure 1 shows a 512*512-pixel matrix, in which all columns and rows are sequentially numbered with binary numbers and in which the letter A is drawn. Reading each of these binary numbers in reverse order leads to renumbering columns and rows and their new placement with the transformation of the entire image (in the second frame). If now part of this frame is painted over or deleted, then reapplying the reverse reading of the binary numbers with the permutation will restore the original image up to identifiable.

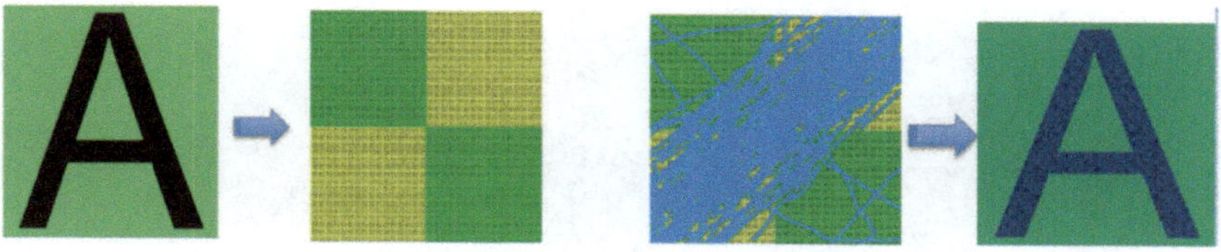

Fig. 1. An example of the bit-reversible holography (from https://habrahabr.ru/post/155471/)

3 Research Results

Let us show the close correspondence of the bit-reversible holography method to the structural foundations of genetic informatics. Genetic information is recorded on DNA molecules as a sequence of four nucleotides: adenine A, guanine G, cytosine C, and thymine T. These four nucleotides are endowed with binary-oppositional molecular traits:

1) two of these nucleotides are purines (A and G), and the other two (C and T) are pyrimidines that give the numerical representing C = T = 1, A = G = 0;
2) two of these nucleotides are keto molecules (T and G), and the other two (C and A) are amino molecules, which gives the numerical representation C = A = 1, T = G = 0.

As it was shown in our previous publications [6, 7], this allows representing DNA alphabets of 4 letters, 16 doublets, 64 triplets, 256 tetraplets, etc. in the form of square tables (Fig. 2), whose columns are numbered with binary indicators "pyrimidine or purine" (C = T = 1, A = G = 0), and rows are numbered with binary indicators "amino or keto" (C = A = 1, T = G = 0). These tables of DNA alphabets are members of the general tensor family of alphabetic matrices [C, A; T, G](n), where (n) is the tensor integer exponent.

	1	0
1	C	A
0	T	G

	11	10	01	00
11	CC	CA	AC	AA
10	CT	CG	AT	AG
01	TC	TA	GC	GA
00	TT	TG	GT	GG

	111	110	101	100	011	010	001	000
111	CCC	CCA	CAC	CAA	ACC	ACA	AAC	AAA
110	CCT	CCG	CAT	CAG	ACT	ACG	AAT	AAG
101	CTC	CTA	CGC	CGA	ATC	ATA	AGC	AGA
100	CTT	CTG	CGT	CGG	ATT	ATG	AGT	AGG
011	TCC	TCA	TAC	TAA	GCC	GCA	GAC	GAA
010	TCT	TCG	TAT	TAG	GCT	GCG	GAT	GAG
001	TTC	TTA	TGC	TGA	GTC	GTA	GGC	GGA
000	TTT	TTG	TGT	TGG	GTT	GTG	GGT	GGG

Fig. 2. Matrices of DNA-alphabets of 4 nucleotides, 16 doublets, and 64 triplets.

The phenomenological fact is known about the binary-oppositional division of the DNA alphabet of 64 triplets - according to their code properties - into two equal sub-alphabets: 32 triplets with strong roots (i.e. triplets starting with 8 doublets CC, CT, CG, AC, TC, GC, GT, GG) and 32 triplets with weak roots (i.e. triplets starting with other 8 doublets) [8, 9]. In the constructed matrices of DNA alphabets of 16 doublets, 64 triplets, and 256 tetraplets (Fig. 2), we mark n-plets with strong roots in black (Fig. 3):

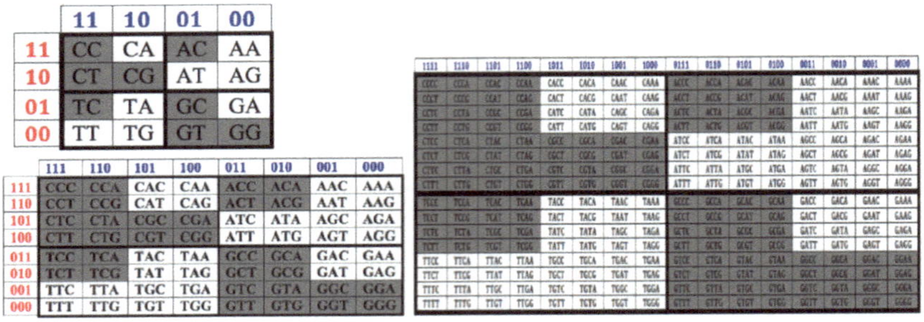

Fig. 3. Mosaic matrices of DNA-alphabets from Fig. 2 (explanation in the text).

Bit-reverse rearrangement of these mosaic matrices in Fig. 3 generates a family of multi-block matrices whose mosaics consist of repetitions of the mosaic (4*4)-matrices of 16 doublets (Fig. 4).

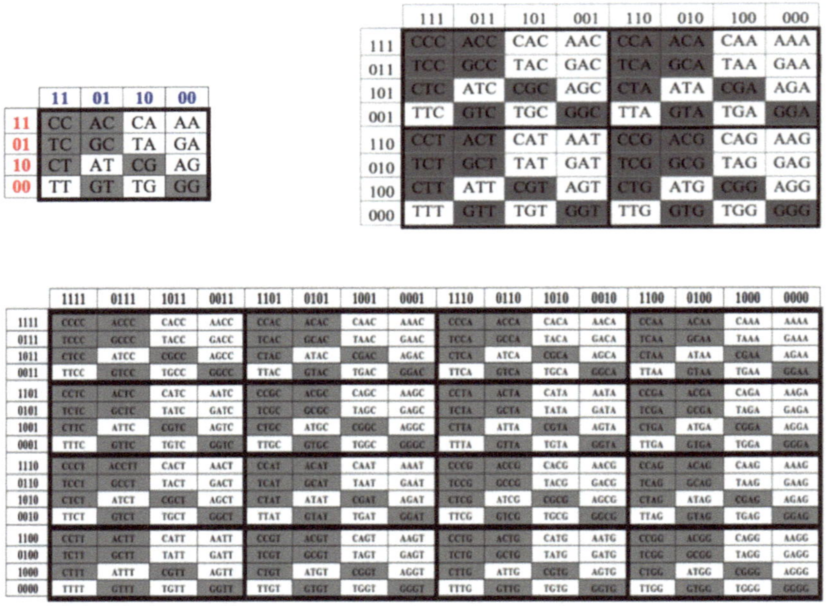

Fig. 4. The result of the bit-reversible rearrangement of the numbering of columns and rows of mosaic matrices of DNA alphabets from Fig. 3.

Black and white cells of the matrices in Fig. 4 reflect the opposition of two subsets of n-plets with strong and weak roots and therefore can be represented by elements +1 and −1 in them. Multi-block mosaics of all emerging numerical matrices of DNA alphabets have the form of "matrix crystals", since they consist of repetitions of a mosaic of the (4*4)-matrix of 16 doublets, which we conditionally term α-matrix (Fig. 5). Does this "universal" repeating block have a meaningful algebraic meaning? Yes, it has.

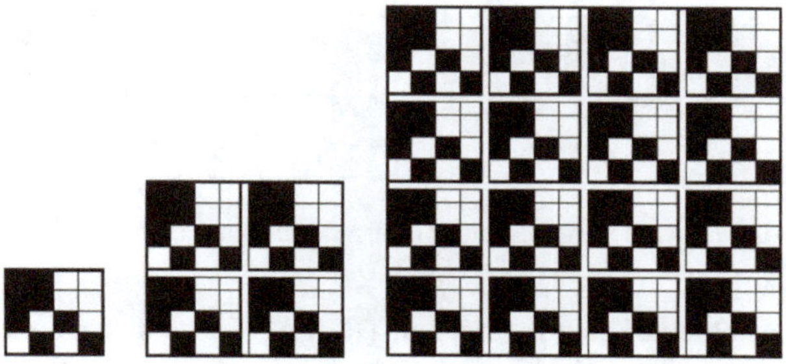

Fig. 5. Numerical representation of mosaic matrices of DNA alphabets from Fig. 4, the black and white cells of the matrices represent elements +1 and −1, respectively.

It turns out that this block is the sum of 4 sparse matrices, the set of which is closed under multiplication and determines the corresponding multiplication table, known as the multiplication table of the algebra of 4-dimensional split-quaternions of Cockle, known since 1849 (Fig. 6). Split-quaternions are used in the Poincaré disk model to describe hyperbolic motions in Lobachevsky's hyperbolic geometry [10]; https://en.wikipedia.org/wiki/Split-quaternion].

$$\begin{vmatrix} 1 & 1 & -1 & -1 \\ 1 & 1 & -1 & -1 \\ 1 & -1 & 1 & -1 \\ -1 & 1 & -1 & 1 \end{vmatrix} = \begin{vmatrix} 1,0,0,0 \\ 0,1,0,0 \\ 0,0,1,0 \\ 0,0,0,1 \end{vmatrix} + \begin{vmatrix} 0,0,-1,0 \\ 0,0,0,-1 \\ 1,0,0,0 \\ 0,1,0,0 \end{vmatrix} + \begin{vmatrix} 0,0,0,-1 \\ 0,0,-1,0 \\ 0,-1,0,0 \\ -1,0,0,0 \end{vmatrix} + \begin{vmatrix} 0,1,0,0 \\ 1,0,0,0 \\ 0,0,0,-1 \\ 0,0,-1,0 \end{vmatrix} = j_0+j_1+j_2+j_3$$

*	j_0	j_1	j_2	j_3
j_0	j_0	j_1	j_2	j_3
j_1	j_1	$-j_0$	j_3	$-j_2$
j_2	j_2	$-j_3$	j_0	$-j_1$
j_3	j_3	j_2	j_1	j_0

Fig. 6. Decomposition of the numerical (4 4)-matrix of the DNA-alphabet of 16 doublets (from Fig. 5) into 4 sparse matrices, the set of which is closed with respect to multiplication. The multiplication table for this set is shown, which matches with the multiplication table of the Cockle split-quaternion algebra used in the Poincaré disk model of hyperbolic geometry. The symbol for the model is shown from https://commons.wikimedia.org/wiki/Category:Poincar%C3%A9_disk_models).

Using the symbol of the Poincaré disk model of hyperbolic geometry from Fig. 6, it is possible to present the obtained block mosaic matrices of structured DNA alphabets in an artistic form to facilitate heuristic associations (Fig. 7).

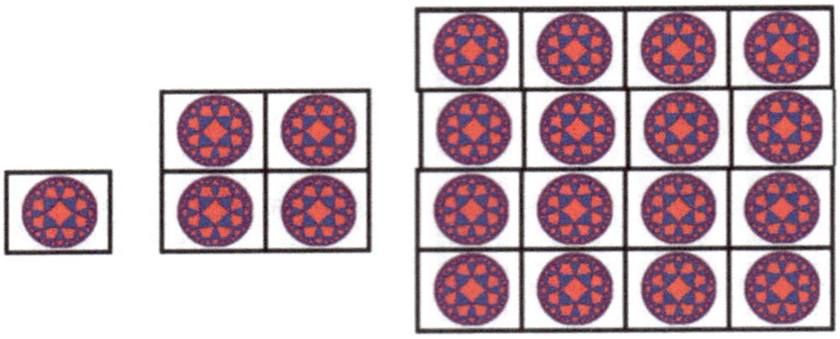

Fig. 7. Artistic representation of mosaic matrices from Fig. 5.

The family of block-unified matrices of DNA alphabets discovered by the method of bit-reversible holography, the blocks of which are conjugated with the disk model of hyperbolic geometry, unexpectedly echoes the theme of "holographic quantum error-correcting codes" developed at the California Institute of Technology in the USA in connection with the same disk model Poincaré and its tilings [11]. This holographic topic is headed by J.Preskill, director of the Caltech Institute for Quantum Information and Matter. The topic also includes the consideration of space-time as a quantum error-correcting code [12]. The following quotes from these works, which describe quantum error-correcting codes with a tensor network structure, clarify these parallels in more detail. "The tensor network is supported on a uniform tiling of a hyperbolic space, known as a hyperbolic tessellation. … We shall focus on examples based on tilings of two-dimensional hyperbolic space, which are specific realizations of uniform hyperbolic tilings known as hyperbolic tessellations" [12]. «Holographic quantum error-correcting codes are constructed by contracting perfect tensors according to a tilling of hyperbolic space by polygons» [11].

Many people are familiar with the images of the Poincaré disk model of the hyperbolic geometry of Lobachevsky from Escher's engravings (Fig. 8).

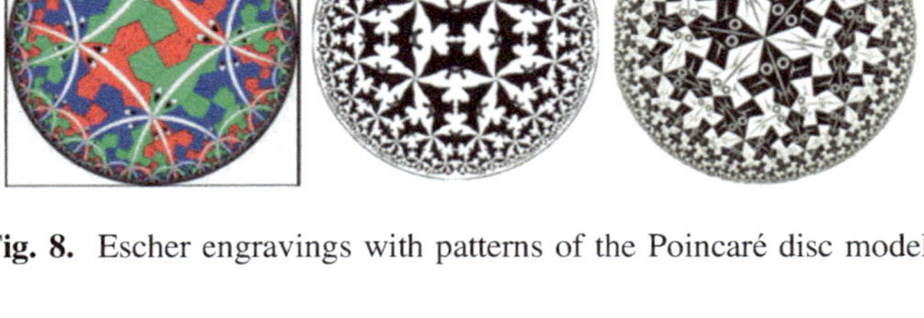

Fig. 8. Escher engravings with patterns of the Poincaré disc model.

Each of the numeric matrices in Fig. 5 can be constructed from the α-matrix by the algorithm (1) based on its tensor product with the matrix [1, 1; 1, 1]:

$$[1,1;1,1]^{(n)}\otimes[1,1,\ -1,\ -1;1,1,\ -1,\ -1;1,\ -1,1,\ -1;\ -1,1,\ -1,1] \quad (1)$$

The matrix [1, 1; 1, 1] used in (1) is the matrix representation of a 2-dimensional hyperbolic number with unit coordinates. Deep structural connections of genetic informatics with 2^n-dimensional hyperbolic numbers are described in [13].

One should else note the actions of the (2^n*2^n)-matrices (Fig. 5) to vectors. Let us consider their actions to voluntary 2^n-dimensional vectors having arbitrary coordinates ai(t), which are not connected to each other and can be, for example, cyclic functions of time. The action of the (4*4)-matrix α = [1, 1, -1, -1; 1, 1, -1, -1; 1, -1, 1, -1; -1, 1, -1, 1] to the vector $\overline{x}_4$ = [$a_0(t)$, $a_1(t)$, $a_2(t)$, $a_3(t)$] with cyclic coordinates unrelated to each other gives a new block-structured vector with interrelated coordinates combined in pairs by a binary-oppositional principle:

$$\overline{X}_4 * \alpha = [v(t),\ w(t),\ -w(t),\ -v(t)] \quad (2)$$

where $v(t) = a_0(t) + a_1(t) + a_2(t) - a_3(t)$, $w(t) = a_0(t) + a_1(t) - a_2(t) + a_3(t)$. In this generated vector, its coordinates v(t) and −v(t) (and also w(t) and −w(t)) are equal to each other up to the sign and their changes are synchronized in time.

The action of the (8*8)-martix α_8 = [1, 1; 1, 1] ⊗ α to the vector $\overline{x}_8$ = [$a_0(t)$, $a_1(t)$, $a_2(t)$, $a_3(t)$, $a_4(t)$, $a_5(t)$, $a_6(t)$, $a_7(t)$] with cyclic coordinates unrelated to each other gives a new block-structured vector with interrelated coordinates combined in pairs by the similar binary-oppositional principle:

$$\overline{X}_8 * \alpha_8 = [v(t),\ w(t),\ -w(t),\ -v(t),\ v(t),\ w(t),\ -w(t),\ -v(t)] \quad (3)$$

where $v(t) = a_0 + a_1 + a_2 - a_3 + a_4 + a_5 + a_6 - a_7$, $w(t)= a_0 + a_1 - a_2 + a_3 + a_4 + a_5 - a_6 + a_7$. This generated vector (3) has a block structure and paired coordinates v(t) and −v(t) (the same for coordinates w(t) and −w(t)), which are equal to each other up to the sign and their changes are coordinated and synchronized in time.

The action of the (16*16)-matrix α_{16} = [1, 1; 1, 1] ⊗ α to a 16-dimensional vector $\overline{x}_{16}$ = [$a_0(t)$, $a_1(t)$, $a_2(t)$, $a_3(t)$, $a_4(t)$, $a_5(t)$, $a_6(t)$, $a_7(t)$, $a_8(t)$, $a_9(t)$, $a_{10}(t)$, $a_{11}(t)$, $a_{12}(t)$, $a_{13}(t)$, $a_{14}(t)$, $a_{15}(t)$] with cyclic coordinates unrelated to each other gives a new block-structured vector with interrelated coordinates combined in pairs by the similar binary-oppositional principle:

$$\overline{X}_{16} * \alpha_8 = [v(t),\ w(t),\ -w(t),\ -v(t),\ v(t),\ w(t),\ -w(t),\ -v(t),\ v(t),\ w(t),\ -w(t),\ -v(t),\ v(t),\ w(t),\ -w(t),\ -v(t)] \quad (4)$$

where

$$v(t) = a_0 + a_1 + a_2 - a_3 + a_4 + a_5 + a_6 - a_7 + a_8 + a_9 + a_{10} - a_{11} + a_{12} + a_{13} + a_{14} - a_{15},$$
$$w(t) = a_0 + a_1 - a_2 + a_3 + a_4 + a_5 - a_6 + a_7 + a_8 + a_9 - a_{10} + a_{11} + a_{12} + a_{13} - a_{14} + a_{15}.$$

This generated vector (4) has a block structure and paired coordinates v(t) and −v(t) (the same for w(t) and −w(t)), which are equal to each other up to the sign and their changes are coordinated and synchronized in time.

In a general case, actions of (2^n*2^n)-matrices $[1, 1; 1, 1]^{(n-2)} \otimes \alpha$ to a voluntary 2n-dimensional vector with cyclic coordinates unrelated to each other gives a new block-structured vector with interrelated coordinates combined in pairs by the similar binary-oppositional principle. If initial vectors $\overline{x}_4, \overline{x}_8, \overline{x}_{16}, \ldots$ are state vectors of systems with unrelated coordinates in configuration spaces (or unrelated behavior in different configuration subspaces), then actions of matrix operators $[1, 1; 1, 1]^{(n)} \otimes \alpha$ transform the system states into new ordered forms with interrelated behavior in different configuration subspaces. For example, if coordinates of an initial state vector of a system are cyclic or wave functions, then the received new state vector have block-structured sets of coordinates whose blocks are equal to each other up to the sign, and their cyclic changes are synchronized in time (that is, cyclic behavior of the considered system in its different configuration subspaces becomes strongly interrelated). Figuratively speaking, these matrix operators (Fig. 5), obtained by analyzing the system of structured DNA alphabets, create order out of chaos.

This algebra-genetic result is essential for the modeling of biological systems since the coordinated cyclic behavior of genetically inherited parts is one of the characteristic features of living bodies. The living organism as a whole is a huge chorus of coordinated cyclic processes. For example, the inherited normal work of the heart is based on the coordinated cyclic work of the parts of the heart muscle; violation of this coordination leads to disease - cardiac arrhythmias. Tibetan pulse diagnosis is based on the relationship of inherited cyclic processes in the body, and so on.

It should be noted holographic-like features of the received multi-block vectors: knowing only one of the blocks of such vector allows restoring the entire vector. For example, knowing two adjacent coordinates v(t) and w(t) in any part of the 16-dimensional vector in (16.4), it is possible to recover the entire vector.

4 Some Concluding Remarks

This article is devoted to applications of algebraic holography methods in genetics for an understanding of inherited holographic-like properties in organisms. The presented author's results show previously unknown abilities and directions for developing genetic informatics on holographic principles in parallels with holographic approaches actively developed in modern theoretical physics and science on brain functions.

For example, let us recall about D.Bohm, who has been described as one of the most significant theoretical physicists of the 20th century and who contributed unorthodox ideas to quantum theory, neuropsychology, and the philosophy of mind [https://en.wikipedia.org/wiki/David_Bohm]. Bohm believed that a separate human consciousness interacts with the consciousness of other people in a holographic way. Human consciousness is part of the universal hologram of the entire human race so that all the data of this great whole can be extracted from a separate consciousness. David Bohm was convinced that the entire universe has a holographic structure and that with its help many events in the universe and human life can be explained. Bohm worked with Pribram on the theory that the brain operates in a manner that is similar to a hologram, by quantum mathematical principles and the characteristics of wave patterns [2]. Nobel laureate in physics Gerard't Hooft has put forward the holographic principle, which is currently being developed by many physicists in string theory, quantum gravity, etc. [https://en.wikipedia.org/wiki/Holographic_principle]. According to this principle, which connects physics with informatics, the description of a volume of space can be thought of as encoded on a lower-dimensional boundary to the region.

Revealing the described connections of the genetic coding system with the Poincaré disk model of the Lobachevski hyperbolic geometry indicates the genetic basis of the known facts of the relationship of physiological phenomena with hyperbolic geometry. Let us recall some. Since the pioneering work [14], articles on non-Euclidean howling geometry of the space of visual perception are published in different countries. The author of the work [15], who carried out particularly careful experiments with the participation of 200 subjects and using 1300 visual patterns, summarized: "Poincaré's model of hyperbolic space ... shows fairly good agreement with the experimental results".

The article [16] on the results of 20 years of research by the author of the locomotions of many animals and humans states that the spatio-temporal organization of locomotions is associated with hyperbolic turns and the geometry of Minkowski. In this regard, Smolyaninov put forward the "locomotor theory of relativity" and wrote about the relativistic brain and relativistic biomechanics.

O. Bodnar studied the regular rearrangements of phyllotaxis lattices during the growth of organisms and, based on the results obtained, stated that living matter is structurally linked to the geometry of Minkowski [17, 18].

Received results can be used in future bioinformatical and technological works of different directions (for example, for developing [19–24]).

In general, the presented results of the author's studies of the structural rules of genetic informatics give pieces of evidence in favor of the effectivity of a model approach to living organisms as quantum-informational algebraic-harmonic essences, which have algebraic-holographic properties related to the Poincaré disk model of the Lobachevski hyperbolic geometry.

Acknowledgments. The authors are grateful to their colleagues M. He, Z. Hu, Yu.I. Manin, I. Stepanyan, V. Svirin, and G. Tolokonnikov for research assistance.

References

1. Belousov, L.V.: Morphomechanics of Development, pp. 3–10. Springer International Publishing, Switzerland (2015)
2. Pribram, K.: Experimental paradoxes and principles in neuropsychology. Englewood Cliffs **6** (1), 442–452 (1971)
3. Lashley, R.S.: Brain Mechanisms and Intelligence, pp. 978–984. University of Chicago Press, Chicago (1929)
4. Talbot, M.: The Holographic Universe. HarperCollins Publishers Ltd, New York (1996)
5. Petoukhov, S.V.: Algebraic rules for the percentage composition of oligomers in genomes. Preprints **3**(1), 60–62 (2021)

6. Petoukhov, S.V.: Matrix Genetics, Algebras of Genetic Code, Noise Immunity, pp. 316–370. RCD, Moscow (2008)
7. Petoukhov, S., He, M.: Symmetrical Analysis Techniques for Genetic Systems and Bioinformatics: Advanced Patterns and Applications. IGI Global (2010). https://doi.org/10.4018/978-1-60566-124-7
8. Rumer, Y.: Condon systematization in the genetic code. Dokl. Akad. Nauk SSSR **183**(1), 225–226 (1968)
9. Fimmel, E., Strüngmann, L.: Yury Borisovich Rumer and his 'biological papers' on the genetic code[J]. Philos. Trans. A Math. Phys. Eng. Sci. **374**(2063), 20150228 (2016). https://doi.org/10.1098/rsta.2015.0228. PMID: 26857664
10. Karzel, H., Kist, Günter.: Kinematic algebras and their geometries. In: Kaya, R., Plaumann, P., Strambach, K. (eds.) Rings and Geometry, pp. 437–509. Springer Netherlands, Dordrecht (1985). https://doi.org/10.1007/978-94-009-5460-1_10
11. Preskill, J.: Stability, topology, holography: the many facets of quantum error correction. APS March Meeting 2016 (2016). http://theory.caltech.edu/~preskill/talks/APS-March-2016-preskill.pdf
12. Pastawski, F., et al.: Holographic quantum error-correcting codes: toy models for the bulk/boundary correspondence. J. High Energy Phys. (2015). https://doi.org/10.1007/JHEP06(2015)149
13. Petoukhov, S.V.: Modeling inherited physiological structures based on hyperbolic numbers. BioSystems **199**(1), 26–47 (2021)
14. Luneburg, R.: The metric of binocular visual space. J. Opt. Soc. Am. **40**(10), 627–642 (1950)
15. Kienle, G.: Experiments Concerning the Non-Euclidean Structure of Visual Space, pp. 386–400. Pergamon Pree, New York (1964)
16. Smolyaninov, V.V.: Spatio-temporal problems of locomotion control. Usp. Fiz. Nauk **170** (10), 1063–1128 (2000)
17. Bodnar, O.Y.: Geometry of phyllotaxis. Rep. Acad. Sci. Ukraine **9**, 9–15 (1992)
18. Bodnar, O.Y.: Golden Ratio and Non-Euclidean Geometry in Nature and Art. Publishing House Sweet, Lviv (1994)
19. Khan, R., Debnath, R.: Human distraction detection from video stream using artificial emotional intelligence. Int. J. Image Graph. Signal Process. **12**(2), 19–29 (2020). https://doi.org/10.5815/ijigsp.2020.02.03
20. Erwin, D.R.N.: Improving retinal image quality using the contrast stretching. Histogram Equal. CLAHE Methods Med. Filters **12**(2), 30–41 (2020)
21. Niaz Mostakim, M., et al.: Design and development of an intelligent home with automated environmental control. Int. J. Image Graph. Sig. Process. (IJIGSP) **12**(4), 1–14 (2020)
22. Arora, N., Ashok, A., Tiwari, S.: Efficient image retrieval through hybrid feature set and neural Network. Int. J. Image. Graph. Sig. Process. **11**(1), 44–53 (2019). https://doi.org/10.5815/ijigsp.2019.01.05
23. Anami, B.S., Naveen, N.M., Surendra, P.: Automated paddy variety recognition from color-related plant agro-morphological characteristics. Int. J. Image. Graph. Sig. Process. **11**(1), 12–22 (2019). https://doi.org/10.5815/ijigsp.2019.01.02
24. Mahtab Ahmed, M.A.H., Akhand, M.M., Rahman, H.: Recognizing Bangla handwritten numeral utilizing deep long short term memory. Int. J. Image. Graph. Sig. Process. **11**(1), 23–32 (2019). https://doi.org/10.5815/ijigsp.2019.01.03

Fuzzy Controller for Automatic Ventilation Control System

Ruslan Bazhenov[1(✉)], Evgeniy Lavrov[2], Nelly Sedova[3], and Viktor Sedov[3]

[1] Sholom-Aleichem Priamursky State University, Birobidzhan 679015, Russia
[2] Sumy State University, Sumy 40007, Ukraine
[3] Maritime State University named after G.I. Nevelskoy, Vladivostok 690059, Russia

Abstract. The paper provides a brief overview of current approaches to modeling intelligent controllers in vent systems of dwellings. The authors propose a model based on fuzzy logic for controlling the speed of rotation of an asynchronous fan engine according to the temperature values inside and outside a housing accommodation and considers the temperature current sanitation and hygiene standards in living accommodations. The model has 2 input linguistic variables (indoor and outdoor temperature) and one output (fan rotor speed). The fuzzy product rule base contains 42 components. 20 fuzzy sets are introduced to describe the model. A graphical and analytical representation is given for each of them. To rate the quality of the proposed fuzzy controller model, the following metrics was used: mean absolute error (MAE), root mean square error (RMSE), and symmetric mean absolute percentage error (SMAPE). The numerical values of the quality metrics based on the test results (MAE = 1.19; RMSE = 2.56; SMAPE = 0.023) indicate that the regulator based on fuzzy logic can adequately control the frequency converter that sets the fan motor speed, and the model showed prospective operability in supply and exhaust ventilation devices.

Keywords: Ventilation · Supply ventilation · Exhaust ventilation · Fuzzy system · Fan frequency control

1 Instruction

A ventilation (air conditioning) system of various accommodations is an assembly of equipment for processing, transporting, supplying and removing air and is divided into natural and mechanical. Natural ventilation is certainly the simplest and the cheapest way to ventilate premises since in this case neither electricity nor special-purpose devices or facilities are required to use. The mechanism of natural ventilation is based on the following natural elements: different and changeable air pressure; wind strength and direction; air aeration. To make a space naturally ventilated, at least one of these changeable natural events is necessary to be regular. It makes such a system too dependent on external factors.

Z. Hu et al. (Eds.): AIMEE 2021, LNDECT 107, pp. 95–104, 2022.
https://doi.org/10.1007/978-3-030-92537-6_9

Mechanical ventilation requires the use of special devices and equipment that allow fresh air to come in from the outside, purify it and, sometimes, even change the temperature over a relatively short time. It hardly depends on natural phenomena needed for natural ventilation, but as opposed to the latter, it requires permanent energy costs.

Mechanical ventilation is supposed to be divided into exhaust, supply, supply and exhaust. Mechanical exhaust ventilation is most often used in industrial buildings and dining facilities since the most critical function of such ventilation is to clean up polluted air in large areas, as well as to cool it. The function of supply mechanical ventilation is to supply air to the place with no pre-conditioning. If necessary, the air masses can be processed with the optional technical facilities, but the supply ventilation does not provide a system for pumping out dirty exhaust air. Depending on the functions and purpose of accommodation, both exhaust and supply ventilation types can be used. However, both systems are more often installed in buildings at once. This is because this type of ventilation combines not only the names of the two most popular air exchange systems but also their most significant and useful properties.

This paper reports on the results of designing a ventilation system with the involvement of a fuzzy controller (FC) used for automatic control of the rotation speed of the supply or exhaust fan. The involvement of that particular fuzzy logic device seems appropriate because widely used controllers based on proportional-plus-integral-derivative controllers have a major drawback. It deals with the fact every time the microclimate of accommodation changes it needs the equipment to be re-configured. The fuzzy controller is self-adaptive to the indoor air condition changes.

2 Fuzzy Controllers Used for Controlling Indoor Ventilation System

Here are some results for controlling the indoor ventilation system that other researchers obtained. So, paper "Environment indoor air quality assessment using fuzzy inference system" presents a system including an FC, used to monitor air quality and a comfortable space temperature [1]. The system is divided into four clusters. Cluster 1 is responsible for monitoring CO and CO_2, cluster 2 is responsible for monitoring NO_2 and O_3, cluster 3 is responsible for monitoring volatile organic compounds and ultrafine particles, and cluster 4, which is responsible for monitoring air humidity and temperature. The system uses the following sensors to obtain this data: a gas analyzer; volatile organic substances counter; moisture and temperature meter. FC is performed in the Matlab environment using the Fuzzy Logic Toolbox. The authors of the study claim that their system is capable of displaying an indoor air quality and thermal comfort index based on the sensor values and FC to be able to solve problems concerned indoor ventilation [1].

The research of Grygierek and Ferdyn-Grygierek presents a method for optimizing mechanical ventilation mainly in private residences in a transitional climate using FC enhanced by a genetic algorithm [2]. To the authors' mind, despite the popularity in space-heating, ventilation and air conditioning systems of proportional-plus-integral-derivative controllers often turn out to be a nonoptimal solution because of the complexities of configuring them. The authors present an FC that regulates the fan speed compared to the temperature outside and in the room. This system is provided in the Matlab and EnergyPlus environments. The studies were conducted to prove the system availability. The authors of the article show that their approach is able to optimize ventilation systems in a separate house and improve the thermal comfort of residents without significant costs [2].

The article "Fuzzy Logic Based Ventilation for Controlling Harmful Gases in Livestock Houses" defines a ventilation system based on fuzzy logic for monitoring toxic gas concentration in the air in shelters to animals [3]. Sensors convey FC information about temperature and humidity, as well as about the carbon dioxide, hydrogen nitride (ammonia) and hydrogen sulfide concentration in the air. The FC regulates the fan rotation speed based on these readings. The system was performed in the Matlab environment using the Fuzzy Logic Toolbox. The authors conclude that it is possible to design an optimal system for monitoring the toxic gases concentration in shelters to animals based on that system [3].

The paper "Impact of Passive Cooling on Thermal Comfort in a Single-Family Building for Current and Future Climate Conditions" is devoted to the study of the influence of passive cooling of accommodation in the current and hotter climate change [4]. The aim of the study was to review the potential of standard passive cooling in providing a sufficient level of thermal comfort and compare it with the method of mechanical ventilation based on an adjustable fan. A detached home designed for a family of four people was taken as an example to put that in context. The layout of the house was transferred to the EnergyPlus program, where the studies were conducted in different weather conditions. The FC for regulating fans was used to simulate the mechanical ventilation of the room. It included additional fans if the indoor air temperature is higher than the predetermined settings of thermal comfort. The study showed that overheating of accommodation will be a problem in buildings with passive ventilation because of global warming. In such conditions, an effective way is to use mechanical ventilation systems with adjustable fans. In case the fans have an appropriate adjustment, they can provide sufficient thermal comfort with low energy consumption.

The article "Fuzzy logic control of air-conditioning system in residential buildings" shows the results of using the FC for controlling air conditioning and heating in accommodations. The sensors of this system transmit the information on the inside temperature and humidity to the FC, it takes into account the received data, and controls the air conditioner operation [5]. The system is performed in the Matlab environment, while the authors conducted a series of experiments comparing a fuzzy controller operation and a proportional-plus-integral-derivative controller in different seasons throughout the year. Finally, it is concluded that the FC can maintain the desired temperature and relative humidity in the room, both under regular and excessive loads, and the proportional-plus-integral-derivative controller cannot control temperatures when the load changes and consumes much more resources than the FC.

The paper "Fuzzy logic controller deployed for indoor air quality control in naturally ventilated environments" denotes a system containing the FC for monitoring the air quality in residential premises, including four sensors that inform about the concentration of carbon dioxide and carbon monoxide gases, air humidity and various odors, which regulate the fan speed [6]. The system is performed in the Matlab environment; studies were conducted under various conditions of gas contamination of the room, which proved the system operability. The authors of the article claim that their system is able to control the number of toxic gases in the room even in emergencies, and is also more energy-efficient in comparison with conventional ventilation control systems.

In the study of author Soleimanzadeh an indoor climate control system based on the FC was developed, aimed at reducing the energy consumed in houses and meeting environmental requirements [7]. Sensors are used to obtain indoor temperature and humidity values, which regulate the operation of a fan, an air washer, a heater and a compressor. The system is equipped with four heaters, two compressors for cooling and one air washer. The following variables are used as input variables: the difference between the indoor temperature and the temperature set by the user, and the difference between the indoor air humidity and the user-set air humidity. The system was performed in the Matlab environment; studies were conducted confirming the system efficiency in cases of a rapid temperature change. The method allows one to prevent frequent switching on and off of equipment and thus reduces power consumption and extends the service life of the system. The authors of the study show that their solution is able to regulate the optimal humidity and temperature of the space while using economical energy consumption in comparison with a standard ventilation, heating and air conditioning system.

The paper "Design of Indoor Temperature and Humidity Monitoring System based on ZigBee and Fuzzy PID Technology" presents a temperature and humidity monitoring system based on the FC and ZigBee protocols, consisting of four sensor nodes, a network gateway and a monitor centre [8]. The ZigBee CC2530 chip is used as the core for sensor nodes, performs real-time monitoring of air temperature and humidity. Communication between the nodes is carried out using the ZigBee protocol and implements a self-organizing network of wireless sensors and automatic aggregation of monitoring data. The indoor temperature and the air humidity are regulated by the FC. A personal computer with the appropriate software is used as a monitor centre. The researchers did an experiment in a laboratory, where it was found out that the network had a good self-correcting capability and high data accuracy. The authors [8] delivered a design that could make a simple and quite inexpensive system for monitoring and regulating indoor air temperature and humidity.

The paper "Modeling, Analysis, and Design of a Fuzzy Logic Controller for an AHU in the S.J. Carew Building at Memorial University" observes the FC review and design for the air handler in the university building [9]. The purpose of the research is to improve modern algorithms for heating, ventilation and air conditioning systems of premises by using the FC. The controller uses data on the temperature of heating elements, atmospheric pressure and the concentration of carbon dioxide in the air, which regulates the fan rotation speed, the hot water supply valve and the operating mode of the fresh air supply fans. The system is designed in the Matlab environment. Testing in it showed that it was able to operate optimally for a long time without any overloads. The researchers [9] claim that the proposed system is more effective than the traditional algorithm, while the FC algorithm provides a durable response and copes with several different parameters better.

The study "Design and Implementation of an intelligent controller for a spilit air conditioner with energy saving" dwells on the implementation of the FC for an air conditioner with energy-saving [10]. The aim of the paper was to reduce the energy consumption of the air conditioner without significantly affecting the thermal comfort of a person since conventional thermostats (water regulators) are not the optimal solution to this problem. This is because they cannot turn off the air conditioner compressor due to minor changes in the indoor temperature, but only when the preset temperature threshold is exceeded. It is assumed that the FC based on the metrics of temperature and humidity sensors can regulate the time of switching on and off the air conditioner better than a conventional thermostat. According to the authors, such an FC can reduce the energy consumption of an air conditioner by 3 kW/h with no marked impact on performance and thermal comfort.

The research "Model predictive and fuzzy control of a road tunnel ventilation system" covers the FC for a pitch-control fan in an underground road [11]. The main aim is to reduce energy consumption by implementing a system capable of analyzing the number of toxic gases in the tunnel and predicting the required number of jet fans in operation for the tunnel exhaust cleaning. The system consists of carbon monoxide sensors that transmit information to FC. It regulates the operation of jet fans necessary for cleaning the tunnel from toxic gases. The proposed scheme has been tested by simulation. The results show that the ventilation response time is much faster than a system controlled by a conventional controller.

The article "Experimental study of an indoor temperature fuzzy control method for thermal comfort and energy saving using wristband device" defines the development result of indoor temperature control based on fuzzy logic and physiological data obtained from a smart watch [12]. The aim of the research was to improve the thermal comfort of the user using his physiological data with reduced energy consumption of an air conditioner for home use. An upgrade method based on a linear regression model is utilized to predict the user's warmth sense. The FC is used to control the comfort temperature indoors and regulate the operating mode of the air conditioner. The control is based on the user's physiological data and the metrics of thermostat sensors. The research findings showed that the control based on forecasting warmth sense can regulate a given temperature value, monitoring the physiological data of the user, without interfering with his routine work. It was also found out that the characteristics of thermal comfort based on forecasting warmth sense are superior to control based on the set temperature values. The researchers claim that their solution will help to avoid hyper cooling of locations. It will also prevent excessive energy consumption and reduce day-time energy consumption by a mean of 20% due to more optimal temperature control in comparison with control based on the set temperature values.

The review of available solutions allowed the authors to conclude that there are many different ways to control ventilation systems, including the involvement of fuzzy logic-based controllers to solve the energy consumption issues and to ensure thermal control of users. On the whole, all the embedding shows significant improvement in outcomes of energy savings without any prejudice as regards the residents' thermal comfort.

One of the shortcomings of the FC-based approaches is a problematic definition of correct values of the membership functions. However, several studies support that the FC can optimize the ventilation process, increasing its energy efficiency with no significant effect on the thermal comfort of the user even based on improper primary data and an intuitive approach. Thus, at present, the FC for the ventilation control system is the optimal solution due to its flexibility, relative simplicity of configuration and economical operation. Also, an additional advantage of fuzzy controllers is the relative simplicity of their hardware implementation [13–15].

3 Controller Model Specification Based on Fuzzy Logic for the Automatic Ventilation Control System

The proposed controller model based on fuzzy logic for the automatic ventilation control system consists of two input linguistic variables. The overall scheme is shown in Fig. 1.

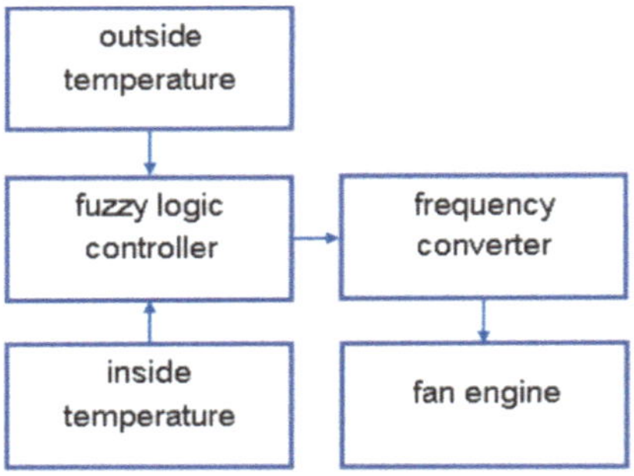

Fig. 1. Flow diagram of ventilation control system

The input parameters involve the values of the temperature inside and outside the room. The values of universal (ground) sets of these variables depend on the probable temperature values. The corresponding terms are generated based on general sanitary requirements. According to them, the most favorable inside temperature should be 20–25 °C, and the allowable temperature variations in warm months are from 20 °C to 28 °C, in cooler and the off-seasons are from 18 °C to 22 °C. For the variables 'Inside Temperature' and 'Outside Temperature', the terms are denoted as follows: NB (Negative Big), NM (Negative Medium), NS (Negative Small), ZR (Zero), PS (Positive Small), PM (Positive Medium), PB (Positive Big).

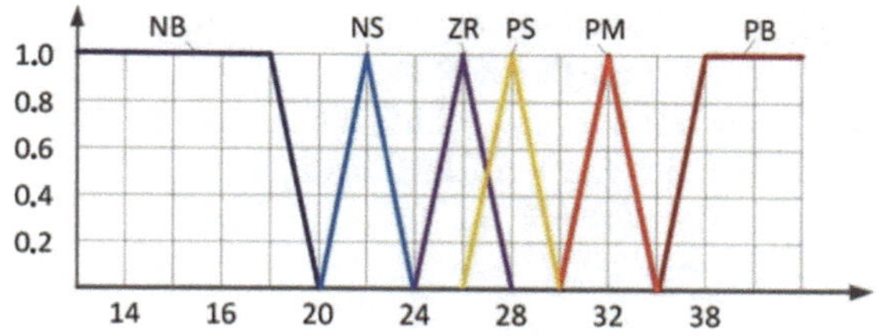

Fig. 2. Membership functions of 'inside temperature' variable

The linguistic variable 'Inside Temperature' (Fig. 2) is characterized by a basic term set consisting of six terms: NB, NS, ZR, PS, PM, PB, set on a universal set from 13 to 42. The term NB of the variable 'inside temperature' has a linear S-shaped function with parameters a = 18, b = 20; the term NS has a triangular (hat) function with parameters a = 20, b = 22, c = 24; the term ZR has a triangular function with parameters a = 24, b = 26, c = 28; the term PS has a triangular function with parameters a = 26, b = 28, c = 30; the term PM has a triangular function with parameters a = 30, b = 32, c = 34; the term PB has a linear S-shaped function with parameters a = 34, b = 38.

The linguistic variable 'Outside Temperature' (Fig. 3) is characterized by a ground term set consisting of seven terms NB, NM, NS, Z, PS, PM, PB, set on a universal set from 0 to 32, while the term NB of the variable 'Outside Temperature' has a linear S-shaped function with parameters a = 2, b = 4; the term NM has a triangular function with parameters a = 4, b = 6, c = 10; the term NS has a triangular function with parameters a = 10, b = 12, c = 14; the term ZR has a triangular function with parameters a = 14, b = 16, c = 18; the term PS has a triangular function with parameters a = 18, b = 20, c = 22; the term PM has a triangular function with parameters a = 22, b = 24, c = 26; the term PB has a linear S-shaped function with parameters a = 26, b = 28.

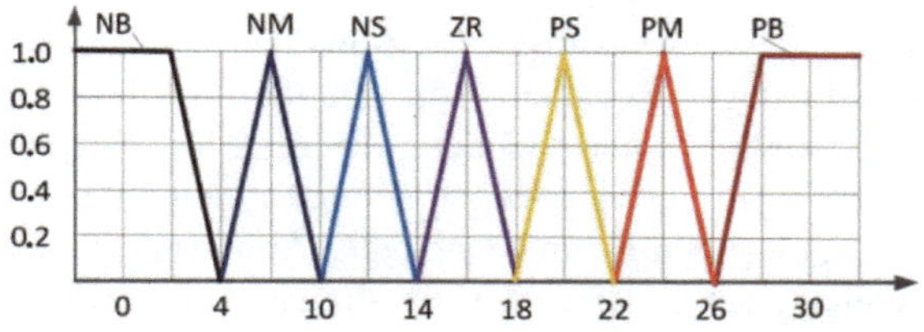

Fig. 3. Membership functions of 'outside temperature' variable

As the assumed output value of the controller, the linguistic variable 'Fan engine frequency' (Fig. 4) is characterized by a ground term set consisting of seven terms NB, NM, NS, Z, PS, PM, PB, set on a universal set from 0 to 75. In this case, the term NB of the output variable 'Fan engine frequency' has a linear S-shaped function with parameters a = 5, b = 10; the term NM has a trapezoidal function with parameters

$a = 10, b = 15, c = 20, d = 25$; the term NS has a trapezoidal function with parameters $a = 20, b = 25, c = 30, d = 35$; the term ZR has a trapezoidal function with parameters $a = 30, b = 35, c = 40, d = 45$; the term PS has a trapezoidal function with parameters $a = 40, b = 45, c = 50, d = 55$; the term PM has a trapezoidal function with parameters $a = 50, b = 55, c = 60, d = 65$; the term PB has a trapezoidal function with parameters $a = 60, b = 65, c = 70, d = 75$.

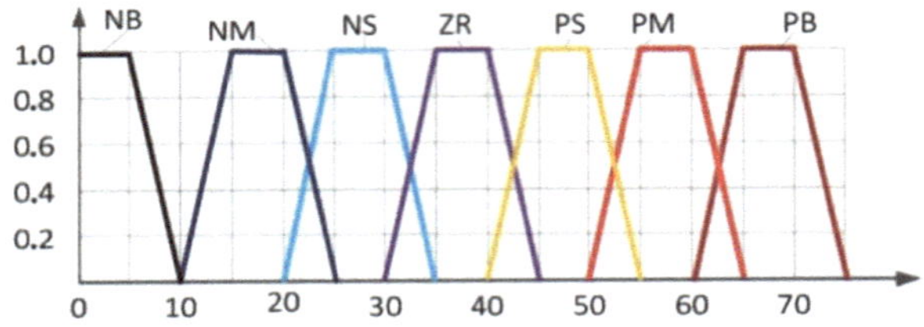

Fig. 4. Membership functions of 'rotation frequency' variable

The fuzzy control rule database includes 42 rules. There are some rules from the base of 'the Rotation Frequency' variable:

Rule #1: If the value of 'inside temperature = Negative Big' and 'outside temperature = Negative Big', then the value of 'frequency of the supply voltage of fan engine = Negative Big'.

…

Rule #21: If the value of 'inside temperature = Zero' and the value of 'outside temperature = Positive Big', then the value of 'frequency of the supply voltage of fan engine = Positive Small'.

…

Rule #42: If the value of 'inside temperature = Positive Big' and the value of 'outside temperature = Positive Big', then the value of 'frequency of the supply voltage of fan engine = Positive Big'.

The controller model based on fuzzy logic for the automatic ventilation control system was performed using a specialized software tool fuzzyTECH, which was used to test the model operability. After conducting a series of computer experiments, a controller model based on fuzzy logic for an automatic ventilation control system was implemented in the C++ programming language. Applications designed in the C++ programming language are faster than most similar applications designed in other programming languages in terms of execution (operating) speed. Qt Creator was used as the programming environment. Since it is one of the best programming environments in C and C++ languages at the moment, it provides better cross-platform compatibility than other programming environments, static code analysis, and deals with various compilers as well as has a fast user interface designer based on the Qt Framework.

Testing was done with the following quality ratings obtained to assess the quality of the proposed controller model based on fuzzy logic for the automatic ventilation control system: Mean Absolute Error (MAE), Root Mean Square Error (RMSE), Symmetric Mean Absolute Percentage Error (SMAPE).

So, the resulting value of MAE is 1.1905, RMSE = 2.5588, and the value of SMAPE = 0.0229. The obtained results prove that the controller based on fuzzy logic controls properly the frequency converter (changer) that sets the rotation frequency of the asynchronous fan engine.

4 Conclusion

This paper presents a model based on fuzzy logic for controlling the rotation speed of an asynchronous fan engine according to inside and outside temperature values and takes into account the current sanitary and hygienic standards for indoor temperature. The model has two input linguistic variables ('Inside Temperature' and 'Outside Temperature') and one output ('Fan rotation speed'). The fuzzy product rule base contains 42 elements. 20 fuzzy sets are introduced to describe the model. A graphical and analytical representation is given for each of them. The conducted testing showed its potential operability in the supply and exhaust ventilation devices. The model is implemented in fuzzy TECH for testing. And it is programmed in C++ language to use it in ventilation systems. The authors' further research will be devoted to increasing the functionality of the model in terms of correspondence to sanitary and hygienic standards depending on the season, fitting air humidity, carbon dioxide concentration, as well as the integration into 'the smart house' system.

References

1. Dionova, B.W., Mohammed, M.N., Al-Zubaidi, S., Yusuf, E.: Environment indoor air quality assessment using fuzzy inference system. ICT Express **6**(3), 185–194 (2020)
2. Grygierek, K., Ferdyn-Grygierek, J.: Multi-objectives optimization of ventilation controllers for passive cooling in residential buildings. Sensors **18**(4), 1144 (2018)
3. Caglayan, N., Celik, H.K., Rennie, A.: Fuzzy logic based ventilation for controlling harmful gases in livestock houses. J. Agric. Mach. Sci. **13**(2), 107–112 (2017)
4. Grygierek, K., Sarna, I.: Impact of passive cooling on thermal comfort in a single-family building for current and future climate conditions. Energies **13**(20), 5332 (2020)
5. Attia, A.H., Rezeka, S.F., Saleh, A.M.: Fuzzy logic control of air-conditioning system in residential buildings. Alexandria Eng. J. **54**(3), 395–403 (2015)
6. Jaradat, M.A.K., Al-Nimr, M.A.: Fuzzy logic controller deployed for indoor air quality control in naturally ventilated environments. J. Electr. Eng. **60**(1), 12–17 (2009)
7. Soleimanzadeh, A.: Designing fuzzy controller for air conditioning systems in order to save energy consumption and provide optimal conditions in closed environments (indoors). J. Artif. Intell. Electr. Eng. **3**(11), 11–18 (2014)
8. Chang, B., Zhang, X.: Design of indoor temperature and humidity monitoring system based on ZigBee and fuzzy PID technology. In: Proceedings of 2018 7th International Conference on Advanced Materials and Computer Science (ICAMCS 2018), pp. 23–30 (2018)

9. Abdo-Allah, A., Iqbal, T., Pope, K.: Modeling, analysis, and design of a fuzzy logic controller for an AHU in the S.J. Carew Building at Memorial University. J. Energy **2018**, 4540387 (2018)
10. Ahilan, C., Kumanan, S., Sivakumaran, N.: Design and Implementation of an intelligent controller for a spilit air conditioner with energy saving. IAENG Int. J. Comput. Sci. **43**(4), 44–65 (2010)
11. Bogdan, S., Birgmajer, B., Kovačić, Z.: Model predictive and fuzzy control of a road tunnel ventilation system. Transp. Res. Part C Emerg. Technol. **16**(5), 574–592 (2008)
12. Li, W., Zhang, J., Zhao, T., Ren, J.: Experimental study of an indoor temperature fuzzy control method for thermal comfort and energy saving using wristband device. Build. Environ. **187**, 107432 (2021)
13. Anand, M.S., Tyagi, B.: Design and implementation of fuzzy controller on FPGA. Int. J. Intell. Syst. Appl. (IJISA) **4**(10), 35–42 (2012)
14. Chen, W., Yuan, H.M., Wang, Y.: Design and implementation of digital fuzzy PID controller based on FPGA. In: IEEE Conference on Industrial Electronics and Application, pp. 393–397 (2009)
15. Erfanian, H.R., Abdi, M.J., Kahrizi, S.: Solving a linear programming with fuzzy constraint and objective coefficients. Int. J. Intell. Syst. Appl. (IJISA) **8**(7), 65–72 (2016)

Development of an Algorithm for Energy Efficient Resource Scheduling of a Multi-cloud Platform for Big Data Processing

Leonid V. Legashev(✉), Lyubov S. Zabrodina, Denis I. Parfenov, and Irina P. Bolodurina

Orenburg State University, Orenburg 460018, Russia

Abstract. The problem of optimal distribution of virtual resources in multi-cloud systems for big data processing is relevant to this day. End users want to access virtual resources as soon as possible using one of the existing content distribution models: virtual desktop as a service, specific software as a service or even infrastructure as a service. The cloud system scheduler can handle incoming requests from users in physical constraints, taking into account the reduction of energy consumption and limited resources of cloud system. Violation of the Service Level Agreement in terms of exceeding the response time can lead to penalty for the service owner. This paper formalized the optimization problem of close to optimal energy efficient scheduling and provided general architecture of cloud platform under consideration. On the basis of the evolutionary annealing simulation algorithm, a cloud system simulator is implemented. Conducted experimental studies have shown the effectiveness of the proposed method in comparison with the greedy algorithm.

Keywords: Big data processing · Energy saving · Evolutionary algorithms · Resource scheduling

1 Introduction

Cloud platforms have been the primary way of delivering computing resources to end users over the past few years across a variety of content delivery models: desktop as a service (DaaS), software as a service (SaaS), infrastructure as a service (IaaS), and more. Cloud service providers need to store virtual machine images all the time, while some computing resources are in idle state, which leads to economic costs. A review of the current state of research shows the relevance of the problem of energy saving and compliance with the quality of service (QoS) and Service Level Agreement (SLA) when placing NFV services (Network Function Virtualization) and virtual computing nodes on the physical infrastructure of a multi-cloud platform. The specified problem is NP-complete; to solve it, one can use optimization algorithms based on the simulated annealing method and a greedy strategy. The general architecture of the cloud platform in terms of the virtual resources allocation is shown in Fig. 1.

Z. Hu et al. (Eds.): AIMEE 2021, LNDECT 107, pp. 105–113, 2022.
https://doi.org/10.1007/978-3-030-92537-6_10

As part of the cloud platform, several servers are presented on which a number of virtual machines can be launched. The Global Monitor module at an arbitrary time moment t send to the Scheduler module information about the number of running virtual machines on each of the available servers, as well as information about the current usage of computing resources of each virtual machine. The scheduler processes incoming requests from end users for renting computing resources; prepared virtual machines images are stored in the VMI storage. In the event of virtual machine downtime, the scheduler can decide to suspend and delete virtual machine images from the cloud storage. At the same time, if a new user request appears, it is necessary to deploy the virtual machine image from scratch, which in turn may lead to a violation of the SLA restrictions on the response time when providing the service.

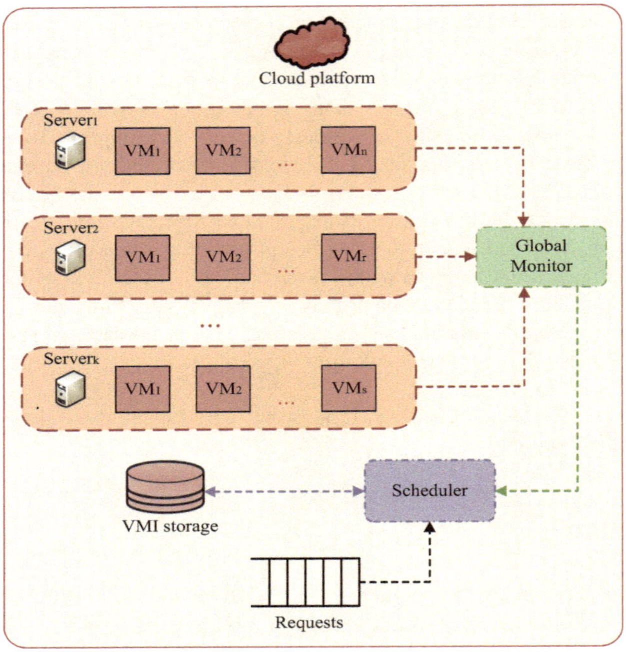

Fig. 1. General architecture of cloud platform

Based on the information received from the Global Monitor module, the Scheduler can decide to permanently store the most frequently used virtual machine images in VMI storage, which will increase the overall energy consumption of the cloud platform. The algorithmic solution developed within the framework of the current paper is in finding a balance between the two presented problems in terms of financial efficiency.

2 Literature Overview

The authors of the paper note that to study the energy saving potential in production management, intelligent algorithms can be used to optimize the total energy consumption on a big data processing platform [1]. The authors of paper use evolutionary algorithms for energy consumption when performing the operation of migrating virtual machines in mobile cloud computing [2]. The book describes the work of a scheduler that optimizes energy consumption while maintaining quality of service [3]. The authors of paper, within the framework of the CloudSim platform, tested the operation of the energy saving algorithm by using the method of dynamic scaling of voltage and frequency for data centers [4]. The authors of the paper reviewed energy saving methods in big data centers, hybrid energy efficiency methods turned out to be the most effective, while it is recommended to use modern machine learning methods to predict computational load [5]. The authors of paper solve the problem of efficient energy consumption in a cloud computing system through live migration of virtual machines based on monitoring CPU utilization, memory load and power consumption [6]. The paper provides an overview of studies of existing energy saving schemes, including energy-efficient management of cluster nodes, energy-efficient data management, energy-efficient resource allocation, and energy-efficient task scheduling [7]. The paper provides a comprehensive analysis of advanced energy saving technologies at the edge devices of the Internet of Things paradigm [8]. The publication describes methods of energy saving and dynamic resource allocation through live migration of virtual machines in a cloud computing environment [9]. The review from paper focuses on the goal of efficient energy consumption while maintaining QoS and reducing service level agreement violations [10]. An approach based on particle swarm optimization was proposed for solving the problem of maintaining a balance between the SLA and the energy consumption [11]. The authors of the publication describe the development and implementation of a strategy for the rapid deployment of NFV services in software-defined networks taking into account energy saving problem [12]. The paper presents the developed architecture focused on the economical use of computing resources and energy for 5G networks [13]. In, an economical scheme for deploying NFV services with guaranteed availability in cloud networks is described in terms of energy consumption and bandwidth [14]. The paper solves the problem of linear programming of load balancing and energy saving in the form of an NFV orchestrator for wireless network environments [15]. NP-hard cloud task scheduling problem in heterogeneous distributed systems is solved using genetic algorithm [16]. Evolutionary algorithm based on ant colony optimization was used to solve scheduling jobs NP-hard problem in grid environment [17].

3 Formalization of Optimization Problem of Close to Optimal Energy Efficient Scheduling

End users of the multi-cloud platform register requests for renting the computing resources for a fixed time. The cloud system scheduler module orchestrates virtual machines launch and VMI storage for the next 24 h in accordance with the current set of requests for processing. Each end-user request can be represented as a tuple of values:

$$R_k = \{VM_k, type, workload, ts_k, d_k, mode\}, k = \overline{1, |R|} \quad (1)$$

where: R – set of incoming requests, VM_k – requested number of virtual machines, *type* = {*small*, *medium*, *large*} – one of the three possible virtual machine configurations. Each configuration is defined by a default three parameters (C_k, M_k, D_k) – requirements for the number of CPU cores, the amount of RAM and disk memory respectively, *workload* = {*low*, *average*, *high*} – virtual machine workload depending on the type of tasks, ts_k – start time in the interval [0, 23], d_k – virtual machine rent duration in hours, *mode* = {*offline*, *online*} – the request mode, depending on whether the virtual machine disk image is in the cloud storage, or is it deployed in real-time at the moment of time ts_k.

For each user k, we introduce the term SLA_k, which is a service level agreement concluded with the k-th user regarding not exceeding a fixed VM response time. Inside each SLA, we introduce the term $vp(SLA_k)$, which is the penalty function for violation of the service level agreement, and the term rt, which is the required response time. The response time of a virtual machine depends on its configuration, workload and storage status: either the disk image of the virtual machine is in the storage, or it needs to be created.

Let's define a set of servers S, each of which is determined by three parameters $(C_i, M_i, D_i), i = \overline{1, |S|}$. The energy used by i-th server to maintain j-th virtual computing node of the physical infrastructure is expressed as a function of the energy consumed by the CPU, the amount of memory consumed, and the amount of disk space, including the amount of data occupied by users working with virtual computing nodes, I/O devices and network card. The total power consumption of virtual machines pool with length n of a physical server S_i is expressed as an additive function of the energy consumed by the server $Serv_j^i$ and the idle power $Idle_j^i$. The total power consumption of the data center of a multi-cloud platform can be represented as a function of the power consumption of the servers. The total energy consumption of the VMI storage is determined as the total power consumption used to store all possible VM disk images. If the end user of the system demands several virtual machines of the same specification, it is necessary to store exactly one disk image of a virtual machine with declared configuration and workload.

Let's define a virtual machine as a tuple of values:

$$VM = \{type, workload, st(type, workload), op(type, workload), dt_s, dt_n\} \quad (2)$$

where: $st(type, workload)$ – function of the storing cost a virtual machine image per hour, depending on its configuration and workload, $op(type, workload)$ is a function of the cost of a virtual machine operation per hour depending on its configuration and workload, dt_s is the time interval for starting a virtual machine if its image is in the VMI storage, i.e. *mode* = *offline*, dt_n is the time interval for starting the virtual machine if its image needs to be created, i.e. *mode* = *online*. Both startup time intervals directly depend on the configuration and workload of the virtual machine, exceeding the threshold value ($dt_n > rt \parallel dt_s > rt$) leads to violation of the SLA conditions and imposition of a penalty $vp(SLA_k)$.

It should be noted that the amount of energy consumed by the server determines the cost of storing a virtual machine disk image, as well as the cost of its operation. Thus, based on the current maintaining cost of a virtual machine, you can estimate the amount of energy consumption. Let’s consider this approach when solving the problem of energy efficient resource scheduling for a multi-cloud platform. Let's introduce the definition of matrix C^{st} of energy consumption, spent on storing a disk image of a virtual machine in storage:

$$C^{st} = \begin{array}{c} \begin{array}{ccc} \text{low} & \text{average} & \text{high} \end{array} \\ \begin{pmatrix} st(1,1) & st(1,2) & st(1,3) \\ st(2,1) & st(2,2) & st(2,3) \\ st(3,1) & st(3,2) & st(3,3) \end{pmatrix} \end{array} \begin{array}{c} \\ \text{small} \\ \text{medium} \\ \text{large} \end{array} \quad (3)$$

The elements of the matrix C^{st} are set manually by the cloud provider based on calculating the function of the cost storing a disk image of a virtual machine per hour.

Let's introduce the definition of matrix C^{op} of energy consumption, spent on virtual machine operation in a cloud system:

$$C^{op} = \begin{array}{c} \begin{array}{ccc} \text{low} & \text{average} & \text{high} \end{array} \\ \begin{pmatrix} op(1,1) & op(1,2) & op(1,3) \\ op(2,1) & op(2,2) & op(2,3) \\ op(3,1) & op(3,2) & op(3,3) \end{pmatrix} \end{array} \begin{array}{c} \\ \text{small} \\ \text{medium} \\ \text{large} \end{array} \quad (4)$$

The elements of the matrix C^{op} are set manually by the cloud provider based on the calculating the function of the cost of running a virtual machine per hour.

In this case, the number of energy consumption modes $st(i, j)$ and $op(i, j)$ $(i = \overline{1,3}, j = \overline{1,3})$ depends on the possible virtual machines configurations and workloads and determines the dimension of the matrices C^{st} and C^{op}.

The multi-cloud platform Scheduler generates an overall virtual machines startup schedule, according to existing servers’ physical constraints and end-user SLAs. The problem of creation an optimal schedule can be represented as a multi-criteria problem of minimizing energy consumption and SLA penalty functions:

$$f(R) = \sum_{k=1}^{|R|} P_{dc} + \sum_{k=1}^{|R|} vp(SLA_k) \rightarrow \min, \tag{5}$$

where:

$$P_{dc} = \begin{cases} \sum_{k=1}^{|R|} (st(type_k, workload_k) \cdot ts_k + op(type_k, workload_k) \cdot d_k \cdot VM_k), \text{if } mode = offline \\ \sum_{k=1}^{|R|} op(type_k, workload_k) \cdot d_k \cdot VM_k, \text{if } mode = online \end{cases}$$

It should also be noted that to appoint virtual machines to the servers of a cloud provider, it is necessary to take into account the resource constraints of each of the servers:

$$\sum_{k=1}^{|R|} C_k \leq \sum_{s=1}^{|S|} C_s, \tag{6}$$

$$\sum_{k=1}^{|R|} M_k \leq \sum_{s=1}^{|S|} M_s, \tag{7}$$

$$\sum_{k=1}^{|R|} D_k \leq \sum_{s=1}^{|S|} D_s, \tag{8}$$

In the case when it is impossible to satisfy the user's request due to resource constraints of cloud system, it is possible to connect the services of an external provider, which, in turn, will affect the renting cost of the services provided.

4 Algorithm of Close to Optimal Energy Efficient Scheduling

Various evolutionary optimization techniques have proven their effectiveness in solving a wide range of NP-complete problems. The stochastic simulated annealing method can be used to solve problem (5). The main idea of the algorithm is to randomly change the current state of the system and a probabilistic estimate of the value of the objective function at each iteration in order to avoid getting into local minima.

Main steps of algorithm for energy efficient resource scheduling of a multi-cloud system can be described as follows:

1) Initialization of the basic parameters of the simulated annealing method: maximum temperature t_{max}, step of decreasing the temperature t_{step}, breakpoint P_{stop}, the probability estimation P_v, and optimal value of the objective function f_{best}.
2) Creation of the initial version of the schedule based on the greedy algorithm.
3) Calculation of the objective function value (5).

4) If the calculated value is less than the specified optimal f_{best}, then take the current value as the optimal one.
5) Otherwise, accept the current value as optimal one based on the probability estimation PV.
6) Stochastic change of the current schedule.
7) Checking the conditions for reaching the breakpoint, returning to step 3 if the temperature has not reached the threshold value.

To create the initial version of the schedule at step 2 of the proposed algorithm, the following greedy technique can be used: the scheduler first assigns virtual machines to operate based on requests from end users, for which the virtual machine configuration and workload will be maximum, i.e. *type* = *large*, *workload* = *high*. By default, all images of virtual machines of hosted applications are created and sent to VMI storage until the ts_k time.

To randomly change the state of the system at step 4 of the proposed algorithm, the following operations can be performed: randomly select *d* number of requests from the set with a given probability. If images of virtual machines are stored in the VMI storage, delete them from there and create an image just before the moment of starting time ts_k. If there are no images, create them in advance and store in VMI storage, making the appropriate energy consumption calculations using Eq. (5).

The simulating annealing algorithm was implemented in the Python programming language to find a close to optimal solution for the problem (5). The main parameters of the simulation are presented in Table 1, the simulation results are presented in Table 2. For each given number of user applications, 10 experiments were carried out and the average percentage of improvement in comparison with the original greedy solution was obtained. Last few iterations results for the case of 50 users' requests are shown in Fig. 2.

Table 1. Parameters of requests generation and simulated annealing algorithm

Parameter	Value
rt, sec	45
vp(*SLAk*), cu	5
dt_s, sec	Uniformly in [5, 55]
dt_n, sec	Uniformly in [30, 60]
VM_k	Uniformly in [1, 5]
ts_k, hours	Uniformly in [1, 18]
d_k, hours	Uniformly in [1, 6]
C^{st}, cu	[0.1, 0.2, 0.3] [0.2, 0.3, 0.4] [0.3, 0.4, 0.5]
C^{op}, cu	[0.2, 0.4, 0.6] [0.3, 0.6, 1.1] [0.7, 1.2, 1.4]
t_{max}	100
t_{step}	0.1

Table 2. Simulations results for different number of requests

Number of requests	Average improvement of function (5)
25	10,89%
50	11,87%
75	09,00%
100	10,63%
125	09,73%
150	08,48%

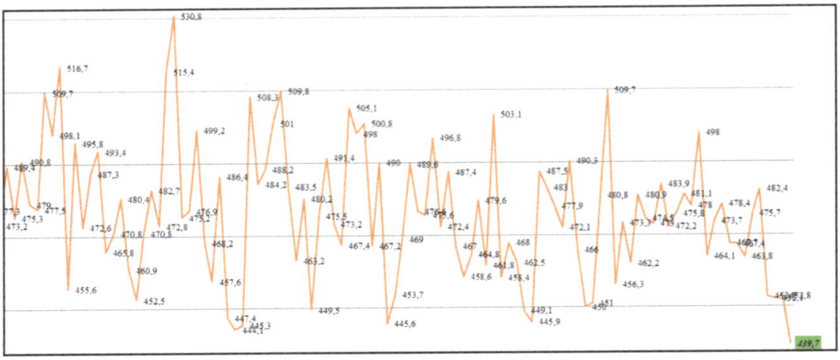

Fig. 2. Demonstration of the f_{best} values distribution

Due to the stochastic nature of the simulated annealing method, in rare cases, the solution found was larger than the original one. An increase in the number of processed requests as a whole does not affect the efficiency of the algorithm. It can be concluded that the usage of the simulated annealing method for solving the problem of energy-efficient scheduling of computing resources in a multi-cloud platform, on average, leads to cost reducing by 10.10%.

5 Conclusion

As a result of the study, the problem of energy-efficient scheduling of computing resources in a multi-cloud system for big data processing was formalized, a method for finding a close to optimal solution based on the evolutionary method of simulating annealing was proposed, and experimental studies of the effectiveness of the presented method were carried out. The simulated annealing method showed good results in reducing computing resources. As part of further research, an in-depth formalization of the problem under consideration is planned, considering the key features of scheduling, use of others evolutionary algorithms such as genetic algorithm and particle swarm optimization, and placing Network Function Virtualization services and virtual computing nodes on the physical infrastructure of a multi-cloud platform for big data processing.

Acknowledgment. The reported study was funded by RFBR (№ 20–07-01065), and by a grant of the President of the Russian Federation (NSh-2502.2020.9).

References

1. Yan, J., et al.: Industrial big data in an industry 4.0 environment: challenges, schemes, and applications for predictive maintenance. IEEE Access **05**, 23484–23491 (2017)
2. Liu, Y., et al.: High-efficient energy saving processing of big data of communication under mobile cloud computing. Int. J. Model. Simul. Sci. Comput. **10**(04), 1–11 (2019)
3. Cordeschi, N., et al.: Energy-saving QoS resource management of virtualized networked data centers for big data stream computing. In: Emerging Research in Cloud Distributed Computing Systems, pp. 122–155. IGI Global (2015)
4. Ahmad, B., et al.: Economic impact of energy saving techniques in cloud server. Cluster Comput. **23**, 611–621 (2020)
5. Arora, S., Bala, A.: A survey: ICT enabled energy efficiency techniques for big data applications. Cluster Comput. **23**, 775–796 (2020)
6. Yang, C.T., Wan, T.Y.: Implementation of an energy saving cloud infrastructure with virtual machine power usage monitoring and live migration on OpenStack. Computing **102**(6), 1547–1566 (2020)
7. Wu, W.T., et al.: Energy-efficient Hadoop for big data analytics and computing: a systematic review and research insights. Future Gener. Comput. Syst. **86**, 1351–1367 (2018)
8. Banka, K., et al.: A study of state-of-the-art energy saving on edges. In: Proceedings of the 2021 ACM Southeast Conference, pp. 224–228 (2021)
9. Yang, C.T., et al.: An energy-efficient cloud system with novel dynamic resource allocation methods. J. Supercomput. **75**(8), 4408–4429 (2019)
10. Qu, Z., et al.: Study QoS optimization and energy saving techniques in cloud, fog, edge, and IoT. Complexity **2020**, 1–16 (2020)
11. Copil, G., et al.: Cloud SLA negotiation for energy saving—a particle swarm optimization approach. In: 2012 IEEE 8th International Conference on Intelligent Computer Communication and Processing, pp. 289–296. IEEE (2012)
12. Tseng, C.W., et al.: NFV deployment strategies in SDN network. Int. J. High Perform. Comput. Netw. **14**(2), 237–248 (2019)
13. Li, X., et al.: Novel Resource and Energy Management for 5G integrated backhaul/fronthaul (5G-Crosshaul). In: 2017 IEEE International Conference on Communications Workshops (ICC Workshops), pp. 778–784. IEEE (2017)
14. Dinh, N.T., Kim, Y.: An efficient availability guaranteed deployment scheme for IoT service chains over fog-core cloud networks. Sensors **18**(11), 3970 (2018)
15. Dalla-Costa, A.G., et al.: Maestro: an NFV orchestrator for wireless environments aware of VNF internal compositions. In: 2017 IEEE 31st International Conference on Advanced Information Networking and Applications (AINA), pp. 484–491. IEEE (2017)
16. Soulegan, N.S., Barekatain, B., Neysiani, B.S.: MTC: minimizing time and cost of cloud task scheduling based on customers and providers needs using genetic algorithm. Int. J. Intell. Syst. Appl. (IJISA) **02**, 38–51 (2021)
17. Sahana, S.K.: An automated parameter tuning method for ant colony optimization for scheduling jobs in grid environment. Int. J. Intell. Syst. Appl. (IJISA) **03**, 11–21 (2019)

Ensemble of Gray Fuzzy Cognitive Maps Application to Solve the Problem of Predicting University Activity

Andrey A. Mikryukov(✉) and Mikhail E. Mazurov

Plekhanov Russian University of Economics, Moscow, Russia
Mikrukov.aa@rea.ru

Abstract. An approach to solving the problem of predicting the performance indicators of a university as a semi-structured system based on cognitive modeling methods is proposed. The university activity model is presented in the form of an ensemble of gray fuzzy cognitive maps, which allows considering the uncertainty of experts' opinions about the state of cognitive map concepts and gives more reliable modeling results. The ensemble of cognitive maps simultaneously uses a finite set of gray cognitive maps built based on expert assessments, the output signals of which are combined into a combined assessment, which is superior in quality to the results obtained using local assessments of the cognitive maps included in the ensembles. Special constructions in the form of interval estimates for assessing the connections between concepts, used in a fuzzy ensemble of cognitive maps, ensured the construction of an adequate cognitive model. The proposed approach allows us to find the most acceptable scenario for planning the increment of baseline indicators with given constraints to target values by identifying hidden factors affecting them and impulse influences on them, ensuring guaranteed achievement of the set goal and obtaining more accurate results.

Keywords: Cognitive modeling · Gray fuzzy cognitive maps · Interval estimates

1 Introduction

The purpose of developing a model of university activities based on a gray fuzzy cognitive map is to build the most preferable scenario for the development of the university. The relevance of the problem being solved is due to the need to develop scientifically grounded proposals to achieve the required values of the target indicators of the university by 2025. In the international QS ranking by calculating the necessary increments of latent factors (indicators), considering their correlations with target indicators. In turn, the main indicator, called the functional F (aka the university rating R), is the sum of the products of the values of the target indicators by their weight coefficients.

To solve this problem, the methodology of fuzzy cognitive modeling was used, designed for analysis and decision-making in poorly defined situations, and a model based on a gray fuzzy cognitive map (FCM), which made it possible to choose the most

Z. Hu et al. (Eds.): AIMEE 2021, LNDECT 107, pp. 114–123, 2022.
https://doi.org/10.1007/978-3-030-92537-6_11

preferable alternative [1]. To obtain a more substantiated solution, the developed model includes an ensemble of gray FCMs, which provides a more accurate and reliable weighted average estimate compared to a separate FCM.

2 Application of Gray Fuzzy Cognitive Maps for Solving the Scenario Forecasting Problem

In the context of the problem under consideration, a scenario is understood as a dynamic sequence of possible events of change in the values of factors - causes and factors - consequences that affect the target indicators [2]. To solve this problem, an approach to scenario forecasting based on a gray fuzzy cognitive map is proposed, a feature of which is the possibility of using a special design that allows one to reduce the uncertainty (spread) of expert estimates of the states of cognitive map concepts. As you know, the FCM is specified using a tuple of sets [3]:

$$\text{FCM} = \langle C, F, W \rangle ,$$

where $C = \{C_j\}$ is a set of concepts - the vertices of the graph, which are the factors that are the most significant in the problem under consideration;

$F = \{F_k\}$ - set of directed arcs of the graph of connections between concepts;

$W = \{W_{ij}\}$ - the set of weights of the FCM links (links can be positive (strengthening, $W_{ij} > 0$) and negative (weakening, $W_{ij} < 0$) the influence of the C_i concept on the C_j concept.

The values of the weights W_{ij} are set using a fuzzy linguistic scale. Each value of the linguistic variable is assigned a certain interval [0,1] for positive relationships or an interval [−1,0] for negative relationships. At an arbitrary discrete time t = 0, 1, 2…, n, the state of the FCM is described by the equation

$$X_i(t+1) = f\left[X_i(t)\sum\nolimits_{j=1}^{n} W_{ij}X_j(t)\right] \quad (i = 1, 2, \ldots, n), \tag{1}$$

where $X_i(t)$ is the value of the state variable of the i-th concept C_j at time $(t + 1)$; n is the number of FCM concepts, f is the concept's nonlinear function. The initial conditions for the calculation are determined by the vector $X(0) = X_1(0), X_2(0), \ldots, X_n(0)^T$.

Each concept is characterized by a term-set of a linguistic variable

$$T_i = \left\{T_1^i, T_2^i, \ldots, T_{mj}^i\right\} \tag{2}$$

where m_j is the number of typical states of the i-th concept. To describe each term T_{ki}, a term is constructed - a set with a membership function $\mu_{Ti}(x)$. The connections between the typical states of each pair of concepts are specified by fuzzy variables described by the corresponding fuzzy sets.

The values of the weights (bond strengths) W_{ij} are set using a fuzzy linguistic scale, which is an ordered set of linguistic values (terms) of bond strength estimates, for example, of the form: LINK_POWER = {Does not affect; Weak; Average; Strong; Very strong}. Each of these values is associated with a certain numerical range belonging to the segment [0, 1] for positive links (Table 1), or the segment [–1, 0] for negative links. In the general case, a weighted digraph with arbitrary values of the weights $W_{ij} \in [-1, 1]$ is described by the dynamics of its state change in time. The state of the digraph (FCM) is determined by the set of states of its concepts C_i, (i = 1, 2,..., n), each of which is described by the state variable X_i (t), which takes values in the interval [0, 1].

Table 1. Assessment of the strength of connection between concepts

Linguistic value	Numerical range	Term designation	Point estimate of bond strength +
No effect	0	Z	0
Very weak	(0; 0,15)	VL	0,12
Weak	(0,15; 0,35)	L	0,23
Medium	(0,35; 0.60)	M	0,47
Strong	(0,6; 0,85)	H	0,72
Very strong	VH	VH	0,93

The strength of the connection between the concept C_i and the concept C_j is chosen by the expert by one of the linguistic values presented in the table, as well as by some "point" estimate of the strength of the connection - a number within this range (if there are several experts, then the weight of W_{ij} is averaged). The disadvantage of this approach is the presence of subjective opinions of experts, which can lead to a significant scatter of estimates of the states of concepts, which can cause a significant decrease in the reliability of the results of the constructed cognitive model due to the incorrect assessment by experts of the weights of connections between concepts.

To eliminate this drawback, it is proposed to use an approach based on special constructions, implemented in fuzzy cognitive models based on gray FCMs, which allow describing the weights of connections between concepts of a cognitive map not by point estimates, but by interval numbers given on fuzzy interval sets [4].

In this case, the equation of state of the FCM (1) can be represented by the expression (3)

$$X_i(t+1) = \left(f \left\{ Xi(t) \bigoplus_{\substack{j=1(j \neq i) \\ i=1,2,\ldots,n),}}^{n} \left[W_{ij} \otimes Xj(t) \right] \right\} \right) \tag{3}$$

where: f is the activation function, and the weights of the connections W_{ji}, as well as the state variables X_i (t + 1), X_i (t) are interval numbers, which are elements of fuzzy interval sets. The operations of addition $\oplus$ and multiplication $\otimes$ of interval numbers are specified on fuzzy interval sets.

The gray set $A \subseteq X$ can be represented as

$$A = \{<x, [\underline{x}, \overline{x}] > |x \in X\} \tag{4}$$

Elements $x \in [\underline{x}, \overline{x}] \leq A$ of the gray set can take values in the range [0,1], where $\underline{x}, \overline{x}$ are the lower and upper bounds of the gray number, X is the universal set. The number $\delta x = (\underline{x}, -\overline{x})$ is called the grayness of the number x, and the number $x^0 = (\underline{x}, -\overline{x})/2$ is the "bleached" (central) value of this number.

The weights of links between gray *FCM* concepts are set as gray numbers $[\underline{W}_{ij}, \overline{W}_{ij}]$. State variables of concepts are also described by gray numbers, the values of which lie in the interval $[\underline{X}_i, \overline{X}_i]$ and are determined by Eq. (3).

In general, the use of the interval approach has several advantages [5]:

- knowledge of the probabilistic characteristics of uncertain factors is not required, which are rarely precisely known in practice;
- with the minimax approach, strict estimates are obtained for the desired quantities themselves, and not for probabilities or mathematical expectations, which is important in the presence of a small number of parameter measurements and one or several realizations
- statistical characteristics cannot guarantee a definite outcome of one particular experience;
- in all cases, guaranteed two-sided approximations of the desired solutions are given.

The accuracy of an interval result is entirely determined by the following factors:

1) Uncertainty in specifying the initial data.
2) The values of rounding off when performing operations that modify or generate interval objects.
3) The approximate nature of the numerical method used.
4) The degree to which the dependencies between the interval objects (variables and constants) participating in the calculation are considered.

3 Scenario Forecasting Cognitive Model of the University Performance Indicators Based on Gray Fuzzy Cognitive Maps Ensemble

At the first stage of the study, the identification and interpretation of the hidden (latent factors) that affect the target indicators and the assessment of their significance using the methods of factor analysis were carried out [6–9].

At the next stage, a cognitive model was developed using the gray FCM, based on which the problem of scenario forecasting of the university performance indicators was solved. The development of a cognitive model is based on the construction of a cognitive map (Fig. 1) in the form of a sign-oriented graph [10–12].

The cognitive map reflects the relationship of latent factors, baseline indicators and functional, considering the values of correlation dependences between the functional and basic indicators obtained based on factor analysis in [9], as well as expert assessments of the mutual influence of latent factors and their impact on target indicators.

The relationship between basic indicators and functionality is determined by formula (5) in accordance with the rules of the international institutional rating QS.

$$R = \sum_{i=1}^{6} w_i x_i, \tag{5}$$

where w_i is the weight of the corresponding indicator; xi- its meaning.

In Fig. 1, the following designations are adopted: F - functional; R - university rating, Basic indicators are set by the rules of the international institutional rating QS [13]: *AR* - academic reputation; *ER* - reputation with the employer; *RS/T* - the ratio of the number of students to the number of teachers; *CT* - indicator of citation of teachers; *IT* - number of international teachers; *IS* is the number of international students.

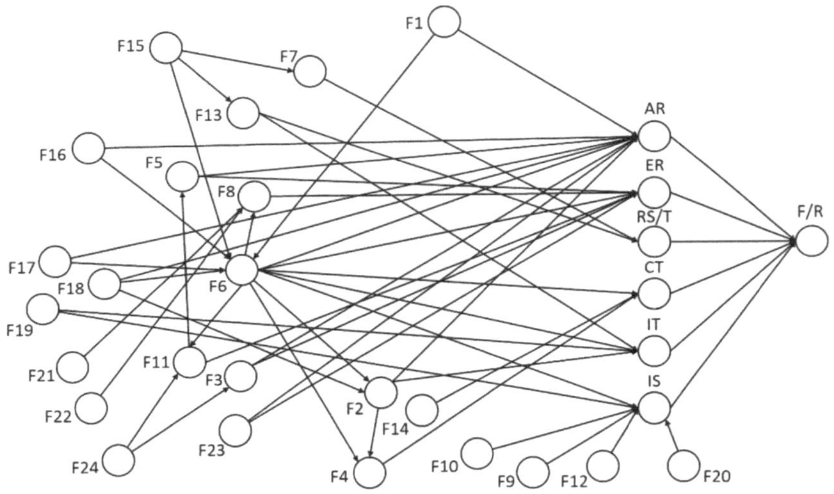

Fig. 1. Cognitive map of the scorecard based on the graph of the relationship of factors, baseline indicators and functionality (Legend: *F* - functional; *R* - rating, (*F1-F24* - latent factors)

Factors are groups of certain variables that correlate with each other more than with the variables included in another factor. Thus, the meaningful meaning of the factors was revealed by examining the correlation matrix of the initial data.

The weighted values of the factors are set on the basis of expert assessments, taking into account the interval scale (lower and upper values, i.e. grayness and "bleached" value of this number).

F1 – factor "Scientific schools and dissertation councils" (0.5–0.7; 0.6); F2 – factor. "Joint research projects" (0.2–0.4; 0.3); F3 – Availability of basic departments (0.1–0.3; 0.2); F4 – Number of publications in the Scopus database (0.5–0.7; 0.6); F5 – Popular areas of training (0.3); *F6* – qualification level of the teaching staff (0.1–0.3; 0.2); *F7* – Number of teaching staff (0.5–0.7; 0.6); *F8* – The level of competence of students (0.4–0.6; 0.5); *F9* – teachers with language training (0.3–0.5; 0.4); *F10* – Places in the hostel (0.1–.3; 0.2); *F11* – Demand for graduates from employers (0.2–0.4; 0.3); *F12* –Areas for educational activities (0.2–0.4; 0.3); *F13* – The level of payment for teaching staff (0.3–0.5; 0.4); *F14* – Stimulating factors (0.1–0.3; 0.2); *F15* – Expansion of the teachers social package (0.2–0.4; 0.3); *F16* – Change in the structure of employment of the teaching staff (0.2–0.4; 0.3); *F17* – Share of teaching staff planning to build an international scientific career (0.1–0.3; 0.2); *F18* – Academic mobility of the teaching staff (0.2–0.4; 0.3); *F19* – Convergence of educational programs with foreign universities (0.3–0.5; 0.4); *F20* – Foreign entrant company (0.2–0.4; 0.3); *F21* – Increase in the number of On-line courses MOOCs (0.2–0.4; 0.3), *F22* – Implementation of individual educational trajectories (0.3–0.5; 0.4); *F23* – Implementation of remote sensing technologies (0.2–0.4; 0.3); *F24* – Close relationship with the employer (0.3–0.5; 0.4).

An approach to scenario forecasting based on a cognitive map includes the following stages [12]: generating scenarios and assessing risks, adjusting scenarios at discrete time intervals (in our case, annually), considering the achieved values of the place in the QS rating of the functional and indicators. To obtain a more substantiated solution of the problem posed, collective methods for solving forecasting problems based on an ensemble of gray FCMs are considered, providing a weighted average estimate of the modeling result, which is superior in accuracy to the result of a single cognitive map [14, 15].

It is known that a combination of several predictions can significantly reduce the overall prediction errors, surpassing individual components. For the formation of the output value of the ensemble at certain states of the outputs of the models, the most common algorithms are based on the following methods: voting; balanced voting; averaging; mixtures of experts; boosting, stacking, bagging, and random subspaces [15].

Considering the specifics of the construction and training of interval forecasting models, the use of an approach based on weighted voting, which is a special case of a mixture of experts, is justified in the work, and allows you to obtain the resulting estimate of modeling based on a predictive ensemble, considering the skill level of experts.

4 Results of University Performance Indicators Scenario Forecasting Based on a Gray Cognitive Maps Ensemble

At the first stage of the study, the identification and interpretation of latent (latent factors) affecting the baseline indicators, and an assessment of their significance using the methods of factor analysis, which is a class of multivariate statistical analysis procedures aimed at identifying latent variables (factors) responsible for the presence of linear statistical relationships (correlations) between the observed variables [16–19]. At the second stage, considering the interval approach, a scenario analysis of the system's behavior on its cognitive model was carried out by means of impulse modeling. A set of realizations of impulse processes is a "development scenario" and indicates possible trends in the development of situations.

The situation in impulse simulation is characterized by a set of all disturbances and control values Q and impulse values X in each simulation cycle.

To form possible strategies for the development of the system, it is necessary, first, to predict its self-development, that is, to study the dynamics of changes in the values of basic reference points in the absence of external control influences.

As a result of such a forecast, a vector of factor values $x = (x_1, x_2, \ldots, x_N)$ at time t is obtained. The values of the initial data that are not quantitatively determined by an expert method. The use of a predictive ensemble based on gray FCMs made it possible to obtain the most preferable values of the increments of factors to achieve the required values of the basic indicators under the conditions of the given resource constraints (Table 2).

Table 2. Results of scenario forecasting based on the ensemble of gray FCMs

No	Indicators (factors)	Gray value of the indicator increments in%
1	Close relationship with the employer	17–19
2	Introduction of distance technologies	14–16
3	Implementation of individual educational technologies	10–12
4	Increase in the number of on-line MOOCs courses	12
5	Foreign applicant company	10–12
6	Convergence is formed. programs with foreign universities	8–10
7	Academic mobility of teachers	11–13
8	The share of the teaching staff, planned build an int. scientific career	7–9
9	Changing the structure of employment of faculty members	5–7
10	Expansion of the teacher's social package	13–15
11	Stimulating factors	21–23
12	Places in the hostel	15–17
13	Places in the hostel	7–9
14	The level of payment of faculty members	9–11
15	Number of research workers	13–15
16	Number of basic departments	11–13
17	Teachers with language training	17–19
18	The level of student's competence	20–22
19	Skill level of the teaching staff	17–19
20	Demand for graduates from the employer	18–20
21	Demand for areas of training	21–23
22	Number of publications in the Scopus database	31–33
23	Joint research projects	24–26
24	Scientific schools and dissertation councils	11–12
25	Required increment of functional F for the year	11–13

Sequentially set "weak" increments in the values of the above factors at the level of 10%, made it possible to assess the sensitivity of the target to control actions in these areas of regulation, based on which the most preferable alternative to scenario forecasting was selected. The preliminary values of the intensity of mutual influence between the measurable factors of the cognitive model were established based on correlation and factor analysis [9]. Further, the coefficients were refined in accordance with the logic of the transition of the system from one stationary state to another because of external impulse influences. Verification of the cognitive model is carried out based on the criterion of completeness and consistency of the influence of factors - causes on factors - consequences and violations of the transitivity of causal phenomena [20–22].

To check the adequacy and accuracy of the cognitive model, it was tested in the retrospective period 2014–2019. Based on the available statistics on the measurable factors of the model. The general correctness of the model based on the predictive ensemble at this stage was confirmed by the closeness of the growth rates of factors calculated on the model to the actual growth rates. At the same time, the accuracy of the model based on the predictive ensemble turned out to be 12% higher than the accuracy of the model based on a single cognitive map.

Scenario analysis of the development of the situation made it possible to choose the most preferable scenario, which, under the conditions of the given restrictions, ensures the achievement of the required planned value of the target indicator with minimal resource consumption for the increment of latent factors.

5 Conclusion

During the study, a cognitive model of scenario forecasting of measures to achieve the required values of the target indicators of the university's activity in the international institutional ranking QS was developed based on an ensemble of gray FCMs, which made it possible to obtain a more accurate result of solving the forecast problem.

To determine the relationship between latent factors and identify the most significant latent factors, the methods of factor analysis were used.

The use of the gray FCMs class made it possible to move from point estimates of expert opinions to interval estimates and to obtain interval results, which made it possible to consider the available data and reduce uncertainty factors with a dispersion of expert opinions more fully.

Building a model based on a predictive ensemble of gray FCMs in the form of a combination of several forecasts made it possible to significantly reduce the overall forecasting errors that arise when using a single gray FCM.

Based on the developed cognitive model, calculations were made and the most preferable scenario for the increment of significant latent factors was selected to achieve the required values of the target indicators of the university development.

The proposed approach allows training the developed model by adjusting the structure of the cognitive map and its parameters based on the results of updating the initial data for its construction.

Acknowledgments. The reported study was funded by RFBR according to the research projects No. 19-07-01137 and 20-07-00926.

References

1. Maksimov, V.I., Kachaev, S.V.: Technologies of the information society in action: application of cognitive methods in business management
2. Axelrod, R.: Decision Structure: Cognitive Maps of Political Elites. Princeton, University Press (2006)
3. Roberts, F.: Discrete mathematical models with applications to social, biological and environmental problems. Nauka (2006)
4. Kosko, B.: Fuzzy cognitive maps. Int. J. Hum. Mach. Res. **24**, 65–75 (1986)
5. Carvalho, J.P., Tom, J.A.B.: Rule-Based Fuzzy Cognitive Maps - Fuzzy Causal Relationships. IOS Press, Amsterdam (1999)
6. Kulba, V., Mironov, P., Nazaretov, V.: Modern approaches to modeling complex socio-economic systems of socio-economic systems using sign digraphs. Autom. Telemech. (7), 45–72 (2001)
7. Karabutov, N.: Structural identifiability of nonlinear dynamical systems under uncertainty. Int. J. Intell. Syst. Appl. (IJISA) **12**(1), 12–22 (2020)
8. Bodyansky, E., Tishchenko, A., Deineko, A.: An evolving neuro-fuzzy system with online learning/self-learning. Int. J. Modern Educ. Comput. Sci. (IJMECS) **2**(7), 1–7 (2014)
9. Vijayalakshmi, V., Venkatachalapati, K.: Comparison of predicting student performance using machine learning algorithms. Int. J. Intell. Syst. Appl. (IJISA) **12**(11), 34–45 (2019)
10. Gorelova, G., Zakharova, E., Rodchenko, S.: Study of Semi-structured Problems of Socio-Economic Systems: A Cognitive Approach. Publishing house of the Russian State University, Rostov (2006)
11. Avdeeva, Z., Kovriga, S., Makarenko, D., Maksimov, V.I.: Cognitive approach in managemen. Probl. Upr. **34**, 2–8 (2007)
12. Bolotova, L.S.: Artificial intelligence systems: models and technologies based on knowledge: textbook/FSBEI RSUITP; FGAU GNĮI ITT "Informatics". Finance and statistics (2012)
13. Borisov, V.V., Kruglov, V.V., Fedulov, A.S.: Fuzzy models and networks, 2nd edn. Stereotype. Hot line - Telecom (2015)
14. Elliott, G., Granger, C.: Timmermann, A.: Handbook of economic forecasting. Elsevier, (12), p. 1324 (2013)
15. Krakovsky, Yu.M., Luzgin, A.N.: Applied aspects of the use of interval forecasting in system analysis. Modern technologies. Syst. Anal. Model. **2**(52) (2017)
16. Goridko, N.P., Nizhegorodtsev, R.M.: Modern economic growth: theory and regression analysis: Monograph. Infra (2017)
17. Draper, N., Williams, I.D.: Applied regression analysis, 912 p. (2019)
18. Sokolov, G.A., Sagiots, R.V.: Introduction to regression analysis and planning of regression experiments in economics: Textbook. Infra (2016)
19. Mikryukov, A.A., Gasparian, M.S., Karpov, D.S.: Development of proposals for promoting the university in the international institutional ranking QS based on statistical analysis methods. Stat. Econ. **17**(1), 35–43 (2020)
20. Belyaev, V.I.: Marketing: fundamentals of theory and practice, 2nd ed., Stereotype. KNORUS, Moscow (2007)

21. Bogomolova, E.: SWOT analysis: theory and practice of application/EV Bogomolova. Econ. Anal. Theory Practice **17**(32), 57–60 (2004)
22. Trakhtengerts, E.A.: Computer support for the formation of goals and strategies. SINTEG (2005)

A Model for Evaluating Remaining Energy of Retired Battery Based on Improved PSO-BP Neural Network

Yongquan Zhu, Jinshan Dai(✉), Mengya Zhang, Xiya Huang, and Feng Zhang

Wuhan University of Technology, Wuhan 430000, China
jinsahn.dai@whut.edu.cn

Abstract. Aiming at the insufficiency of the remaining energy detection methods for decommissioned batteries specified in the current national standards, it is proposed to use neural networks to build a model for evaluating the remaining energy of decommissioned batteries to provide a basis for the subsequent recycling and reuse of decommissioned batteries. In view of the problems of traditional BP neural network being easy to fall into local optimal solutions, it is proposed to increase the momentum term and the learning rate adjustment function at the same time; for the problems of slow convergence of the PSO algorithm, the linear decreasing weight method is used as the weight update strategy, and the learning factor is adjusted at the same time. The improved PSO algorithm is combined with the improved BP neural network to construct a PSO-BP neural network model to evaluate the residual energy of decommissioned batteries. It has been verified that the PSO-BP model has a faster convergence speed and a higher accuracy during the training process, and the accuracy rate of the residual energy evaluation result of the retired battery is 98%, which has high practical application value.

Keywords: Improved BP algorithm · PSO algorithm · PSO-BP neural network

1 Introduction

With the large-scale application of new electric vehicles, the power batteries used by them have also ushered in the peak of retirement. When reclaiming the power battery, the residual energy of the retired battery needs to be tested to facilitate the classification and reuse of the subsequent battery. Compared with the residual energy detection method of retired batteries stipulated by the current national standard, the method of evaluating the residual energy of retired batteries using neural network model combined with battery charging and discharging test data has the advantages of high accuracy and low energy consumption.

According to the existing national standards of battery residual energy detection, it is necessary to fully charge and discharge the retired battery many times to evaluate the residual energy of the retired battery. This method has the following problems:

Z. Hu et al. (Eds.): AIMEE 2021, LNDECT 107, pp. 124–133, 2022.
https://doi.org/10.1007/978-3-030-92537-6_12

(1) When charging and discharging the battery, it is necessary to fully charge and discharge the battery several times and measure the relevant parameters, which takes a long time, and the long-term occupation of test equipment is not conducive to large-scale assembly line operation.
(2) When charging and discharging the battery, a professional must take care of the whole process. In addition, due to the current battery residual energy test standard, the current must be limited during the test of the decommissioned battery. This kind of residual energy detection method consumes a lot of human and material resources.
The use of more automated measuring instruments for the charging and discharging test of decommissioned batteries can alleviate the current situation that the measurement methods specified by the national standard have a large demand for human resources. At the same time, considering energy consumption and battery loss, it is proposed to use the BP neural network model to evaluate the residual energy of retired batteries based on a small amount of charge and discharge test data.

In the view of the fact that the initial parameters of the BP neural network are randomly generated, the randomness is large, and it is easy to fall into the local optimal solution during the model training process, the PSO (Particle Swarm Optimization) algorithm is introduced to optimize the initial parameters of the BP neural network [1–3]. The search of PSO algorithm is based on the direction of the particles, by tracing the historical optimal position of each particle and the historical optimal position of the population to achieve the purpose of fast global optimization. The global search feature of PSO algorithm can make up for the defect that BP neural network search is easy to fall into local optimal solution, and at the same time enhance the generalization of the model. Based on BP neural network, PSO algorithm is introduced to expand the parameter search space and avoid improper parameter selection, which may lead to the local optimal solution during the model training process.

Based on the above research basis, improved BP algorithm and PSO algorithm are proposed to improve neural network model. Based on two improved algorithms, PSO-BP neural network model is put forward and applied to the residual energy detection and evaluation of decommissioned batteries [4].

2 The Improvement of BP Neural Network and PSO Algorithm

In order to solve the problems that the traditional BP neural network convergence speed is too slow, the convergence speed is greatly affected by the initial threshold and bias. The following methods are proposed to solve the problems.

2.1 Improved BP Neural Network

The multi-layer perceptron that takes the S-type activation function as the neuron transformation function and uses the BP (Back Propagation) algorithm as the connection weight learning algorithm is called the BP neural network. BP neural network is a multi-layer feedforward neural network trained according to the error back propagation algorithm.

BP neural network has received widespread attention due to its good generalization ability but limited by the shortcomings of BP algorithm that it is easy to fall into the local optimal solution and cannot obtain the global optimal solution, its use in certain fields is greatly restricted. In response to these problems, based on existing research, the following methods are proposed to improve the defects of BP neural network:

(1) Adding the Momentum Term

The BP algorithm is based on the gradient descent method when the weight is adjusted. During the training process, the weight is easy to oscillate, which leads to a slower convergence speed. In order to improve the convergence speed of the algorithm, a momentum term is introduced in the weight adjustment strategy. The weight and threshold adjustment formulas are as follows:

$$\begin{cases} \Delta W(t+1) = (1-\alpha)\eta\delta X + \alpha\Delta W(t) \\ \Delta\theta(t+1) = (1-\alpha)\eta\delta + \alpha\Delta\theta(t) \end{cases} \tag{1}$$

In the formula, t is the number of training times, W represents the weight matrix of a certain layer, X represents the input vector of a certain layer, θ represents the threshold matrix of a certain layer, η is the learning rate, δ is the output error of a certain layer, α is the momentum coefficient, $\alpha \in (0, 1)$. The momentum term can integrate the learning rate and the weight at that moment. When the error surface fluctuates greatly, it can reduce the oscillation and improve the convergence speed. Adjusting the momentum coefficient according to the following formula:

$$\alpha = \begin{cases} 0, E(t+1) > 1.05E(t) \\ 0.93, E(t+1) < E(t) \\ \alpha, \text{other} \end{cases} \tag{2}$$

$E(t+1)$ is the mean square error at time $t+1$, while $E(t)$ is the mean square error at time t.

(2) Adaptive Learning Rate

The learning rate η is also called the step size. At the beginning of the network training, a larger learning rate is needed to achieve a rapid drop in error, and a smaller learning rate is needed to prevent overfitting during fast convergence [5]. To realize the dynamic change of the learning rate, the following adjustment methods of η are proposed:

Introduce a scale factor $\eta_1 \in (0, 1]$ to adaptively change the learning rate, the formula is as follows:

$$\eta(t+1) = \begin{cases} \frac{\eta(t)}{\eta_1}, E(t+1) < E(t) \\ \eta_1\eta(t), E(t+1) > E(t) \end{cases} \tag{3}$$

When $E(t+1) < E(t)$, divide η by η_1 to improve learning rate and speed up convergence; while $E(t+1) > E(t)$, η multiply by η_1 to decrease learning rate and fix the weight in time.

The addition of the momentum term and adaptive adjustment of the learning rate can effectively prevent the model from falling into a local minimum during training, and improve the convergence speed and learning efficiency, and to a certain extent make up for the defects of the BP algorithm.

2.2 Improved PSO Algorithm

The PSO algorithm solves the high-dimensional optimization problem by simulating the foraging process of a flock of birds. The search space of the optimization problem is regarded as the flight space of birds, and each bird is regarded as a particle. The optimization process is the search of the flock of birds. Food process. The food location is regarded as a feasible solution of the optimization problem. The feasible solution is continuously searched, and the obtained feasible solutions are compared, and the optimal solution that meets the requirements is finally obtained.

Supposing that in a d-dimensional target search space, there are m particles in the population. The position vector of the $i-$ th particle is $X_i = (\mathrm{x}_{i1}, \mathrm{x}_{i2}, \cdots, \mathrm{x}_{id})$ with the velocity vector $V_i = (\mathrm{v}_{i1}, \mathrm{v}_{i2}, \cdots, \mathrm{v}_{id})$ respectively, where $i = 1, 2, \cdots, m$. The particle velocity and position update formula is as shown:

$$V_{id}(t+1) = wv_{id}(t) + c_1r_1(P_{id}(t) - x_{id}(t)) + c_2r_2(P_{gd}(t) - x_{id}(t)) \tag{4}$$

$$x_{id}(t+1) = x_{id}(t) + v_{id}(t+1) \tag{5}$$

In the formula, t is the current iteration number of the algorithm, and w is the inertia weight. c_1 and c_2 are learning factors, usually $c_1 = c_2 = 2$. r_1 和 r_2 are uniformly distributed random numbers between 0 and 1. The position of the $i-$ th particle, also called the current local optimal value, is denoted as P_{best}, $P_{best} = (P_{i1}, P_{i2}, \cdots, P_{id})$; the current optimal position found by the entire particle swarm is the global optimal value g_{best}, $g_{best} = (P_{g1}, P_{g2}, \cdots, P_{gd})$.

Aiming at complex optimization problems, traditional PSO algorithm has defects such as premature convergence, slow convergence speed, low search accuracy and easy to fall into local optimal values. In view of the above-mentioned shortcomings, the PSO algorithm is optimized from the following aspects in combination with the related parameters of the particle swarm [6, 7].

(1) Inertial Weight Update Strategy

To accelerate the convergence speed of the algorithm, in the initial stage of the particle search, a larger inertia weight should be set to improve the algorithm's global search ability, and in the later stage of the search, the inertia weight should be reduced to improve its local search ability. The most used weight update strategy is linear declining weight method (LDW), and its principle is as follows:

$$w = w_{\max} - \frac{t}{t_{\max}}(w_{\max} - w_{\min}) \tag{6}$$

In the formula, $w_{\max}$ is the maximum weight. $w_{\min}$ is the minimum weight.$t_{\max}$ is the maximum number of iterations. w determines the impact of the velocity of the particle in the previous iteration on the current velocity. However, the linear decrement method has certain defects. As the number of iterations increases, the algorithm's global search capability will decrease. Therefore, it is proposed to increase the inertia weight on the basis of LDW, and at the same time adopt the inertia weight decay function, which can adapt the weight change to the search process, and achieve the purpose of improving the global search ability in the initial stage and the local search ability in the final stage. The improved weight update formula is as shown in formula (7):

$$w(t+1) = w(t) + \gamma r \tag{7}$$

In the formula, $w(t)$ and $w(t+1)$ are the inertia weights after iterations t and $t+1$ respectively. γ is the introduced disturbance factors, and are random numbers between [0, 1].

(2) Learning Factors

The learning factor c_1 represents the maximum step length when the particle reaches the optimal position of the individual, and c_2 represents the maximum step length when the particle reaches the global optimal position. A large value of c_1 will cause the particles to search only in a local area, and a small value of c_2 will cause the particles to converge prematurely. The asynchronous change method is used to determine the value of the learning factor, and the iterative formula of the learning factor is shown in Eq. (8).

$$\begin{cases} c_1 = c_{1i} - (c_{1i} - c_{1f})(\frac{h}{t_{\max}}) \\ c_2 = c_{2i} - (c_{2f} - c_{1i})(\frac{h}{t_{\max}}) \end{cases} \tag{8}$$

In the equation, the initial values of c_1 and c_2 are c_{1i} and c_{2i}, and the end values are c_{1f} and c_{2f} respectively. h represents iterations. Through the standard function test, the improved PSO algorithm has better global search ability and convergence.

3 Detection Method of Retired Battery Residual Energy Based on Improved PSO-BP Neural Network

The improved PSO algorithm replaces the gradient descent method in the BP neural network to optimize the initial parameters of the network. Using optimized weights and thresholds to construct a neural network can effectively prevent the model from falling into a local minimum during the training process, improve the training speed and learning ability of the network, and optimize the performance of the network. The implementation process of the improved PSO-BP neural network algorithm is as follows.

Step 1: Determining the network structure of the BP neural network.
Step 2: Initializing the weights and thresholds of the BP neural network and encoding them to get the initial population.
Step 3: Initializing the parameters of the particle swarm algorithm: population size, particle dimension, learning factor, inertia weight, maximum number of iterations, speed, and position.
Step 4: An appropriate fitness function is chosen to calculate the fitness value of each particle. The calculation formula is as follows:

$$f(i) = \sum_{i=1}^{n} (\hat{y}_i - y_i)^2, i = 1, 2, \cdots, N \tag{9}$$

In the equation, n is the number of training samples, N is the population size of the swarm.

Step 5: Finding the individual optimal position P_{best} and the global optimal position g_{best} according to fitness.
Step 6: Updating the velocity and position of the particles.
Step 7: Calculating the fitness value of the new particle and updating P_{best} and g_{best} according to the fitness value.
Step 8: When the number of iterations is less than the target value, return to step 4, otherwise end the iteration.
Step 9: According to the obtained optimal solution, that is, the optimal weights and thresholds optimized by the improved PSO algorithm, a neural network is constructed for learning and training.

The algorithm flow chart is shown in Fig. 1.

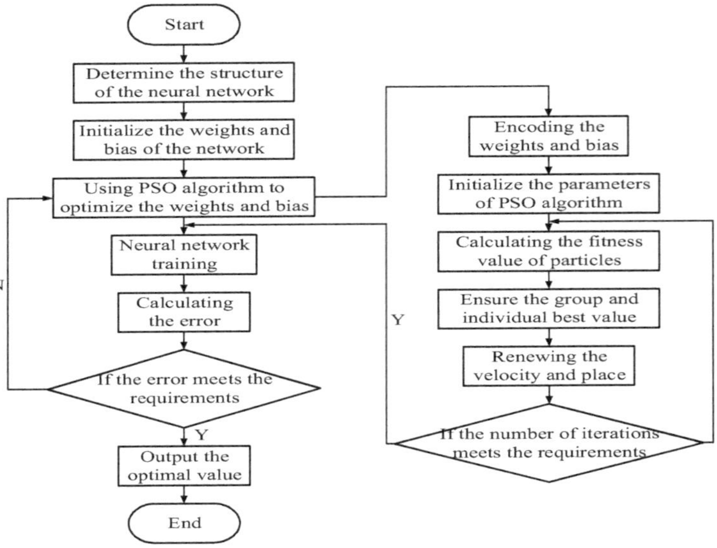

Fig. 1. The flowchart of improved PSO-BP algorithm

4 The Verification and Analysis of the Improved Algorithm

The neural network to be tested adopts a basic three-layer network structure. The 10-dimensional feature vector obtained after the battery test data is processed is used as input, and the output layer is the six classification results of battery residual energy. The classification results use One-Hot Encoding. Therefore, the number of neurons in the output layer is 6. According to the empirical formula $l = \sqrt{m+n} + a$ (a is a constant between 1–10), setting the number of hidden layer neurons to 11, the topology of the neural network to 10-11-6, the improved PSO algorithm parameters are set as follows: the population size is 50, the maximum number of iterations is 400, and the maximum particle velocity is 5, initial learning factors c_{1i} and c_{2i} are both 0.5. The parameters of the improved BP algorithm are set as follows: momentum coefficient is 0.9, error accuracy is 0.0001, maximum number of iterations is 400, learning rate range is [0.01, 0.8], hidden layer and output layer activation functions adopt tansig and purel in functions, respectively [8].

4900 randomly selected samples are used to train the traditional neural network, the improved BP neural network and the improved PSO-BP neural network. The training process is shown in the Figs. 2, 3 and 4:

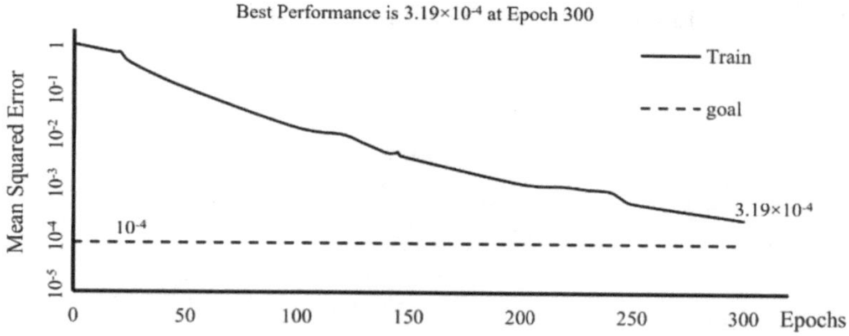

Fig. 2. The training progress of BP neural network

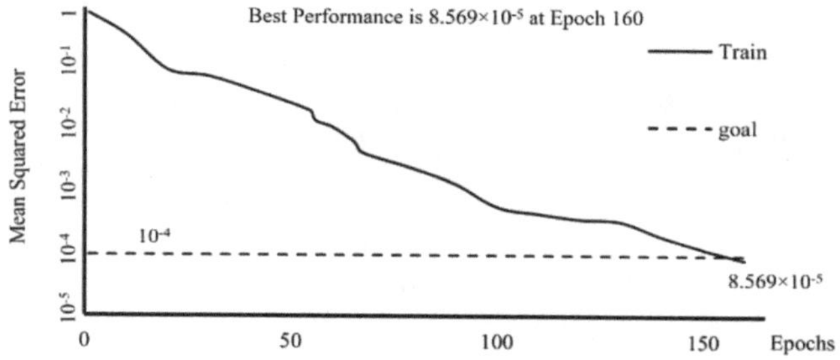

Fig. 3. The training progress of improved BP neural network

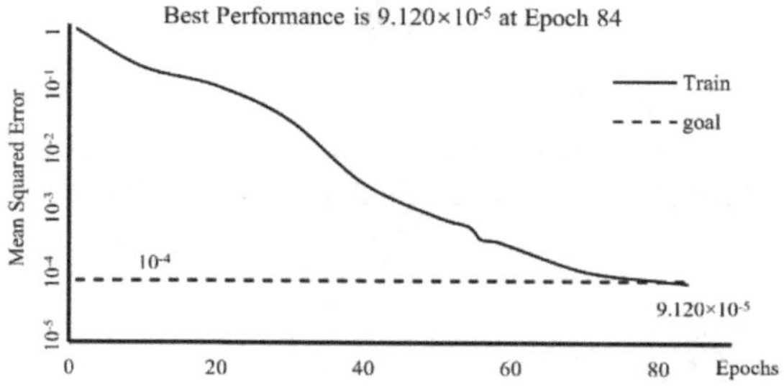

Fig. 4. The training progress of improved PSO-BP neural network

It can be seen from the above training process that when the error is large, the error reduction speed of the improved PSO-BP neural network is better than the other two algorithms; when the error is small, the error reduction speed of the improved PSO-BP is significantly reduced, and overfitting is less likely to occur. And after 84 iterations of the improved PSO-BP neural network, the error has been less than the target value of 0.0001.

To verify the effectiveness of the improved PSO-BP neural network, the battery sample verification set is applied to the trained improved PSO-BP neural network model. There are 100 samples in the validation set, and the test results are shown in the Table 1:

Table 1. Test result of improved PSO-BP neural network

Residual energy level	Number of samples	Accurate identification number
1	15	15
2	10	9
3	15	16
4	20	21
5	25	24
6	15	15

Acc (Accuracy), Se (Sensitivity) and Sp (Specificity) are used as indicators of battery residual energy grading effect. The calculation formulas of the three indicators are as follows:

$$\text{Acc} = \frac{\text{TP}}{\text{TP} + \text{FN}} \tag{10}$$

$$\text{Se} = \frac{\text{TN}}{\text{TN} + \text{FP}} \tag{11}$$

$$\text{Sp} = \frac{\text{TN} + \text{TP}}{\text{TN} + \text{TP} + \text{FN} + \text{FP}} \tag{12}$$

According to Table 1 and Eq. (10), the improved PSO-BP neural network has an accuracy rate of 98% in the verification set, and the classification effect is significant. The improvement of BP algorithm and PSO algorithm mentioned above has significant effects [9, 10].

5 Summary and Conclusion

Based on the improvement of PSO algorithm and BP algorithm, an improved PSO-BP neural network is proposed and applied to battery residual energy detection. The momentum term and adaptively adjusting the learning rate are introduced to improve the BP neural network; the PSO algorithm is improved in two aspects: the inertia weight strategy and the learning factor.

Based on the above improvements, the PSO algorithm is used to optimize the initial weights and thresholds of the BP neural network. The processed battery residual energy data is used to train the traditional neural network, the improved BP neural network and the improved PSO-BP neural network, and the trained neural network is used to separately evaluate the residual energy of the retired battery. The results show that the improved PSO-BP neural network has shorter training time, faster convergence speed and higher classification accuracy. Compared with the traditional BP neural network and the improved PSO-BP neural network has a higher accuracy rate. Based on its higher accuracy, shorter training time and better generalization compared with traditional neural network, improved PSO-BP has a promising future in the field of battery residual energy detection.

References

1. Wang, Y.D., Xiao, H., Jianxin, X., Wang, H.: Prediction model of PSO-BP neural network on coliform amount in special food. Saudi J. Biol. Sci. **26**(6), 1154–1160 (2019)
2. Huang, J., He, L.: Application of improved PSO - BP neural network in customer churn warning. Procedia Comput. Sci. **131**, 1238–1246 (2018)
3. Wen, L., Yuan, X.: Forecasting CO2 emissions in Chinas commercial department, through BP neural network based on random forest and PSO. Sci. Total Environ. **718**, 137194 (2020)
4. Huang, J., He, L.: Application of improved PSO-BP neural network in customer churn warning. Procedia Comput. Sci. **131**, 1238–1246 (2018)
5. Sang, J.: Research on pump fault diagnosis based on pso-bp neural network algorithm. In: IEEE 8th Joint International Information Technology and Artificial Intelligence Conference (ITAIC), pp. 1748–1752 (2019)
6. Zhang, Y., Cui, N., Feng, Y., Gong, D., Hu, X.: Comparison of BP, PSO-BP and statistical models for predicting daily global solar radiation in arid Northwest China. Comput. Electron. Agricu. **164**, 104905 (2019)
7. Khan, K., Sahai, A.: A comparison of BA, GA, PSO, BP and LM for training feed forward neural networks in e-learning context. Int. J. Intell. Syst. Appl. **4**(7), 23–29 (2012)
8. Sreedhara, B.M., Kuntoji, G., Manu, S.M.: Application of particle swarm based neural network to predict scour depth around the bridge pier. Intell. Syst. Appl. **11**, 38–47 (2019)
9. Gupta, S.L., Baghel, A.S., Iqbal, A.: Threshold controlled binary particle swarm optimization for high dimensional feature selection. Int. J. Intell. Syst. Appl. (IJISA) **10**, 75–84 (2018)
10. Zhou, D.: Optimization modeling for GM(1,1) model based on BP neural network. Int. J. Comput. Network Inf. Secur. (IJCNIS) **4**(1), 24–30 (2012)

Evolution Management of Multistage Manufacturing Based on Evidence Theory and Methods of Interval Analysis

Boris V. Paliukh and Alexander N. Vetrov(✉)

Tver State Technical University, Tver 170026, Russia

Abstract. The article substantiates the principles of managing the operational reliability of continuous production through the use of soft measurements and calculations. Instead of the traditional system of technical diagnostics based on the analysis of instantaneous values of diagnostic variables, a new approach is proposed. It is based on the analysis of the intervals of changes in diagnostic variables in a given time interval. The use of the proposed approach makes it possible to reduce the sensitivity to noise interference and delays in the dynamic channels between the input and output of a continuous technological process. An intelligent decision-making system based on the Dempster-Shafer theory is substantiated. The combination of these approaches makes it possible to create a more efficient system for managing the operational reliability of continuous production compared to traditional methods. The proposed approaches are illustrated by setting specific tasks with solutions of control examples.

Keywords: Safety of complex engineering systems · Methods of interval analysis · Dempster-Shafer theory

1 Introduction

The methods of managing the operational reliability of continuous production mainly contain a mathematical apparatus that is of little use for reasonable diagnostics in conditions of unreliability of measurements [1]. The classical problems of parametric diagnostics are solved by rapid calculation of diagnostic parameters through the measured instantaneous values of the measured technological variables, such as temperature, pressure, flow, and the like [2–5]. However, in this case, the calculated parameters may have a very large error, so the diagnostic system can often issue false alarms or skip defects [2, 6]. This is due to the following factors [6]. First, the instantaneous values of the measured variables are too sensitive to noise interference and can randomly change by a significant amount. Secondly, the measured variables can relate to both the input and output of the diagnostic object, and therefore their values will differ by the amount of the lag of the dynamic channel between the input and output. These difficulties can be overcome by applying soft measurements and calculations. Reducing the sensitivity to noise interference is achieved by using instead of the traditional technical diagnostics system based on the analysis of instantaneous values o f dagnostic parameters, such a diagnostic system that analyzes the intervals of

Z. Hu et al. (Eds.): AIMEE 2021, LNDECT 107, pp. 134–143, 2022.
https://doi.org/10.1007/978-3-030-92537-6_13

changes in diagnostic parameters for a given period of time [7, 8]. For continuous production, usually the regions of operable O and defective D states overlap significantly, therefore, it is fundamentally impossible to avoid erroneous decisions [8]. The likelihood of erroneous decisions can be reduced by using interval analysis [9]. Let there be an interval of values of the diagnostic parameter:

$$X = [\underline{x}; \overline{x}] = \{x | \underline{x} \leq x \leq \overline{x}; \quad \underline{x}; \overline{x} \in R\} \tag{1}$$

where R is the set of real numbers, and $\underline{x}; \overline{x}$ respectively, the upper and lower values of the diagnostic parameter. We denote

$$T = [\underline{t}; \overline{t}] = \{t | \underline{t} \leq t \leq \overline{t}; \quad \underline{t}; \overline{t} \in R\} \tag{2}$$

the specified time interval during which the diagnostic variable is measured Here $\underline{t}$; $\overline{t}$ are the lower and upper bounds of the time interval measurement interval.

Theorem 1. The probability of a false alarm of a continuous production diagnostic system built on the principle of analyzing the instantaneous value of a diagnostic parameter is greater than the probability of a false alarm of a diagnostic system built on the principle of analyzing the interval of values of a diagnostic parameter.

Theorem 2. The probability of missing a defect in a continuous production diagnostics system built on the principle of analyzing the instantaneous value of a diagnostic parameter is greater than the probability of missing a defect in a diagnostics system built on the principle of analyzing an interval of values of a diagnostic parameter. The proof of the Theorems 1 and 2 is given in [7]. Thus, if we analyze not the instantaneous values of the diagnostic parameter, but the intervals of its change for a given time interval, then the probability of a false alarm and missing a defect decreases. The differences in the values of the input and output measured variables of the diagnostic object associated with the lag of the dynamic channel between the input and output can be overcome by choosing the time interval T in this way; so that $\overline{t} = t_0 + \tau$. Here τ is the delay time of the dynamic channel between the input and output of the object.

2 Evaluation of Calculated Diagnostic Indicators

Diagnostic variables are divided into observable – z and – x calculated variables. During the selected time interval, the intervals of the observed technological variables $Z = [\underline{z}; \overline{z}]$ are measured. It is necessary to find a guaranteed estimate of the calculated diagnostic indicators in the form of interval $X = [\underline{x}; \overline{x}]$. The mathematical model of the diagnostic object is usually given in the form of a system of algebraic or differential equations. In our case, the equations will be with interval variables and coefficients. Let's consider the solution of this problem on a specific example of the production of weak nitric acid [2]. Let there be a system of five algebraic equations with five calculated diagnostic indicators: X_1 - the degree of conversion, X_2 - the degree of absorption, X_3 - the content of ammonia in fractions of one, X_4 - the temperature of the ammonia-air mixture, X_5 - the pressure in the catalytic reactor. The system of equations has the form

$$\frac{Z_3}{0.02041 Z_1 X_3} = \frac{355.56}{X_1 Z_2}; \tag{3}$$

$$X_3 = \frac{1.03}{Z_4/Z_3 + 1}; \tag{4}$$

$$273 + Z_5 = 1055 + 177 X_5; \tag{5}$$

$$273 + X_3 = -1055 + \{(Z_5 + 273) - (X_4 + 273)\} 0.8533 \times 10^{-3}; \tag{6}$$

$$X_1 = 0.97 + 0.0046 X_5 - X_5^2 \tag{7}$$

Let's consider an example of the solution of this system. Input observed technological variables:

Z_1 = [53.3; 55.5] - concentration of the production acid, %;
Z_2 = [16.24; 16.44] - production acid consumption, m^3/h;
Z_3 = [5200; 5300] - consumption of gaseous ammonia for the ammonia-air mixture (ABC), m^3/h;
Z_4 = [48000; 49000] - air flow for ABC, m^3/h;
Z_5 = [898; 902] - the temperature of the nitrous gas under the grids of the catalytic reactor, K.

Results of solving a system of interval algebraic equations:

X_1 = [0.957; 0.958] - the degree of conversion, unit fraction.;
X_2 = [0.958; 0.993] - the degree of absorption, unit fraction.;
X_3 = [0.959; 0.0994] - the ammonia content in ABC, unit fraction;
X_4 = [515; 523] - the temperature of ABC, K;
X_5 = [0.655; 0.678] - pressure in the catalytic reactor, MPa.

Experimental verification of the calculation results showed that the actual values of unobserved diagnostic indicators are guaranteed to be included in the calculated intervals.

As an example of constructing parametric models in the form of interval ordinary differential equations, we take a model of the transformation of ammonia X_6 in the catalyst layer:

$$\frac{dX_6}{dt} = 6750 (\frac{Z_4}{4})^{0.56} (\frac{Z_5}{4.5})^{0.59} e^{-793(\frac{1}{Z_3} - \frac{1}{1123})} (1 - X_6),\ X_6(0) = 0 \tag{8}$$

The values of the measured variables are presented:

Z_1 = [460; 470] - volume of ammonia per 1 ton of product, m^3/t;
Z_2 = [3880; 3950] - air volume per 1 ton of product, m^3/t;
Z_3 = [1123; 1133] - temperature of nitrous gases, K;

Z_4 = [1.9; 1.1] - gas velocity in the catalytic reactor, m/sec.

Solving this equation by two-way numerical methods, we obtain the dependence of the total degree of transformation of ammonia X_6 in the catalyst layer on time t. The calculated interval values of the diagnostic parameters allow us to reliably solve the problem of the state of individual units of continuous production. The conducted studies allow us to conclude that it is necessary to build diagnostic systems for continuous production not on the traditional principles of analyzing instantaneous values of diagnostic parameters, but on the analysis of the intervals of changes in these parameters for a certain time interval. Therefore, it is most expedient to solve the problems of managing the operational reliability of continuous production using soft measurements and calculations. To control the operational reliability of continuous production, the most promising method is parametric diagnostics based on the control of permissible deviations of technological variables.

The use of parametric diagnostics has the following features:

- it is possible to control unobservable diagnostic indicators by calculating through measured variables;
- it is possible to predict the occurrence of gradual (parametric) failures;
- monitoring of diagnostic indicators is carried out for the most part not continuously, but discretely with a certain period, therefore, the measured or calculated indicator characterizes the diagnostic object at a certain time interval.

When using traditional methods of parametric diagnostics, use special techniques for accounting for errors and complex mathematical methods for converting diagnostic data. It is possible to get rid of these shortcomings by applying interval analysis methods, which make it possible to automatically take into account the error in the measured initial data and facilitate decision-making on diagnostics [2].

3 Principles of Application of Interval Analysis

Let's consider the principles of using interval analysis for parametric diagnostics. Let x_i be the value of the i - th technological change on a serviceable diagnostic object. Over the time of operation of the object under consideration, the variable x_i will deviate by some small value δx_i. In order to diagnose the defect early, it is necessary to measure and analyze the entire set of small deviations of the object variables. The values of x_i are recorded by measuring devices. Let the measurement error be included in the value δx_i. To find a defect, the relations between the deviations of these variables are analyzed using mathematical models of objects, according to which linear diagnostic matrices are compiled [1, 2]. However, this problem can be solved without the use of diagnostic matrices, if you apply interval analysis directly to mathematical models of objects. The solution of such a problem is possible if we represent small deviations of each i - th diagnostic indicator in the form of an interval $X = [\underline{x}; \overline{x}]$.

If the value of the i - th technological variable x_i increases by the value δx_i during operation of the equipment, then the boundaries of the interval are defined as

$$\underline{x} = x_i; \quad \overline{x}_i = x_i + \delta x_i \tag{9}$$

If x_i, decreases by δx_i, then

$$\underline{x} = x_i - \delta x_i; \quad \overline{x}_i = x_{ii} \tag{10}$$

Similarly, you can take into account the time intervals (frequency) of monitoring. If the measurements of a technological variable are carried out periodically with a time period τ, then any measurement of the variable characterizes the diagnostic object over the entire j-th time interval $T_j = [\underline{t}_j; \overline{t}_j]$. Here $\underline{t}_j$, $\overline{t}_j$ is the beginning and end of the time interval in which the j - th measurement was performed.

$$\overline{t}_j = \underline{t}_j + \tau; \quad j = 1, 2, \ldots \tag{11}$$

Figure 1 shows a functional block diagram of an automated system for managing the operational reliability of continuous production. Each production facility has measuring devices that control the observed technological variables. Unobservable diagnostic indicators are calculated by parametric diagnostic methods.

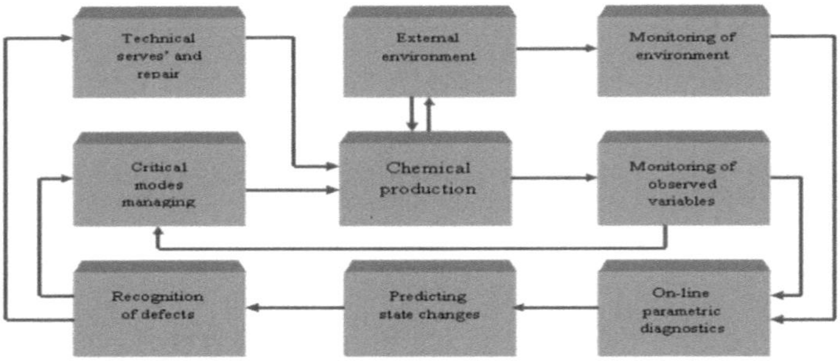

Fig. 1. Functional block diagram for managing the operational reliability of continuous production

According to the predicted values of the indicators, production defects are recognized. When a pre-emergency state occurs, the operational reliability management system is activated, which can work in the "adviser" mode or in real time, depending on the structure of the automation equipment.

The most important issue in the construction of interval parametric models is the method of finding the minimum interval distances. The method is based on the following principles.

1) For early diagnostics of production, small deviations of technological variables inherent in the forms of incipient defects should be measured and analyzed.
2) The use of the method of small deviations involves the analysis of the production operation in a steady-state mode.

3) As a parametric diagnostic model, it is recommended to use the equations of the material and thermal balances of the technological process, presented in the form of a closed system of nonlinear algebraic equations with interval variables and coefficients; this system can be supplemented with equations of kinetics, hydro - and gas dynamics.
4) The analytical form of each equation of the system should be transformed in order to reduce the width of the natural interval expansion.
5) To strictly guarantee that the solution belongs to the obtained intervals, it is necessary to take into account computer rounding errors, but in practical calculations these errors are small, and they can be ignored.
6) When using models in practice, it is necessary to use the same measuring circuits to measure the observed technological variables at different time intervals in order to eliminate systematic measurement errors.
7) Due to the delay of the dynamic channel between the input and output of the diagnostic object, the time interval $T_j = [\underline{t}_j; \overline{t}_j]$ of measuring technological variables should be set so that $\underline{t}_j = t_0$ and $\overline{t}_j = t_0 + \tau$ where t_o is the beginning of the measurement of the interval of change in the technological variable; τ is the delay time of the dynamic channel between the input and output of the object.

4 Using the Theory of Evidence to Solve the Problems of Industrial Diagnostics

To make decisions on operational reliability management within the framework of the task, it is necessary to use artificial intelligence methods that allow you to assess the likelihood and take into account the uncertainties in the initial information. Let there be a set of objects C and a set of all possible properties S that they can have. It is necessary to determine objects with the specified properties by a known set of features P. Then, in accordance with [10, 11], a fuzzy set of objects C_k is assigned to each k - th feature P_k

$$C_k = \{c_i; \mu_k(c_i)\}; \quad C_k \in C, \tag{12}$$

where c_i are objects that could have the specified properties in accordance with the value of the k - th attribute); $\mu_k(c_i)$ is the membership function of the element c_i to the set C_k (equal to the degree of confidence reduced to the range [0; 1] in the presence of the specified property by the k - th attribute on the object c_i), C is the complete set of objects.

Let the current situation correspond to the set of registered features - $P^* = \{P_k | P_k = 1\}$. Let's denote the set of objects that actually contain the specified properties as C^* . Then the procedure for determining objects with the specified properties is reduced to solving the following logical equation

$$\text{F: } P^* \rightarrow C^*, \tag{13}$$

where F is a predicate function of the form

$$F = \underset{P_k=1}{\wedge} (\underset{c_i \in C_k}{\vee} c_i); \quad C_k \in C \tag{14}$$

The solution of Eq. (14) can be found in the form

$$C^* = \bigcap_{P_k \in P^*} c_i | c_i \in C_k. \tag{15}$$

If the presence of the property $s \in S$ on the object c_i is a necessary and sufficient condition for the appearance of all signs of P_k, such that $c_i \in C_k$, then the solution (14) could be found using Boolean or Bayesian approaches [6, 9]. In practice, this means the requirement of mutual independence and clear of relations between properties and objects.

Currently, there are methods that allow you to mitigate the requirements of mutual independence and clear of relationships between properties and objects. So, when using the fuzzy sets approach, this is achieved by expanding the C_k set by including unlikely hypotheses at the expert classification stage. This creates a certain information redundancy in solving the problem.

When finding a solution to Eq. (14) in the form of (15) by the fuzzy sets method, the degree of belonging of the object c_N to the set of specified properties is determined as

$$\mu^*(c_N) = \max\{\mu_i(c_N), \mu_j(c_N)\}, \quad i \neq j, \tag{16}$$

where $\mu_i(c_N)$ is the degree of belonging of the object c_N to the set C_i, activated by the appearance of the sign $P_i = 1$. However, neither Boolean nor Bayesian approaches, nor even the method of fuzzy sets [9] allow using the procedure (16) without taking into account the requirements of mutual independence and clear of relations between properties and objects to obtain a reliable and complete solution, since the uncertainty associated with decision-making in artificial intelligence problems is far from being exhausted by the fuzziness of expert assessments. It can be caused by the stochastic nature of the processes, the instability of the parameters, the noise of the measurements, etc.

To solve such problems, the most promising approaches are based on the use of improbability uncertainty measures that have intuitively justified minimal properties. This approach was developed in [12, 13], and used to solve diagnostic problems in the works [14–16].

The main idea of this approach is that a certain measure of probability can be attributed not only to individual elements of a set of events in the subject area, but also to a subset of this set as a whole. Moreover, a more detailed distribution of this partial probability measure over a subset is unknown. At the same time, some probability measures related directly to the elements of the set of events may also be unknown.

The considered methods based on the theory of evidence (evidence intervals) can be used to solve the tasks set. At the same time, the assignment of the general probability measure to subsets of the set of objects with specified properties can occur for the following reasons. 1 – to the presence of uncertainty in the expert classification.

That is, the expert cannot unambiguously determine the presence of a connection between diagnostic indicators and technological equipment. 2 – a inaccuracy of determining the value of the diagnostic feature. And, thirdly, when assigning a common level of belief for all signs. Despite the unknown specific probability measures assigned to individual objects with specified properties, it is possible to draw some conclusions based only on the known distribution of probability measures. Therefore, it is most expedient to make decisions for the task on the basis of the Dempster-Schafer theory of evidence [14, 17, 18].

Let's consider the application of the Dempster-Shafer theory on a specific example. Suppose you have 5 of technological units for continuous production (agents' a_1, a_2, a_3, a_4, a_5) and two diagnostic indicators x1, x3. As a result of processing the data of measuring diagnostic indicators, normalized values of the probability of equipment malfunction were obtained.

$$m_{x1} = \{a_1, a_2, a_5, A\} = \{0.13; 0.25; 0.48; 0.25\} \tag{17}$$

$$m_{x1} = \{a_1, a_2, (a_3, a_4), A\} = \{0.32; 0.18; 0.10; 0.40\} \tag{18}$$

According to Dempster's rule, the combination of various evidence, with probability distributions m_1 and m_2, is performed as follows:

$$m(A_N) = \frac{1}{1 - m(\varnothing)} \sum_{C_i \cap C_j} m_1(C_i) * m(C_j) \tag{19}$$

Where, $m(A_N)$ is the function of combining the certificates $m_1(C_i)$ and $m_2(C_j)$ corresponding to the presence of the object property A_N; $\varnothing$ - is an empty set.

The degree of probability that falls on an empty set

$$m(\varnothing) = \sum_{C_i \cap C_j} m_1(C_i) * m_2(C_j) \tag{20}$$

For the considered example

$$m(\varnothing) = 0.08 + 0.12 + 0.02 + 0.07 + 0.04 = 0.35 \tag{21}$$

The combinations of various m_{xi}, based on the Dempster rule, are presented in Table 1.

Based on the new basic probabilities, two values, belief (bel) and plausibility (pl) are calculated for all the hypotheses under consideration. For the alternatives under consideration, we have intervals bounded by two values, belief and plausibility (belief $\leqslant$ plausibility). Solution result: a_1[0.26; 0.41], a_2[0.28; 0.43], a_1[0.03; 0.18], a_3[0.03; 0.18], a_4[0.03; 0.18], a_5[0.29; 0.44], A [1.0; 1.0].

Table 1. An example of combining different evidence

	a_1 0.13	a_2 0.25	a_5 0.48	A 0.25
a_1 0.32	a_1 0.04	$\varnothing$ 0.08	$\varnothing$ 0.12	a_1 0.08
a_2 0.18	$\varnothing$ 0.02	a_2 0.04	$\varnothing$ 0.07	a_2 0.04
$a_3 \vee a_4$ 0.10	$\varnothing$ 0.01	$\varnothing$ 0.02	$\varnothing$ 0.04	$a_3 \vee a_4$ 0.02
A 0.40	a_1 0.05	a_2 0.10	a_5 0.19	A 0.10

The approach to taking into account uncertainty in the diagnosis of the state of complex technical systems provides a number of advantages compared to the Bayesian approach or the approach based on the use of fuzzy sets.

In particular, the application of the Dempster rule allows you to transform the results of many estimates in such a way as to make them consistent and use them to combine with the results of mathematical modeling. Also, the use of evidence theory allows to reduce the conditions of necessity and sufficiency by redistributing part of the confidence over the entire set of possible outcomes of the diagnosis.

5 Conclusion

In accordance with the principles of operational reliability management presented above, uncertainty in the tasks of technical diagnostics is taken into account by using the theory of evidence and the apparatus of interval mathematics. In this case, the probability of the statement A is defined as the interval [bel(A), pl(A)], where bel (A) $\in [0; 1]$ is the probability of the evidence of the statement A (sufficiency). The value pl(A) is the probability of the likelihood of the statement A (necessity) and is calculated as pl(A) = 1 − bel(A). The difference $u(A) = pl(A) - bel(\overline{A})$ represents the uncertainty of the diagnosis. Here $\overline{A}$ A is the negation of A. Thus, the proof of confirmation of the hypothesis does not negate its negation, as it happens with the Bayesian approach, which is more in line with the structure of human thinking. The Bayesian approach arises as a special case when bel(A) = pl(A) is equal. The proposed principles of managing the operational reliability of continuous production allow us to quickly recognize technological and mechanical failures, predict their development at a given time interval, and bring the technological process out of the pre-emergency state.

Acknowledgment. This project is supported by Russian Foundation for Basic Research (project no. 20-07-00199).

References

1. Horst, C.: Handbook of Technical Diagnostics. Springer, Heidelberg (2013). https://doi.org/10.1007/978-3-642-25850-3
2. Kafarov, V.V., Ivanov, V.A., Palyukh, B.V., Babij, I.I.: Problemy obespecheniya bezopasnosti i ekspluatacionnoj nadezhnosti himicheskih proizvodstv. In: Itogi nauki i tehniki. Processy i apparaty himicheskoj tehnologii. VINITI, Moscow (1992)
3. Teslyuk, V., Beregovskyi, V., Denysyuk, P., Teslyuk, T., Lozynskyi, A.: Development and implementation of the technical accident prevention subsystem for the smart home system. Int. J. Intell. Syst. Appl. (IJISA) **10**(1), 1–8 (2018)
4. Shamim Hasan, A.S.M., et al.: Multi-sensor approach for monitoring pipelines. Int. J. Eng. Manuf. (IJEM) **7**(6), 50–58 (2017)
5. Alguliyev, R.M., Aliguliyev, R.M., Imamverdiyev, Y.N., Sukhostat, L.V.: An anomaly detection based on optimization. Int. J. Intell. Syst. Appl. (IJISA) **11**(9), 87–96 (2019)
6. Fedulov, A.A., Fedulov, Yu.G., Cygichko, V.N.: Vvedenie v teoriyu statistiki nenadezhnyh reshenij. Statistika, Moscow (1979)
7. Kafarov, V.V., Palyukh, B.V., Perov, V.L.: Reshenie zadachi tehnicheskoj diagnostiki nepreryvnogo proizvodstva s pomoshyu intervalnogo analiza. Dokl. AN SSSR **311**(3), 677–680 (1990)
8. Palyukh, B.V.: Upravlenie ekspluatacionnoj nadezhnostyu himicheskogo proizvodstva v usloviyah neopredelennosti ishodnoj informacii. Teor. osnovy him. Texnol. **28**(5), 514–518 (1994)
9. Truxaev, R.I.: Modeli prinyatiya reshenij v usloviyah neopredelennosti. Nauka, Moscow (1981)
10. Borisov, A.N., et al.: Obrabotka nechetkoj informacii v sistemah prinyatiya reshenij. Radio i svyaz, Moscow (1989)
11. Borisov, A.N., Krumberg, O.A., Fedorov, I.P.: Prinyatie reshenij na osnove nechetkix modelej: Primery` ispolzovaniya. Zinatne, Riga (1990)
12. Dyubua, D., Prad, A.: Teoriya vozmozhnostej. Radio, Moscow (1990)
13. Marefat, M., et al.: A Dempster-Shafer approach for recognizing machine features from CAD models. Pattern Recogn. **36**(6), 1355–1368 (2003)
14. Simon, C., Weber, P., Evsukof, A.: Bayesian networks inference algorithm to implement Dempster Shafer theory in reliability analysis. Reliab. Eng. Syst. Saf. **93**, 950–963 (2008)
15. Ballal, P., Lewis, F.: Condition-based maintenance using dynamic decisions by Petri nets and Dempster—Shafer theory: a matrix-based approach. Trans. Inst. Measur. Control **31**, 323–340 (2009)
16. Fang, H., Chakrabarty, K., Wang, Z., Gu, X.: Diagnosis of board-level functional failures under uncertainty using Dempster-Shafer theory. IEEE Trans. Comput. Aided Des. Integr. Circ. Syst. **31**(10), 1586–1599 (2012)
17. Jiang, W., Zhan, J.: A modified combination rule in generalized evidence theory. Appl. Intell. **46**(3), 630–640 (2017). https://doi.org/10.1007/s10489-016-0851-6
18. Xu, H., Deng, Y.: Dependent evidence combination based on Shearman coefficient and Pearson coefficient. IEEE Access **6**(1), 11634–11640 (2018)

Consideration of Expert Judgement by the Morphological Matrix Formation

Dmitry L. Rakov(✉)

Mechanical Engineering Research Institute of the Russian Academy of Sciences (IMASH RAN), Moscow 101000, Russia

Abstract. The article discussed with the creation of new innovation systems and engineering solutions based on the morphological approach. Researches belong to the field of knowledge discovery and management, as well as knowledge engineering. Particular attention is paid to the procedures for taking into account the judgement of experts. The peculiarities of the proposed approach are to take into account all options for solutions synthesized by experts. Subsequently, the variants are clustered and obtained estimates. The introduction of average normalized evaluations allows to analyze both experts individually and the whole group as a whole. The application of the morphological approach is shown on the example of decision selection. The classification of additive technologies is given as an example. A morphological matrix has been generated. The combinatorial number of possible solutions is tens of thousands of variants. The use of cluster analysis, system theory, set theory, set of rules allows to identify new innovative technologies.

Keywords: Knowledge discovery and management · Knowledge engineering · System theory · Morphological matrix · Expert judgement · Engineering solution · Additive technologies

1 Introduction

The field of research belongs to the intensively developing knowledge management and engineering [1]. Knowledge management is recognized as an important weapon for sustaining competitive advantage and improving performance [2]. Knowledge sharing has become an important tool in every organization's success story today [3].

The developed approach is intended for the solutions synthesis at the stage of conceptual design of innovative systems and the search for new engineering solutions (ES). This stage is characterized by the fact that it elaborates the possible principles of implementation of the goal, the conditions of application, manufacturing and operation of engineering systems [4, 5]. Possible schematic diagrams, general views, layouts and technical characteristics of the design object are also identified. An error in the choice of scheme and appearance cannot be corrected at the following stages. Therefore, it is necessary to review all real options, the number of which can reach several hundred. For systems of new types, the effectiveness of design research largely affects the course of further development, so at the conceptual stage it is advisable to fully implement new methods that increase the reliability and quality of the results.

Z. Hu et al. (Eds.): AIMEE 2021, LNDECT 107, pp. 144–151, 2022.
https://doi.org/10.1007/978-3-030-92537-6_14

The proposed approach was developed with a specific goal: to enable the researcher to navigate in all the variety of combinations of elements of the object. It also increases the confidence that the selected structural combinations are really close to the best possible in a number of indicators.

The proposed method of ES synthesis is a development of F. Zwicky's "morphological box" method [6–8]. Fritz Zwicky formulated: "The goal of morphological research is to see the perspective of a complete "field of knowledge" about a subject" [6]. Morphological approaches are used in the study of complex subject areas. These approaches are the most rational for solving problems, for which it is impossible or difficult to give an exact and correct mathematical formulation.

The effectiveness of the method is measured by the degree of correspondence of the found engineering solution to the conditions of the problem and the time for which the solution is obtained. Morphological methods can be used to create artificial intelligence systems as well as to teach students [9, 10].

2 Consideration of Expert Judgement

The following approach can be proposed to take into account the expert opinions. In the proposed morphological method, conceptual design is conducted in next steps (Fig. 1) [11]:

- synthesis of the morphological matrix,
- definition of a system of criteria,
- weighting of options,
- selection of reference variants,
- generation,
- selection of variants using estimates of each variant and comparison with others,
- clustering of variants based on similarity measure and creation the solution space,
- analysis of clusters and solutions,
- analysis of the design risk, variants and selection,
- synthesis of anticipation models, parametric modeling and optimization stage.

After the general stage 1 and 2, the experts get a morphological matrix and a set of criteria. The morphological matrix and the set of criteria are fixed and do not change further.

Based on this information, the experts at further stages evaluate the criteria and evaluate the options. The formation of incompatible options and reference variants can be assigned to experts or formed in advance. For example, after the generation of variants we get four variant sets A, B, C, D of options formed by the experts (Fig. 2, 3 and 4).

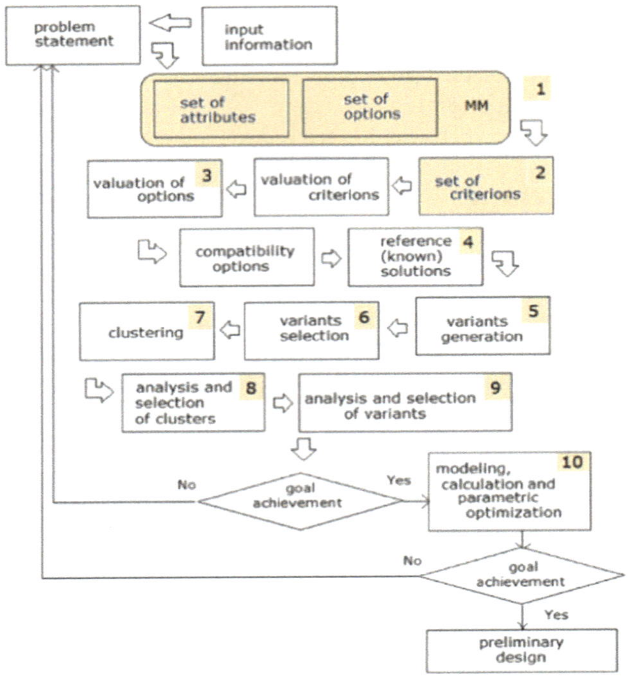

Fig. 1. Block diagram of work execution

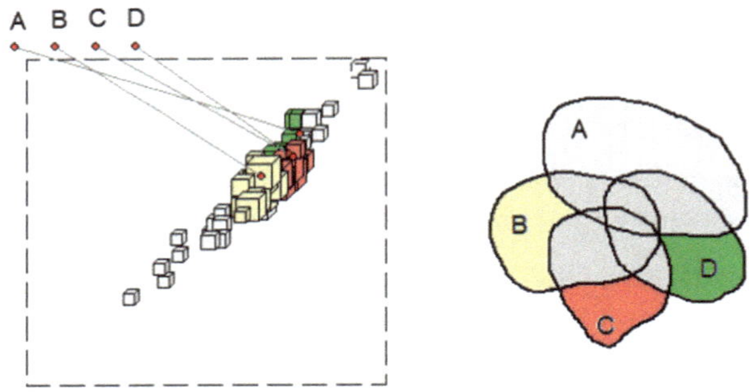

Fig. 2. Expert sets of variants in the decision space (screenshot)

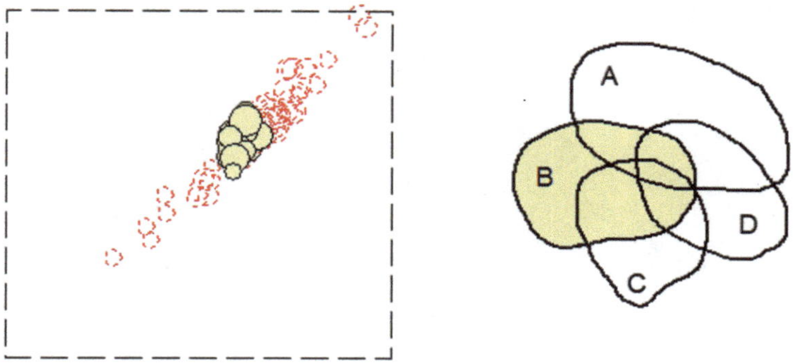

Fig. 3. Variants B in the solution space (screenshot)

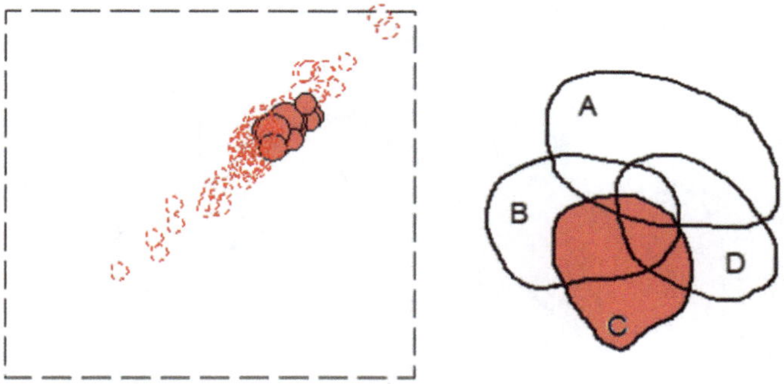

Fig. 4. Variants C in the solution space (screenshot)

After generating variants, the set of all variants E is synthesized (Fig. 5):

$$E = A \cup B \cup C \cup D \qquad (1)$$

And the set F consisting of general variants (Fig. 6):

$$F = A \cap B \cap C \cap D \qquad (2)$$

Further we cluster the variants and calculate the average evaluations of the variants and clustering (Fig. 7). The introduction of average normalized evaluations allows to analyze both experts individually and the whole group as a whole (Table 1).

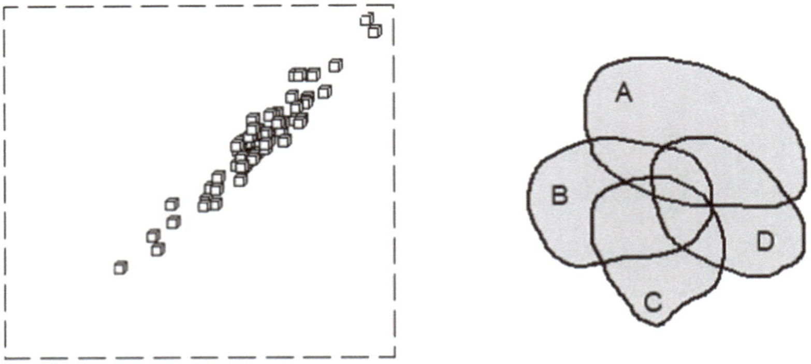

Fig. 5. Set E of all variants (screenshot)

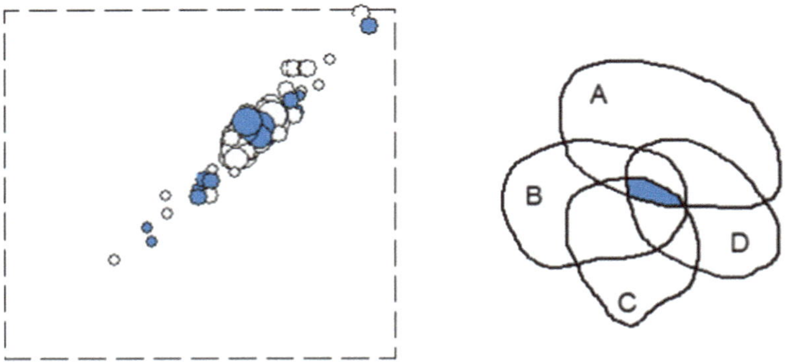

Fig. 6. General variants F in the solution space (screenshot)

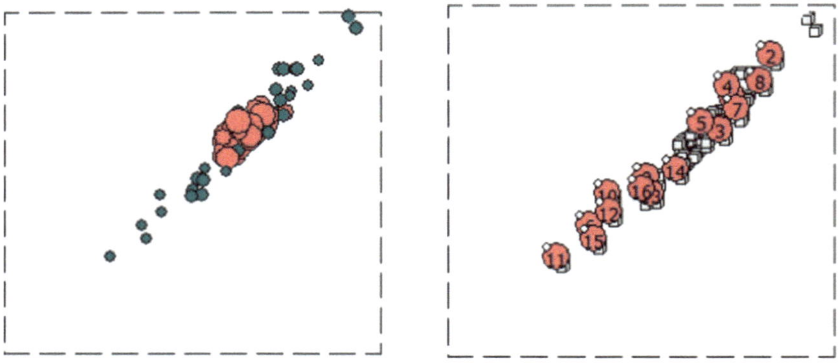

Fig. 7. Clustering centers and reference variants in solution space

Table 1. Normalized expert evaluations

	Set	Average normalized evaluation of the variants	The highest normalized variant score in the set
1	A	1,12	1,36
2	B	0,87	1,08
3	C	0,96	1,42
4	D	1,23	1,1
5	E	1	1,36
6	F	0,85	1,35

3 Analysis of Additive Technologies Based on the Morphological Approach

One of the main development areas of Industry 4.0 is the use of new materials and technological processes in the production of parts and products, in particular additive technologies (Fig. 8) [12].

Fig. 8. Engine combustion chamber parts

A number of works have analyzed and classified additive technologies [13]. Society for Materials and Testing (ASTM) together with the International Organization for Standardization (ISO) created the international standard on additive technologies ISO/ASTM 52900:2015.

Additive technologies according to the methods of obtaining objects are classified according to the following attributes:

- material extrusion,
- material jetting,

- binder jetting,
- sheet lamination,
- vat photopolymerization,
- powder bed fusion,
- directed energy deposition

For the tasks of synthesis of technological processes, it is reasonable to have attributes as these processes are performed, which increases the clarity and logicality of the work (Fig. 9). On the basis of the analysis, a morphological matrix was built [14].

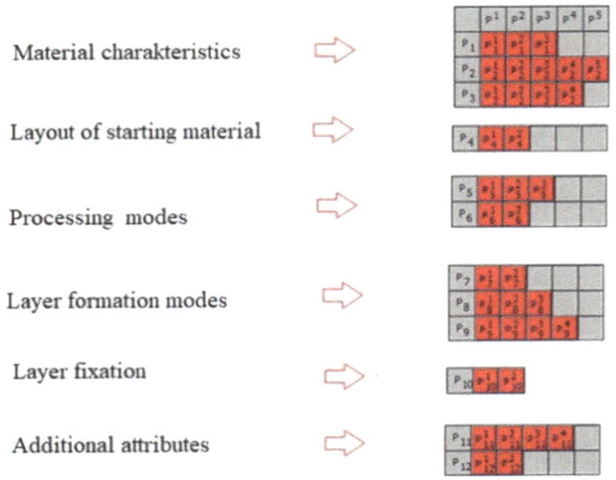

Fig. 9. Morphological matrix feature grouping

4 Summary and Conclusion

The article considers the problems in the synthesis of innovative systems and engineering solutions based on the morphological approach. The considered approach provides clarification and streamlining of the problem.

The use of all options generated by the experts allows increasing the degree of validity of the decisions made and expanding the variant field. This allows selecting the most rational solutions.

The classification of additive technologies is considered as an example. The power of the combinatorial set is tens of thousands of variants. The use of cluster analysis, system theory, set theory, set of rules allows to identify new innovative technologies.

In the future, the number of systems under study is expected to increase. This will make it possible to formulate general rules and trends for systems.

The proposed system can also be considered as a multi-intelligent agent system [15].

References

1. Filemon, A., Uriarte, J.: Introduction to Knowledge Management. ASEAN Foundation, Jakarta (2008)
2. Abdel, N., Zaied, G., Hussein, M.: The role of knowledge management in enhancing organizational performance. Int. J. Inf. Eng. Electron. Bus. **4**(5), 27–35 (2012)
3. Keishing, V., Renukadevi, S.: A review of knowledge management based career exploration system in engineering education. Int. J. Mod. Educ. Comput. Sci. (IJMECS) **8**(1), 8–15 (2016)
4. Pahl, G., Beitz, W.: Engineering Design-A Systematic Approach. Springer, Heidelberg (1996)
5. French, M.J.: Conceptual Design for Engineers. Springer, Heidelberg (1985). https://doi.org/10.1007/978-3-662-11364-6
6. Zwicky, F.: Discovery, Invention, Research-Through the Morphological Approach. The Macmillan Company, Toronto (1969)
7. Bardenhagen, A., Pecheykina, M., Rakov, D.: Advanced morphological approach for knowledge-based engineering (KBE) in aerospace. In: Zhengbing, H., Petoukhov, S.V., He, M. (eds.) Advances in Artificial Systems for Medicine and Education II, pp. 495–505. Springer International Publishing, Cham (2020). https://doi.org/10.1007/978-3-030-12082-5_45
8. Ritchey, T.: General morphological analysis as a basic scientific modelling method. Technol. Forecast. Soc. Chang. **126**, 81–91 (2018)
9. Alvarez-Dionisi, L., Mittra, M., Balza, R.: Teaching artificial intelligence and robotics to undergraduate systems engineering students. Int. J. Mod. Educ. Comput. Sci. (IJMECS) **11**(7), 54–63 (2019)
10. Njenga, S., Oboko, R., Omwenga, E., Maina, E.: Use of intelligent agents in collaborative mlearning: case of facilitating group learner interactions. Int. J. Mod. Educ. Comput. Sci. **9**(10), 18–28 (2017). https://doi.org/10.5815/ijmecs.2017.10.03
11. Bardenhagen, A., Rakov, D.: Analysis and Synthesis of Aircraft Configurations During Conceptual Design Using an Advanced Morphological Approach. Deutscher Luft- und Raumfahrtkongress, Berlin (2019)
12. Kablov, E.: At the intersection of science, education and industry. Sci. Technol. **15**(941) (2016)
13. Ngo, T., Kashani, A., Imbalzano, G., Nguyen, K., Hui, D.: Additive manufacturing (3D printing): a review of materials, methods, applications and challenges. Compos. Part B Eng. **143**, 172–196 (2018)
14. Rakov, D., Sukhorukov, R.: Classification and analysis of additive technologies based on the morphological approach. Prob. Mech. Eng. Autom. **4**, 80–86 (2020)
15. Kadhim, M., Afshar Alam, M., Kaur, H.: A multi-intelligent agent system for automatic construction of rule-based expert system. Int. J. Intell. Syst. Appl. **8**(9), 62–68 (2016)

Applying Machine Learning Algorithms to Automatically Classify Emergency Messages

Alexander Shukhman[1(✉)] and Elena Shukhman[2]

[1] Orenburg State University, 460018 Orenburg, Russia
[2] Orenburg State Pedagogical University, 460014 Orenburg, Russia

Abstract. The rapid development of machine learning and natural language processing leads to the possibility of improving the efficiency of unified duty and dispatch services. It is possible to develop algorithms for the automatic classification of emergency messages to three classes: ambulance service, fire service, and police service. In this paper various methods of text classification were studied and compared with each other. The transfer learning method shows the best accuracy with F1-score exceeds 95%. Based on the experimental study a website prototype for handling emergency calls was developed. The site saves information about calls to emergency services, displays them on an interactive map. Authorized operators can add, view, edit, and delete calls. The site supports the REST API to add and view emergency calls. The Yandex Alice voice assistant can be used to call the service.

Keywords: Emergency messages · Natural language processing · Classification

1 Introduction

To unify the method of calling emergency services in the Russian Federation, the Unified Duty and Dispatch Service (UDDS) is created, operating at 112. The creation of a single service solves the problem of a comprehensive assessment of the situation and ensuring the interaction of various structures in cases where the solution to the problem requires the involvement of several services at the same time. In emergency situations, the UDDS-112 system should provide information interaction of the following services: fire and rescue service, police, ambulance, gas service, anti-terror service.

There are problems when, in emergency situations, people cannot immediately get through to the dispatcher, due to the long connection with the operator, which can cost them their lives. To solve this problem, it is possible to develop automated systems based on artificial intelligence, providing a guaranteed answer to the victim's call, even in cases when all service operators are busy. The development of such system requires the use of algorithms that, based on the received voice message from the victim, determine the list of services to which it is necessary to transmit the received information.

A small number of publications are devoted to the classification of emergency messages. The authors of paper [1] use natural language processing based filtering for

Z. Hu et al. (Eds.): AIMEE 2021, LNDECT 107, pp. 152–160, 2022.
https://doi.org/10.1007/978-3-030-92537-6_15

segregation of high priority emergency messages. For message classification support vector machines (SVM) was used. In another research [2] a comparison study of metrics for evaluating relevance and accuracy of emergency messages in social media was carried out. A similar study [3] was carried out to classify Facebook messages of US police departments using support vector machines and convolutional and recurrent neural networks. The authors of paper [4] present a new classification system based on the characteristics of social medial emergency messages about typhoon disasters with according category labels. The research [5] is devoted to development of a methodology to summarize emergency text messages for real-time disasters.

Analysis of publications and practice of the UDDS allows us to conclude that currently no algorithms have been developed for the classification of emergency messages in Russian.

2 Classification Methods of Emergency Messages

An automatic service detection by the emergency message is a special case of the document classification problem. A document classification is a problem to assign a document to one of several predefined categories (classes) based on the document content.

To solve the document classification problem, supervised learning methods are used. A set of decision rules for the classifier is calculated automatically based on the training dataset containing sample documents for each class [6].

The solution to the document classification problem consists of three sequential stages.

1. Preprocessing and vectorization of documents.
2. Building a classifier and training it using machine learning methods.
3. A quality assessment of the classification.

At the first stage of text preprocessing a tokenization is carried out. It selects tokens in text such as words, numbers, punctuation marks, formulas, etc. After that, it is necessary to remove semantically neutral words such as conjunctions, prepositions, articles, etc. [6].

Next, morphological analysis is carried out marking by parts of speech, stemming or lemmatization is performed. Stemming is the process of finding the stem of a word. Lemmatization is the process of bringing a word to its normal (dictionary) form. This allows you to significantly reduce the number of several tokens. As a result, all the significant words found in the document are used as the features.

Document vectorization is the construction of a certain numerical model of the text. It translates the text into a vector representation that is convenient for further processing. For example, the bag-of-words model represents each document as a vector in a multidimensional space, coordinates correspond to word numbers, values are weights. Another vectorization Word2Vec converts each word to a vector that contains information about the context words [7]. Also for vectorization n-grams (sequences of n adjacent words) can be taken into account [7].

For classification, we used both classical machine learning methods based on linear separation of objects (logistic regression, support vector machine SVM [6]) and various neural network classification methods [7]. These methods are widely used in solving other problems of text classification, for example, sentiment analysis [8, 9].

The SVM views a document dataset as a set of points in a multidimensional space. A set of points is called linearly separable if points can be separated into different classes by a hyperplane. In this case all the points of one class lie on one side of this hyperplane, and all the points of another class are on the opposite side. Then, in order to classify unknown points, it will be enough to look at which side of the hyperplane they will be. It is best to choose the hyperplane that is as distant as possible from the points. In SVM the distance between a hyperplane and the nearest point from the point set is maximized. In practice, it is very rarely possible to construct a separating hyperplane. It means that it is impossible to guarantee linear separability of the points. There may be documents that the algorithm assigns to one class, but in reality they should belong to the opposite. These points are called outliers, and they create method bias. Another linear method, logistic regression, is often used [10].

Linear method benefits are of good quality and that a small training dataset is sufficient. The disadvantage of the methods is outlier volatility in source data.

The basic architecture of neural network classifiers is a multilayer perceptron. A multilayer perceptron includes an input layer, an output layer and intermediate layers: the signal goes sequentially from the input layer through intermediate layers to the output.

In recurrent neural networks the neurons of the subsequent layers have connections with the neurons of the previous layers. This architecture leads to the possibility of considering the results of the neural network at the previous stage for processing the input vector at the next stage of the network functioning. The most commonly used architectures of recurrent neural networks are LSTM and GRU [7].

Neural networks provide high accuracy with a successful selection of parameters. The disadvantage of neural networks is the low learning rate and the need for a large amount of training data to achieve high accuracy.

Currently, the highest accuracy of text classification is provided by the transfer learning technique, when a pretrained network is trained on a specific dataset. The basic network is most often taken from the BERT network (bi-directional multilingual model with transformer architecture). We considered the RuBERT network, developed at the Neural Systems and Deep Learning Laboratory of the Moscow Institute of Physics and Technology [11].

The RuBERT network includes a sequence of the following blocks:

- the input embeddings layer forms 768-byte vector representation of a token;
- the encoder consists of 12 blocks of transformers including an attention layer, fully connected layers, and normalization layers;
- the pooler - the final fully connected layer.

RuBERT was trained on the Russian-language section of Wikipedia and news sites and uses a dictionary of about 120 thousand tokens. The network has about 180 million parameters, its total size is 711 MB.

Initially RuBERT includes an output layer that predicts masked words in text. To solve the classification problem, it was replaced by an output layer with three outputs.

3 Experimental Study of Classification Methods of Emergency Messages

For an experimental study of text classification methods based on the research of audio recordings of calls to the UDDS-112 of the Orenburg region, a dataset was created, including messages for three services – ambulance service, fire and rescue service and police service. In total, the dataset included 1200 messages ranging in length from 10 to 50 words. The average message length is 18 words. Each message is located in a separate text file in utf-8 encoding. Messages for the ambulance service are placed in the Ambulance folder, for the fire service in the Fire folder, for the police service in the Police folder.

The main criteria for assessing the quality of the classification are the combination of precision and recall. The precision of classification within a class is the proportion of documents found by the classifier that actually belong to this class, relative to all documents that the system has assigned to this class. The recall of a classification is the proportion of documents found by the classifier that actually belong to a class, relative to all documents of this class in the test sample. Let us introduce the notation TP - true positive decision; TN - true negative decision; FP - false positive decision; FN is a false negative decision. The precision is now determined as follows:

$$p = \frac{TP}{TP + FP} \tag{1}$$

The recall is determined as follows:

$$r = \frac{TP}{TP + FN} \tag{2}$$

F-score is a characteristic of the quality of the algorithm, which combines information about the precision and recall:

$$F_\beta = \frac{(\beta^2 + 1)pr}{\beta^2 p + r}, \; where\, 0 \le \beta < \infty \tag{3}$$

At $0 \le \beta < 1$, precision is more important.

For $\beta = 1$, precision and recall are equal, then $F_\beta = \frac{2pr}{p+r}$.

For $1 \le \beta < \infty$, recall is more important.

Comparison of classification methods is a rather difficult task due to the fact that different input data can lead to different results. To compare different classifiers, it is necessary to construct them and calculate their efficiency on the same sets of documents for training and testing phases. For the experimental study, we used the Python 3.8, and scikit-learn, keras, and pytorch machine learning libraries. Before training the classifiers, the dataset was loaded into memory and divided into training and test samples in a 3: 1 ratio using the train_test_split function.

At the first stage it was necessary to construct vector embeddings of messages for classical linear classifiers. The text embeddings were constructed based on the CountVectorizer class (the bag-of-words model).

For neural network methods, the embeddings were formed using the input layer of neural network based on a list of dictionary numbers of message tokens.

After that, the transformed texts were used to train the following classifiers:

1) Logistic regression LogReg [12].
2) Support vector machine SVM [12].
3) Multilayer perceptron MLP (with two hidden layers) [7].
4) Recurrent neural network LSTM [7].
5) Recurrent neural network GRU [7].
6) Pretrained neural network RuBERT [11].

The optimal values of the model parameters were selected to obtain the maximum value of the F1-score. So, for the multilayer perceptron, the sizes of the hidden layers were selected as follow: 200 and 80 neurons. The experimental results are shown in the Table 1.

Table 1. Classifiers comparison results

Model	Precision	Recall	F-score	Accuracy
LogReg	88,4%	83,6%	85,2%	90,0%
SVM	90,1%	86,1%	87,5%	91,1%
MLP	88,9%	88,2%	88,3%	90,8%
LSTM	84,3%	82,5%	83,2%	87,8%
GRU	88,3%	83,2%	84,4%	87,5%
RuBERT	96,2%	94,4%	95,2%	96,3%

It can be concluded that linear methods for relatively small datasets give a high quality of classification results, comparable or superior to the results of neural networks. The best quality is achieved with the transfer learning method based the RuBERT pretrained neural network.

4 Development of a Website Prototype for Handling Emergency Calls

Based on experimental study, a website prototype for automate handling emergency calls was developed. For development were used the following tools: the Python 3.8, the Flask web-framework [13], and the PostgreSQL database management system.

The site saves information about calls to emergency services, displays them on an interactive map (Fig. 1). For an unauthorized user, only call map view, registration and login are available. The call map is made using the Yandex Map API, it displays all emergency calls for which the status is not "completed". The map is automatically scaled when the page is refreshed to display all calls. Various specially selected icons are used to call the police, the ambulance and fire brigade.

Authorized operators can add, view, edit, and delete calls. An authorized operator can click the call icon and view a link to open the call edit page in a new tab.

When adding a call, only the message and address are filled in. The called service is automatically detected based on the pre-trained classifier.

Addresses are checked for their existence using the Yandex Geocoder API.

The time in the database is stored in UTC, the operator's time zone is determined automatically based on the time in his/her browser using the flask-moment library. On the pages, the time is displayed in the operator's time zone.

The address in the call list is displayed as a geo-link; when you click on the address, a window with a map will pop up (Fig. 2).

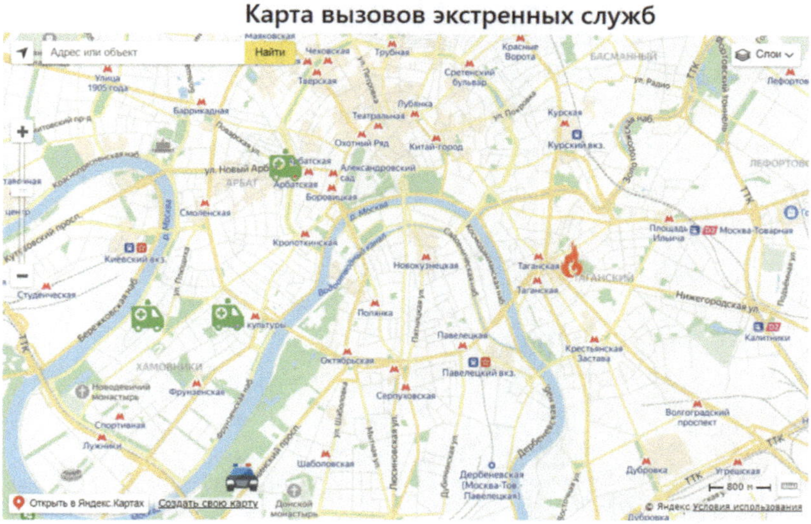

Fig. 1. The interactive call map

The button in the call line on the right opens the call editing page.

The call editing page displays full information about it, you can change all the fields, including the call service, status, address, and message. The call editing page also displays a map using the Yandex Map API (Fig. 3).

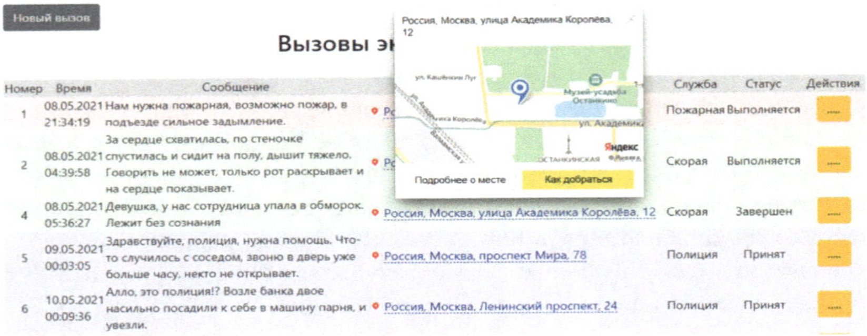

Fig. 2. The call list

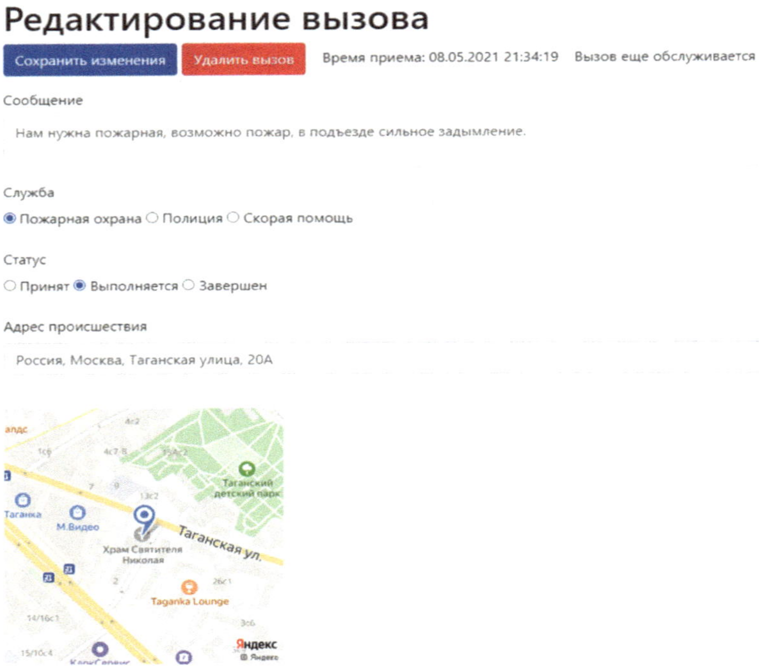

Fig. 3. The call editing page

The site supports REST API access to add and view emergency calls. The Yandex Alice voice assistant can be used to call the service.

The skill for the Yandex Alice introduces itself and offers to report what happened. If the number of tokens in the message is less than 8, then it is proposed to talk about the incident in more detail. Then you are asked to provide the address. If Alice did not find the geo object, it contains incomplete information, or the address did not pass verification through the geocoder API, then it is suggested to repeat it.

If the message and address are received, then the call is added to the database and information about the call is sent.

5 Summary and Conclusion

Studying the work of UDDS operators made it possible to define the problem for automation - recording a user's message and automatically determining an emergency service for processing such message. Based on the study of educational and scientific sources, the analysis of text classification algorithms was carried out; their advantages and disadvantages are highlighted. To train text classification algorithms, a dataset of 1200 emergency messages was generated. An experimental comparison of the operation of six classification algorithms on the created dataset was carried out, the most accurate algorithm based on the transfer learning method is selected, for which the F1-score exceeds 95%.

Based on the experimental study a website prototype for handling emergency calls was developed. The site saves information about calls to emergency services, displays them on an interactive map. Authorized operators can add, view, edit, and delete calls. The site supports the REST API to add and view emergency calls. The Yandex Alice voice assistant can be used to call the service.

It is planned to supplement the dataset for emergency municipal services and make the system integration with digital telephony systems.

Acknowledgment. This project is supported by the RFBR (№ 20-07-01065) and by the grant of the President of the Russian Federation (NSH-2502.2020.9).

References

1. Bhattacharjee, S., et al.: Best-effort delivery of emergency messages in post-disaster scenario with content-based filtering and priority-enhanced PRoPHET over DTN. In: 2016 8th International Conference on Communication Systems and Networks (COMSNETS), pp. 1–7 (2016)
2. Kikin, P.M., Kolesnikov, A.A., Panidi, E.A.: Social media data processing and analysis by means of machine learning for rapid detection, assessment and mapping the impact of disasters. Int. Arch. Photogrammetry Remote Sens. Spatial Inf. Sci. **43**, 1237–1241 (2020)
3. Pogrebnyakov, N., Maldonado, E.: Identifying emergency stages in Facebook posts of police departments with convolutional and recurrent neural networks and support vector machines. In: 2017 IEEE International Conference on Big Data (Big Data), pp. 4343–4352 (2017)
4. Yu, J., Zhao, Q., Chin, C.S.: Extracting typhoon disaster information from VGI based on machine learning. J. Marine Sci. Eng. **7**(9), 318 (2019)
5. Nelson, C., Pottenger, W.M.: Optimization of emergency response using higher order learning and clustering of 911 text messages. In: 2013 IEEE International Conference on Technologies for Homeland Security (HST), pp. 486–491 (2013)
6. Zhang, T., Oles, F.J.: Text categorization based on regularized linear classification methods. Inf. Retrieval **4**(1), 5–31 (2001)

7. Goldberg, Y., Hirst, G.: Neural Network Methods in Natural Language Processing. Morgan & Claypool Publishers, San Rafael (2017)
8. Rahman, A., Sadat, M., Siddik, S.: Sentiment analysis on twitter data: comparative study on different approaches. Int. J. Intell. Syst. Appl. **13**(4), 1–13 (2021)
9. Shaheen, M., Awan, S.M., Hussain, N., Gondal, Z.A.: Sentiment analysis on mobile phone reviews using supervised learning techniques. Int. J. Modern Educ. Comput. Sci. **11**(7), 32–43 (2019)
10. Al-Tahrawi, M.M.: Arabic text categorization using logistic regression. Int. J. Intell. Syst. Appl. **7**(6), 71 (2015)
11. Kuratov, Y., Arkhipov, M.: Adaptation of deep bidirectional multilingual transformers for Russian language. In: 2019 Annual International Conference on Computational Linguistics and Intellectual Technologies, pp. 333–339 (2019)
12. Raschka, S., Mirjalili, V.: Python Machine Learning: Machine Learning and Deep Learning with Python, Scikit-Learn, and TensorFlow. Packt Publishing, Birmingham (2017)
13. Grinberg, M.: Flask Web Development: Developing Web Applications with Python. O'Reilly Media, Newton (2018)

Categorical Model of a Plant as a System

George K. Tolokonnikov[1(✉)], V. I. Chernoivanov[1], Yu. Kh. Shogenov[2], and A. S. Dorokhov[1]

[1] Federal Scientific Agro-Engineering Center VIM, Russian Academy of Sciences, Moscow, Russia
[2] Russian Academy of Sciences, Moscow, Russia

Abstract. In this work, on the basis of the categorical theory of systems, using the apparatus of categorical splices, the construction of a mathematical model of a plant as a categorical functional system has begun, at the present level of development of the system approach. The categorical theory of systems formalizes the intuitive constructions of functional systems and, for the case of plants, makes it possible, on a strictly mathematical basis, to implement the remaining undeveloped approach to plants as functional systems, analogous to a large extent to the construction for humans and animals performed by P.K. Anokhin and his school. A general framework of a categorical model of a plant as a system is given and further steps for its development are outlined.

Keywords: Categories · Splices · Categorical systems · Functional systems · Biomachsystems · Artificial intelligence

1 Introduction

The work is devoted to the modeling of plants, considered as systems, in the framework of the categorical theory of systems, which to a large extent formalized the theory of functional systems of P.K. Anokhin. The study of higher nervous activity on the basis of functional systems [1] and categorical systems theory serves as an outlet for modeling strong artificial intelligence. It is possible that the study of plant intelligence within the framework of plant neurobiology [2, 3] and a systematic approach to plant neurobiology will also be useful for modeling artificial intelligence.

Outstanding advances in the theory of functional systems led to a full-scale revision of the traditional organ approach and the construction of human and animal physiology on the basis of functional systems [4]. However, for such living organisms as plants, a similar construction of their physiology has not been carried out, since many systemic mechanisms of plants are insufficiently studied, and the existing studied mechanisms do not always fit naturally into the ideology of the theory of functional systems. Categorical systems cover functional systems as special cases (categorical functional systems), describe other known systems, and also give examples of possible other systems, in particular, they should contain models of systems that correspond to plants as integral organisms. Considering the individual cells of a plant as a system, we represent the plant itself as a system with subsystems in the form of these individual cells. However, the systemic hierarchy of a plant is much more complicated, for

Z. Hu et al. (Eds.): AIMEE 2021, LNDECT 107, pp. 161–169, 2022.
https://doi.org/10.1007/978-3-030-92537-6_16

example, a plant contains tissues (educational, integumentary, conductive, mechanical, and others), the first classification of which was proposed in [5], and tissues, as proved in [6], are histions, that is, the selected subsystems of cells, formed according to the laws of the theory of the division of labor [6]. In the categorical theory of systems, as described below, a system of future subsystems is formed with the help of a convolution operation that has a strict mathematical definition, formalizing the system-forming factor according to P.K. Anokhin, while in ordinary expositions the authors use intuitive ideas about the concept of a system, argue about assembling a system from a subsystem in intuitive terms. In particular, understanding that a multicellular organism cannot do without the ordered interaction of its parts, one speaks about the integration of plant organs, about the coordination of their activity [7]. The carriers of coordination and integration are hormones (auxins, gibberellins, peptides, and other hormones) and bioelectric signals (action potentials of APs and resting, variable potentials) [8]. In the categorical theory of systems, these hormones and bioelectrosignals are modeled by splices (for electrical signals, they are modeled by directional connections, similar to the models of neurons in neural networks modeled by convolutional polycategories of the crown type [9]). The intuitive concept of an organism is rather vague, so in [10] plants, although they are declared "multicellular eukareotic systems," are not considered organisms.

APs were discovered by Bose at the beginning of the 20th century on mimosa, his work was continued in the second half of the 20th century and the beginning of the 21st century [11–16], APs are being intensively studied in our time [17]. It is interesting to note that Bose concluded that plants have an electromechanical pulse, a nervous system, a form of intelligence and are able to memorize and learn, which remains controversial until now, but is supported within the framework of plant neurobiology [2, 3].

Many environmental influences require a quick reaction from the plant (caterpillars eating leaves, fire, etc.), therefore, first of all, coordination and integration should be manifested in the fastest signals in plants, these are the action potentials of APs, but there are unstable data in the form of experimental results observation of ultrafast APs of animals comparable to APs [18, 19]. This is a fundamental question, since slow single APs cannot transfer information about the nature of the impact, which is provided in animals by packs of spikes.

As in animals, APs is generated due to ion fluxes through ion channels (in the plasmalemma), there are differences, the question is under study. In [8], the presence of any information processing center in the plant is not presupposed, and the APs, spreading through the plant, are themselves part of the plant's response to the stimulus, "transfer stress", which excites the response of other parts of the plant to the stimulus. According to [8], the role of the spreading AP is reduced to the coordination of various types of responses to stimuli available in the plant.

In the cited articles, there is no account of the main systemic artifact of the categorical theory of systems, the theory of functional systems and biomachsystems [20], the presence of a system-forming factor as a result useful for the organism. It is not discussed where in the plant the information about the result necessary for the.

plant is formed and stored, when the reaction to the stimulus is carried out, the actions of the plant are not discussed if the reaction that is essential to them did not lead to the result that the plant needs.

The same drawback occurs in the article [21], where the author refers to N. Wiener, and not to P.K. Anokhin, whose works were already in the 1930s (30 years before N. Wiener, who publicly recognized the priority on cybernetics for P.K. Anokhin), nevertheless, in [21], a step was made to consider the plant as a system that necessarily has receptors, efferent and afferent pathways of excitation, central regulatory elements, effectors and feedback elements between receptors and effectors.

"Command centers of integration" do not have an established understanding of their nature and location in plants. These are apical meristems of the shoot and root, nodes of the stem, apexes of the shoot and root.

There is a hypothesis about the presence in a plant of several relatively autonomous subsystems in the form of phytomers: "In the process of initiation and development of phytomere elements (node, internode, leaf, bud, root), elements of the control system are formed in the form of receptors, pathways for conducting excitation, central regulatory elements, executive elements and elements of feedback between receptors and effectors" [7].

It is proposed to consider sclerenchyma cells as central regulatory elements, which are endowed with a similar function in addition to traditional mechanical properties. There is even an outward resemblance of some types of these cells to neurons (see Fig. 1).

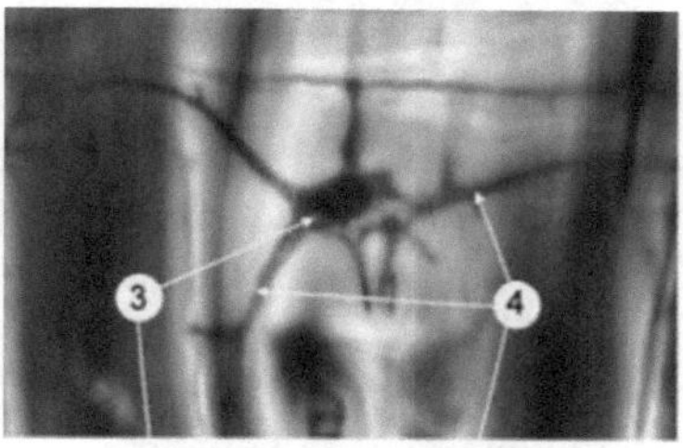

Fig. 1. Similarity of the sclerenchymal cell to the neuron: 3 - cell body, 4 - long processes of the cell

In the recently formed science of plant neurobiology [2, 3], the study of the functions of electrical signals at large plant-scale distances and their role in the regulation of plant responses, the role of homologous plant molecules, like neuroreceptors and neurotransmitters, in the nervous system of animals is defined as the main tasks. Ultimately, the challenge is to gain an understanding of how plants receive and integrate information in order to coordinate responses that affect the entire plant as a whole and as a system. Such tasks are of direct interest for crop production (see, for example, [12–15]). In plants, among the metabolic neurotransmitters are acetylcholine, catecholamines, histamines, serotonin, dopamine, melatonin, GABA (g-aminobutyric acid) and glutamate, which in the nervous system of animals play a role in perception, movement, vision, information processing and development. The tasks of plant neurobiology include elucidating the metabolic or signaling role of these compounds in plants [22]. In particular, there is convincing evidence that glutamate is a signaling molecule in plants due to the discovery of glutamate receptors in plants [23]. Moreover, genes similar to the genes of glutamate receptors in the nervous system of animals were found in plants, for example, Arabidopsis has 20 such genes [24]. Auxin, which plays an important role in plant metabolism, which is transported from cell to cell, also has characteristics that resemble neurotransmitters, in particular, there is a similarity between the release of auxin and neurotransmitter from neuronal cells [25]. It is known that auxin induces rapid electrical reactions when applied extracellularly [19], the role of auxin can be seen in a new light from the point of view of plant neurobiology. One hypothesis is that auxin molecules secreted through auxin-enriched vesicles induce electrical responses in neighboring cells within a few seconds [19]. This electrical activity resembles signaling molecules with neuro-like properties.

These ideas of Bose about the intelligence of plants and the well-known ideas of Charles Darwin, who called plateau eating plants "smart animals", have been developed up to attributing consciousness to plants [26], which is undoubtedly interesting from the point of view of artificial intelligence, but we do not stop there.

An important modern direction of modeling plants as systems is the issues of networks, they are also close to many methods of modern artificial intelligence [27–31].

The systems themselves are often described in the form of networks, for example, in an extensive survey [32], a system is called, referring to L. Bertalanffy, a connected graph (a set of vertices and pairs of vertices, a pair of vertices is called an edge): "… the system is represented as a set of nodes, which are connected by edges that indicate relationships or interactions between these elements. By definition, all nodes of these networks must be connected to at least one other node by at least one edge so that the systems represent a "single whole". The way in which edges connect nodes in a given network is called network topology… As von Bertallanffy showed in his general systems theory, all systems are connected to networks". L. Bertalanffy's approach to the concept of a system is of an intuitive nature and at the same intuitive level was subjected to exhaustive criticism by P.K. Anokhin [33], which does not allow calling by systems, those systemic formations in biology that L. Bertalanffy discusses. The main weakness of L. Bertalanffy's approach is the absence, apart from a number of other systemic attributes, the main attribute, the system-forming factor.

As we have indicated, the intuitive approach of PK Anokhin, including the system-forming factor, is formalized within the framework of categorical systems theory [33].

It is usually believed that the above definition also describes artificial neural networks, which are not really the above graphs and for which a model without inaccuracies of the engineering level of rigor was found in [9] in the form of a convolutional polycategory with convolutions of the corona type. In the next section, precise definitions of categorical splices are given and, on their basis, a definition of the categorical system is given. The third section deals with the categorical plant model. In the conclusion, the results are summed up and further steps for the development of this direction are outlined.

2 Categorical Splices and Categorical Systems

Consider a countable number of varieties with variable and functional projection symbols $x^{(i)}{}_k$, $C^{(j)}$, $i,j,k = 1,2,\ldots$. We introduce functional symbols ξ_{ij} with properties $\xi_{ij}\xi_{ji}\, x^{(i)}{}_k = x^{(i)}{}_k$. For each i, we denote $C^{(j)}{}_i = \xi_{ij}C^{(j)}\xi_{ji}$ and introduce the axioms $C^{(k)}{}_i$ $C^{(j)}{}_i = C^{(l)}{}_i$. Thus, the function symbols form an algebra, with the multiplication table corresponding to the last axioms.

Introduce the notation for the formulas $Ob^{(i)}(x^{(i)}{}_k) = (x^{(i)}{}_k =_i C^{(i)}(x^{(i)}{}_k))$ as well as variables $a^{(i)}{}_k$, $k = 1,2,\ldots$ of sort i, bounded by scope $Ob^{(i)}(x^{(i)}{}_k)$. We introduce equality predicates of the form $(x^{(i)}{}_k =_{ij} x^{(j)}{}_k) = (x^{(i)}{}_k =_i \xi_{ij}\, x^{(j)}{}_k)$. For each sort, we introduce a set of basic functional symbols $\mu^{(i)}{}_l$, $\nu^{(i)}{}_l$, $l = 1,2,\ldots$ called n-ary operations, $n = 0,1,2,3,$ … and a set of ***identities*** as equalities of terms of the same kind $t^{(i)}{}_m = t^{(i)}{}_k$. We enter the countable number $\alpha = 1,2,3,\ldots$ of the specified sets of sorts, to indicate it we add an additional index $x^{(i,\alpha)}{}_k$, $C^{(i,\alpha)}$, $\mu^{(i,\alpha)}{}_l$, $=_{i,\alpha,i',\alpha'}$, $\xi^{(\alpha)}{}_{ij}$.

Definition. A signature is a set of functional letters of projections and operations $C^{(j)}$, $\mu^{(i)}{}_l$.

We introduce copies of varieties.

$$x_k^{(i,\alpha)}, C^{(i,\alpha)}, \mu_l^{(i,\alpha)}, \ldots, =_{i,\alpha,i',\alpha'}, \xi_{ij}^{(\alpha)}, x_k'^{(i,\alpha)}, C'^{(i,\alpha)}, \mu_l'^{(i,\alpha)}, \ldots, ='_{i,\alpha,i',\alpha'}, \xi_{ij}'^{(\alpha)}, x^{(i,\alpha)}{}_k,$$
$$C^{(i,\alpha)}, \mu^{(i,\alpha)}{}_l, \ldots, =_{i,\alpha,i',\alpha'}, \xi_{ij}^{(\alpha)}, x'^{(i,\alpha)}{}_k, C'^{(i,\alpha)}, \mu'^{(i,\alpha)}{}_l, \ldots, ='_{i,\alpha,i',\alpha'}, \xi'^{(\alpha)}_{ij},$$
$$x'^{(i,\alpha)}{}_k, C'^{(i,\alpha)}, \mu'^{(i,\alpha)}{}_l, \ldots, ='_{i,\alpha,i',\alpha'}, \xi'^{(\alpha)}_{ij}, x^{(i,\alpha)}{}_k, C^{(i,\alpha)}, \mu^{(i,\alpha)}{}_l, \ldots, ='_{i,\alpha,i',\alpha'}, \xi'^{(\alpha)}_{ij}.$$

Definition. A complete base splice is a formula made up of conjunctions of formulas $a^{(i,\alpha)}{}_k =_i C^{(i,\alpha)}(x^{(i,\alpha)}{}_k)$, $a'^{(i,\alpha)}{}_k = '_i C'^{(i,\alpha)}(x'^{(i,\alpha)}{}_k)$, $a'^{(i,\alpha)}{}_k = '_i C'^{(i,\alpha)}(x'^{(i,\alpha)}{}_k)$, $a^{(i,\alpha)}{}_k =_i C^{(i,\alpha)}(x^{(i,\alpha)}{}_k)$ and equals are any of the symbols of the specified sorts, the equals sign has the corresponding indices omitted here.

Definition. We will distribute the conjunctions for each formula of full splices in four cells of the 2×2 table, in cell (1,1) we place the formulas of outer splices. In cell (1,2) we place the formulas of outer convolutions. In cell (2.1) we place the formulas of inner splices. In cell (2,2) we place the formulas of inner convolutions:

u	v'
$\overline{w}$	$\overline{r'}$

By the construction of formulas for complete splices, the equality subformulas contain only those letters of the variables that occur in the subformulas of the form $a = C(x)$. This allows the equality signs (formulas with equal variables) to be excluded from the table, replacing the equality formulas with dotted lines connecting the letters of the variables included in the equalities.

Let us define inference rules specific for categorical splice M_t, M_r, M_l, M_b, which translate some formulas of complete splice into others by moving the formulas between cells of the table of full splice. M_b moves to the bottom cells from the top those conjunctions from formulas that have connections, when moving, a change in varieties is specified that meets the definition on M_b. M_t moves to the upper cells from the lower ones those formulas that have compounds. M_l moves linked vertically and within a pair from right to left cells. M_r moves connected pairs from left to right cells. The rule of derivation Ev is introduced, and its inverse Ev^{-1}, based on the function symbols of operations $\mu^{(i,\alpha)}{}_l$ and the placement of = signs in the corresponding places. The rules of projection derivation Pr_{ij}, i,j = 1,2 are introduced, transforming the formula for full splices into the formula for ij-cells with the removal of dashes connecting the cells. The application of the convolution available in cell (1,2) of the splices table consists of the sequential application of the inference rules M_b, M_l, Ev. The reverse operation of the indicated one consists of sequential application of the rules Ev^{-1}, M_t, M_r, inference rules.

Definition. A complete splice is called a basic splice in which the variables x (with indices) are replaced by the terms of the formal signature language of the functional letters of operations and projections.

Definition. The procedure for applying convolution, indicated above for full base splices, is literally applicable to full splices and gives an inductive construction of formula terms: full base splice is a formula term; if there is a formula term, then the application of the convolution operation to it is also a formula term and there are no other terms.

Now we can define the categorical splices theory.

Definition. The formal theory of categorical splices is a signature, a set of identities together with a set of formula terms constructed above.

With the help of the formal theory of categorical splices, we define categorical systems as follows.

When considering models of categorical splices, as in the case of categories [4], it is possible to emphasize the difference between the axiomatic definition and the model by the prefix "meta", that is, to call, for example, the meta-splice for formal theory, and to use the term splice for the model.

Definition. Let categorical splices be given as a model of formal categorical meta splices. Then any splice of a categorical splices is called a categorical system or simply a system. A system is called composite if it is representable in the form of convolutions of other systems, called subsystems of the original system. If a system cannot be represented as a convolution of other systems, then it is called a simple system.

The system-forming factor for each system K in a given categorical splice is determined by the result, that is, by the system R and the convolution Δ, which acts on the set of systems K_i, i = 1,2,… and gives the system K, which has a subsystem R.

3 Categorical Model of a Plant to be taken as a System

In contrast to the above definition of a plant as a system according to L. Bertalanffy on the basis of networks [32], we will use strict concepts to consider the initial stages of constructing a model of a plant as a categorical system free from the above disadvantages in the approaches in [32].

As simple (non-composite) systems, at the first step, we take individual cells of the plant selected for modeling. In the future, it is supposed to continue the hierarchy of systems, decomposing individual cells into subsystems. Note that a step in this direction is the categorical theory of chemical molecules proposed in [35], which includes biomolecules of DNA and RNA.

The plant cells themselves form a set in which the interpretation of the formal theory is given. A set of copies of this set is selected to implement the interpretation of the required number of sorts of variables. Next, it is necessary to define interpretations for the functional symbols of the projections. To do this, it is necessary to specify which connections and inputs-outputs for directional cell connections should be specified. Among those already discovered and studied, these links include hormones (auxins, gibberellins, peptides, and other hormones) and bioelectric signals (action potentials of AP and rest of PP, variable potentials of EP). The appearance of new links and new properties or the refinement of the properties of links in this list will make the model more adequate to a real plant. After describing simple systems as cones of categorical splice, one should proceed to the description of convolutions from which system-forming factors for subsystems of various levels will be formed in particular, the convolution for the formation of histions corresponding to the tissues available in the plant according to the theory of division of labor for cells [6]. These steps require further study of the subject area of plant physiology, for example, histions have been studied mainly for animal tissues, and for plants there is only fragmentary information.

At the initial stage, it is possible not to take into account the subordination of the plant, as a system, to the system-forming factor of the superior system and, ultimately, the system-forming factor of the population, which is also considered as a categorical functional system.

Detailed approaches in the theory of functional systems for animals and humans can be used as a guiding analogue in constructing a model. Similarly to the constructed functional systems of nutrition, respiration and other subsystems for the plant model, it is necessary to construct categorical functional systems characteristic of plants (respiration, plant nutrition and other subsystems). At the same time, based on the study of the subject area for constructing the dynamics of plant development, sets of subsystems are distinguished that implement in a categorical form a set of individual systems quanta according to K.V. Sudakov [4], which are intermediate categorical functional systems for achieving the next stage result. On the discussed path of constructing a model, the lack of information allows us to speak only of approximate models, which for certain practical purposes, for example, in crop production, may be sufficient.

In the plant growing industry, many machines and mechanisms are used, and the plants themselves become subsystems of the biomachsystem "man-machine-living" [20], the higher biomachsystem begins to influence the plant, changing its system-forming factor accordingly.

4 Conclusion

The work began to build a mathematical model of a plant as a categorical functional system, at the current level of development of a systemic approach. The model corresponds to the implementation of the remaining unconstructed approach to plants as functional systems, largely carried out by P.K. Anokhin and his school for humans and animals. A general framework of a categorical model of a plant as a system is given and further steps for its development are outlined. The most important stage in the development of the model is the elucidation of the nature of electrical signals in the plant, in particular, in the direction of the experimental and theoretical works, and the functions that they perform for the plant organism in its life.

References

1. Tolokonnikov, G.K.: Functional systems integrated with a universal agent of artificial intelligence and higher neurocategories[J]. Adv. Intell. Syst. Comput. **2006**, 3–12 (2021)
2. Trewavas, A.: The foundations of plant intelligence. Interface Focus **7**, 20160098 (2017). https://doi.org/10.1098/rsfs.2016.0098
3. Brenner, E.D., et al.: Plant neurobiology: an integrated view of plant signaling. Trends Plant Sci. **11**(8), 413–419 (2006)
4. Sudakov, K.V.: Physiology. Fundamentals and functional systems. Medicine, 784 (2000)
5. Haberlandt, G.: Physiological plant anatomy[M]. Macmillan, London (1914)
6. Savostyanov, G.A.: Structural foundations of developmental biology and 3D histology: a new approach to the study of carcinogenesis. Petersburg, Lema, 832 (2020)
7. Stepanov, S.A.: The nervous system of plants: hypotheses and facts. Bul. Bot Garden Sarat. **15**(4), 31–56 (2017)
8. Pyatygin, S.S.: Comparative characteristics of action potentials in animals and higher plants. J. General Biol. **69**(1), 72–77 (2008)
9. Tolokonnikov, G.K.: Convolution polycategories and categorical splices for modeling neural networks. In: Hu, Z., Petoukhov, S., Dychka, I., He, M. (eds.) Advances in Computer Science for Engineering and Education II, pp. 259–267. Springer International Publishing, Cham (2020). https://doi.org/10.1007/978-3-030-16621-2_24
10. Gamaley, Y.: Supercellular plant organization. Russ. J. Plant Physiol. **44**(6), 706–730 (1997)
11. Gunar, I.I., Sinyukhin, A.M.: Propagating excitation wave in higher plants. DAN SSSR **142** (4), 954–956 (1962)
12. Shogenov, Y., et al.: Mathematical modeling of electrical signals in the conducting system of a plant. Izvestiya TSKhA **2**, 114–128 (1999)
13. Vasilyev, V.A., et al.: Light induced electrical activity of green plants. Biophysics **48**(4), 662–671 (2003)
14. Shogenov, Y.K., et al.: Effect of monochromatic electromagnetic irradiation in the wavelength range of 330–3390 nm on plant bioelectric activity. Russ. J. Plant. Physiol. **46**(5), 697–703 (1999)
15. Mironova, E.A., et al.: Bioelectric responses of plants to the low-intensive irradiation in the visible and infrared ranges. In: Proceedings of SPIE (Society of Photo-Optical Instrumentation Engineers), Moscow, pp. 349–352 (1998)
16. Opritov, V.A., Pyatigin, S.S., Retivin, V.G.: Bioelectrogenesis in higher plants. In: Nauka, M., p. 213 (1991)

17. Zimmermann, M.R., Mithöfer, A., Will, T., Felle, H.H., Furch, A.C.U.: Herbivore-triggered electrophysiological reactions: candidates for systemic signals in higher plants and the challenge of their identification. Plant Physiol. **170**(4), 2407–2419 (2016). https://doi.org/10.1104/pp.15.01736
18. Volkov, A.G., Collins, D.J., Mwesigwa, J.: Plant electrophysiology: pentachloro- phenol induces fast action potential in soybean. Plant Sci. **153**(2), 185–190 (2000)
19. Steffens, B., et al.: The auxin signal for protoplast swelling is perceived by extracellular ABP1. Plant J. **27**(6), 591–599 (2001)
20. Chernoivanov, V.I.: Biomachsystems: emergence, development of perspective. Biomachsystems **1**(1), 7–58 (2017)
21. Zubkus, O.P.: Peculiarities of generation of electrical impulses by plants. Izvestiya. Sib. dep. Acad. Sci. USSR. Ser. Biol. Sci. **5**, 120–124 (1986)
22. Brenner, E.D.: Drugs in the plant. Cell **109**(6), 680–681 (2002)
23. Lam, H.M., et al.: Glutamate-receptor genes in plants. Nature **396**(6707), 125–126 (1998)
24. Lacombe, B., et al.: The identity of plant glutamate receptors. Science **292**, 1486–1487 (2001)
25. Baluška, F., Volkmann, D., Menzel, D.: Plant synapses: actin-based adhesion domains for cell-to-cell communication. Trends Plant Sci. **10**(3), 106–111 (2005)
26. Taiz, L., et al.: Plants neither possess nor require consciousness. Trends Plant Sci. **24**(8), 677–687 (2019)
27. Dharmajee Rao, D.T.V., Ramana, K.V.: Winograd's inequality: effectiveness for efficient training of deep neural networks. Int. J. Intell. Syst. App. **10**(6), 49–58 (2018). https://doi.org/10.5815/ijisa.2018.06.06
28. Hu, Z., Tereykovskiy, I.A., Tereykovska, L.O., Pogorelov, V.V.: Determination of structural parameters of multilayer perceptron designed to estimate parameters of technical systems. Int. J. Intell. Syst. App. (IJISA) **9**(10), 57–62 (2017)
29. Awadalla, M.H.A.: Spiking neural network and bull genetic algorithm for active vibration control. Int. J. Intell. Syst. App. **10**(2), 17–26 (2018). https://doi.org/10.5815/ijisa.2018.02.02
30. Abuljadayel, A., Wedyan, F.: An approach for the generation of higher order mutants using genetic algorithms. Int. J. Intell. Syst. App. (IJISA) **10**(1), 34–35 (2018)
31. Kasihmuddin, M.S.M., Mansor, M.A., Sathasivam, S.: Bezier curves satisfiability model in enhanced hopfield network. Int. J. Intell. Syst. App. **8**(12), 9–17 (2016)
32. Lucas, M., Laplaze, L., Bennett, M.J.: Plant systems biology: network matters. Plant. Cell. Environ. **34**, 535–553 (2011)
33. Anokhin, P.K.: Fundamental Questions of the General Theory of Functional Systems, pp. 5–61. Princ. Syst. Organ. Funct., Moscow (1973)
34. Tolokonnikov, G.K.: Informal theory of categorical systems. Biomachsystems **2**(4), 41–144 (2018)
35. Tolokonnikov, G.K.: Modelling biomolecular structures in categorical systems theory. EPJ Web. Conf. **248**, 1–6 (2021)

Advances in Medical and Technological Approaches

Determination of Race Using the FFRACE Software Package

Victor N. Zvyagin[1], Elena E. Fomina[2](✉), Olga I. Galitskaya[1], Nina V. Narina[1], and Lyudmila L. Usacheva[1]

[1] Russian Center of Forensic Medicine of the Ministry of Health of Russia, Moscow 125284, Russian Federation

[2] Tver State Technical University, Tver 170026, Russian Federation

Abstract. Information technologies are used in all spheres of public life. They are also effectively introduced into the work of forensic experts and criminologists. The use of specialized software can significantly increase the efficiency of solving practical problems and the scientific validity of the conclusions obtained as a result of an expertise. This article describes the algorithm and software package "FFRACE", designed to determine the race of an individual (Caucasoid or Mongoloid) by eighteen external identification features of the head. The proposed algorithm is based on the step-by-step Bayes procedure modified by A.A. Genkin. The software complex, developed on the basis of the constructed algorithm, includes a module for inputting initial information, a module for fixing external identification features of the head, and a module for racial diagnostics. The result of the race determination in the software package is an expert opinion containing a conclusion about the race of the individual. The software package is intended for use both in laboratory research and at a scene of an accident.

Keywords: Software · Bayesian procedure · Race determination · External identification features of the head

1 Introduction

Information technologies are actively used in all spheres of public life. Forensic medicine and criminalistics are no exception [1, 2].

The introduction of specialized software into the practice of forensic experts will increase both the efficiency of solving practical problems and the scientific validity of the conclusions obtained as a result of an expertise [3, 4].

One of the important tasks in forensic medicine and forensic science is the task of describing the general characteristics of a person's appearance [5]. In particular, an analysis of the external appearance is necessary when registering corpses or living persons, as well as when searching for missing persons or, when restoring a face from a skull. Along with the determination of gender, age, body size, an important stage is the establishment of race.

Race is assessed by anthropologists, forensic scientists, and police experts. Races differ from each other in anatomical, physiological, biochemical and other

Z. Hu et al. (Eds.): AIMEE 2021, LNDECT 107, pp. 173–183, 2022.
https://doi.org/10.1007/978-3-030-92537-6_17

characteristics. Races are interconnected by intermediate types, which make it difficult to diagnose.

Several racial diagnostic techniques are currently known in forensic medicine.

The paper [6] describes a method for establishing Caucasian-Mongoloid affiliation by size and head markers, based on one-dimensional discriminant analysis for both intact and fragmented skulls. The disadvantages of this method include the complexity of calculations associated with the lack of software implementation, as well as the lack of practical application with an incomplete set of features included in the discriminant functions [6].

The work [7] discusses the validity of the classification of people into separate racial groups on the basis of their skeletal characteristics. At the same time, only local ethnic races were studied, which significantly limits the possibilities of the proposed method.

In [8], the authors proposed a method of racial diagnosis based on morphological analysis of the craniofacial region, within the framework of which the sum of characteristics was determined and compared with known samples. But, since not all traits can be strongly reflected in one particular person, the results obtained are more subjective.

The software implementation has a method for determining the race by the size of the angles of profiling of the face, which characterizes the degree of its flattening [6]. The computer program allows you to measure with sufficient accuracy the angles of the profiling of the face, but does not provide a way to determine the race on these grounds. The training sample included only 18 skulls, which raises doubts about the statistical reliability of the results, the assessment of which is not provided by the authors.

The authors of this article have developed the algorithm and a unique software package that makes it possible to determine a race (Caucasoid or Mongoloid) by external identification features of the head, to obtain reasonable conclusions and form an expert opinion. The methodology is based on the study of 42 physiognomic characteristics of Caucasoids (2881 men; 1319 women) and Mongoloids (12650 men; 4617 women).

We formed a complex of morphological racial-diagnostic characteristics, the assessment of which will make it possible to distinguish between Mongoloids and Caucasians.

This article is aimed to describe the algorithm and main characteristics of the software package "FFRACE", designed to determine the Caucasoid or Mongoloid race of an individual based on the external identification features of the head [9].

2 The Algorithm of Race Determination Underlying the Software Package

The software package is based on an algorithm of determining Caucasoid-Mongoloid races by external identification feature of the heads, which includes the following stages (Fig. 1) [6].

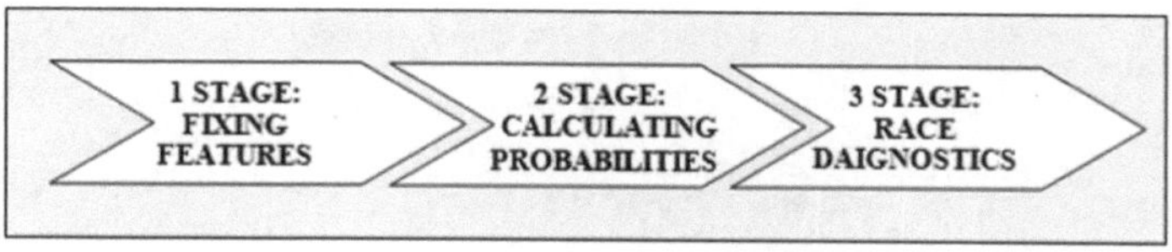

Fig. 1. Algorithm steps

1) *Stage 1.*
An expert fixes 18 main external identification features of the head (X_1-X_{18}), which are characterized by the major racial shifts (Table 1), which are characterized by the major racial shifts (Table 1). Fixation is possible both using a picture and examining the appearance of an individual. It is a scoring of features in the order corresponding to their significance. The significance of the features was calculated on the basis of the Kullback criterion (Table 1). At least 12 signs should be evaluated to make a correct and reasonable conclusion about Caucasoid or Mongoloid race. If a feature is impossible to evaluate, it is skipped. If an ambiguous situation has occurred during the assessment of a feature, then it is recommended to conduct several examinations with different values of the feature.

Table 1. Features

Name	*Variable*	*Feature meaning*	*Racial-diagnostic significance of the feature, calculated on the basis of the Kullback criterion*
Epicanthus	X_1	0-absent	4,059
		1-weak	
		2-medium	
		3-strong	
Eye color	X_2	0-light	3,385
		1-mixed	
		2-dark	
Horizontal face shaping	X_3	1-weak	2,973
		2-medium	
		3-strong	
Hair color	X_4	0-blond	2,318
		1-light brown	
		2-fair-haired	
		3-dark blond	
		4-black	
Upper eyelid fold	X_5	0-absent	1,815
		1-weak	
		2-medium	
		3-strong	
Height of the nose bridge	X_6	1 low	1,586
		2-medium	
		3-high	

(*continued*)

Table 1. (*continued*)

Name	*Variable*	*Feature meaning*	*Racial-diagnostic significance of the feature, calculated on the basis of the Kullback criterion*
Upper lip thickness	X_7	1-thin	1,214
		2-medium	
		3-thick	
		4-bloated	
Beard growth	X_8	1-weak	1,009
		2-medium	
		3-strong	
Transverse profile of the nose bridge	X_9	1-flat	0,912
		2-medium	
		3-protruding	
Lower lip thickness	X_{10}	1-thin	0,796
		2-medium	
		3-thick	
		4-bloated	
Protrusion of cheekbones	X_{11}	1-weak	0,725
		2-average	
		3-strong	
Position of the base of the nose	X_{12}	1-raised	0,460
		2-horizontal	
		3-omitted	
Upper lip profile	X_{13}	1-proheilia	0,352
		2-orthocheilia	
		3-opistocheilia	
Position of the nose tip	X_{14}	1-raised	0,323
		2-horizontal	
		3-omitted	
Upper lip height	X_{15}	1-small	0,093
		2-medium	
		3-large	
Nosal dorsum contour	X_{16}	1-concave	0,042
		2-straight	
		3-convex	
		4-winding	
Height of the wings of the nose	X_{17}	1-small	0,035
		2-medium	
		3-large	
Expressiveness of the wing grooves	X_{18}	1-weak	0,031
		2-medium	
		3-strong	

2) *Stage 2.*

Calculate the probabilities of belonging of the studied individual to the Mongoloid or Caucasian race (formulas 1 and 2) according to Bayes' formulas modified by A.A. Genkin [10–12].

Prior probabilities of the occurrence of each variant of features ($X_1\_X_{18}$) in Caucasoids and Mongoloids are described in [6, 13, 14].

$$P_k(e./xi_j) = \frac{P_k(e.) * P(xi_j / e.)}{P_k(м.) * P(xi_j / м.) + P_k(e.) * P(xi_j / e.)}, \quad (1)$$

$$P_k(м./xi_j) = \frac{P_k(м.) * P(xi_j / м.)}{P_k(м.) * P(xi_j / м.) + P_k(e.) * P(xi_j / e.)}, \quad (2)$$

- i - the total number of features, $i = 1,\ldots,18$;
- j - the ordinal number of the variant of the feature (may be in the range from 1 to 4);
- k - the iteration number in the step-by-step Bayes procedure ($1 \leq k \leq 18$);
- xi_j - the j value of the feature Xi score;
- $P_k(e./xi_j)$, $P_k(м./xi_j)$ - the probabilities that the individual belongs to the Caucasoid or Mongoloid race at the current step of the iterative process;
- $P_k(e.)$, $P_k(м.)$ - the probabilities that the individual belongs to the Caucasoid or Mongoloid race at the previous step of the iterative process.
- For $k = 1$: $k = 1$: $P_k(e.) = P_k(м.) = 0{,}5$;
- for $k > 1$: $P_k(e.) = P_{k-1}(e./xi_j)$, $P_k(м.) = P_{k-1}(м./xi_j)$);
- $P(xi_j/e.)$ and $P(xi_j/м.)$ - the probabilities that the feature X_i took the j-th value among Caucasoids and Mongoloids, respectively [6].

3) *Stage 3.*

Determination of an individual's race, which is carried out according to the following rules:

1) The individual is considered to be authentically Caucasoid (Mongoloid) if, when calculating by iterative formulas (1)–(2) at any of the steps, it was found that $P_k(e./xi_j)$ or $P_k(м./xi_j)$ is equal to 1, or the probability of an individual belonging to the Caucasoid (Mongoloid) race, calculated at the last step, is greater than 0.95.
2) The individual is considered probably Caucasoid (Mongoloid) if the probability of an individual belonging to the Caucasoid (Mongoloid) race obtained at the last step is in the range from 0.75 to 0.95.
3) The race of an individual is considered undefined if the probabilities of attributing the individual to the Caucasoid and Mongoloid races obtained at the last step are less than 0.75.

The software package "FFRACE" was developed to solve the problem of determining Caucasoid-Mongoloid races based on the external identification features of the head [9].

3 Description of the Software Package

3.1 Structure

The software package "FFRACE" consists of the following modules:

- *Module 1. Module of initial information input;*
- Module 2. Module of fixing the external identification features of the head;
- *Module 3. Race identification module.*

Figure 2 shows the scheme of the software package.

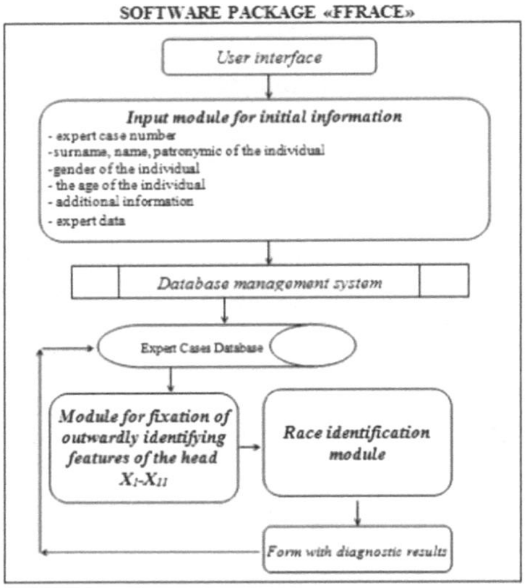

Fig. 2. Software modules

3.2 Input Module of Initial Information

Each expert case is associated with the input of the following information: the serial number of the case, surname, name and patronymic of the studied individual, gender and age of the individual (if they are known), additional information; data of the expert who makes the assessment. All information is stored in a database.

3.3 Module of Fixing the External Identification Features of the Head

It is intended for fixing the specific values of the external identification features of the head of an individual. To fix a particular feature, the expert goes through the information tabs containing reference information, such as definition, detailed description of the point scale values, and a set of pictures showing different expressiveness of a particular feature (Fig. 3). At least 12 signs should be recorded to make a reasonable conclusion about the race of an individual.

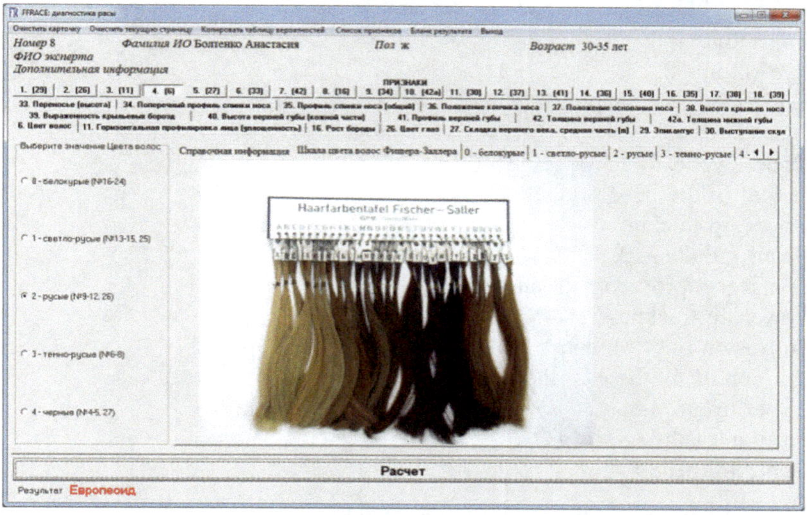

Fig. 3. Module of fixing external identification features of the head

3.4 Race Identification Module

Race identification module is based on the algorithm described in the previous section.

After calculating the probabilities, there will be generated a form containing a table with the individual's personal information; variations of each fixed feature; probabilities of occurrence of this variant in Caucasoids and Mongoloids; calculated probabilities of belonging of the studied individual to a particular race at each step of the iterative process; and a final conclusion about the race of the studied individual. The diagnostic result is also saved in the database.

It should be noted that if the racial identity of the individual was not established as a result of the diagnostics, the expert may determine the predominance of certain features in the appearance using the data in the table.

4 An Example of the Application of the Algorithm and Software Package

Let's consider an example of the practical use of the algorithm and software package.

An expert, analyzing a photograph or directly the head of an individual, captures eighteen outwardly identifying features. If it is impossible to fix a feature, it is skipped. At the same time, for a reasonable conclusion, it is necessary to determine the values of at least twelve signs from the list under consideration (Table 1).

Let us assume that the expert has obtained the following scoring of the features:

1) Epicanthus – 0.
2) Eye color – 2.
3) Horizontal face shaping – 3.
4) Hair color – 2.
5) Upper eyelid fold – 1.
6) Height of the nose bridge – 2.
7) Upper lip thickness – 1.
8) Beard growth – 0.
9) Transverse profile of the nose bridge – 3.
10) Lower lip thickness – 2.
11) Protrusion of cheekbones – 1.
12) Position of the base of the nose – 2.
13) Upper lip profile – 2.
14) Position of the nose tip – 2.
15) Upper lip height – 1.

The values of the signs in the order of their racial-diagnostic significance, as well as the relevant information of the expert case, are entered into the database and the racial diagnostics module is launched, as a result of which the following form will be obtained (Table 2).

The form will contain a point score for the recorded traits, the probability of occurrence of the fixed value in Caucasians and Mongoloids, the likelihood of the studied individual belonging to a particular race at each step of the iterative process for each trait.

In the example under consideration, at the last step, the probability of belonging to the Caucasian race was 0.99998. Thus, in the line with the expert opinion, the conclusion will be written “Authentically Caucasian”.

Table 2. Blank of result

RESULT FORM

Determination of race (Caucasoid-Mongoloid) of an individual based on the results of the assessment of racial-diagnostic features of appearance and the calculation of posterior probabilities

Number: 8 *Full name*: unknown *Sex*: M *Age*:-

Expert: Ivanov I.I. *Additional information*: no

№	*Feature*	*Feature meaning*	$P(xi_j / м.)$	$P(xi_j / e.)$	$P_k(м./ xi_j)$	$P_k(e./ xi_j)$
1	Epicanthus	0	0.289	0.997	0.22473	0.77527
2	Eye color	2	0.853	0.068	0.7843	0.2157
3	Horizontal face shaping	3	0.055	0.281	0.41579	0.58421
4	Hair color	2	0.001	0.217	0.003269	0.99673
5	Upper eyelid fold	1	0.301	0.314	0.0031341	0.99687
6	Height of the nose bridge	2	0.624	0.758	0.0025815	0.99742
7	Upper lip thickness	1	0.138	0.525	0.00067985	0.99932
8	Beard growth	0			0	0
9	Transverse profile of the nose bridge	3	0.057	0.248	0.00015634	0.99984
10	Lower lip thickness	2	0.440	0.645	0.00010666	0.99989
11	Protrusion of cheekbones	1	0.166	0.194	9.1263E-005	0.99991
12	Position of the base of the nose	2	0.385	0.579	6.0686E-005	0.99994
13	Upper lip profile	2	0.568	0.800	4.3088E-005	0.99996
14	Position of the nose tip	2	0.427	0.549	3.3513E-005	0.9997
15	Upper lip height	1	0.088	0.129	2.2862E-005	**0.99998**
16	Nose bridge profile	0				
17	Height of the wings of the nose	0				
18	Expressiveness of the wing grooves	0				

RESULT: *AUTHENTICALLY CAUCASIAN*

5 Summary and Conclusions

The authors of the article proposed a complex of morphological racial-diagnostic characters, the assessment of which will allow distinguishing between Mongoloids and Caucasians.

An algorithm and a software package have been developed to solve the problem of establishing the race (Caucasoid-Mongoloid) of an individual by eighteen outwardly identifying features of the head.

The algorithm has a high representativeness of training samples, independence from the complete list of features, it has developed clear decision rules.

The presented software package has no analogues and allows solving the following tasks:

1) To unify the assessment of racial diagnostic signs by the operator, using visual prompts and reference information.
2) Receive an original form for assessing the development of the appearance signs of an individual, which verbally characterizes his uniqueness and can be used in comparative studies.
3) Create databases containing personal information and the values of external identification signs of the human head.
4) Carry out the definition of race, based on quantitative criteria for determining the racial characteristics of the external appearance and assessing their complex.
5) Mathematically substantiate the reliable or probable assignment of an individual to Caucasians or Mongoloids, or a conclusion about his mixed origin.

The software complex has been tested by experts from the Russian Center for Forensic Medicine of the Ministry of Health of Russia and has shown high efficiency of diagnostic results in comparative and identification studies.

The developed software can be used both in laboratory studies to establish the racial identity of a person, and at the scene of an accident during mass disasters.

References

1. Kopanitsa, G., Hildebrand, C., Englmeier, K.H., Stausberg, J.: Visualization of medical data based on EHR standards. Methods Inf. Med. **52**(1), 43–50 (2013)
2. Gusev, A.V., Pliss, M.A., Levin, M.B., Novitsky, R.E.: Trends and forecasts for the development of medical information systems in Russia. Doct. Inf. Technol. **2**, 38–49 (2019)
3. Belyshev, D.V., Guliev, Y.I., Mikheev, A.E.: The development of medical information systems at the present stage - how to combine many years of experience and new technologies? Doct. Inf. Technol. **4**, 40–51 (2017)
4. Stolyar, V.P., Krainyukov, P.E., Rybakov, Y., Gukasov, V.M.: Digital medicine: theory, state, prospects for implementation and practical application. Med. High Technol. **4**, 5–16 (2018)
5. Krivtsova, A.G.: The role of information technology in modern forensics. Innov. Technol. Mech. Eng. Educ. Econ. **7**, 4–1(6), 5–7 (2017)

6. Zvyagin V.N., Kovalev A.V., Galitskaya O.I., Narina N.V., Usacheva L.L., Fomina E.E.: The method of establishing Caucasoid-Mongoloid affiliation of an individual according to external identifying signs of the head. Invention patent number 2661075 of 06 Feb 2017
7. New method for estimating race/ethnicity and associated disparities where administrative records lack self-reported race/ethnicity. https://www.researchgate.net/publication/5372963_A_New_Method_for_Estimating_RaceEthnicity_and_Associated_Disparities_Where_Administrative_Records_Lack_Self-Reported_RaceEthnicity. Accessed 03 Sep 2019
8. The assessment of ancestry and the concept of race. https://www.researchgate.net/publication/281747419_The_Assessment_of_Ancestry_and_the_Concept_of_Race. Accessed 03 Sep 2019
9. Zvyagin, V.N., Kovalev, A.V., Galitskaya, O.I., Narina, N.V., Usacheva, L.L., Fomina, E. E.: "FFRACE". Certificate of state registration of computer programs No. 2015617098 dated 30 Jun 2015
10. Terekhov, S.A.: Introduction to Bayesian Networks. Neuroinformatics-2003: scientific. Session MIFI-2003. V All-Russian scientific tech. conf. M.: MIFI Publishing House, Part 1, pp. 149–187 (2003)
11. Eremeev, A.P., Khaziev, R.R., Zueva, M.V., Tsapenko, I.V.: The prototype of the diagnostic decision support system based on the integration of Bayesian trust networks and the Dempster – Schaefer method. Softw. Syst. **1**(101), 11–16 (2013)
12. Fatima, K., Ahmad, S.P.: Bayesian approximation techniques of inverse exponential distribution with applications in engineering. Int. J. Math. Sci. Comput. **4**(2), 49–62 (2018)
13. Hamd, M.H., Ahmed, S.K.: Biometric system design for iris recognition using intelligent algorithms. Int. J. Mod. Educ. Comput. Sci. **10**(3), 9–16 (2018)
14. Obuhuma, J.I., Okoyo, H.O., McOyowo, S.O.: Driver behaviour profiling using dynamic bayesian network. Int. J. Mod. Educ. Comput. Sci. **10**(7), 50–59 (2018)

Automated System for Modeling the Casting Process of Non-ferrous Metals

Elena E. Fomina(✉)

Tver State Technical University, Tver 170026, Russian Federation
f-elena2008@yandex.ru

Abstract. Improving the quality of products is the main task of any production, including metallurgy industry. Carrying out experiments to select the optimal values of technological parameters on a real unit is associated with a number of technical difficulties and material costs. To solve this problem, mathematical modeling can be used, which will significantly reduce the cost of carrying out full-scale experiments, analyze various casting modes and describe the process of solidification of the billet. The article presents a mathematical model used to describe the technological process of continuous and discrete-continuous casting of cylindrical billets from non-ferrous metals. The issue of the mathematical description of the quality of cast blanks is considered separately. Optimality criteria are proposed. To solve the constructed system, an algorithm was developed based on the SIMPLER algorithm, taking into account the features of the problem being solved. The description of the automated process control system developed on the basis of the constructed algorithm is given. The results of modeling the process of casting copper cylindrical billets are presented. Namely, the influence of the casting speed and stopping time during discrete-continuous casting on the billet solidification process has been investigated.

Keywords: Casting of nonferrous metals · Mathematical modeling · Casting packages

1 Introduction

The range of products manufactured by enterprises of the metallurgical complex is quite wide. The task of improving the quality and competitiveness of products is one of the main tasks facing engineers and designers.

This question is especially relevant when designing the technological process of production of billets from non-ferrous metals by continuous and discrete-continuous casting, where conducting experiments on a working unit is associated with a large number of technical difficulties, material and time costs [1–3].

To solve this identified actual problem, it is necessary to develop and implement automated decision support systems based on mathematical modeling methods [4–6].

Z. Hu et al. (Eds.): AIMEE 2021, LNDECT 107, pp. 184–195, 2022.
https://doi.org/10.1007/978-3-030-92537-6_18

The program should visualize the crystallization and solidification process, give an idea of the shrinkage processes and pores that can form in the billet. Using the program will allow avoiding technical difficulties caused by small or large dimensions of the test piece, high temperatures, and will also save resources and improve product quality.

Currently, models are being actively developed and improved for the mathematical description of the continuous casting process. As a rule, they include the equation of continuity, equations of motion and energy, as well as boundary and initial conditions. The following works can be distinguished, in which the features of modeling the process of continuous vertical and horizontal casting of non-ferrous metals are described in sufficient detail [7–10]. However, it should be noted that when solving the modeling problem, simplifications are often used, namely, stationary models are considered, not all models take into account flow turbulence, which has a significant effect on the crystallization process.

To date, a number of software products have also been developed that allow the metallurgist to analyze and visualize the casting process. Among the most famous are such casting packages as PolyCast [11], Polygon [12], LVMFlow [13], ProCAST [14], MAGMASoft [15], FLOW3D [16].

Each of them is a powerful modeling tool, but most of them are mainly intended for injection molding (PolyCast, Polygon, LVMFlow, ProCAST, MAGMASOFT) and therefore cannot be used for continuous casting modeling. In turn, those software products that can be used to simulate continuous casting processes (FLOW3D) must be adapted for a specific process, setting a number of thermophysical parameters, which can be difficult to determine.

Analyzing the requirements for casting packages, we can conclude that the modeling system should contain databases of characteristics of the cast metals and equipment, software, i.e., casting process models, software modules for inputting initial data, a process calculation module and modules for generating simulation results in the form of tables, graphs and a temperature profile.

In addition, it should be noted that there is no methodological apparatus and software systems for optimizing the modes of obtaining blanks of various nomenclature [2, 17].

This article is devoted to the actual problem of developing a mathematical model and a software package for the study and optimization of processes occurring in the production of cylindrical billets from non-ferrous metals by continuous and discrete-continuous casting.

2 Mathematical Model

The scheme of the technological process of casting cylindrical billets from non-ferrous metals, which is considered in this article shows on Fig. 1. The liquid melt is continuously (or with periodic stops) fed into a cylindrical crystallizer. The feed is carried out either through the entire inlet or through the annular wall openings. The second option of feeding the melt will contribute to a more uniform cooling and the formation of the correct crystal structure. Additionally, it is possible to heat the melt with an electron beam, which will allow taking into account the features of each mode.

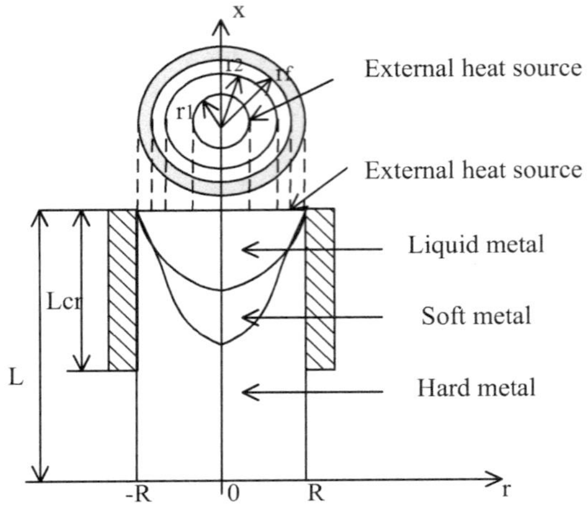

R, L - radius and length of the crystallizer,
Lcr - length of the crystallizer, r1, r2 - inner and outer radius of the ring for additional heating,
rf - inner radius of the ring for casting

Fig. 1. Scheme of the technological process of continuous casting with billet drawing down

The model provides for the possibility of extracting the solidified ingot from the crystallizer either at a constant speed (continuous casting) or with periodic stops.

The proposed mathematical model includes the following equations written in cylindrical coordinates: continuity Eq. (1); equations of motion (2), (3); energy Eq. (4); boundary and initial conditions.

Continuity equation:

$$\frac{\partial r\rho}{\partial t} + \frac{\partial}{\partial x}(r\rho(u + u_{cast})) + \frac{\partial}{\partial r}(r\rho v) = 0. \tag{1}$$

Equations of motion:

Along the Ox axis:

$$\begin{aligned} &\rho\frac{\partial u}{\partial t}+\frac{1}{r}\left[\frac{\partial}{\partial x}(r\rho(u+u_{cast})u)+\frac{\partial}{\partial r}(r\rho uv)\right] \\ &=-\frac{\partial p}{\partial x}+\frac{1}{r}\left[\frac{\partial}{\partial x}(r\mu\frac{\partial u}{\partial x})+\frac{\partial}{\partial r}(r\mu\frac{\partial u}{\partial r})\right]+Au+\rho g \end{aligned}, \tag{2}$$

Along the or axis:

$$\begin{aligned} &\rho\frac{\partial v}{\partial t}+\frac{1}{r}\left[\frac{\partial}{\partial x}(r\rho(u+u_{cast})v)+\frac{\partial}{\partial r}(r\rho v^2)\right] \\ &=-\frac{\partial p}{\partial r}+\frac{1}{r}\left[\frac{\partial}{\partial x}(r\mu\frac{\partial v}{\partial x})+\frac{\partial}{\partial r}(r\mu\frac{\partial v}{\partial r})\right]-\frac{2\mu v}{r^2}+Av \end{aligned}. \tag{3}$$

Energy equation:

$$\begin{aligned} &\rho\frac{\partial T}{\partial t}+\frac{1}{r}\left[\frac{\partial}{\partial x}(r\rho(u+u_{cast})\phi)+\frac{\partial}{\partial r}(r\rho v\phi)\right] \\ &=\frac{1}{r}\left[\frac{\partial}{\partial x}(rk\frac{\partial T}{\partial x})+\frac{\partial}{\partial r}(rk\frac{\partial T}{\partial r})\right] \\ &-\frac{1}{r}\left[\frac{\partial}{\partial x}(r\rho(u+u_{cast})\Delta H)+\frac{\partial}{\partial r}(r\rho v\Delta H)\right] \end{aligned}. \tag{4}$$

Boundary conditions along the length of the ingot are specified in the form of the following differential equations describing the heat flux at the boundary:

A) *On the side walls (r = R):*
   If $T<T_{sol}$: $-k\frac{\partial T}{\partial r}\Big|_{r=R}=\sigma\varepsilon_s(T^4-T^4_{form})\Big|_{r=R}$, $x\geq(L,\ u(x;R)=0,\quad v(x;R)=0,$
   or $-k\frac{\partial T}{\partial r}\Big|_{r=R}=\sigma\varepsilon_s(T^4-T^4)\Big|_{r=R}$, $x<(L-L_{cr})$, $u(x;R)=0,\quad v(x;R)=0,$
   If $T\geq T_{sol}$: $-k\frac{\partial T}{\partial r}\Big|_{r=R}=\alpha_c(T-T_{form})\Big|_{r=R}$, $u(x;R)=-u_{sol},\quad v(x;R)=0,$

B) *Bottom border of the form ($x=0$):* $\frac{\partial T}{\partial x}\Big|_{x=0}=0$, $u(0;r)=0,\quad v(0;r)=0.$

C) *Top border of the form ($x=L$):*
   If $r_1\leq r\leq r_2$, то $k\frac{\partial T}{\partial x}\Big|_{x=L}=q_b-\sigma\varepsilon_l(T^4-T^4_{env})\Big|_{x=L}$.
   If $r<r_1$ or $r_1<r<r_f$, $-k\frac{\partial T}{\partial x}\Big|_{x=L}=\sigma\varepsilon_l(T^4-T^4_{env})\Big|_{x=L}$.
   If $r_f<r<R$, then $T(L;r_f<r<R)={}_{melt}$, $\mu\frac{\partial v}{\partial x}\Big|_{x=L}=\frac{\partial r}{\partial T}\frac{\partial T}{\partial r}$, $0\leq r\leq R$, $u(L,r)=0$.

The RNG $k-\varepsilon$ model was used to simulate turbulence [2].

All variables used in the model are described in Table 1.

Table 1. Variables used in the model

Variable	Description	Units
u_{cast}	Constant casting speed	*m/s*
ρ	Metal density	kg/m^3
$\boldsymbol{u}$	Speed along the *Ox* axis	*m/s*
$\boldsymbol{v}$	Speed along the *Or* axis	*m/s*
$\boldsymbol{r}$	Distance from the *Or* axis to the current point of the ingot	*m*
$\boldsymbol{x}$	Distance from the *Ox* axis to the current point of the ingot	*m*
$\boldsymbol{t}$	Time	*s*
p	Pressure	$kg/m{\cdot}s^2$
μ	Effective viscosity	$kg/m{\cdot}s$
$A = -C(1-\lambda)^2/\lambda^3$	Parameter λ, which is equal to 1 in liquid, 0 in solid and varies linearly with temperature in the soft zone	
C	Darcy coefficient is equal 10^8	*1/s*
g	Acceleration of gravity	m/s^2
c	Heat capacity	$J/kg{\cdot}K$
ϕ	Enthalpy	$J{\cdot}K/kg{\cdot}s$
T	Temperature	*K*
k	Thermal conductivity	$W/m{\cdot}K$
$\Delta H_f, \ \Delta H = \lambda \Delta H_f$	Total latent heat	*J/kg*
$\boldsymbol{\sigma}$	Stefan-Boltzmann constant, $5{,}667{\cdot}10^{-8}$	$W/m^2{\cdot}K^4$
ε_s	Radiation on a solid surface	
T_{form}	Mold surface temperature	*K*
T_{env}	Ambient temperature	*K*
$L, \ L_{cr}$	Length of mold surface and crystallizer	*m*
α_c	Heat transfer coefficient at contact between the crystallizer and the ingot	$W/m^2{\cdot}K$
q_b	Electron beam heat flux	W/m^2
γ	Surface tension	
r_f	Inner radius of the casting ring	*m*
ε_l	Emissivity of the environment surrounding the free surface of the melt	

The solution of the presented system of equations was carried out using the finite volume method and the SIMPLER algorithm, which was refined and modified [18].

When developing the model, special attention was paid to the problem of choosing the optimal values of technological parameters and managing the quality of the resulting products.

The concept of «ingot quality» is quite capacious and includes a large number of indicators. Measurement of these indicators during the production of castings is associated with technical difficulties or even impossible.

The paper proposes to use indirect quality indicators, called optimality criteria. These criteria must satisfy the following conditions: they must to some extent characterize the quality of the castings and must be related to adjustable process parameters.

The following criteria were considered as optimality criteria (Table 2), which are functionals of the temperature field [19].

Table 2. Optimality criteria

№.	Criterion	Description
1	$I_1 = \int_0^H \int_\xi^a \sqrt{T_x^2 + T_r^2} dxdr^{\rightarrow min}$	The criterion is introduced based on the requirement of small temperature gradients over the thickness of the crust
2	$I_2 = \int_0^H \int_\xi^a \lvert T_x - T_r \rvert dxdr^{\rightarrow min}$	The criterion determines the uniformity of the heat sink in the cross section inside the ingot crust
3	$I_3 = \int_0^H \int_\xi^a \lvert T_{xr} \rvert dxdr^{\rightarrow min}$	The criterion is introduced based on the recommendations on the desirability of an approximate straightness of the distribution of temperature curves in the longitudinal section of the solidifying part of the ingot
4	$I_4 = \int \int \left\lvert \frac{\partial^2 T}{\partial x \partial r} \right\rvert dxdr^{\rightarrow min}$	The criterion is introduced based on the requirements of equality of the cooling rate of the ingot surface at each time of the cooling rate of the newly solidified ingot (this contributes to an approximation to such a cooling mode when no thermal stresses occur)
5	$I_5 = \int \int \left\lvert \frac{grad_x T}{grad_r T} \right\rvert dxdr^{\rightarrow min}$	The criterion is the solidification rate for a stationary front, in the solid part of the ingot determines the direction of growth and the location of crystals

The task of controlling the continuous casting process was posed as follows: it is necessary to obtain a continuously cast ingot, the temperature field of which corresponds to the specified criteria (I_1–I_5) with the specified restrictions on the technological parameters.

To solve the optimization problem, the method of random search with self-learning and adaptation of the working step was used.

3 Software Package

On the basis of a mathematical model and an algorithm, a software package «Modeling and optimization of processes for obtaining crystalline billet from non-ferrous metals by continuous casting methods» [20] was developed, which can be used in technological practice for analysis, visualization, and also for calculating the main indicators of the process of continuous and discrete-continuous casting of cylindrical billets from non-ferrous metals.

The software package includes the following modules (Fig. 2).

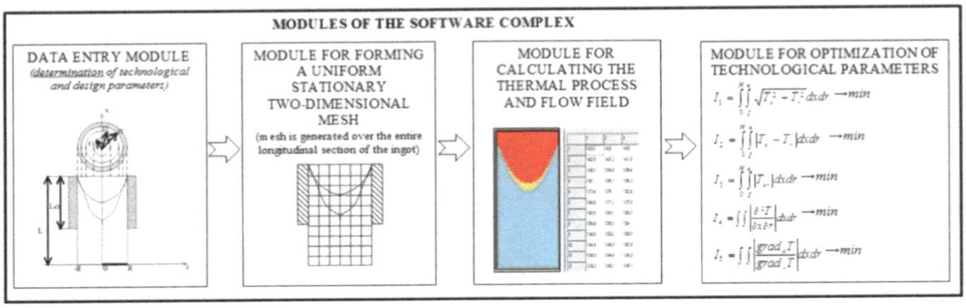

Fig. 2. Modules of the software complex

After the development of the model and the software package, their adequacy was checked. For this, Fisher's criterion was used, which was calculated for the depth of the solidification zone. The depth of the solidification zone was calculated in the software package. The obtained data were compared with those calculated experimentally [21–24]. The fulfillment of the adequacy condition allows the use of a mathematical model and a software package developed on its basis for modeling the technological process of obtaining crystalline blanks from non-ferrous metals by the method of continuous casting.

4 Research Results and Discussion

The developed software package was used to solve the following set of tasks:

1) Modeling of technological modes of casting in order to visualize the processes occurring inside the ingot;
2) Conducting research in order to identify the patterns of the process and prepare recommendations for technologists and designers;
3) Optimization of technological parameters.

A series of computational experiments made it possible to describe and visualize the processes occurring in the ingot during its cooling and further solidification, both during continuous and discrete-continuous casting. In particular, the article presents the results of modeling the technological process of casting copper billets.

It was found that the temperature profile includes three zones: the molten metal is red, the soft metal is yellow, and the hardened metal is blue (Fig. 3).

In Fig. 2 shows the temperature profiles along the longitudinal section of the ingot for different values of the casting speed in the case of continuous casting with pulling the billet downward. As can be seen, an increase in the casting speed leads to an increase in the solidification zone. This effect can lead to a loss of product quality when the molten metal leaves the crystallizer.

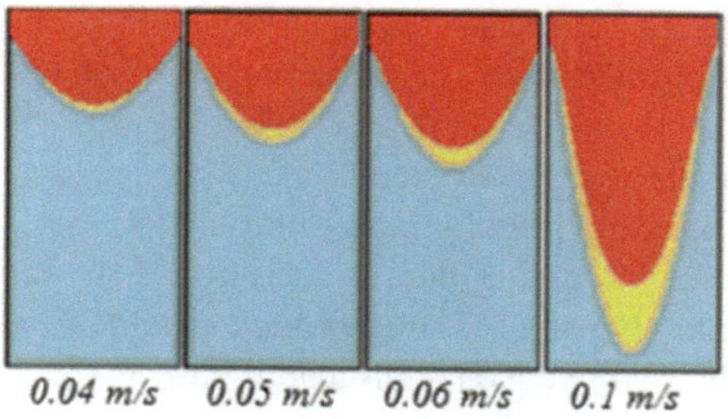

Fig. 3. Temperature profile when casting at different casting speeds

In Fig. 4 shows a temperature profile corresponding to a discrete-continuous casting with a billet pulling down at a speed of 0.07 m/s. Analysis of the temperature profile shows that with this casting mode, the formation of zones of liquid and soft metal in the solidified ingot is possible, which can lead to the formation of shrinkage cavities.

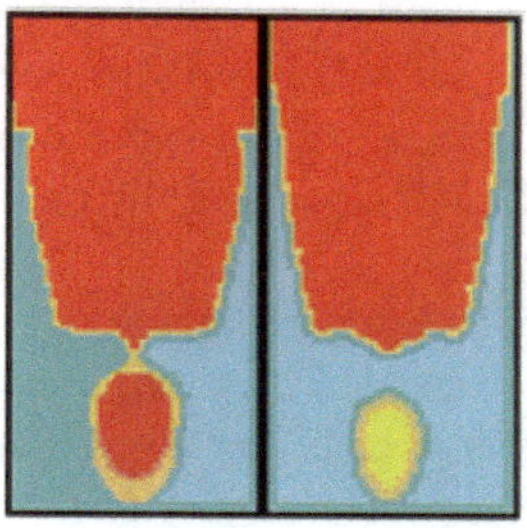

Fig. 4. Temperature profile at the end of the drawing and stopping phase for a speed of 0.07 m/s

In Fig. 5 show fragments of tables with the values of the temperature and velocity fields.

The software package has the ability to obtain such tables for other variables that characterize the process.

The temperature field

	1	2	3
1	1425,1	1430	1430
2	1402,5	1407,3	1411,5
3	1390,1	1394,9	1399,4
4	1381	1385,7	1390,3
5	1373,4	1378	1382,6
6	1366,6	1371,2	1375,9
7	1360,5	1365,1	1369,7
8	1354,9	1359,4	1364

The velocity field

	1	2	3
1	0,032	2,744238E-5	2,744238E-5
2	0,032	3,658087E-5	2,744541E-5
3	0,032	3,653741E-5	2,752577E-5
4	0,032	3,63607E-5	2,752681E-5
5	0,032	3,618885E-5	2,75266E-5
6	0,032	3,601864E-5	2,752638E-5
7	0,032	3,584994E-5	2,752615E-5
8	0,032	3,568259E-5	2,752593E-5

Fig. 5. A fragment of a table of values

The studies carried out served as the basis for identifying patterns and made it possible to prepare recommendations for the technologist on the choice of optimal technological modes.

First, it was found that the melt temperature, casting speed and billet radius have the greatest influence on the cooling and crystallization process.

Secondly, the issue of increasing the productivity of the process by increasing the casting speed and maintaining the required product quality has been comprehensively analyzed. Increasing the casting speed significantly affects the temperature profile. At a casting speed of less than 0,1 m/s, the phase transition has time to complete in the zone of the crystallizer; however, at high speeds, an incompletely solidified ingot leaves the crystallizer. In the case of real production, the speed limit is determined experimentally on a case-by-case basis.

The stability problem is one of the most important and difficult to analyze. This is due not only to technological parameters, but also to fluctuations in their values. However, the geometric parameters of the temperature profile (for example, such as the depth of the solidification zone, the width of the solidification zone) can be used as a basic characteristic of the stability of the process. Thus, it is possible to determine some critical values of the parameters at which, on the one hand, the maximum possible performance of the process is ensured, and on the other, there is no loss of process stability. The temperature profile will show that the phase transition is completed in the zone of the mold.

Using the software package, lines of equal values of the depth of the solidification zone were obtained in the coordinates T_{melt} - u_{cast} and R - u_{cast} with the remaining technological parameters of the copper billet being fixed.

Having a set of such data, the technologist can, within the limits of the melting temperature and the casting rate, increase the productivity of the process while maintaining constant values of the geometric characteristics of the temperature profile.

With the help of the software package, the problem of optimization of technological parameters was solved.

The software package was used to solve the problem of optimizing technological parameters. The following parameters were taken as the optimized technological parameters of casting: casting speed; melt temperature; cooling water temperature; mold length; stop time for discrete-continuous casting.

The analysis of the results of calculations of the optimal technological parameters leads to the following main conclusions: 1) the casting speed, shutdown time and the length of the mold have the greatest influence on the value of all criteria, while the temperature of the melt and the temperature of the cooling water have a much smaller effect; 2) if we consider the criteria from the point of view of the productivity of the casting process, defined as the casting speed, then the highest productivity both when casting downward and upward is provided by criterion I_3; 3) in the general case, a large overheating of the metal at the entrance to the mold is not required.

Along with the casting speed, the stop time in discrete-continuous casting has a great influence on the process productivity and product quality.

Using the software package, a series of computational experiments were carried out to optimize the stopping time for the critical mode of casting a copper billet at a speed of 0.07 m/s, in which the stability of the process was violated (Fig. 4). The stop time limits were set as follows: $1 \leq dt \leq 30$ (sec). The calculation results are presented in Table 3 and Fig. 6.

Table 3. Calculation of the optimal stopping time

Criterion	I_1	I_2	I_3	I_4	I_5
Optimal stop time (s)	11,6	11,6	6,5	10,1	16,95

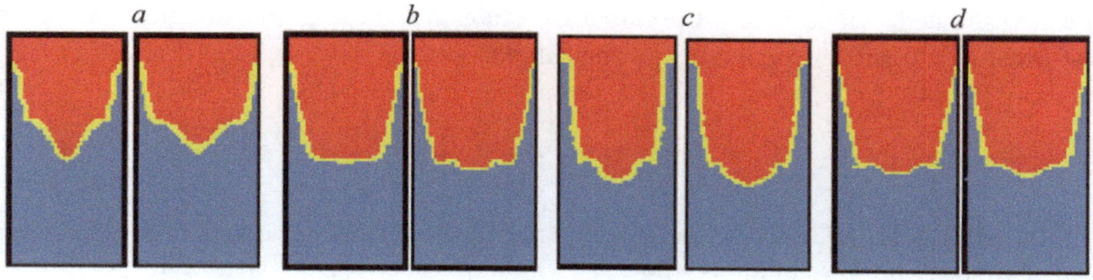

Fig. 6. Temperature profiles at the end of the draw and shutdown phases: a) criterion I_1, I_2, b) criterion I_3, c) criterion I_4, d) criterion I_5

The calculated optimal stopping time for discrete-continuous casting at a speed of 0.07 m/s made it possible to simulate a process in which the continuity of the medium is not violated during solidification, and the process of solidification is completed in the zone of the mold, in contrast to the process shown in Fig. 3. This fact allows us to conclude that the software package can be used to calculate the optimal values of technological parameters without field experiments.

Thus, computer modeling makes it possible not only to visualize the casting process, but also to calculate its characteristics, as well as to control it based on the specified optimality criteria.

5 Summary and Conclusion

This article describes a comprehensive study of the problem of modeling the process of continuous casting of billets from non-ferrous metals.

A non-stationary mathematical model of the process of continuous and discrete-continuous casting of cylindrical billets made of non-ferrous metals is proposed, which makes it possible to take into account additional heating of the melt, casting through the annular wall openings of the mold, as well as an algorithm based on the SIMPLER algorithm, intended for calculating the process, which makes it possible to reduce computational costs when modeling.

The description of the modules and capabilities of the developed software package for visualizing the processes of continuous and discrete-continuous casting of cylindrical billets from non-ferrous metals, calculating the characteristics of the temperature profile, pressure fields, velocity and temperature in the longitudinal section, as well as optimizing the technological and design parameters, which allows a new technological process in production to replace a physical experiment with a computational one, which leads to savings in material and time costs.

The results of computational experiments on the influence of technological and design parameters on the solidification process of a continuously cast billet are presented, allowing the technologist and designer to be guided by them in the design and production of castings.

The technique and software for optimizing the technological parameters of casting are described, the features of solidification of the billet under optimal casting conditions are investigated, which made it possible to establish the process parameters that have the greatest impact on the quality of the billets.

References

1. Batyshev, K.A., Semenov, K.G., Svinoroev, Y.A.: Modern Technologies for the Production of Castings from Non-ferrous Alloys, p. 150. Publishing House «First Volume», Moscow (2020)
2. Zhiganov, N.K., Volnov, I.N., Fomina, E.E.: Modeling of continuous casting processes of non-ferrous metals and their alloys, p. 222. Tver State Technical University (2007)
3. Monastyrskiy, V.P.: Modern state and problems of computer modeling of casting processes. In: The Collection: Progressive Foundry Technologies. Proceedings of the X International Scientific and Technical Conference, Moscow, pp. 438–444 (2020)
4. Kukushkina, V.A., Alexandrov, G.I.: State and prospects of application of digital technologies in casting production. In: The collection: Topical Issues of the Development of the Machine Tool Industry. Collected Works of the I International Scientific and Practical Conference, pp. 128–134 (2018)
5. Skuratov, A.P., Gorokhov, Y.V., Potapenko, A.S.: Application of cad systems for modeling thermal processes in a continuous casting and non-ferrous metal pressing plant. In: The Collection: Modeling and High-tech Information Technologies in Technical and Socio-Economic Systems. Proceedings of the IV All-Russian Scientific-Practical Conference with International Participation, pp. 154–158 (2016)
6. Aktar, S., Kawser, M.A.: Asymptotic solutions of a semi-submerged sphere in a liquid under oscillations. Int. J. Math. Sci. Comput. **4**(3), 66–79 (2018)
7. Skuratov, A.P., Potapenko, A.S.: Computer model of heat exchange in unit for continuous casting and pressing of non-ferrous metals. J. Siber. Federal Univ. Ser.: Eng. Technol. **10**(8), 1019–1030 (2017)
8. Shyy, W.: Modeling of turbulent transport and solidification during continuous ingot casting. Heat Mass Transf. **35**(5), 1229–1245 (1992)
9. Kryukov, I.Y., Naumova, M.G., Vdovin, K.N., Larina, T.P.: Development of a mathematical model of thermal state of crystallizing rectangular billet in a horizontal semi-continuous casting machine. Basic Res. **10–2**, 306–311 (2016)
10. Minakov, A.V., Pervukhin, M.V., Platonov, D.V., Khatsyuk, M.Y.: Mathematical model and numerical simulation of the casting and crystallization of aluminum in a magnetic field taking into account a free surface. J. Comput. Math. Math. Phys. **55**(12), 2094 (2015)
11. System of computer modeling of foundry processes POLYCAST. http://www.foundry.spb.ru/html/customers/polycast.shtml. Accessed 09 Feb 2021

12. Poligonsoft. https://poligonsoft.ru/. Accessed 10 Feb 2021
13. Lvmflow. http://lvmflow.ru/. Accessed 01 Feb 2021
14. Casting Simulation Software. https://castsoft.ru/. Accessed 01 Feb 2021
15. Magma. https://www.magmasoft.com/. Accessed 03 Feb 2021
16. FLOW3D. http://www.flow3d.com/. Accessed 05 Feb 2021
17. Albu, A.F., Zubov, V.I.: About the influence of plant parameters on control of the process of crystallization of substance in casting. J. Comput. Math. Math. Phys. **53**(2), 238 (2013)
18. Fomina, E.E., Zhiganov, N.K.: About modification of the simpler algorithm for solving the problem of modeling the process of continuous casting of non-ferrous metal blanks. Voronezh State Tech. Univ. Bull. **12**, 32–35 (2016)
19. Zhiganov, N.K., Fomina, E.E.: Optimization of the processes of discrete-continuous casting of non-ferrous metals and their alloys. Bull. Tver State Tech. Univ. **15**, 22–27 (2009)
20. Modeling and optimization of processes for obtaining crystalline billets from non-ferrous metals by continuous casting methods. Certificate of state registration of a computer program No. 2013616906 dated 25 July 2013
21. Shchipanov, K.A.: Development and research of mathematical models, creation of software for control of objects in metallurgy. Dissertation. Ekaterinburg, p. 161 (2007)
22. Yunusa, U., Usman, B., Ibrahim, M.B.: Adsorptive removal of basic dyes and hexavalent chromium from synthetic industrial effluent: adsorbent screening, kinetic and thermodynamic studies. Int. J. Eng. Manuf. **10**(4), 54–74 (2020)
23. Nazimuddin, A.K.M., Ali, M.S.: Application of the flow curvature method in Lorenz-Haken model. Int. J. Math. Sci. Comput. **6**(1), 33–48 (2020)
24. Ramshankar, Y., Deivanathan, R.: Development of machine vision system for automatic inspection of vehicle identification number. Int. J. Eng. Manuf. **8**(2), 21–32 (2018)

Evaluation and Analysis of Cross-border Logistics Security Risk Between China and Kazakhstan Based on SVM Under the Background of "One Belt and One Road"

Lei Zhang[1,2](✉) and Zhongquan Wang[3]

[1] School of Logistics, Wuhan Technology and Business University, Wuhan, China
[2] School of Management, Wuhan University of Technology, Wuhan, China
[3] Beijing Branch of China Post Technology Co., LTD, Beijing, China

Abstract. In the context of the Belt and Road Initiative, Kazakhstan, as the bridgehead of the belt and Road initiative on land, will surely play an increasingly important role. In recent years, china-Kazakhstan oil project cooperation has been growing and has made great progress. With the increasing scale of cross-border logistics, logistics security has been paid more and more attention. Cross-border logistics security between China and Kazakhstan is more affected by political risks, economic regional risks, industrial market risks and other factors in Kazakhstan. Therefore, based on the collection, induction and analysis of these factors, this paper constructs the basic framework of the comprehensive evaluation system of cross-border logistics security, and uses the support vector machine (SVM) method to evaluate, and puts forward corresponding countermeasures and suggestions, in order to expect to make an exploration in this aspect.

Keywords: One Belt and One Road · Cross-border logistics · Support vector machine

1 Introduction

"One Belt and One Road" was put forward by the top government in 2013. The purpose of its strategic vision is to build a community of security, responsibility and destiny for the common development of the Eurasian region through economic means [1]. It aims to build a free and open economic system including countries along the Belt and Road through more smooth and convenient circulation of production factors. In the "One Belt and One Road" circulation system, logistics is not only one of the most important connotations, but also an indispensable and important means to realize the "One Belt and One Road" construction [2, 3]. Logistics is the guiding demand of economic development. If information flow and capital flow are the "nerve" and "center" of economy, logistics is the "blood" of economic development flow. Through the "One Belt and One Road" trans-Eurasian trade channel and industrial channel, the fragmented regional economy will be connected, and through the increase of mutual

Z. Hu et al. (Eds.): AIMEE 2021, LNDECT 107, pp. 196–205, 2022.
https://doi.org/10.1007/978-3-030-92537-6_19

trade and investment among countries along the belt and Road and the acceleration of industrial transfer, the logistics flow along the belt and Road will be greatly increased. At the same time, the smooth flow and scale expansion of modern logistics industry can also make all regions better integrate into "One Belt and One Road", promote the further integration of factor market, industrial chain and industrial cluster in Eurasia, so as to form a more unified and closely connected economic space [4].

"One Belt and One Road" involves 66 countries around the world, whose national conditions, environment, culture, geography and resources are very different. They face great security risks in capital flow, information flow and logistics, among which the security risk of logistics is the most important and intuitive embodiment [5]. In order to realize the long distance and large space and time transfer of goods from space and time, its security risks exist in every link and every moment of logistics. Thus to promote the construction of "area" orderly implementation, it is vigorously advancing along the country in the construction of each project, must be prospectively conducted in view of the "area" of the logistics security risks of the project, that is, from logistics safety risk identification, risk assessment, risk response from several aspects, such as system design and research, it is very necessary, It's very urgent. This is undoubtedly of practical significance for the construction of "One Belt and One Road", for the construction of engineering projects in countries along the Belt and Road, and for enterprises undertaking specific international engineering projects to reduce safety risks, promote construction progress, reduce losses and improve economic benefits.

Various risk factors affecting the implementation of international engineering logistics are constantly changing, and even the transportation of each batch and each kind of equipment will have different risk factors, and the change of these risk factors will bring various losses to the implementation of foreign contracted engineering projects [6]. However, the previous researches on the risk management of engineering logistics tend to be a static mode, which carries out comprehensive analysis and discussion through the process of risk management, and pays less attention to the dynamic changes. The international engineering logistics discussed in this paper is regarded as an independent operating system in the process of execution of contracted projects. In the implementation of foreign contracted projects, it is operated under high risk and complex conditions, and the risk source may be transformed with the implementation of foreign contracted projects [7].

Many Chinese engineering companies focus on the undertaking and execution of the project in the implementation of the foreign contracted engineering project, but lack the correct understanding of the international engineering logistics and the possible consequences caused by the risk factors. Compared with other costs of foreign contracted projects, the international engineering logistics is only a small part of foreign contracted projects. Because the projects contracted by Chinese engineering companies are mainly concentrated in Asia, Africa, Latin America and other countries and regions with relatively backward industrial foundation, and the industrial foundation of these regions is weak. Therefore, during the implementation of the project, a large amount of equipment, materials and other materials need to be purchased from China or a third country and transported to the project site through international engineering logistics. The risks faced by international engineering logistics in the process of execution are very complex and changeable. In the process of project execution, many Chinese engineering companies have not formed a dynamic monitoring mechanism for the risk factors affecting the execution of international engineering logistics, as well as the prediction and evaluation of the overall operation trend of international engineering logistics under the influence of risks. As a result, when some Chinese engineering companies encounter an emergency of international engineering logistics during the implementation of foreign contracting projects, they have no plans and then take timely remedial measures, leaving them in a state of laissez - laire and out of control.

Therefore, its risk management should be dynamic, from the inside to the outside of its comprehensive monitoring, take timely measures to deal with the possible loss caused by the change of risk factors.

Kazakhstan is an energy-rich country with huge reserves of oil and natural gas. In recent years, China's total imports of oil and natural gas from Kazakhstan have shown a steady growth trend [8].

With China's oil demand set to grow further and a greater share of its oil coming from overseas, China-Kazakhstan oil projects are becoming increasingly important. The risks and safety of future projects and the logistics risks and safety related to them determine the basis and guarantee for the smooth progress of oil projects. Then how to quantitatively evaluate the risk and safety degree of the logistics of Chinese and Kazakstan projects?

This paper establishes the oil project logistics security risk evaluation system, and to evaluate based on Support Vector Machine (SVM) theory, because of the SVM compared with the Analytic Hierarchy Process (AHP) and fuzzy evaluation method to, in addition to the more subjective compared with other methods such as neural network methods, simple classification of surface, high fitting precision [9–11]. Moreover, it can analyze the inconsistent results according to the needs [12]. Therefore, it is evaluated on the basis of support vector machine theory, in order to achieve the purpose of evaluating the security of cross-border logistics of petroleum projects [13, 14].

2 Support Vectory Machine Theory

Statistical learning theory is a machine learning method with solid theoretical foundation developed on the basis of traditional statistics. Since the 1990s, other learning methods such as neural network have encountered difficulties. It has formed a relatively perfect theoretical system-statistical learning theory, and put forward a new pattern recognition method-support vector machine, so this research has been paid more and more attention.

SVM is a new machine learning method proposed by V. Vapnik et al. based on Statistical Learning Theory (referred to as SLT). It shows many unique advantages in solving the problems of small sample, nonlinear and high-dimensional pattern recognition, and can effectively realize the accurate fitting of high-dimensional nonlinear system based on small sample [15–17]. Good results have been achieved in pattern recognition, function approximation and probability density estimation. The regression function can be expressed as follows:

$$y_i[(w \cdot x_i) + b] - 1 \geq 0 \tag{1}$$

In this case, the classification interval is equal to $2/\|w\|$, so that the maximum interval is equivalent to the minimum. The classification surface that satisfies the condition and minimizes is called the optimal classification surface, and the training sample point on H1 and H2 is called support vector. By using the Lagrange optimization method, the above optimal classification surface problem can be transformed into its dual problem, the maximum values of the following functions can be solved.

$$\mathrm{Q}(a) = \sum\nolimits_{i=1}^{n} a_i - \frac{1}{2}\sum\nolimits_{i,j=1}^{n} a_i\, a_j y_i y_j \left(x_i \cdot x_j\right) \tag{2}$$

For the Lagrange multiplier corresponding to each sample, the corresponding sample is the support vector, and the optimal classification function is obtained.

$$f(x) = \mathrm{sgn}\{(w^* \cdot x) + b^*\} = \mathrm{sgn}\left\{\sum\nolimits_{i=1}^{n} a_i^*\, y_i\left(x_i \cdot x_j\right) + b^*\right\} \tag{3}$$

For the nonlinear problem, it can be transformed into a linear problem in a high-dimensional space by nonlinear transformation, and then the optimal classification plane in the space can be transformed. Select the appropriate kernel function to meet the Mercer condition to transform into a high-dimensional space, and the objective function is transformed into:

$$\mathrm{Q}(a) = \sum\nolimits_{i=1}^{n} a_i - \frac{1}{2}\sum\nolimits_{i,j=1}^{n} a_i\, a_j y_i y_j K\left(x_i \cdot x_j\right) \tag{4}$$

The corresponding classification function becomes:

$$f(x) = \mathrm{sgn}\{(w^* \cdot x) + b^*\} = \mathrm{sgn}\left\{\sum\nolimits_{i=1}^{n} a_i^*\, y_i K\left(x_i \cdot x_j\right) + b^*\right\} \tag{5}$$

Gaussian radial basis kernel function commonly used in support vector machines.

$$K\left(x_i, x_j\right) = \exp\left(-\gamma\|x_i - x_j\|^2\right) \tag{6}$$

Among these kinds of kernel functions, the radial basis kernel function is the most widely used, which has a wide convergence domain and is an ideal classification basis function. Gaussian radial basis kernel function is used in this study.

3 System Construction

The complexity of the countries involved in the Belt and Road makes China face various risks in its foreign energy investment, such as international politics, economy, market, etc. The same is true for Kazakhstan, as well as the impact on the logistics security of the projects it executes.

3.1 Political Risk

There are some political risks discussed below.

The first one is the risk of government intervention. Kazakhstan interferes in the investment behavior of enterprises in the investing countries based on the protection of its own interests, which is mainly caused by the restriction of investment area, investment proportion or the setting of additional laws and regulations. Such intervention is deformed and concealed.

The second one is the risk of a change of government. The risk of government change refers to the uncertainty brought about by a series of circumstances such as political instability and regime change. The power change will bring hidden risks to Chinese energy companies' overseas assets in the country and have a huge impact on logistics.

The third one refers to the risk of war. War risk refers to the possibility of the outbreak of war within the country where the energy is located, or the possibility of losses caused by the coup d 'etat or civil strife between different parties, ethnic groups, religious factions, etc. At the present stage, the coup and civil strife between ethnic and religious factions in Kazakhstan are mainly concentrated in the high-risk areas with unstable political situation.

The fourth risk is about policy and law changes. Policy change risk refers to the possibility of loss caused by the policy change of the investee country. These include major changes in the investment policies of some host countries, rather than the direct implementation of restrictive policies. These may be the biggest non-economic risks encountered.

3.2 Economic Risks

In terms of economic risks, there are mainly risks caused by changes in operation, exchange rate and logistics property rights risk.

The first economic risk is operational risk.

Due to the low price of oil, gas, coal and other commodities, the economic decline of energy economies, the emergence of fiscal deficit crisis, and the rise of sovereign credit risk, the logistics security risk of related projects increases accordingly.

The second one is exchange rate risk.

Exchange rate risk is the loss caused by uncertainty or possibility caused by exchange rate fluctuations. Income generated by overseas investment of energy companies is very easily affected by exchange rate fluctuations, while logistics is more directly affected by exchange rate fluctuations.

The third one is logistics property rights risk.

Property rights risks include the forms of property rights held by foreign investment, including equity participation, holding and sole proprietorship, etc. At present, energy enterprises in Kazakhstan are dominated by equity participation, which directly causes that Chinese enterprises do not have the decision-making right and the final right of speech, and also brings risks to the transfer of logistics property rights.

3.3 Industry Market Risks

Industry risks mainly come from the uncertainty in the market between the investing country and the receiving country, such as market competition, product market, industry technology and other risks. Competitive risk refers to the measures taken by the same industry to deal with environmental changes. For example, in order to ensure market share and domestic fiscal balance, oil-resource countries have not cut production to counter falling prices, and they have also adopted cost control levels to eliminate competitors. The influx of American oil and gas into the international market is bound to break the balance between supply and demand. See Table 1 and Fig. 1.

Table 1. Evaluation index system for the construction of cross-border logistics security system

Target layer	Criterion layer	Index layer	Serial number
Cross-border logistics security system	Political risk	Risk of government intervention	C1
		The risk of a change of government	C2
		The danger of war	C3
	The economic risk	Operational risk	C4
		Currency risk	C5
		Logistics Property Rights Risk	C6
	Industry market risk	Competition risk in the market	C7
		Product market risk	C8
		Industry technology risk	C9

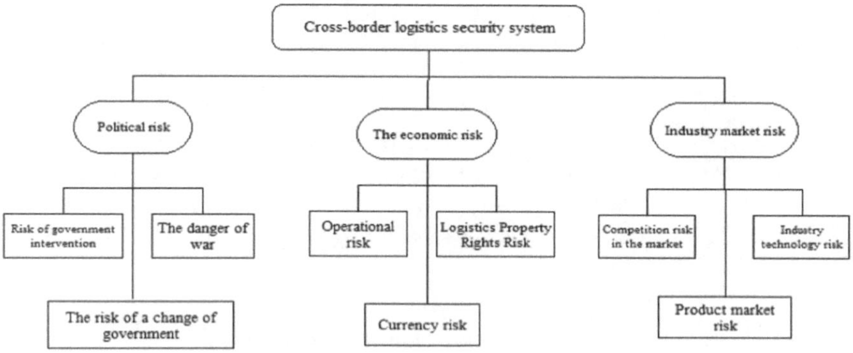

Fig. 1. The construction of cross-border logistics security system

According to the basic situation of the construction of cross-border logistics security system, the index evaluation system is established. After the evaluation index system is established, the safety grade is divided into five levels, namely, unqualified, qualified, medium, good and excellent. The range of the safety grade is [0, 0.2], (0.2, 0.4], (0.4, 0.6), (0.6, 0.8] and (0.8, 1]. From this, the set of evaluation criteria is determined. See Table 2.

Table 2. Evaluation grade standard of cross-border logistics security system

Grade	Level	Grade interval
I	Excellent	[1.0, 0.8)
II	Good	[0.8, 0.6)
III	Secondary	[0.6, 0.4)
IV	Qualified	[0.4, 0.2)
V	Unqualified	[0.2, 0.0]

4 Evaluation of Logistics Security Based on SVM

Sample data came from data collection in 2020. Experts collected 50 training samples and 50 prediction samples, and established the initial data table of logistics safety indicators, as shown in Table 3.

Through the joint discussion of the expert group, the feature items of transnational logistics security level are classified, deleted, merged and reconstructed, and finally nine evaluation indexes of logistics security level are obtained.

Table 3. Evaluation index data and evaluation grade

Sample	Serial number	C1	C2	C3	C4	C5	C6	C7	C8	C9	Overall grade
Train sample	1	0.4	0.7	0.5	0.3	0.6	0.4	0.2	0.7	0.5	High
	2	0.5	0.3	0.6	0.4	0.7	0.5	0.1	0.3	0.6	Secondary
	…										…
	50	0.6	0.7	0.5	0.3	0.5	0.3	0.6	0.7	0.5	Low
Test sample	1	0.4	0.2	0.5	0.5	0.3	0.6	0.4	0.2	0.5	High
	2	0.7	0.5	0.4	0.2	0.5	0.3	0.6	0.5	0.4	Secondary
	…										…
	50	0.1	0.4	0.2	0.5	0.3	0.6	0.4	0.4	0.2	Low

After the training set is determined, knowledge acquisition, decision function construction and result verification of support vector machine are realized by programming on Matlab7.01 platform and SVMlab1.5 toolbox, and risk prediction results are output. To verify the practicability and accuracy of SVM model evaluation, through repeated training and learning, the training effect has been obtained.

Training model using Matlab7.0 language programming to realize training SVM model, and select the RBF kernel function model, a wide range of domain of convergence, is an ideal classification basis function, so it USES gaussian radial basis kernel function, and to determine the two parameters: the penalty parameter and kernel parameter, and selection parameters by using the method of cross validation. Taking 50 training samples as the learning set, when the penalty parameter Cure 1.16387 and kernel function K = 0.6604 were taken, the error was the smallest and The SVM model was the optimal mode.

According to the basic process of SVM, as the post-information recognition system of the prediction model, it calculates the training samples by recognizing the running results of the training samples and obtains the results.

After the training of the SVM model, the optimized SVM model was tested using the untrained samples. Through testing the past 50 groups in Table 3, the obtained test results, and the coincidence rate grade between the test level and the actual is 91.1465%, indicating that it is feasible to use support vector machine to evaluate the level of intelligent logistics, the evaluation grade is qualified. It has good application and popularization value in practical evaluation.

The kernel function of RBF is an ideal basis function for classification. Therefore, gaussian RBF is used for operation. Two parameters are determined: penalty parameter based on cross validation and kernel parameter selection. 30 training samples are taken as the learning set. When the parameter value is the penalty parameter CURE 1.1673 and the kernel function K = 0.6704, the error is the smallest and the SVM model is the optimal model.

According to the post-SVM information recognition system as the prediction model, the training samples are calculated and the running results of the training samples are obtained through the cognition of the running results of the training samples.

After training the SVM model, untrained samples were used to test the optimized SVM model. Through the past 30 groups of tests in Table 3, the test results, and the test level and the actual coincidence rate grade is 91.1546%, indicating that it is feasible to use support vector machine to evaluate the level of logistics security risk, qualified assessment level. It has good application and popularization value in practical evaluation.

5 Conclusion

Under the background of "One Belt and One Road", China-Kazakhstan energy cooperation has entered a new development stage. It is necessary to ensure the smooth and long-term stable operation of the oil and gas field cooperation projects between the two countries, and logistics security is the basic guarantee of the project cooperation.

Based on the reality of sino-Kazakhstan economic cooperation and logistics development, this paper establishes a logistics security evaluation system from the aspects of politics, economy and market, and from the three dimensions of target level, standard level and index level, and uses support vector machine to evaluate it.

With the deepening of bilateral cooperation projects, it is also very important and necessary to dynamically screen corresponding indicators at different development stages. It is also necessary for the security of cross-border logistics in the project cooperation between the two countries to evaluate and analyze accordingly as objectively and dynamically as possible and establish an evaluation system.

The project cooperation between China and Kazakhstan has broad prospects, with both advantages and challenges, but the advantages outweigh the challenges. China should take corresponding measures to maintain the sound political and economic cooperative relations with Kazakhstan, enhance mutual trust and push the china-Kazakhstan comprehensive strategic partnership to a new level.

This topic for logistics security risks, especially for "area" in the way of the first countries-kazakhstan logistics security risk, political risk, economic risk, market as the main aspects, the construction of the security system, to the two countries on building the security system of logistics security risks, security risks for multinational logistics to achieve the purpose of evading and resist, Taking China and kazakhstan as a case, it plays an exemplary role of taking the point with the line and the point with the surface, so it is taken as an exploration and topic.

On the basis of drawing on many domestic and foreign literature, combined with the actual situation and certain data of China and Kazakhstan, this topic puts forward a set of transnational logistics security risk system suitable for "The Belt and Road". On the macro level, it provides ideas for solving the problems of "One Belt and One Road" development, which is conducive to the long-term development of transnational logistics and can drive the internationalization of the whole economy; On the micro level, it can provide theoretical and practical reference for enterprises to carry out transnational logistics and establish security risk system and mechanism, so as to adopt effective logistics risk control methods and improve economic and social benefits.

Acknowledgement. This paper is supported by: (1) Doctoral Research Foundation of Wuhan Institute of Industry and Technology: Analysis and Countermeasures of Cross-border Logistics Security of Petroleum projects in China and Kazakhstan based on "Belt and Road Initiative" (D2019004); (2) Hubei Business Service Development Research Center Fund project: Hubei Intelligent Logistics Development support system Research (2020Z01).

References

1. Reeves, J.: China's silk road economic belt initiative: network and influence formation in central Asia. J. Contemp. China **27**(112), 502–518 (2018)
2. Mitrovic, D.: The belt and road: China's ambitious initiative. China Int. Study **59**, 76 (2016)
3. Ferdinand, P.: Westward ho. the China dream and one belt, one road: Chinese foreign policy under Xi Jinping. Int. Aff. **92**(4), 941–957 (2016)
4. Dadi, Z.: China's energy situation and new energy industry development trend and bright spots. China Econ. Trade Guide **1**, 39–40 (2017). (in Chinese)
5. Kim, K., Kim, H., Kim, S.K., et al.: i-RM: an intelligent risk management framework for context-aware ubiquitous cold chain logistics. Expert Syst. Appl. **46**, 463–473 (2016)
6. Vinay, D.R., Ananda, B.J.: A novel secure data hiding technique into video sequences using RVIHS. Int. J. Comput. Netw. Inf. Secur. (IJCNIS) **3**(2), 53–65 (2021)

7. Dadi, Z.: China's energy revolution faces many challenges. Manag. State-owned Enterp. **2**, 18–19 (2018). (in Chinese)
8. Huiqing, K.: Research on the Risk of Foreign Investment of Chinese Energy Enterprises under the One Belt and One Road Initiative. University of International Business and Economics, Beijing (2018). (in Chinese)
9. Chatterjee, S., Jose, P.G., Datta, D.: Text classification using SVM enhanced by multithreading and CUDA. Int. J. Mod. Educ. Comput. Sci. (IJMECS) **11**(1), 11–23 (2019)
10. Desai, P., Kulkarni, G.R.: Use of API's for comparison of different product information under one roof: analysis using SVM. Int. J. Inf. Technol. Comput. Sci. (IJITCS) **10**(6), 11–22 (2018)
11. Thomas Scaria, L., Christopher, T.: Microarray gene retrieval system based on LFDA and SVM. Int. J. Intell. Syst. Appl. (IJISA), **10**(1), 9–15 (2018)
12. Vapnik, V.N.: The Nature of Statistical Learning Theory. Springer-Verlag, NY (1995)
13. Xuegong, Z.: On statistical learning theory and support vector machine. Acta Autom. Sin. **26** (1), 32–42 (2000)
14. Ronghe, Y., Zonghui, L., Xiangqian, W.: Study on safety management risk prediction model of coal mine enterprises-based on SVM and RS. East China Econ. Manag. **28**(10), 159–163 (2014). (in Chinese)
15. Ahmad, M., Aftab, S.: Analyzing the performance of SVM for polarity detection with different datasets. Int. J. Mod. Educ. Comput. Sci. (IJMECS) **9**(10), 29–36 (2017)
16. Haiyan, W., Jianhui, L., Fenglei, Y.: Overview of support vector machine analysis and algorithm. Appl. Res. Comput. **31**(5), 1281–1286 (2014)
17. Dong, X., Zheng, W.: System Analysis and Design Based on MATLAB-Neural Network. Xidian University Press, Xi 'an (2002)

Simulation and Optimization of Flight Scheduling in Dual-Runway Cargo Airport for Vaccine Transport Based on Artificial Bee Colony Algorithm

Jinshan Dai, Yao Zhang, Mengya Zhang(✉), and Mengjian Wang

School of Logistics Engineering, Wuhan University of Technology, Wuhan 430063, China
jinshan.dai@whut.edu.cn

Abstract. The mass production of COVID-19 vaccine is gradually mature, transportation and vaccination of it is imminent. To meet the requirement of COVID-19 vaccine freight, the rapid response of the corresponding freight airport is important, any flight delay during the process will influence the quality of COVID-19 vaccine, hence the entry/departure scheduling of the airport needs further optimization. Therefore, taking the aircraft entry/departure process as the target, the paper analyzes and compares the current situation of international cargo airports, and conducts a research on the aircraft entry/departure scheduling process for vaccine transportation. The model in this paper, is based on the dual-runway airport, adds the vaccine transport flight delay loss function to optimize the objective function, uses Python language to compel the discrete Artificial Bee Colony Algorithm to calculate, and the comparison with the first come first serve method which used widely in airport scheduling is made to verify its effect. The simulation result shows that the overall flight optimization effect is 48% and the flight delay time of COVID-19 vaccine is reduced by 50%, which can be regarded as an accepted result of scheduling optimization.

Keywords: COVID-19 vaccine freight · Aircraft scheduling problem · Multi-objective optimization · Artificial bee colony algorithm

1 Introduction

COVID-19 has attracted high attention from various countries by its fast spread speed, wide spread range and serious infection consequences. The continuous upgrading of the epidemic situation has not only made a huge impact on the China economy and Chinese society, in the short term, but also caused indelible fluctuations on the global economy and society in the long term. In order to ensure the rapid response of COVID-19 vaccine transportation, major airports around the world (especially cargo airports) have built facilities and management teams for vaccine transportation, and implemented support services such as priority entry and priority departure for vaccine transportation.

Z. Hu et al. (Eds.): AIMEE 2021, LNDECT 107, pp. 206–218, 2022.
https://doi.org/10.1007/978-3-030-92537-6_20

1.1 Analysis of the International Cargo Airports

Following the approval of an emergency permit for the use of COVID-19, FedEx at Memphis Airport used priority night service and began shipping vaccines and vaccine management supplies packages under FedEx Priority Alert Advanced Monitoring. From 2020, Louisville Airport's "UPS World Port" became the air transportation hub for the delivery of the first COVID-19 vaccine to all parts of the United States. DHL Global Freight has provided organizations and governments with dedicated airlift capacity of 100 tons per week to transport testing kits and medicines during the COVID-19 period, and has launched charter flights connecting Asia Pacific, Europe and the United States to meet customer needs. Facing the medical supplies and vaccine transportation pressure from COVID-19 outbreak, Liege Airport as Europe's hub for the transportation of vital medical supplies and one of eight new WHO hubs around the world, has installed a new "pop-up" cargo handling facility to speed up the transport and storage of medical equipment, vaccines and related medical supplies.

1.2 Major Challenges for Air Transportation of COVID-19 Vaccine

The main difficulties in recent air freight for COVID-19 vaccine distribution lies in three aspects: firstly, airline cold chain transportation resources (such as RFID and sensor networks) for COVID-19 vaccine are inadequate [1]. In the context of the COVID-19, there are not only huge air transportation demand for COVID-19 vaccine, but also have demands for affiliated medical supplies and equipment, resulting in the gradual saturation of traffic flow to the airport and the corresponding airspace. Secondly, impact of air transport delay. The contradiction between the limited air transport facilities and the continuously increasing air transport demand is directly manifested by the high level of flight delays, which will greatly increase the risk of cold-chain break for COVID-19 vaccine transportation [2]. Thirdly, traditional scheduling methods are not effective. The traditional scheduling method is FCFS or experience based scheduling. Such scheduling methods are not effective enough, the effect is difficult to evaluate, and difficult to meet the requirements of COVID-19 vaccine airline transportation.

1.3 Influencing Factor Analysis for Flight Entry and Departure Scheduling

Judging from the current airport development situation, most mainstream airports adopt the design of double parallel runway. The configuration of the dual-runway can mainly be divided into three categories by location difference: parallel dual-runway, overlap dual-runway and V-type dual-runway. The entry/departure stage is essentially the comprehensive designation of the departure time and runway occupation queue according to the differences between airport resources and flight attributes. The busier the large airports, the denser the flights are, so the more the flight sorting affects the flight capacity of the entire airport [3–5].

Scheduling is an important work integrating many technologies and involving many contents, which needs special attention and is worthy of in-depth research [6–9].

In order to systematically and structurally analyze the aircraft entry/departure problem, this paper mainly analyzes the influencing factors of flight entry/departure scheduling from two perspectives of certainty and uncertainty, based on the MECE principle (Mutually Exclusive Collectively Exhaustive), the summary extracts 12 influencing factors. The main analysis results are shown in Table 1 below.

Table 1. Influencing factors of flight entry and departure scheduling

Deistic factors	Uncertainty factors
Flight delay time	Weather factors
Total flight delay costs	Airport Facilities issues
Runway capacity	Passenger and postal delay
Safety interval	Airline service delays
Aircraft model	Airspace traffic control failure
Flight Priority	Flight accident

Further research shows that the aircraft flight scheduling problems mainly focus on the objective function, constraints and solving algorithms. The objective functions include but are not limited to: the maximization of runway utilization, the minimization of flight scheduling completion time, the minimization of flight delay cost and the maximization of flight scheduling fairness. Constraints include but are not limited to: exclusive constraints of runway for incoming and departing flights, unique constraints of runway allocation for incoming and departing flights, constraints of maximum runway capacity and constraints of safe time interval between take-off and landing flights, etc. Solving algorithm can be generally divides into tow categories of accurate algorithm and intelligent algorithm. Accurate solving algorithms include branch and bound method and column generation algorithm. Intelligent algorithms include genetic algorithm, ant colony algorithm, bee colony algorithm, Wolf swarm algorithm, immune algorithm and particle swarm algorithm.

In order to improve the airline cold chain transport efficiency of COVID-19 vaccine, it is very important to study and improve the flight scheduling and entry/departure sequencing algorithm for COVID-19 vaccine transport.

2 Literature Review

Scholars have studied many cases of single and dual-runways scheduling problem. Beasley gives a mixed integer 0–1 formula for the single-runway problem and extends it to multi-runway problems, and strengthens the linear programming relaxation of these formulae by introducing additional constraints [10]. Salehipour proposes a simple heuristic strategy for the airport start and landing scheduling problem. The algorithm features the decomposition of problems into a smaller, easier to solve problem chain [11]. Messaoud proposed an exact method to solving the static situation of the aircraft landing problem on a pair of runways. Computational experiments using data from Carthage Airport in Tunisia show that the proposed method has certain advantages over the air traffic control rules applied to relevant airports [12]. Zhou proposes a flower pollination algorithm embedded in runway balance to solve the problem and designed cognitive learning and runway balance strategies in context to improve its search capabilities [13]. Abdullah proposed a harmonious search algorithm for multi-runway landing scheduling; each aircraft has its predefined target landing time within one time window and the separation time between each continuous aircraft pair; the goal is to minimize the deviation between each aircraft and target landing time under a set of constraints [14]. Celikbilek presents a genetic algorithm based on fuzzy clustering to solve the problem of landing ordering on the runway (ALS); the model comprehensively considers the optimal target landing time for each aircraft based on their transport type and fuel loss [14]. Rezaei proposed an intelligence-based aircraft landing scheduling approach to balance airline agents (flights) to minimize the deviation between actual landing and target landing times and air traffic entity agents to maximize the use of the runway targets [15]. Kam Hung Ng proposed a constructive heuristic algorithm based on Artificial Bee Colony Algorithm to solve aircraft speed control in near terminal zone delay and waiting mode, increasing conservatism when handling air and terminal traffic to maintain high runway system elasticity, enhance the robustness of landing plans, and reduce the workload of air traffic controllers [16]. Risikat developed an adaptive dynamic scheduling algorithm based on artificial bee colony for vehicular traffic control which can minimizing the average waiting time at the cross intersection [17]. R K Jena proposed a novel multi-objective artificial bee colony algorithm based framework and optimized the operational modes of sensor nodes along with clustering schemes and transmission signal strengths [18]. Pengli Zhao have proposed a new approach to quickly solve the multi-target aircraft landing problem, which combs the different criteria for the aircraft landing problem and selects the average scheduling time, maximum flow time, and maximum delay time as the multi-objective [19]. Kulida studied the problem of aircraft landing order optimization. For formal description, main limitation and target function of aircraft landing optimal queuing problem are presented. For real time solution, optimization methods for the initial sequence is presented and a heuristic algorithm for the approximate solution of the problem [20].

From insights of the literature, in this paper, the mathematical model of entry/departure scheduling for COVID-19 vaccine cold chain transportation in dual-runway airport is made, combining with the optimization of the objective function and constraints under COVID-19 vaccine transportation feature, establish COVID-19 vaccine transport dual-runway entry/departure scheduling method, and tried to use discrete artificial bee colony algorithm to calculate and verify its effect.

3 Modeling of Flight Entry and Departure Scheduling Problem

3.1 Common Strategies for Flight Entry and Departure Scheduling

When coordinating the flights, determine the scheduling strategy for entry/departure flights is important [21]. The Flight entry/departure scheduling strategy is summarized into three categories by flight priority: entry priority strategy, departure priority strategy and entry/departure joint scheduling strategy. The optimization results will change greatly under different strategy.

1) The entry priority strategy, which landing flights are in high priority in the runway time slot allocation. Priority runway time slot for incoming flights. The mobilization priority strategy is conducive to saving the airline fuel costs and ensuring the safety of the aircraft in the flight state.
2) The departure priority strategy, which departure flights have a high priority in runway time slot allocation. Priority runway time slot for departure flights. The departure priority strategy is often used for the saturation of airport parking capacity. In order to avoid airport congestion caused by excessive aircraft waiting on the ground, priority is arranged to release the airport capacity.
3) The entry/departure joint scheduling strategy, which entry/departure flights were scheduled regardless of priority. The entry/departure flights are scheduled through the optimization algorithm, and the airport runway time slot is uniformly allocated to the entry/departure flights. Under this strategy, a relatively good entry/departure order can often be obtained. Reasonably allocate the entry/departure time flights accelerates the release of the runway resources, and improve the utilization rate of the runway. Therefore, this paper mainly studies the airport entry/departure scheduling strategy based on dual-runway airports.

3.2 Joint Scheduling Mathematical Model of Dual Runway Airport

3.2.1 Mathematical Model Hypothesis of Dual-Runway Joint Scheduling

In a certain time, window Δt, there are n incoming flights and M departing flights.

1) The airport has two runways;
2) The type of incoming and departing flights is known, and the estimated arrival time is known;
3) Departure flights will take off on time according to the assigned time slot, and the errors are ignored;
4) The incoming flight lands on time according to the assigned time slot, and the errors are ignored;
5) Flight arrivals and departures can arrive at the monitoring points on time, and the errors are ignored.

3.2.2 Definitions of Symbols Related to Dual-Runway Joint Scheduling Problem

See Table 2.

Table 2. Symbols and definitions

Parameters	Description
A_{ir}^{v}	A collection of airlines, $A_{ir}^{v} = \{A_{ir}^{1}, A_{ir}^{2}, A_{ir}^{3}, \ldots A_{ir}^{v}\}$
F_C	Entry and departure flight collection, $F_C = \{f_1, f_2, f_3, \ldots f_c\}$ and $F_C = F_A^a \cup F_D^d$
F_A^a	Access flight Collection, $F_A^a = \{f_A^1, f_A^2, f_A^3, \ldots f_A^n\}$
F_D^d	Departure Flight Collection, $F_D^d = \{f_D^1, f_D^2, f_D^3, \ldots f_D^m\}$
R	Parallel, independent runway collections, $R = \{1, 2, 3, \ldots r\}$
STA_i	Actual arrival time of the incoming flight f_A^n
ETA_i	Estimated arrival time for incoming flights f_A^n
STD_i	Actual departure time of departure flight f_D^m
ETD_i	Expected arrival time for departure flights f_D^m
AAT_{max}	Maximum advance time allowed for entry flight
ADT_{max}	Maximum delay time allowed for entry flight
DAT_{max}	Maximum advance time allowed for departure flight
DDT_{max}	Maximum delay allowed for departure flight
R_A^r	$R_A^r = 1$, Flight access on runway r $R_A^r = 0$, no flights entered on runway r
R_D^r	$R_D^r = 1$, departure on runway r $R_D^r = 0$, no flights leaving on runway e
S_s^c	$S_s^c = 1$, indicating that the time slot is assigned to the flight f_c $S_s^c = 0$, indicating that the time slot is not assigned to the flight f_c
θ_{ij}	Minimum safety interval for two consecutive flights
T	Target function with the minimum total delay time
C	Function of COVID-19 Vaccine value loss over time
W	Total delay time and loss function

3.2.3 Modeling of Dual-Runway Entry and Departure Joint Scheduling Problem

The absolute difference between the actual and estimated entry/departure times of all flights is the total delay time [22–25]. In this model, all advance and delay of flight is taking into consideration.

The objective function of the flight in time period Δt, which takes the total delay time as the measurement index, is shown in Eq. (1):

$$\min T = \sum_{i=1}^{n} |ETA_i - STA_i| + \sum_{i=1}^{m} |ETD_i - STD_i| \tag{1}$$

First of all, based on the analysis of the influencing factors of flight approach and departure sequencing above, this paper mainly models the deterministic factors and ignores the uncertain factors. The aircraft type and safety interval are relatively determinable, so they are not reflected in the modeling equation. The runway capacity has been determined in the assumption process that only one aircraft is allowed to take off and land on each runway; Therefore, the next step is to comprehensively consider the waiting time of flight delay, the total cost of flight delay and flight priority.

In the traditional model, the objective function is determined by minimizing the delay and waiting time. However, due to the strict requirements on transportation temperature, transportation packaging and transportation conditions, the damaged transportation packaging or flight delay may lead to the failure of the vaccine of an entire transportation unit. Therefore, to some extent, it is impossible to arrange the arrival and departure of COVID-19 vaccine transportation flights with the minimum total delay time within a simple time period. Therefore, new constraint conditions are introduced in this paper and the solved algorithm is adjusted to improve the deficiencies in the basic model.

(1) Loss function of the COVID-19 Vaccine transportation flight delay.

Define the loss of the delay to enter the airport airspace or leave the hanger until leaving the airport airspace or entering the hanger. The delay cost can be present as in the function (2):

$$V = \sum_{v}^{i=1} f_i \cdot C_i \tag{2}$$

$C_i = e^{d_i t_i}$ is the function of value loss of the COVID-19 vaccine over time, and d_i is the flight delay loss coefficient of the COVID-19 vaccine, t_i is the delay time of COVID-19 Vaccine transportation flight General freight and civil air traffic are considered not to be affected by delay.

(2) Target function in the background of the COVID-19 Vaccine transportation.

Considering the need for the control tower to make comprehensive decisions based on the actual situation of the airport, and the loss caused by the delay of the COVID-19 vaccine should be minimized. The objective function (3) puts the two functions in the same order of magnitude by making the compensation factor β, and then assigns the p, q these two weight coefficients according to the actual situation, transforming the multi-object optimization problem to a single-object optimization problem. The constraints on the two subjective weight coefficients are shown in formula (4).

$$W = min(p\beta T + qC) = \\ min\left[p\beta\left(\sum_{i=1}^{n}|ETA_i - STA_i| + \sum_{i=1}^{m}|ETD_i - STD_i|\right) + q\left(\sum_{i=1}^{v} e^{d_i t_i} \cdot f_i\right)\right] \tag{3}$$

$$p + q = 1 \tag{4}$$

For the cross dual-runway, V-type dual-runway and some close distance parallel dual-runway airport, two runways need adjusted entry/departure of flight in two different runway according to the type of the aircraft and the condition of the airport. To simplify the model, this paper mainly targets the far distance parallel dual-runway model, and each runway can independently operate without interference with each other.

Accordingly, the parallel dual-runway entry and exit joint scheduling constraints under the transportation background of COVID-19 vaccine distribution are as follows:

(3) Constraints

$$\theta_{ij} \leq \begin{cases} STA_j - STA_i \\ STA_j - STD_i \\ STD_j - STA_i \\ STD_j - STD_i \end{cases} \tag{5}$$

$$\sum_{a=1}^{n} S_s^{F_A^a} + \sum_{a=1}^{m} S_s^{F_D^d} \leq 2 \tag{6}$$

$$R_A^1 + R_D^1 \leq 2 \tag{7}$$

$$\begin{cases} STA_i - ETA_i \geq -AAT_{max} \\ STD_i - ETD_i \geq -ADT_{max} \end{cases} \tag{8}$$

$$\begin{cases} STA_i - ETA_i \leq DAT_{max} \\ STD_i - ETD_i \leq DDT_{max} \end{cases} \tag{9}$$

Formula (5) Represents the constraint between the safety time and the safe distance of the flight, namely, the minimum safety interval between the front and rear two flights must be less than or equal to the actual landing time or distance interval of the two flights. formula (6) (7) indicates the time slot and runway resource allocation constraints. For the remote parallel dual-runway entry/departure sorting process, because each runway can operate independently without interference with each other, its time slot and runway resources can be allocated to two flights for activities of entrance and departure. Formula (8) Represents the time constraint of the arrival flight landing in the entry/departure scheduling. Formula (9) Represents the time constraint for the departure flight in the entry/departure scheduling.

4 Model Calculation Based on Artificial Bee Colony Algorithm

4.1 Design of Discrete Artificial Bee Colony Algorithm

Since the data of aircraft approach and departure process belong to discrete state, this section mainly focuses on this model and adopts the Artificial Bee Colony Algorithm (ABC) to carry out an example analysis on the parallel double runway approach and departure joint sorting under the background of COVID-19 vaccine. The algorithm flow is shown in Fig. 1.

4.2 Simulation Analysis of Entry and Departure Joint Scheduling

Randomly Selected Memphis Airport's 10 flight information the departure time. CAAC has three types of aircraft types, Small (S), Large (L) and Heavy (H). The ten flights are being numbered as HF1, LF2, HF3, HF4, SF5, HF6, LF7, HF8, HF9, SF10. The encoded letters represent the type of the aircraft. The main data are shown in Table 5 below.

Flight HF3 and HF8 were being randomly selected as the COVID-19 vaccine transport flights.

For easy computing, p = q = 0.5. The minimum time intervals of different types of aircraft are represented by a time matrix S in the matrix (10).

$$S = \begin{bmatrix} 1.0 & 1.5 & 2.0 \\ 1.0 & 1.5 & 1.5 \\ 1.0 & 1.0 & 1.0 \end{bmatrix} \tag{10}$$

Table 3 is the basic data of the calculation example. Table 4 and Table 5 represent the operation results of the present paper model using FCFS and the swarm algorithm. Table 6 mainly shows the operational effect analysis of the two algorithms.

Mathematical model of parallel dual-runway joint sorting is performed by the use of discrete Artificial Bee Colony Algorithm. It can be seen from Table 6 that the overall delay time from 12.5 min solved to FCFS algorithm to 6.5 min, improvement efficiency of 48%; and the overall delay time of COVID-19 flight was reduced from 2 min to 1 min, improvement effect of 50%; although the overall calculation time increases to some extent, the overall calculation time is within 3 s.

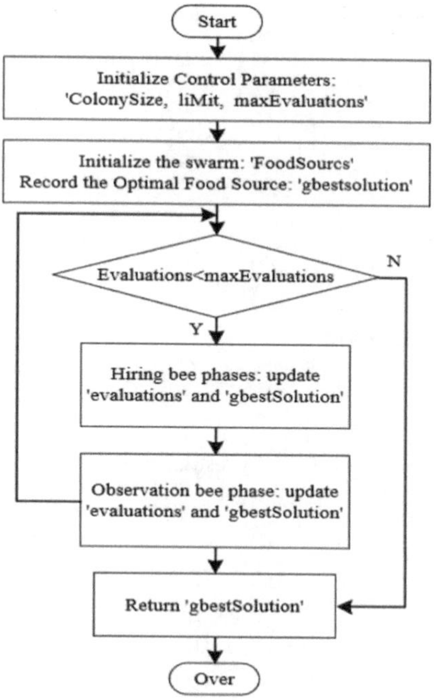

Fig. 1. Flow chart of discrete bee colony algorithm

Table 3. Accounting example flight start and drop information table

No	Runway # 1, ETA,/min	Runway # 2, ETA,/min	COVID-19 flights
HF1	11.0	10.0	No
LF2	15.0	17.0	No
HF3*	6.0	7.0	Yes
HF4	6.0	7.0	No
SF5	9.0	12.0	No
HF6	7.0	6.0	No
LF7	15.0	17.0	No
HF8*	6.0	7.0	Yes
HF9	6.0	7.0	No
SF10	9.0	12.0	No

Table 4. FCFS simulation results table

Flights number	Runway # 1			Flights number	Runway # 2		
	ETA	STA	AAT		ETA	STA	AAT
HF3*	6.0	6.0	0.0	HF6	6.0	6.0	0.0
HF4	6.0	7.0	1.0	HF1	10.0	10.0	0.0
HF8*	6.0	8.0	2.0				
HF9	6.0	9.0	3.0				
SF5	9.0	11.0	2.0				
SF10	9.0	12.0	3.0				
LF2	15.0	15.0	0.0				
LF7	15.0	16.5	1.5				

Table 5. Simulation results of ABC algorithm

Flights number	Runway # 1			Flights number	Runway # 2		
	ETA	STA	AAT		ETA	STA	AAT
HF2*	6.0	6.0	0.0	HF5	6.0	6.0	0.0
HF7*	6.0	7.0	1.0	HF3	6.0	7.0	1.0
SF4	9.0	9.0	0.0	HF8	6.0	8.0	2.0
SF9	9.0	10.0	1.0	HF0	10.0	10.0	0.0
LF1	15.0	15.0	0.0				
LF6	15.0	16.5	1.5				

Table 6. Effect analysis of two solution methods

Algorithm	Optimal solution/min	Worst solution/min	Average solution/min	Average calculation time/s
FCFS	12.5	12.5	12.5	0.988
ABC	6.5	7.5	7	2.856

5 Conclusion

5.1 Simulation and Results

Basing on the analysis of current air cargo freight for vaccine, this paper mainly focuses on the modeling and analyzing of the dual-Runway flight sorting process in the background of COVID-19 vaccine transportation, and proposes the loss function of the transportation to optimize the target function of the original model. Finally, the discrete artificial bee colony algorithm and Python programming were used to solve the model. The result were compared with the FCFS method result, the overall flight optimization effect is 48%, namely decreasing from 12.5 min to 6.5 min, and the flight delay time of COVID-19 vaccine is reduced by 50%, which is depressed from 2 min to 1 min and can be regarded as an accepted result of scheduling optimization.

5.2 Limitation and Implication

Aiming at the optimization research of airport vaccine scheduling based on bee colony algorithm, this paper has studied the flight entry/departure sequencing to some extent, but in view of time and capacity limitations, there are still some problems worth further improvement in the future, mainly including:

1) First, the priority of flight arrival and departure is dynamic. In this article, only for the volume of an airplane and simple new crown to set the priority vaccine transport flight, the aircraft carries the pharmaceutical goods are not discussed, and according to the cargo capacity, cargo volume and cargo type and time sensitive factors such as dynamic intelligent on the priority setting can better flight into out optimized scheduling process in the future.
2) Second, the model of parallel double runway is extended. In this article only on the part of the deterministic index analysis and discussion in modeling, the actual include fuel consumption in the process of airport operation, aircraft deterministic problem such as the passenger comfort and other uncertainty index value are analyzed, through the analysis of the diversified index in the future can be more accurate more reasonable on the whole optimization.
3) Thirdly, the application of the algorithm can be further optimized. This paper to solve the problem in the process of the main using a simple discrete artificial colony algorithm, the quantum artificial colony algorithm are popular now, and other including genetic algorithm, and the idea of narrative of quantum approximation algorithm and other algorithms to solve the problem under the effect of no validation and discussed. Further experiments can be carried out to determine a more appropriate and accurate method.

References

1. Ting, P.-H.: An efficient and guaranteed cold-chain logistics for temperature-sensitive foods: applications of RFID and sensor networks. Int. J. Inf. Eng. Electron. Bus. **6**, 1–5 (2013)
2. Li, Z.: Transportation capacity of novel coronavirus vaccine in Chinese civil aviation needs to be improved. China Aviation News 02(006) (2021). (in Chinese)
3. Nazini, H., Sasikala, T.: Simulating aircraft landing and takeoff scheduling in distributed framework environment using Hadoop file system. Clust. Comput. **22**(6), 13463–13471 (2019)
4. Khassiba, A., Bastin, F., Cafieri, S., et al.: Two-stage stochastic mixed-integer programming with chance constraints for extended aircraft arrival management. Transp. Sci. **54**(4), 897–919 (2020)
5. Xu, B.: An efficient ant colony algorithm based on wake-vortex modeling method for aircraft scheduling problem. J. Comput. Appl. Math. **317**, 157–170 (2017)
6. Biswas, D., Samsuddoha, M.: Determining proficient time quantum to improve the performance of round robin scheduling algorithm. Int. J. Mod. Educ. Comput. Sci. (IJMECS) **11**(10), 33–40 (2019)
7. Soulegan, N.S., Barekatain, B., Neysiani, B.S.: MTC: minimizing time and cost of cloud task scheduling based on customers and providers needs using genetic algorithm. Int. J. Intell. Syst. Appl. (IJISA) **13**(2), 38–51 (2021)

8. Mishra, A., Ahmed, A.O.: Simulation of cpu scheduling algorithms using poisson distribution. Int. J. Math. Sci. Comput. (IJMSC) **6**(2), 71–78 (2020)
9. Ankita, Sahana, S.K.: An automated parameter tuning method for ant colony optimization for scheduling jobs in grid environment. Int. J. Intell. Syst. Appl. (IJISA) **11**(3), 11–21 (2019)
10. Beasley, J.E., Krishnamoorthy, M., Sharaiha, Y.M., et al.: Scheduling aircraft landings - the static case. Transp. Sci. **34**(2), 180–197 (2000)
11. Salehipour, A.: An algorithm for single-and multiple-runway aircraft landing problem. Math. Comput. Simul. **175**, 179–191 (2020)
12. Messaoud, M.B., Ghedira, K., Harizi, R.: The multiple runway aircraft landing problem: a case study for Tunis Carthage airport. In: 2017 IEEE International Conference on Systems, Man, and Cybernetics (SMC), pp. 2802–2807. IEEE (2017)
13. Zhou, G., Wang, R., Zhou, Y.: Flower pollination algorithm with runway balance strategy for the aircraft landing scheduling problem. Clust. Comput. **21**(3), 1543–1560 (2018)
14. Abdullah, O.S., Abdullah, S., Sarim, H.M.: Harmony search algorithm for the multiple runways aircraft landing scheduling problem. J. Telecommun. Electron. Comput. Eng. (JTEC) **9**(3–7), 59–65 (2017)
15. Çelikbilek, Y.: A fuzzy cluster based genetic algorithm approach for the aircraft landing scheduling problem. Istanbul Gelişim Üniversitesi Sosyal Bilimler Dergisi **5**(1), 127–148 (2018)
16. Rezaei, H., Crespo, A.M.F., Chen, M., et al.: An iterative bidding approach applied to cost reduction in the context of aircraft landing problem. In: 2018 IEEE 22nd International Conference on Computer Supported Cooperative Work in Design, pp. 1–6. IEEE (2018)
17. Risikat, F.A., Kabir, A.A., Abdoulie, M.S.T., Busayo, H.A.: Management of vehicular traffic system using artificial bee colony algorithm. Int. J. Image Graph. Signal Process. **11**, 18–28 (2017)
18. Jena, R.K.: Artificial bee colony algorithm based multi-objective node placement for wireless sensor network. Int. J. Inf. Technol. Comput. Sci. **6**, 25–32 (2014)
19. Ng, K.K.H., Lee, C.K.M.: Aircraft scheduling considering discrete airborne delay and holding pattern in the near terminal area. In: Huang, D.-S., Bevilacqua, V., Premaratne, P., Gupta, P. (eds.) Intelligent Computing Theories and Application. LNCS, vol. 10361, pp. 567–576. Springer, Cham (2017). https://doi.org/10.1007/978-3-319-63309-1_50
20. Zhao, P., Zhang, J., You, L.: A composite dispatching rule-based method for multi-objective aircraft landing problem. In: COTA Conference International Conference of Transportation Professionals (CICTP2019), pp. 4902–4913 (2019)
21. Kulida, E.L., Lebedev, V.G., Egorov, N.A.: Study of the effectiveness of the algorithm to optimize the flow of aircraft on landing. Problemy Upravleniya **6**, 63–69 (2019)
22. Xu, Z., Yu, J., Luo, Y., Xu, X.: Research on approach and departure scheduling of long-distance parallel runways based on independent operation. J. Civ. Aviat. Univ. China **33**(05), 1–6 (2015). (in Chinese)
23. He, P.: Research on inbound and departure flight scheduling problem of single runway airport under non-emergency condition. Jilin University, Changchun (2020). (in Chinese)
24. Samà, M., D'Ariano, A., Palagachev, K., et al.: Integration methods for aircraft scheduling and trajectory optimization at a busy terminal manoeuvring area. OR Spectr. **41**(3), 641–681 (2019)
25. Daneshvar, S., Yazdi, M., Adesina, K.A.: Fuzzy smart failure modes and effects analysis to improve safety performance of system: case study of an aircraft landing system. Qual. Reliab. Eng. Int. **36**(3), 890–909 (2020)

The Role of Artificial Intelligence Technology in Improving the Resilience of Supply Chain During COVID-19

Zhong Zheng[1,2], Genge Zhang[1], Yun Lin[3], Yanfang Pan[1], and Yandong He[4(✉)]

[1] Nanning University Nanning, Nanning 530200, China
[2] King Mongkut's Institute of Technology Ladkrabang, Bangkok 10520, Thailand
[3] Chongqing University, Chong Qing 400030, China
[4] Shenzhen Zhongsheng Intelligent Technology Co., Ltd., Shenzhen, China
ydhe602@cqu.edu.cn

Abstract. The outbreak of COVID-19 has posed continuous and huge challenges to the global supply chain. Difficulties in resuming work, border blockade, and sharp decline in market sales have increased the brittleness of the supply chain, which has put forward new and urgent demands on the resilience of the supply chain. In this context, this paper studies the impact of the epidemic on the supply chain, summarizes the thinking of enterprise representatives in typical industries on improving supply chain resilience under the epidemic situation, studies the specific strategies of using artificial intelligence technology to improve supply chain resilience, studies the correlation between supply chain resilience and artificial intelligence technology, analysis of artificial intelligence in improving supply chain resilience mechanism, specific analysis of artificial intelligence through what key technologies to improve supply chain resilience. On the basis, this paper construct artificial intelligence in the supply chain resilience technical framework, and construct a preliminary scheme for the integration of artificial intelligence and supply chain resilience, in order to provide for the outbreak to improve supply chain resilience reference and guiding significance.

Keywords: Supply chain resilience · Artificial intelligence · COVID-19

1 Introduction

Since the outbreak of COVID-19, it has spread around the world, causing heavy casualties. According to the latest data on the WEBSITE of the World Health Organization (WHO), as of 17:00 central European time on August 20, the total number of confirmed cases of COVID-19 in the world has reached 20,9876,613, and the total number of deaths has reached 4,400,284. The current outbreak is still moving at a very rapid pace.

Z. Hu et al. (Eds.): AIMEE 2021, LNDECT 107, pp. 219–232, 2022.
https://doi.org/10.1007/978-3-030-92537-6_21

The epidemic has brought a huge impact on the global economy. As an important Index reflecting the manufacturing industry, the Purchasing Managers' Index (PMI) can better reflect the operation of the economy. As shown in Fig. 1 and Fig. 2. The impact of COVID-19 is extremely sudden and devastating, due to the impact of the epidemic, the PMI Index of China's manufacturing industry in February 2020, The manufacturing PMI was only 35.7%, and the non-manufacturing business activity index was only 29.6%, falling off a cliff and creating the lowest level in statistical history. With the control of the epidemic and the resumption of work and production in China, the manufacturing PMI also kept recovering. Meanwhile, according to the Global Manufacturing Purchasing Managers' Index (PMI) (shown in Fig. 2), the global impact of COVID-19 has had a significant impact on economic activity, with the PMI hitting a low of 39.5% in April 2020, demonstrating the disruption of the pandemic to the global economy.

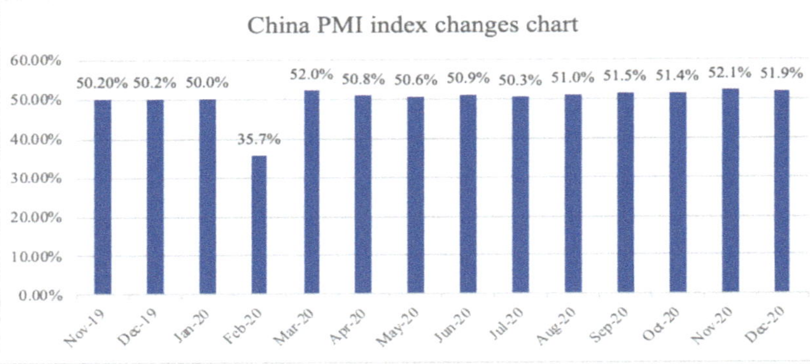

Fig. 1. Changes in China's PMI index under the impact of the epidemic

COVID-19 has had a severe impact on manufacturing and business activities, particularly the disruption of supply chains. As the epidemic continues to spread around the world, countries around the world are gradually upgrading their prevention and control measures. Restrictions on human flow and logistics have severely impacted the global division of labor, and the global industrial chain is facing a huge risk of disruption [1]. Due to different epidemic prevention and control policies in different countries, international logistics is limited and logistics costs are rising. Residents are isolated at home, factories are difficult to start, and the upstream and downstream supply chains are affected, resulting in a disconnect between the front and back ends. 94% of Fortune 1,000 companies have experienced supply chain disruptions during the pandemic, according to an assessment by Dun & Bradstreet。 According to Capgemini Research Institute (CRI) survey data, more than 80% of companies report that their supply chain has been negatively affected by the COVID-19 crisis, and the vast majority are experiencing difficulties in all aspects of their operations.

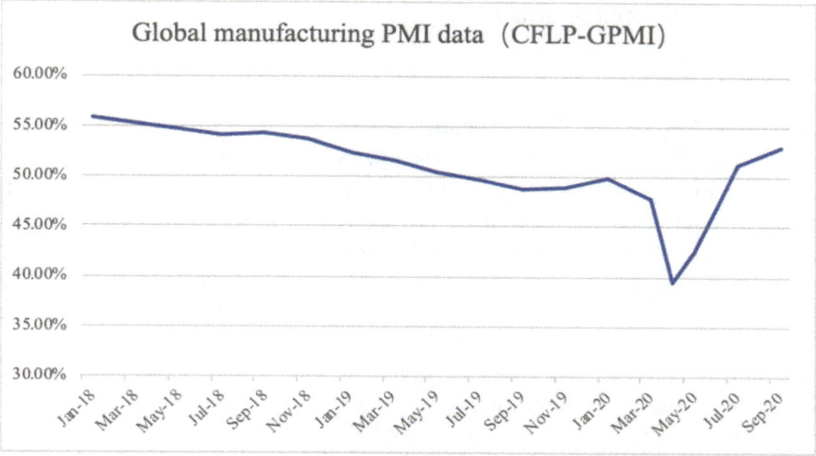

Fig. 2. Global PMI index changes under the impact of the epidemic

COVID-19 is just an emergency in this era. Many experts believe that the world is gradually entering the era of VUCA (Volatile, Uncertain, Complex and Ambiguous), where trade protectionism is on the rise, anti-globalization undercurrents are surging, global trade frictions are intensifying, and the global industrial chain is at high risk of breaking off and restructuring [2, 3]. Trade friction leads to a significant increase in the brittleness and uncertainty of supply chain, so the importance of supply chain resilience (SCR) is becoming more and more prominent. How to improve the supply chain resilience strategy has become an urgent problem concerned by both theoretical and practical circles. It is more and more difficult to manage the supply chain through traditional forecasting and planning. Reshaping the supply chain architecture by changing the management mode and improving the application effect of technology has become an important challenge to improve supply chain resilience. Faced with the impact of the pandemic, many businesses are responding to the supply and demand shock by looking for alternative suppliers and logistics, and some companies with simpler supply chain structures, such as food and beverage, are shifting production to domestic or closer to home (nearshore). For more complex supply chains, such as pharmaceuticals or electronics, it is difficult to cope with the crisis simply by finding alternative companies. Complex supply chains put more emphasis on resilience management and technology improvement, the application of advanced technologies is an important response measure pursued by enterprises, such as blockchain, artificial intelligence (AI), Internet of Things and other technical tools that can solve the vulnerability and disruption of the supply chain caused by the epidemic [3].

Therefore, this paper focuses on the following issues.

Question1: Analyze the factors that affect the resilience of the supply chain, and what are the key strategies to improve the resilience of the supply chain?

Question 2: Research the role of AI technology in supply chain management, and study how AI technology can improve the resilience of supply chain, especially when confronted with emergencies such as COVID-19.

2 Theoretical Background

2.1 The Conception of SCR

The key to understanding supply chain resilience is to understand the concept of "resilience", which is derived from physics and mechanics. Resilience theory describes how an object moves or deforms under the action of external forces and then returns to its original size and shape. The concept of resilience has been continuously integrated into economics and sociology and gradually used to describe the behavior of organizations or individuals. In sociology, the study of resilience originates from the developmental theory of social psychology, the concept of resilience is directly related to important issues such as ecological and social vulnerability, political and psychology of disaster recovery, and risk management under increasing threat [4]. Expert concluded that resilience has three main properties: A: The amount of change that can occur when the structure and function control of the system remain unchanged. B: The degree to which the system can self-organize and develop without interference from external factors. C: The degree to which the system develops learning and adaptive capacity to cope with interference and self-recovery [5, 6].

Supply chain is a complex and important form of organization, the concept of resilience has been applied more and more widely in the field of supply chain. The definition of supply chain resilience has been defined by many experts and scholars [6–9], typically shown in Table 1.

Table 1. Typical definitions of supply chain resilience.

Authors	Definition
Barroso et al. (2010)	SCR refers to the ability of the supply chain to respond to the negative effects of disturbances occurring at a given moment in time, thus maintaining the objectives of the supply chain
Ponomarov and Holcomb (2009)	SCR refers to the adaptive ability of a supply chain to better respond to emergencies, respond to supply chain disruptions, and restore operations by maintaining the desired level of connectivity and structural and functional control
Datta (2007)	SCRES is not only the ability to maintain control over performance changes in the face of interference, it is also adaptive and able to continuously respond to sudden and significant changes in the environment in the form of uncertain requirements
Christopher, M., & Peck, H. (2004)	SCR is the ability of the supply chain system to quickly restore to the original state or move to a new and more ideal state after being disturbed. Supply chain resilience, to a large extent, means the flexibility and agility of the supply chain
Kamalahmadi and Mellat-Parast (2016)	SCR means that the supply chain can reduce the probability of sudden interference, resist the diffusion of interference by maintaining the control of structure and function, recover and respond by precise and effective response plan, so as to transcend the interference and restore the supply chain to a robust operation state

The concept of "Supply Chain Resilience" was first proposed by Professors Rice and Caniato in 2003, and its formal definition was first proposed by Professors Christopher and Peck in 2004. Supply chain resilience is defined as "the ability of supply chain to return to its original state or more ideal state after disturbance". Subsequently, other definitions of supply chain resilience were put forward one after another. Currently, the more widely used definition is "the adaptability of supply chain to prepare in advance for potential emergencies and to respond quickly to and recover from disruptions" [4, 6].

Expert provides a hierarchical description of supply chain resilience by dividing it into three stages (readiness, response, and recovery) and a total of 11 factors [4]. Massachusetts Institute of Technology professor David Simchi-Levi mentioned that this global crisis is a great challenge to the overall stability of the global supply chain, and proposed two key indicators of supply chain stress test, TTR (Time to Recovery, Recovery Time) and TTS (Time to Survive) [11–13].

By analyzing the meaning of typical definitions and experts' understanding of key indicators of supply chain resilience, it can be concluded that supply chain resilience focuses on preparing for disaster, responding to disaster and recovering from disaster, and that adequate preparation can effectively respond to disasters and reduce losses in the upstream and downstream of the supply chain and quickly recover from supply chain disruptions caused by disasters.

2.2 Entrepreneurs' Thoughts on Supply Chain Resilience Under the Epidemic

The flow of goods has been disrupted over the past year as the coronavirus pandemic has spread around the world, forcing many countries to close borders, close workplaces or restrict exports. The spread of a more contagious variant of the Delta virus has renewed fears of supply disruptions. At the end of February 2020, a survey of 212 parts and components companies conducted by China Automobile Association showed that 16% of parts and components companies suffered losses of 20 million to 50 million yuan, with the highest losses reaching 2 billion yuan.

The survey was conducted by the Economist Intelligence Unit (EIU). In February and March 2021, the agency surveyed 175 supply chain managers, more than 70 percent of whom were based in Asia. In addition to automobiles, respondents were from five other industries: footwear and apparel; food and beverage; manufacturing; information technology, technology and electronics; healthcare; pharmaceuticals and biotechnology. About 51.7% of auto industry respondents said supply chain disruptions were "very serious," the highest percentage among the six industries. Footwear and clothing came in second, with 43.3% of respondents describing supply disruptions as "very serious". The survey found that the pandemic has prompted some companies to rethink their long-term supply chains, with about a third of respondents overhauling their supply chains. Among them, increasing supply chain flexibility to deal with risks is the key direction to consider.

In order to better understand the impact of the epidemic on the supply chain and countermeasures, the paper summarizes the opinions of enterprise leaders in typical industries as shown in Table 2 below.

Table 2. Views from representatives of enterprises in typical industries

Interviewee	Profile of interviewee	Industry involved	The impact on supply chains resilience	Key point for coping strategy
Hawk Xu	Chief Transformation Officer, Global Supply Chain, Lenovo	Computer Manufacturing	Delayed resumption of work, stagnant production capacity and difficulties in recruiting workers have disrupted the normal operation of the global supply chain, hindered delivery of goods and raised the risk of supply chain disruption	COVID-19 has elevated the stability and sustainability of supply chains to a more important level. It is also important to enhance sustainability and transparency of supply chains in response to similar extreme events
Xin Wang	Senior Manager of purchasing Department of BMW Brilliance	Automobile Manufacturer	A car is composed of about 28,000 parts. If any part is stockout, we cannot complete the production smoothly. If the vehicle cannot be delivered on time due to the shortage of parts, it will cause huge losses to the enterprise. The epidemic has brought huge uncertainties to the automotive parts supply chain	Enhancing flexibility is the key for supply chain to deal with market fluctuations and risks, while enhancing transparency through digital and intelligent supply chain is an important strategy to improve efficiency and deal with risks
Philip Johnson	Chief strategy officer, QA Chef	Food Industry	Lack of unified traceability system safety supervision, terminal distribution becomes complicated and difficult, food sales decline, enterprises face difficulties in business, more enterprises due to lack of funds and bankruptcy	Traceability and transparent process management of the food supply chain have become weak links and hidden worries, and terminal distribution faces huge challenges. The digital food supply chain led by automation and intelligent technology has become a trend

(*continued*)

Table 2. (*continued*)

Interviewee	Profile of interviewee	Industry involved	The impact on supply chains resilience	Key point for coping strategy
Jim Bingmeng	Executive Vice President, SAP Worldwide	Software Industry	The epidemic has intensified the needs of e-commerce, telecommuting and other fields. There will also be a spurt of digital demand related to building supply chain resilience	Enterprises should accelerate digital transformation in order to adapt customer requirements. Digital and intelligent supply chain can be more agile, flexible and resilient to respond to emergencies and adjust
Tony McKie	General Manager of MemsStar	Equipment Manufacturer	Goods cannot move freely, greatly extending the production and transportation cycle of factories. At the same time, as MEMS devices become more complex, supply chain disruptions caused by COVID-19 are becoming more serious and the entire manufacturing process is affected to varying degrees	After the epidemic, more information technology will be used to make the upstream and downstream of the supply chain more transparent and smooth business communication

Summarize the views of typical entrepreneur representatives of different industries, it shows that the COVID-19 pandemic has brought a severe test to the global supply chains. The outbreak highlights the instability, uncertainty, complexity and lack of transparency of the traditional supply chain, which has sounded the alarm for supply chain management. The massive crisis has reminded global supply chain leaders that enterprises should pay attention to the resilience and stability of supply chains in addition to pursuing low cost and efficiency. Many experts believe that after the painful lessons brought by the outbreak, many "hidden diseases" in the traditional supply chain have been concentrated and magnified in front of everyone. The outbreak first affected the emergency supply chain, but it will certainly affect other supply chains to varying degrees for a long time to come. In the final analysis, these problems also lie in the lack of maturity of supply chain infrastructure and need to be further improved [15–17]. If companies can get through the whole supply chain, and coordinate each link, the realization of the sharing of data real-time synchronous on-demand production and dispatch in time according to the latest data analysis, such as the combination of artificial intelligence (AI) technology can achieve intelligent operation, as well as for unmanned machine operation, can greatly promote the collaborative efficiency, will encounter similar problems later, may be able to better respond to treatment [13, 19, 20].

2.3 Approach to Increase Supply Chain Resilience

In terms of studying how to improve supply chain resilience, many experts made quantitative analysis on resilience by establishing mathematical models and adopting certain strategies to analyze the changes of resilience indicators before and after, and many experts made meaningful explorations [17, 22–27], as shown in Table 3.

Table 3. Example of approaches for SCR

Authors	Modeling approach	Key drivers of SCR
Chowdhury and Quaddus (2017)	Dynamic capability theory	Agility, Flexibility
Day, J. M. (2014)	Complexity system theory	Network layout, Preparation, Cooperation
Hosseini and Khaled (2016)	Hybrid ensemble and AHP	Geographic segregation, Raw material replacement, Budget resource, Surplus inventory, Reliability of supplier
Haobin Li (2016)	The hree-echelon dynamic models	Information sharing, Risk management, Cooperation relationship
Liu et al. (2018)	Structural equations modeling	Agility, Capacity,Integration, Technology
Ivanov (2018)	Dynamic control	Backup supplier, Surplus inventory, Capacity buffer, Multiple sourcing

From these studies, we can learn that many experts have conducted quantitative analysis of supply chains, as well as the key factors driving supply chain resilience. The analysis shows that although many researches focus on supply chain resilience, most of the current researches focus on relatively simple scenarios and idealized states, mainly analyzing the resilience of a single enterprise, and rarely involve the studies on the resilience of inter-enterprises in the upstream and downstream of the supply chain. At the same time, the mathematical models in the current literature on SCR seldom consider the dynamics of resilience and the large and lasting impact of COVID-19, but this epidemic is a huge disaster to the global supply chain, and it must be studied more deeply and widely. At the same time, most of the literature did not consider the impact of technology on supply chain resilience in the model.

Many experts believe that there are two main ways to increase the resilience of the supply chain: one is the strategic level, the other is the strategic level. At the strategic level, resources are often taken out, backup resources are added, and supply chain layout is changed to enhance the resilience of supply chain. The strategic or execution level focuses on specific ways to increase supply chain resilience, such as advanced information systems, early warning mechanisms and emergency processes, partner communication plans, etc.

2.4 Core Technology of Artificial Intelligence

Since Turing put forward the concept of artificial intelligence in the middle and late 20th century, artificial intelligence research has attracted more and more attention. Experts have made a thorough and comprehensive analysis of its definition and principle, and conducted a lot of research from principle to application. Scholars have different definitions of artificial intelligence. For example, artificial intelligence is defined as: show intelligent programs, algorithms, systems, and machines; A robot that represents human intelligence, an automated machine capable of imitating human behavior; Ability to flexibly accomplish specific goals and tasks through learning and interpreting external data. Still, artificial intelligence is a fusion and fuzzy concept, and many of the questions surrounding it remain open. Fundamentally, it is because artificial intelligence is a constantly evolving technology, its connotation, form, and function are constantly changing with the development of technology. Turing proposed an operational definition of artificial intelligence, the Turing test, through the Turing test machine is referred to as artificial intelligence, the Turing test mainly include natural language processing, knowledge expression, robotics, machine vision, human-computer interaction, etc., these characteristics define the artificial intelligence also expresses the core technology of artificial intelligence. Much of the current AI research is around these core technologies [3, 28–30].

After comprehensive analysis of existing research results and applications, the core technologies of artificial intelligence are summarized in eight aspects: Machine Learning, Natural Language Processing, Human-Computer Interaction, Computer Vision, Knowledge Graph, Virtual Reality/Augmented Reality, Biometrics Recognition [3, 29, 31]. The details are shown in the Fig. 3.

According to a 2018 survey conducted by Adobe, 15% of businesses are already using AI, while 31 percent plan to do so in 2019. Areas that generate revenue in this process include research and development, product innovation, supply chain operations, and customer service. According to a Deloitte Survey of 2,737 IT and business-line executives in nine countries conducted in August 2020, more than 93 percent of the respondents believe that AI is a positive and strategic driver of their companies' business outcomes. Machine learning, deep learning, language processing and computer visualization are the most used. At the same time, the field of artificial intelligence is developing rapidly, and more and more operations in the supply chain management, accelerate the process of supply chain digitalization, to improve the resilience of the supply chain plays an important role. McKinsey & Company, the research firm, estimates that companies could gain between $1.3 trillion and $2 trillion annually in economic value from using A.I. in their supply chains. In addition,

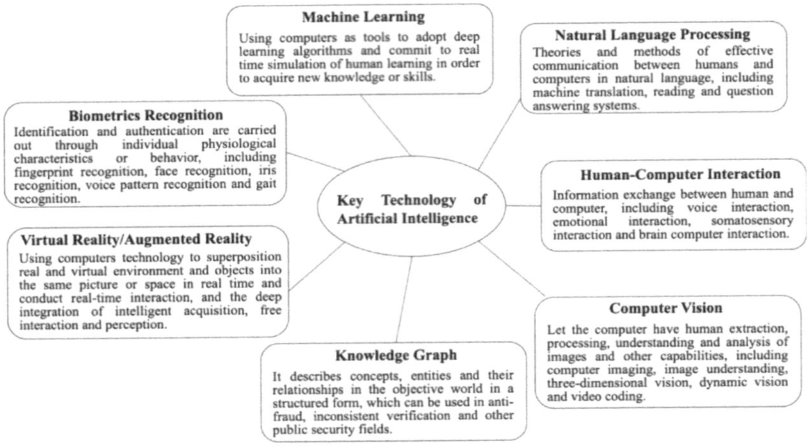

Fig. 3. The Key technology of artificial intelligence (Data from White Paper on Standardization of Artificial Intelligence (2018 Edition))

PricewaterhouseCoopers also said that ai is expected to contribute nearly \$15.7 trillion in value to the global economy by 2030. As the supply chain becomes increasingly complex, AI technology continues to improve, and the application foundation and environment continue to improve, the application of AI in the supply chain will be more and more extensive.

3 Application Mechanism of AI in SCR

According to McKinsey, the application of AI in supply chains is one of the most important areas for generating future financial value. First, data provided by AI can boost sales and satisfaction by supporting real-world product manufacturing marketing and personalized services, thereby increasing revenue. Second, data also has the ability to adjust bottom line results by simplifying and automating workflow and decision-making, identifying inefficiencies, and controlling supply chain costs. Just as importantly, by integrating data streams from third parties, companies can gain the critical ability to predict outages and take corrective action to shape future demand.

Artificial intelligence technology, supported by digital intelligent supply chain by constructing real-time information cloud platform, the entire supply chain logistics information can be real-time monitoring and management, enterprise can also for huge amounts of logistics data layer, flow and phases of statistics and analysis, and to provide customers with a full range of real-time information sharing service, At the same time, the application of artificial intelligence can make full use of the existing information to forecast and analyze the supply and demand situation of the product market, which greatly improves the resilience of the supply chain [3].

The specific mechanism of artificial intelligence in improving supply chain resilience is shown in Fig. 4.

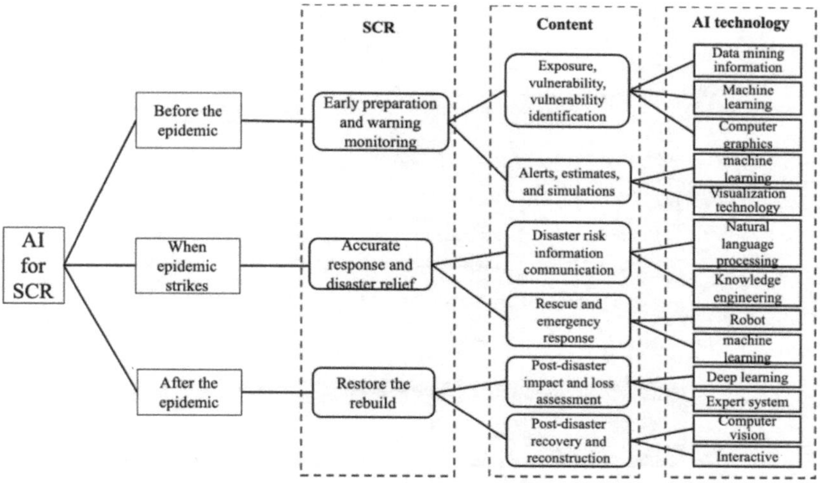

Fig. 4. Mechanism of artificial intelligence in improving supply chain resilience

According to the characteristics of AI technology and the performance needs of supply chain resilience, the application technology framework of AI technology in supply chain resilience is designed as shown in Fig. 5.

Before the disaster, by collecting all kinds of monitoring data and through a variety of deep learning algorithm for accurate prediction, so as to know the supply chain upstream and downstream proper preparation and reserves, at the same time through digital map information, market changes did, and so on and so forth in advance to backup or supply chain security layout, fully prepared for supply chain to deal with disaster. When disaster strikes, AI technology also by sensing external information, automatic access to information, to judge the damage, the relief at the same time, by providing scientific plan, reasonable allocate resources as a whole, to provide the supply of accurate and efficient tool, through the wisdom of the auxiliary even fully automated equipment, as far as possible to reduce or slow down the chain of supply chain is broken. After the disaster, the supply chain needs to recover quickly. Artificial intelligence technology helps enterprises find the best recovery plan through depth algorithm and edge algorithm, and provides professional guidance for supply chain recovery through expert system. Therefore, artificial intelligence technology can better improve supply chain resilience.

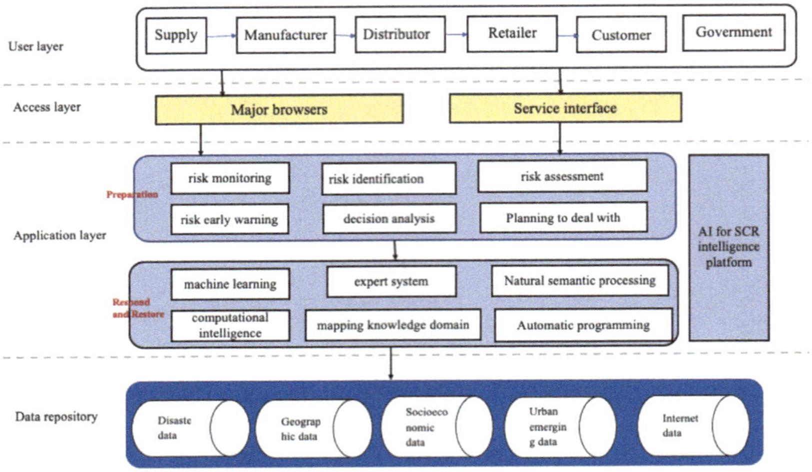

Fig. 5. The technical framework of artificial intelligence in supply chain resilience

4 Conclusion

With the pandemic still raging around the globe, this paper based on the analysis of epidemic situation of industrial economy, especially under the background of a huge impact on the global supply chain, realize the importance of improve supply chain flexibility, through literature review and collect the typical entrepreneur's point of view, on the one hand, epidemic factors to the impact of supply chain is analyzed, the key influence factors, Provided guidance to find countermeasures, analysis of the shortcomings of the current literature research on supply chain flexibility, hence the necessity of using AI technology to improve supply chain flexibility and urgency, through the analysis of the characteristics of the technology of artificial intelligence, the resilience of supply chain characteristics and combined artificial intelligent technology, and research its concrete operation mechanism, On this basis, the technical application framework of artificial intelligence in supply chain resilience system is established so as to provide some exploration for improving supply chain resilience.

Acknowledgment. This project is supported by Guangxi philosophy and social science planning research project(17FJY014) and Guangxi young and middle-aged teacher's basic ability enhancement project (2018KY0744).

References

1. Modgil, S., Gupta, S., Stekelorum, R., Laguir, I.: AI technologies and their impact on supply chain resilience during COVID-19. Int. J. Phys. Distrib. Logist. Manage. **13**(3), 69–83 (2021)

2. Nayal, K., Raut, R.D., Queiroz, M.M., et al.: Are artificial intelligence and machine learning suitable to tackle the COVID-19 impacts? An agriculture supply chain perspective. Int. J. Logist. Manage. 35–52 (2021)
3. Belhadi, A., Kamble, S., Fosso Wamba, S., et al.: Building supply-chain resilience: an artificial intelligence-based technique and decision-making framework. Int. J. Prod. Res. 1–21 (2021)
4. Ponomarov, S.Y., Holcomb, M.C.: Understanding the concept of supply chain resilience. Int. J. Logistics Manage. **19**, 11–23 (2009)
5. Longo, F., Oren, T.: Supply chain vulnerability and resilience: a state-of-the-art overview. In: Proceedings of European Modeling & Simulation Symposium, pp. 17–19 (2008)
6. Caniato, F.F.A., Rice, J.: Building a secure and resilient supply chain. **8**(1), 22–30 (2003)
7. Christopher, M., Peck, H.: Building the resilient supply chain. Int. J. Logistics Manage. **15** (2), 1–13 (2004)
8. Barroso, A.P., Machado, V.H., Barros, A.R., et al.: Toward a resilient supply chain with supply disturbances. In: 2010 IEEE International Conference on Industrial Engineering and Engineering Management, IEEE pp. 245–249 (2010)
9. Kamalahmadi, M., Parast, M.M.: A review of the literature on the principles of enterprise and supply chain resilience: major findings and directions for future research. Int. J. Prod. Econ. **171**, 116–133 (2016)
10. Ponis, S.T., Koronis, E.: Supply chain resilience? definition of concept and its formative elements. J. Appl. Bus. Res. **28**(5), 921–935 (2012)
11. Ivanov, D., Dolgui, A.: Viability of intertwined supply networks: extending the supply chain resilience angles towards survivability. a position paper motivated by COVID-19 outbreak. Int. J. Prod. Res. **58**(10), 2904–2915 (2020)
12. Belhadi, A., Kamble, S., Jabbour, C.J.C., et al.: Manufacturing and service supply chain resilience to the COVID-19 outbreak: lessons learned from the automobile and airline industries. Technol. Forecast. Soc. Change **163**, 120447 (2021)
13. Hosseini, S., Ivanov, D., Blackhurst, J.: Conceptualization and measurement of supply chain resilience in an open-system context. IEEE Trans. Eng. Manage. **79**(2), 60–72 (2020)
14. Gu, M., Yang, L., Huo, B.: The impact of information technology usage on supply chain resilience and performance: An ambidextrous view. Int. J. Prod. Econ. **232**, 107956 (2021)
15. Sabahi, S., Parast, M.M.: Firm innovation and supply chain resilience: a dynamic capability perspective. Int. J. Logistics Res. Appl. **23**(3), 254–269 (2020)
16. Li, Y., Zobel, C.W., Seref, O., et al. Network characteristics and supply chain resilience under conditions of risk propagation. Int. J. Prod. Econ. **223**, 107529 (2020)
17. Zheng, Z., Lin, Y., Li, L., Lu, L., Pan, Y.: The application of data-driven technologies to enhance supply chain resilience in the context of COVID-19. In: Hu, Z., Zhang, Q., Petoukhov, S., He, M. (eds.) ICAILE 2021. LNDECT, vol. 82, pp. 238–253. Springer, Cham (2021). https://doi.org/10.1007/978-3-030-80475-6_24
18. Han, Y., Chong, W.K., Li, D.: A systematic literature review of the capabilities and performance metrics of supply chain resilience. Int. J. Prod. Res. **58**(15), 4541–4566 (2020)
19. Surasma Surung J., Agung Bayupati, I.P., Agung Ayu Putri, G.: The implementation of ERP in supply chain management on conventional woven fabric business. Int. J. Inf. Eng. Electron. Bus. (IJIEEB) **12**(3), 8–18 (2020)
20. Anitha, P., Patil, M.M.: A review on data analytics for supply chain management: a case study. Int. J. Inf. Eng. Electron. Bus. (IJIEEB) **10**(5), 30–39 (2018)
21. Ali, M.H., Suleiman, N., Khalid, N., et al.: Supply chain resilience reactive strategies for food SMEs in coping to COVID-19 crisis. Trends Food Sci. Technol. (37) 18–31 (2021)
22. Chowdhury, M.M.H., Quaddus, M.: Supply chain resilience: conceptualization and scale development using dynamic capability theory. Int. J. Prod. Econ. **188**, 185–204 (2017)

23. Day, J.M.: Fostering emergent resilience: the complex adaptive supply network of disaster relief. Int. J. Prod. Res. **52**(7), 1970–1988 (2014)
24. Hosseini, S., Al Khaled, A., Sarder, M.D.: A general framework for assessing system resilience using Bayesian networks: a case study of sulfuric acid manufacturer. J. Manuf. Syst. **41**, 211–227 (2016)
25. Li, H., Pedrielli, G., Lee, H., et al.: Enhancement of supply chain resilience through inter-echelon information sharing. Flex. Serv. Manuf. J. **29**(2), 260–285 (2017)
26. Xiaotong, L., Kai, Z., Bokui, C., et al.: Analysis of logistics service supply chain for the one belt and one road initiative of China. Transp. Res. Part E: Logistics Transp. Rev. **117**, 23–39 (2018)
27. Ivanov, D.: Revealing interfaces of supply chain resilience and sustainability: a simulation study. Int. J. Prod. Res. **56**(10), 3507–3523 (2018)
28. Turing, A.M.: Computing machinery and intelligence. In: Epstein, R., Roberts, G., Beber, G. (eds.) Parsing the Turing Test, pp. 23–65. Springer, Dordrecht (2009). https://doi.org/10.1007/978-1-4020-6710-5_3
29. Dhuri, U., Jain, N.: Teaching assessment tool: using AI and secure techniques. Int. J Educ. Manage. Eng. **6**(08), 12–21 (2020)
30. Mohammadi, N., Zangeneh, M.: Customer credit risk assessment using artificial neural networks. Int. J. Inf. Technol. Comput. Sci. (IJITCS) **8**(3), 58–66 (2016)
31. Al Shibli, M., Mathew, B.: Artificial intelligent machine learning and big data mining of desert geothermal heat pump: analysis, design and control. Int. J. Intell. Syst. Appl. **4**(8), 1–13 (2019)

Analysis and Countermeasures of Inland Shipping Logistics in Fujian Province

ZhiZhong Zhuang[1], Wen Zhang[2(✉)], and Yong Gu[2]

[1] Fujian Chuanzheng Communications College, FuZhou 350007, China
[2] School of Transportation and Logistics Engineering, Wuhan University of Technology, Wuhan 430063, China
zhangwen99@whut.edu.cn

Abstract. China's rivers are rich in superior natural conditions. After years of sustainable development, an inland river shipping system with four major water systems as the backbone has been formed. There are many good ports along the coast of Fujian, and the port group has developed rapidly. In recent years, the modern port group has accelerated its rise. Inland river shipping has received extensive attention due to small investment and high economic benefits. Firstly, aiming at the development status and existing problems of Fujian inland waterway shipping, this paper uses PEST analysis method to analyze Fujian inland waterway shipping logistics from four aspects of politics, economy, society and technology, and then constructs an exponential smoothing prediction model to predict and analyze the freight throughput and waterway freight volume of Fujian inland waterway shipping logistics in the next few years, Finally, it puts forward the corresponding countermeasures for the logistics development of inland shipping in Fujian.

Keywords: Inland shipping · PEST analysis · Exponential smoothing method · Forecast · Countermeasures

1 Instruction

The steady development of inland shipping not only makes transportation very convenient, but also brings rich economic benefits. In order to ensure the sustainable development of inland shipping in Fujian, starting from the current situation, this paper explores the existing problems of inland shipping, studies the political, economic, social and technical factors of inland shipping in Fujian Province by using PEST analysis method, forecasts and analyzes the development prospect of inland shipping in Fujian Province by exponential smoothing method, and finally puts forward the inland shipping policy of Fujian Province Based on the analysis.

1.1 Development Status of Inland Shipping Logistics in Fujian

China's rivers run from east to west and from north to south. The navigable mileage of inland waterways is 125000 km, accounting for 29% of the total length of rivers. After years of continuous development, the inland shipping system with the Yangtze River,

Z. Hu et al. (Eds.): AIMEE 2021, LNDECT 107, pp. 233–243, 2022.
https://doi.org/10.1007/978-3-030-92537-6_22

Pearl River, Huaihe River, Heilongjiang and Songhua River as the backbone has basically taken shape. Among them, Fujian inland waterways are distributed in the Pearl River system.

Fujian waterway is mostly planned from coast to inland, with many ports, and the radiation range of coastal ports is expanded. In recent years, the modern port group has accelerated its rise, and the economic operation of port and shipping has improved year by year. Since 2014, cargo throughput, container throughput, waterway cargo volume and waterway cargo turnover have shown an overall growth trend, and the development situation is very considerable (Table 1).

Table 1. 2014–2020 Economic operation of port and shipping in Fujian Province

Year	Cargo throughput (MT)	Container throughput (Million TEU)	Waterway freight volume (MT)	Waterway cargo turnover (Gt-km)
2014	492	12.71	257.82	365.57
2015	503	13.64	293.71	430.80
2016	508	14.40	316.68	484.64
2017	520	15.65	334.53	542.98
2018	558	16.47	368.54	620.94
2019	595	17.26	422.63	713.56
2020	621	17.20	450.18	781.17

Fujian inland river shipping only needs to plan the channel with the help of rivers, with large traffic volume, low cost, less investment and high economic benefit. However, the external connectivity of inland shipping in Fujian is poor, which can only be used for transportation in the province. Except Minjiang River, other rivers are short and small, which is not conducive to the development of inland navigation (Fig. 1).

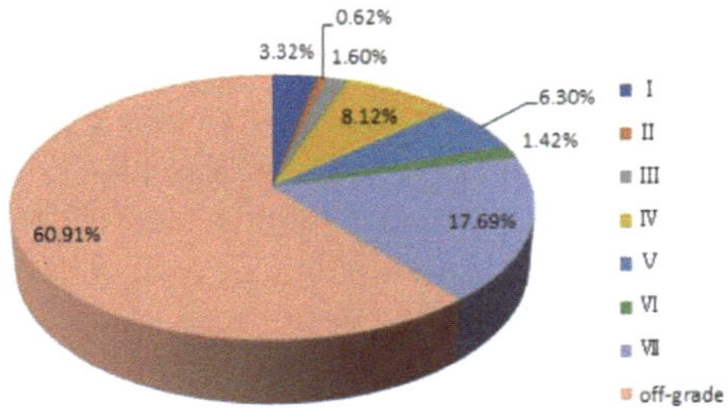

Fig. 1. The composition of navigable mileage of inland waterways in Fujian Province in 2020

By the end of 2020, the navigation mileage of inland waterways in the province was 3245.28 km. The grade channel is 1268.65 km, accounting for 39.1% of the total mileage. Among them, 443.8 km of class IV and above high-grade channels account for 13.7% of the total mileage. The proportion of specific navigable mileage of channels at all levels is shown in the figure.

1.2 Literature Review

In order to put forward the optimization countermeasures according to the current situation of inland shipping in Fujian Province, it is necessary to understand the theories of PEST analysis, prediction analysis and green shipping.

With regard to shipping analysis, Chen et al. (2016) conducted a strategic study on the development of Wuxi inland waterway based on PEST analysis [1]. Huang (2007) used the combination forecasting method to forecast the railway freight volume [2]. Wen et al. (2007) used combined forecasting for optimization to obtain air cargo volume [3]. Li Jun and Wan Licheng (2017) accurately predicted the freight volume of the Yangtze River trunk line, which is of great significance to the Yangtze River freight policy [4]. Many scholars used exponential smoothing model to predict [5, 6].

Recently, for the development of shipping, Lian et al. (2011) analyzed the development opportunities and challenges of inland shipping in Fujian, He put forward the concept of leapfrog development of inland shipping in Fujian earlier [7]. In addition, many scholars put forward opinions on promoting shipping development from multiple angles [8–12]. For example, Li Yue and Wang Na (2020) put forward their views on the green development of inland water transport economy [13].

2 PEST Analysis of the Development of Fujian Inland Shipping Logistics

2.1 Problems Existing in the Development of Fujian Inland Shipping Logistics

With the rapid development of inland shipping in Fujian, more and more enterprises choose shipping. However, there are still some problems in the development of shipping industry.

(1) The relevant national laws and regulations are not perfect
The protection of inland waterways in the existing national legal system is mainly the "Waterway Management Regulations" promulgated by the State Council in 1987, which no longer meets the needs of reasonable protection of shipping resources and management of shipping development.
(2) Insufficient construction investment and poor navigation environment
At present, the construction and maintenance of inland waterways in Fujian Province only rely on the provincial finance and the Ministry of Transport subsidies, and the amount is limited, which makes it difficult to carry out the maintenance and dredging of the waterways and the navigation environment is not smooth.
(3) The function of industry regulation and guidance is weakened, and market competition is disorderly
Private individual ships and shipping companies are the main force in Fujian's inland waterway shipping industry. They are small in scale and lack corporate

management functions. The freight rate is negotiated and decided by the shipowner and the shipper, and vicious competition in freight rate often occurs.

(4) The construction of transportation informatization is lagging behind, and the technological content is not high
The healthy and sustainable development of inland shipping is inseparable from advanced information services. At present, the docking of cargoes at the Fujian Inland River Terminal relies on the face-to-face communication between the shipowner and the cargo owner. The timeliness is poor and the docking efficiency of cargo ships is low.

2.2 PEST Analysis on the Development of Inland Shipping Logistics in Fujian

2.2.1 Political Environment Analysis

In the politics aspect, it analyzes the development of inland shipping logistics in Fujian from the aspects of the 14th five year plan.

(1) Inland Waterway Shipping Is Focused by the "14th Five-Year" Water Transport Plan
Inland shipping is a weak link in our country's comprehensive transportation system During the "14th five year plan" period, the development of inland shipping in Fujian Province will accelerate the construction of high-grade channel network and promote the smart, green and safe development of inland ports with the goal of accelerating the completion of short boards. At the same time, a special inland river innovation project was established to build the core area of the 21st century Maritime Silk Road.

(2) Construct Modern Comprehensive Transportation
By vigorously promoting the development of Minjiang River shipping, Fujian Province has achieved the steady development of inland river shipping, and a relatively perfect modern comprehensive transportation system has been basically completed in 2020. In addition, the Fifth Plenary Session of the 19th CPC Central Committee proposed to "improve the comprehensive transportation channel, comprehensive transportation hub and logistics network". During the "14th five year plan" period, through the implementation of port and waterway upgrading project and comprehensive hub efficiency improvement project, Fujian Province has accelerated the establishment of a leading area of transportation power and built a modern comprehensive transportation hub.

2.2.2 Economic Environment Analysis

In the economic aspect, it analyzes the development of inland shipping logistics in Fujian from the aspects of industry and free trade zone.

(1) Developed Economy Supports Inland Shipping
Located in the southeast coast, Fujian is an important window of foreign trade. Its economy has developed rapidly in recent years. As a pilot free trade zone and the

core area of the 21st century Maritime Silk Road, the developed economy has laid a solid foundation for its construction and development of inland shipping.

(2) The Development of Industries along the River Brings Vitality
Fujian's private economy is developed, and most industries along the river are industrial enterprises. Therefore, some bulk raw materials are transported through inland rivers. The rapid development of key industries in Fujian Province and the continuous upgrading of industrial structure along the river can promote the development of Fujian's overall regional economy.

(3) The Establishment of Free Trade Zone Promotes the Development of Inland Shipping
Fujian pilot free trade zone can not only promote Fujian Taiwan cooperation, but also expand the scale of foreign trade and promote the growth of GDP in Fujian Province. Policies such as promoting shipping liberalization in the construction plan of the free trade zone also play a positive role in inland shipping.

(4) The Construction of Port Logistics Park Provides Guarantee for Inland Shipping
There are many port logistics parks in Fujian Province. Port logistics parks are an important platform for foreign trade. Inland shipping in Fujian can achieve greater economic benefits by combining these port logistics parks with other modes of transportation.

2.2.3 Social Environment Analysis

In terms of social, this paper analyzes the development of inland shipping logistics in Fujian from the aspects of urbanization.

(1) The demand of two oriented society drives the development of inland shipping
Two oriented society refers to "resource-saving society and environment-friendly society". China's development of green economy has identified two oriented society as a strategic task in development. Compared with other modes of transportation, inland shipping occupies less land, has low cost, large traffic volume, low pollution and high safety, which is in line with the strategic goal of the development of two oriented society.

(2) New urbanization and transportation modes promote each other
Transportation can promote Fujian's new urbanization. The new pattern with urban agglomeration as the main form requires Fujian to build a more open comprehensive transportation system, and inland water transportation also plays an enhanced role in adjusting regional transportation structure and promoting regional industrial development.

(3) Urban development promotes water transportation
At present, with the rapid development of economic construction in Fujian, a large number of building materials are needed for urban infrastructure construction and real estate development. If we can give full play to the advantages of inland water transportation and undertake transportation tasks, we can reduce freight rates and alleviate urban traffic pressure. Therefore, urban development provides opportunities for the development of inland water transportation.

2.2.4 Technical Environment Analysis

In terms of technology, this paper analyzes the development of inland shipping logistics in Fujian from the aspects of intelligence and green ships.

(1) Intellectualization and informatization promote the transformation and upgrading of terminals
In recent years, new technologies such as cloud computing and big data have been continuously applied to the operation of terminals. Many ports in Fujian Province are introducing automation equipment, building port comprehensive information platform, and paying attention to the intelligence and informatization of various operation links, in order to improve the overall operation ability of ports and promote the transformation and upgrading of ports.

(2) Green ship is the focus of green wharf
Under the trend of common pursuit of environmental protection in today's world, the Ministry of transport recently put forward that green construction should focus on "protection first and prevention first". Green ships are concerned with low-carbon, environmental protection and pollution-free, which has become a vital link in the green construction of ports.

3 Forecast and Analysis of Inland Shipping Logistics in Fujian Province

3.1 Historical Data of Inland Shipping Volume in Fujian Province

By collecting historical data, it is available to analyze the inland shipping volume in Fujian Province.

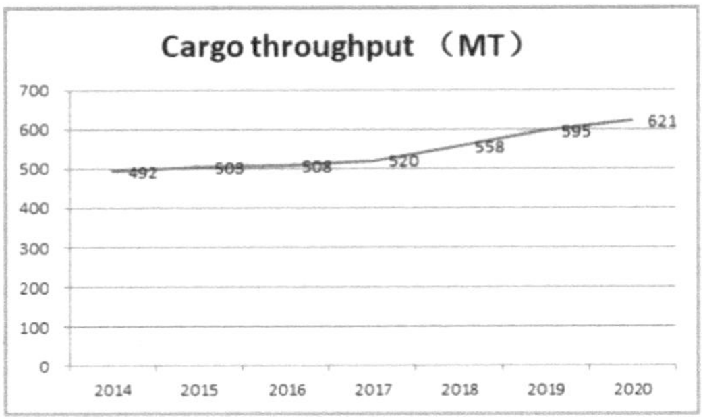

Fig. 2. Fujian cargo throughput (MT)

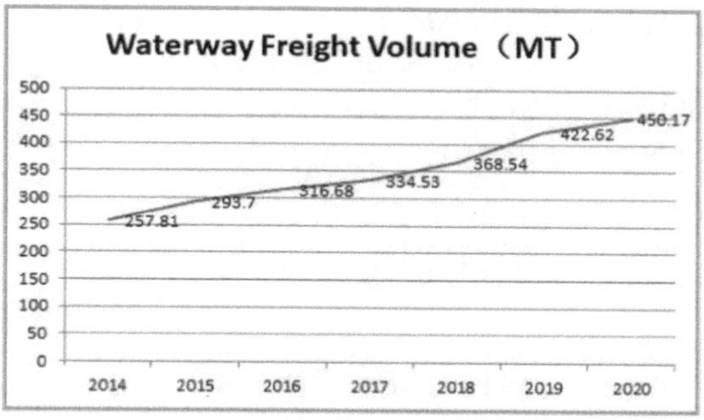

Fig. 3. Fujian waterway freight volume (MT)

According to the statistics of Fujian Province, it can be concluded that the cargo throughput of Fujian Province from 2014 to 2020 is shown in Fig. 2, and the waterway freight volume of Fujian Province is shown in Fig. 3. It can be seen from the figures that the cargo throughput and freight volume of Fujian Province have gradually increased in recent years. This chapter will use the exponential smoothing method to predict these two data.

3.2 Forecast of Shipping Volume Based on Exponential Smoothing

3.2.1 Exponential Smoothing

Compared with the simple full-period averaging model, which assigns equal weight to all past data, and the moving average model, which assigns more weight to recent data without considering the longer-term data, the exponential smoothing method sets the advantages of both, which not only considers the long-term data, but also gives the data a gradually weakening influence factor from near too far, and gradually converges to zero.

The three basic formulas of exponential smoothing are as follows:

$$S_t = \alpha \frac{X_t}{I_{t-L}} (1 - \beta)(S_{t-1} + b_{t-1}) \tag{1}$$

$$b_t = \gamma(S_t - S_{t-1}) + (1 - \gamma)b_{t-1} \tag{2}$$

$$I_t = \beta \frac{X_t}{S_t} + (1 - \beta)I_{t-L} \tag{3}$$

Among them, L is the season length, S is the stable component, b is the linear component, I is the seasonal component, and α, β, and γ are smoothing parameters.

The forecast formula is:

$$f_{t+m} = (s_t + b_t m)I_{t+m-L} \tag{4}$$

Among them, m is the selected period, f_{t+m} is the predicted value.

3.2.2 Cargo Throughput Forecast

When the smoothing parameter is 0.8, the smallest MSE can be obtained, and the forecast of inland shipping cargo throughput from 2021 to 2025 is as follows (Fig. 4):

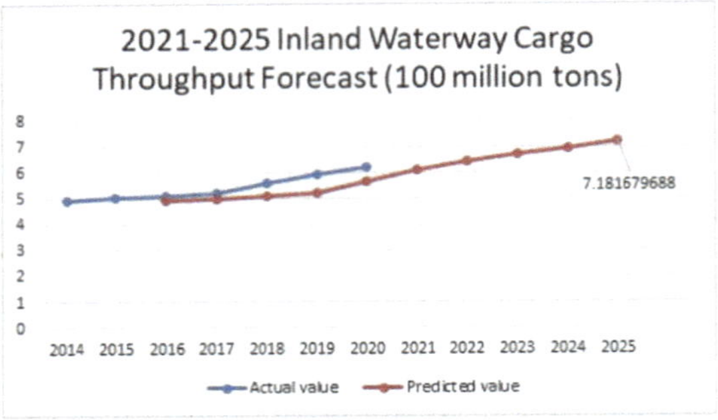

Fig. 4. Comparison of predicted value and actual value

An analysis of the cargo throughput of inland water transport in Fujian Province from 2014 to 2020 can predict that the cargo throughput of inland water transport in Fujian Province will reach 7.181679688 million tons by 2025.

3.2.3 Waterway Freight Volume Forecast

On the basis of collecting relevant data of waterway freight volume in Fujian Province over the years, the exponential smoothing method is used to gradually calculate the predicted value of waterway freight volume in each year, and it is calculated that the MSE value is the smallest when the smoothing coefficient is 0.8, and the optimal value can be obtained at this time. Result. Forecasts indicate that by 2025, the freight volume of Fujian inland waterway shipping will reach 64,8125,000 tons (Fig. 5).

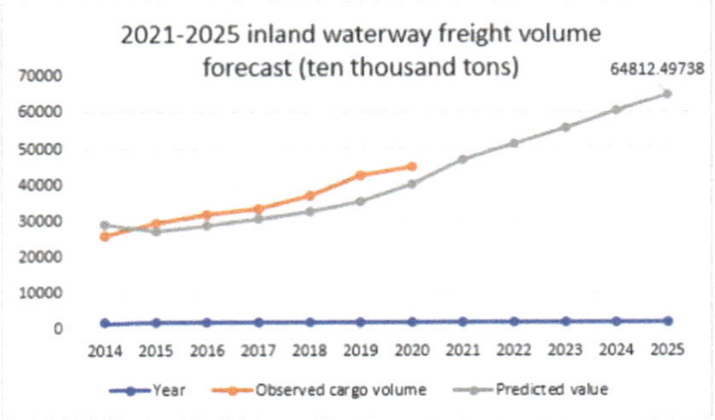

Fig. 5. Comparison of predicted value and actual value

3.3 Forecast Result Analysis

According to the above prediction, it can be found that the logistics volume of inland waterway shipping in Fujian Province is increasing year by year. It is expected that the cargo throughput will reach nearly 720 million tons and the waterway freight volume will reach 648.125 million tons by 2025. Facing this situation, Fujian inland shipping should actively change its development mode to meet the increasing transportation pressure. In addition, the forecast results of the year are generally smaller than the actual value, which shows that the shipping infrastructure construction and shipping related enterprises in Fujian Province have developed to a certain extent, so the overall actual value can be larger than the forecast results.

4 Countermeasures for the Development of Fujian Inland Waterway Logistics

4.1 Strengthen the Construction of Water Transportation Infrastructure

The construction of inland shipping needs to promote infrastructure construction from the long-term interests. Combined with the construction of water conservancy projects, establish a planning system for the comprehensive utilization of water resources in the basin to ensure the implementation of the comprehensive development of inland navigation; Plan and build a comprehensive transportation system, give full play to the advantages of various transportation modes, and form an efficient and perfect inland river transportation system.

4.2 Optimize the Industrial Layout Along the River

The growth rate of emerging industries should be accelerated, and the layout of industries along the river should be optimized. Special plans should also be issued in support of the development and expansion of a new generation of information technology and other industries. Optimizing the regional layout of Fujian Province requires strengthening the overall coordination at the provincial level to form an industrial development pattern with a clear division of labor and interconnection.

Carry out gradient training for industries along the river. Fujian should focus on strategic emerging industries to implement the gradient cultivation project of small and medium-sized enterprises. It should also improve the precision service quality according to different development stages of enterprises and accelerate the formation of advantageous enterprise echelons.

4.3 Develop Green Inland Shipping Logistics

Green design and green management should be implemented in the construction of inland waterways. China should build ecological slope protection along the waterway to prevent soil erosion, improve the landscape of the inland waterway to promote the harmonious development of the waterway and the city, encourage inland water transport to take advantage of the resource saving, green and low-carbon advantages of inland waterways, Strengthen the supervision of ship pollutant discharge, and improve the supporting facilities for energy saving and emission reduction in the channel. At the same time, the construction of inland waterways should also strengthen the awareness of serving the people, to improve the service level of the waterway as the core, and expand the service scope of the waterway.

4.4 Promote Shipping Transformation and Upgrading

Under the new pattern of port competition, information technologies such as Internet of things and cloud computing must be introduced to speed up the construction of smart port. Inland ports should introduce intelligent loading and unloading equipment and develop port production operation and management system to realize the intellectualization of port operation. At the same time, inland ports should build an integrated inland shipping information service platform for terminals and cooperative enterprises, so as to optimize supply chain management.

For the sustainable development of inland shipping, the ship shall be equipped with corresponding anti pollution facilities and equipment, and the ship type design shall be able to improve the utilization rate of navigation facilities and reduce transportation costs. We should also speed up the transformation of existing ships and complete the transformation of pollution prevention facilities.

5 Conclusions

Based on the development status of Fujian inland waterway shipping, this paper conducts a PEST analysis on the development of Fujian inland waterway shipping logistics, and uses exponential smoothing method to predict Fujian shipping cargo throughput and waterway cargo volume from 2014 to 2020, and obtains Fujian inland waterway shipping volume in the next five years. The trend of change. It can be seen from the forecast results that the inland waterway shipping in Fujian will continue to increase in the next five years.

To sum up, in the development of inland waterway, Fujian inland waterway should make full use of its own advantages and external environment, effectively combine the development trend of science and technology and the requirements of regional economic development, and implement a number of strategies such as the comprehensive development strategy of municipal waterway, the promotion strategy of waterway service level, the implementation strategy of "green" waterway and the improvement strategy of waterway information system.

References

1. Yuan, C., Lijie, D., Shulin, F.: Research on Wuxi inland waterway development strategy based on PEST analysis. Mod. Transp. Technol. **13**(04), 77–80 (2016). (in Chinese)
2. Huang, F., Shen, Y., Tao, S.: The application of combined forecast method in predicting freight volume of railway. In: First International Conference on Transportation Engineering (ICTE 2007), no. 4, pp. 3347–3352 (2007)

3. Wen, J., Wang, L.: Combined forecast of air freight volume based on approximate optimal non-negative weight. In: International Conference on Computational and Information Sciences, pp. 1232–1235 (2010)
4. Li, J., Wan, L.: Study on the sub-period forecast of the Chang Jiang River trunk freight volume. In: Proceedings of 2017 World Transportation Conference, p. 9 (2017)
5. Wu, X.: China Southern Airlines, Shi Fucun, Wang Qian. Research on cargo volume prediction of three gorges ship lock based on exponential smoothing model. China Water Transp. **21**(5), 13–15 (2021). (in Chinese)
6. Filimon, A.M., Mashaka, M., Mohamed, R.: Trend analysis and forecasting of water level in mtera dam using exponential smoothing. Int. J. Math. Sci. Comput. (IJMSC) **6**(4), 26–34 (2020)
7. Datong, L.: Analysis on the leapfrog development of inland river shipping in Fujian. China Water Transp. **11**(01), 48–50 (2011). (in Chinese)
8. Li, P., Shao, Q., Shen, J.: Brief discussion on the development status and Countermeasures of green channel in China. China Highway Society, vol. 6 (2018). (in Chinese)
9. Deng, A., Tao, B.: The coordinated development of inland shipping and comprehensive transportation system. In: The 4th Environmental Technology and Knowledge Transfer (ICET 2012), pp. 754–758 (2012)
10. Min, D., Yin, X.: The Study of improving china's shipping soft power: from the new perspective of cruise industry development. In: 2010 International Conference on Management and Service Science, pp. 4357-4360 (2010)
11. Anitha, P., Malini, M.P.: A review on data analytics for supply chain management: a case study. Int. J. Inf. Eng. Electron. Bus. (IJIEEB) **10**(5), 30–39 (2018)
12. Zine, B., Ghalem, B., Abdelkader, N.: A Cost measurement system of logistics process. Int. J. Inf. Eng. Electron. Bus. (IJIEEB) **10**(5), 23–29 (2018)
13. Yue, L., Na, W.: Preliminary study on green development of inland waterway economy. Financ. Circles **26**, 10–11 (2020). (in Chinese)

On the Development of Macao Film Industry Based on "New Diamond Model" from a Systematic Perspective

Yinan Lin[1,2] and Xingru Li[1,2](✉)

[1] Business School of Macao University of Science and Technology, Macao, China
[2] Beijing International Studies University, Beijing, China

Abstract. Since its return to the motherland in 1999, Macao and its film industry have developed rapidly with the support of the SAR government and domestic policies. In view of this, from a perspective of systematic theory, this paper uses the diamond model to explore the "diamond system" affecting the industrial competitive advantage. Based on the "new diamond model", the paper combs the main advantageous resources of Macao film industry, and explores the industrial advantages and industrial development of Macao film industry. The paper analyzes the resources of Macao's film industry from six aspects: development opportunities, government policies, channel resources, talent resources, capital resources and market demand; studies and judges the future development of Macao's film business, puts forward that under the development policy of "innovation" and "combination", Macao's film industry should be based on its own unique resource advantages, promote the virtuous cycle of industry helping cultural construction and creating industrial benefits with culture, and probes into the future development path of Macao's film industry.

Keywords: Diamond model · System theory · Macao film industry · Resource advantage

1 Introduction

As a place where Eastern and Western cultures collide and blend in recent 500 years, Macao has formed a special culture with both Chinese Lingnan style and Portuguese southern European characteristics. Its visually recognizable historical sites, local customs and practices continue to provide a unique source of inspiration for film creators all over the world [1–3]. Since Macao's return in 1999, film business in Macao has got support from the SAR government, and then achieved development greatly and rapidly. From 2009 to now, the Macao SAR government has successively issued a number of policies to support the film industry and further promote the development of Macao's film industry. The construction of Guangdong Hong Kong Macao (or, G-H-M, for short) Great Bay area, the technological revolution and the rapid development of film and television industry have brought historical development opportunities to Macao's film industry. Based on the "new diamond model", this study combs and analyzes the main advantageous resources of Macao's film industry, and analyzes the future

Z. Hu et al. (Eds.): AIMEE 2021, LNDECT 107, pp. 244–254, 2022.
https://doi.org/10.1007/978-3-030-92537-6_23

development path of Macao's film industry. To analyze the development of Macao's film industry from a systematic point of view is to look at the composition, current situation and development trend of the industry from a large perspective, and clarify the future development direction [4–6]. To analyze the development of Macao's film industry from a systematic point of view is to look at the composition, current situation, and development trend of the industry from a large perspective, and clarify the future development direction [7, 8].

2 Diamond Model and System Theory

It is understandable to take Macao film industry as the research object of this paper. It is a typical social system, and conforms to the basic characteristics of the general system, such as integrity, relevance, hierarchy, timing, etc. These are not only the basic ideas of the system, but also the fundamental principles of the system method. In this sense, system thought is the most basic starting point of this paper.

The "Diamond Theory" used in this paper, also known as "Diamond Model" or "Diamond Theory", was proposed by Michael Porter in 1990. It is a classic method to understand the national or regional industrial competitiveness [9]. According to Porter's research, there are four factors that have the greatest and most direct impact on the industry, namely, production factors, such as human resources, knowledge resources, capital resources, infrastructure, etc.; Demand conditions, i.e., the market demand for a certain type of products or services; Related and supporting industries, or upstream industries and complementary industries associated with a business; Enterprise strategy, structure and horizontal competition, which implies how an enterprise establishes, organizes and manages an enterprise in a specific environment, and what kind of competitors it faces [10–12]. These four key factors may accelerate the cultivation of industrial competitive advantage, or cause the stagnation of industrial development, which is known as the "Diamond System" affecting industrial competitive preponderance. In addition, two important variables, opportunity and government, may have a significant impact on industrial competitive edge. These six factors together constitute the classical model of Porter's "Diamond Theory", shown in Fig. 1.

Porter's diamond model is a typical system model, which consists of many elements; There are countless connections between those elements; The system exists in the environment and is inevitably affected by various environmental factors.

In the decades after Porter put forward the "Diamond Theory", many scholars have supplemented and improved the "Diamond Model" based on different entry angles, research perspectives and regional characteristics. For example, British economist Dunning believes that Porter's "Diamond Theory" underestimates the impact of product and market globalization on national competitive advantage, so he adds the influencing factor of "business activities of multinational corporations" to the classical theoretical model [13]; Professor Zhao Dongcheng of Seoul University, Korea, combined with the characteristics of industrial development in Korea, put forward the "nine factor model" to better evaluate the factors acting the creation of international competitiveness of underdeveloped countries [14].

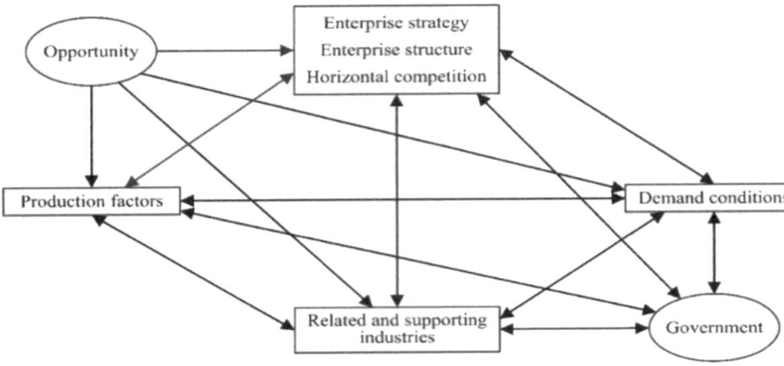

Fig. 1. Porter's "Diamond Model"

Professor Rui Mingjie of Fudan University, China, added the core of knowledge absorption and innovation ability to the classic "Diamond Model", forming the "new diamond model", shown in Fig. 2 [15]. He believes that this core is the driving force for the sustainable competitiveness of the industry. If China's relevant industries want to seek development, they should first cultivate their own knowledge absorption and innovation capability, and secondly actively participate in the industrial division system, so as to maintain and develop their own industrial competitiveness.

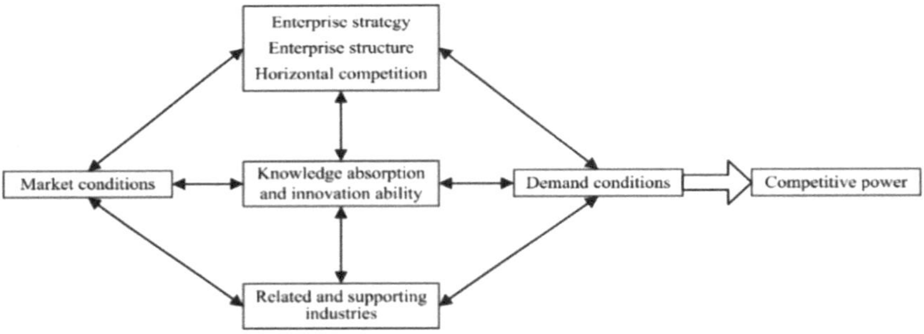

Fig. 2. Rui Mingjie's "new diamond model"

As an industry with high intelligence, high input and high output supported by high and new technology, the film industry is an important part of the cultural business with great vitality and freshness [16–18]. The ability of knowledge absorption and innovation has a vital impact on industrial development. Therefore, this study selects the "new diamond model" as the main theoretical basis to explore the industrial dominant position and industrial development of Macao's film business.

3 Advantageous Resources of Macao's Film Industry Based on the "New Diamond Model"

From a systematic perspective, it is needless to say that compared with the mainland, Macao's film industry is still in a relatively immature development stage, but from another perspective, this is the development advantage of Macao's film industry. This business, which is growing synchronously with the construction of Guangdong, Hong Kong and Macao Bay area, has a lot of development experience and development resources that can be used for reference [19]. These industrial resources are the important factors in the composition of the system and constitute the core competitiveness of Macao's industry in the future.

3.1 Development Opportunities

Development opportunities refer mainly to construction of G-H-M Greater Bay Area which brings historical opportunities to Macao film business.

In 2016, the guiding opinions on deepening regional cooperation in the Pan Pearl River Delta issued by the State Council first put forward the concept of Guangdong Hong Kong Macao Bay area. In this instruction, Guangzhou and Shenzhen are required to gang up with Hong Kong and Macao to jointly carry G-H-M Bay area and build a world-class urban agglomeration. From the historical perspective, the construction of this Great Bay area is based on the Pan Pearl River Delta cooperation, and the regional scope is basically consistent with the original Pan Pearl River Delta urban agglomeration.

The structure of G-H-M Bay area is not only the construction of economic community, but also the construction of cultural community. From the perspective of human geography, Guangdong, Hong Kong, and Macao Greater Bay Area belongs to Lingnan cultural circle and has a certain common cultural foundation. How to inherit the original foundation, explore and innovate it, so as to further form the unique culture belonging to Greater Bay Area and build a solid regional cultural identity is an important factor for the smooth realization of the national strategy of G-H-M Greater Bay Area. Because of its cultural product attribute, film plays an irreplaceable role in constructing the cultural identity of Guangdong, Hong Kong and Macao Bay area and helping its cultural construction. It can be said that the construction of G-H-M Bay area provides a major opportunity for the development of Macao's film industry.

3.2 Government Policy

Government policy refers to the support of Macao's film industry through the superposition of multiple means.

Porter believes that government policies are important resources that will have an impact on industry. Scientific and correct policies of government can help industrial development definitely. On the contrary, unreasonable policies may form obstacles to industrial development. Macao has always regarded the gambling industry as its unique feature. At the same time, the SAR government has been supporting the film industry

[20]. In order to promote the diversified development of economic structure and avoid excessive dependence on gambling and tourism, after Macao's return to China, the SAR government signed the mainland Macao Closer Economic and trade relationship arrangement, in which the terms on audio-visual services clearly allow Macao service providers “in the form of joint venture, Engage in the distribution business of audio-visual products (no more than 70% controlling interest) and operate Cinemas (no more than 75% controlling interest) “and” films made in Macao can be distributed in the mainland and are not limited by the quota of imported foreign films” [21]. With the support of the above policies, the policy framework for the development of Macao's cultural industry has been improved, and institutions such as the Department for the promotion of cultural and creative industries, the cultural industry committee and the cultural industry fund have been established successively, which has played a role in managing and supporting the cultural industry, including the film industry.

The Macao Cultural Bureau helps the local film industry by means of financial subsidies, building a platform and reducing the entry threshold. First of all, the Macao Cultural Bureau adopts a variety of funding methods, such as funding local directors to produce feature films through the “film feature film production support program”, providing financial support for the “Macao Film Festival”, “Original Video Exhibition”, and funding films to go abroad to participate in film festivals. In addition, the Macao Cultural Bureau and other government departments jointly launched the “co-ordination service of shooting application” for institutions and departments applying for shooting services.

The photographer only needs to apply to one department of the Cultural Bureau, and the coordination of the other nine departments is completed by the Cultural Bureau. This measure greatly simplifies the preliminary application procedures for film shooting in Macao, reduces the administrative threshold for local and other film makers to shoot in Macao, and enhances the attraction of Macao as a film location and production place.

In addition to Macao's local policy, the state film administration of China has also issued five measures to support the development of Macao's films in the mainland, including: there is no quantitative limit on the participation of Hong Kong and Macao people in mainland film production; There are no restrictions on the proportion of actors and mainland elements in co productions between the mainland and Hong Kong and Macao; Cancel the collection of project application fees for co productions with the mainland; Hong Kong and Macao film and film workers can sign up for mainland film awards such as Huabiao award and Golden Rooster Award, as well as Hundred Flowers Award; Various awards can be applied for the distribution and promotion of excellent domestic films in Hong Kong, Macao, Taiwan and overseas [22]. It can be said that Macao's film industry has policy resource advantages at the national and local government levels, and this resource superiority can be continued for a long time.

3.3 Channel Resources

Channel resources are mainly used to build a platform for film industry communication through the organization of various festival activities.

The holding of festival activities has multi-directional and multi-level positive significance for the film industry. Conducting a film festival can praise and affirm the local films in the host place and bring high attention to the host place. At present, the most influential film festival hosted by Macao is the Macao International Film Festival, which has been successfully held for 11 times since 2009. During the festival, activities such as "international film, television and animation products fair" have been held, and professional services and international film and television copyright trading platform have been provided. Through film exhibitions and various supporting activities, combined with Macao's tourism resources and entertainment elements, a film carnival with the participation of world filmmakers and the general public will be created. The Golden Lotus awards one belt, one road and another, has been inspiring and nurturing a number of excellent domestic and foreign films. In recent years, the influence of the film festival has been increasing with the help of the activities [23].

3.4 Human Resources

Talent resources are to establish a powerful talent team through self-cultivation and talent introduction.

Building a talent team with reasonable structure, complete categories, clear division of labor and excellent quality is an important factor to promote the good development of the film industry. In recent years, Macao has made a lot of efforts in the cultivation of film talents. Currently, the universities offering film related majors in Macao are mainly the School of Art, Macao Polytechnic Institute and the film School of the Macao University of Science and Technology. In addition to higher education, Macao also began to try to introduce film talents from other regions.

3.5 Capital Resources and Market Demand

The advantages in terms of capital resources and market demand are mainly reflected in: the gambling industry is an important source of funds, which has a great impact on the performance of the mainland market.

From the perspective of the film market, there were 15 screens in Macao in 2016, and the average number of film viewers per screen was 77297; In 2017, there were 16 screens in Macao, and the average number of film viewers per screen was 62770. The number of film viewers per year is about one million. From 2014 to 2017, the film box office was between 100 million and 140 million every year, and the local audience market was small [24–26].

The mainland film market provides a broad development space for Macao's film industry. In recent five years, the income of the gambling industry and the performance of the mainland film market have a direct impact on the investment of relevant funds in the film industry.

Table 1 presents the participation of Macao's gambling industry funds in film investment from 2015 to 2019. In 2015 and 2016, the revenue of Macao's gambling industry is considerable, and coincided with the development period of mainland film and television industry. National policy support and sustained economic growth made the market demand for cultural products rise. The gambling industry has increased its investment year by year. In 2015, the main types of films invested were small and medium-cost commercial films and Hong Kong and Macao films produced by independent filmmakers. In 2016, it began to increase its investment in large-scale commercial films, the box office of investment films increased explosively by 53 times, of which the box office of two major films, Macao situation 3 and Mekong River operation, accounted for nearly 90% of the box office of all investment films. Therefore, the investment in 2017 continued the strategy of annual investment in more than two large-scale commercial films, supplemented by the production of Hong Kong and Macao co productions at small and medium costs. In that year, the mainland film market gradually entered a rational stage, and the box office revenue decreased. In 2018, due to the decrease of gambling revenue and gambling funds, the investment in film and television industry, and the mainland film market entered the adjustment period, the box office revenue of investment films decreased by 83% year-on-year. In 2019, there were nearly 40 million Macao tourists, which has reached an all-time high. The revenue of the gambling industry has improved, and the mainland film and television market has also begun to pick up. The large-scale films invested by the relevant capital layout have been released one after another, making the investment and box office record high.

Figure 3 shows Macao's box office revenue and annual growth from 2015 to 2019.

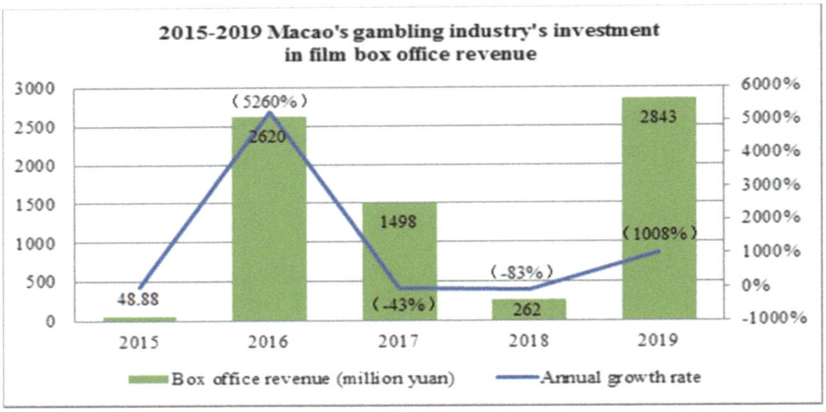

Fig. 3. Macao box office revenue and annual growth from 2015 to 2019

3.6 Absorption and Innovation Capability

Absorption can be understood as the learning and digestion of existing knowledge. Innovation refers to breaking the thinking set and putting forward new ideas. These two abilities are required.

As a special zone under the guidance of the "one country, two systems" policy, Macao's cultural environment has the characteristics of more diversity and inclusiveness. Both collision of multiple cultures and the open environment have given Macao's film industry great creative space. With the construction of Guangdong, Hong Kong and Macao Bay area, the mainland film market has been opened to Macao films; The

Table 1. Macao's gambling industry's participation in film investment from 2015 to 2019

Year	Movie title	Box office revenue
2015	ANNIVERSARY	12 million
	Guia In Love	0.207 million
	Paris holiday	14.8 million
	Robbery	1.84 million
	Little Big Master	20.04 million
In 2015, the total box office revenue of investment films was 48.88 million		
2016	Nessun Dorma	2.15 million
	Heartfall Arises	110 million
	Call of Heroes	160 million
	House of Wolves	27.54 million
	From Vegas to Macau 3	1117 million
	Operation Mekong	1184 million
In 2016, the total box office revenue of investment films was 2620 million. An increase of 5260% over the previous year		
	A Nail Clippers Romance	4.5 million
2017	Perfect Couple	24.292 million
	Love Contractually	59.40 million
	Always be with You	46.3493 million
	The big call	55.3128 million
	The Brink	67.7 million
	Love Off the Cuff	170 million
	Paradox	515 million
	Chasing the Dragon	575 million
In 2017, the total box office revenue of investment films was 1498 million, 43% less than the previous year		
Year	Movie title	Box office revenue
2018	When Sun Meets Moon	0.325 million
	House of the Rising Sons	1.905 million
	Keep Calm and Be a Superstar	39.00 million
	The Leaker	71.00 million
	Big Brother	150 million
In 2018, the total box office revenue of investment films was 262 million, 83% less than the previous year		
2019	IP Man 4	1190 million
	A Home with a View	33 million
	Chasing the Dragon II	310 million
	The White Storm 2: Drug Lords	1310 million
In 2019, the total box office revenue of investment films was 2843 million. An increase of 1008% over the previous year		

cultivation of local excellent film talents has also brought more possibilities for the exploration and improvement of Macao's film creation. Therefore, the innovation and development of Macao films in content has great potential.

The development of new technology also provides an innovative driving force for the development of Macao's film industry. At present, Macao's online film and television business is gradually increasing. ITunes, Netflix, TVB anywhere and Youtube can provide both paid and free film viewing services.

4 Research and Judgment on the Future Development of Macao's Film Industry

Through the above systematic and comprehensive analysis of Macao's film industry, it can be seen that the main advantageous resources of Macao's film industry are mainly reflected in policies, opportunities, funds and innovation ability. Although Macao's local film market is small, the Greater Bay area of Guangdong, Hong Kong and Macao has ranked first in the national film "ticket warehouse" ranking for 14 consecutive years. In addition to the mainland market can provide support for the demand of Macao's film market, Portuguese speaking countries, especially the large population of Portuguese speaking countries, should also become the advantageous market of Macao's films. Nowadays, the population of Portuguese speaking countries around the world is about 250 million, distributed in Europe, Asia, Africa, Latin America and other continents. Among them, Brazil has developed into the world's tenth largest film market in 2013, with 149.5 million film viewers and an annual output of more than 100 films. Macao can take advantage of its unique preponderances to assume the function and responsibility of the communication center between Chinese films and Portuguese films.

Based on this situation, Macao's film industry can start from the two directions of "innovation" and "cooperation" at the same time to promote the further development of the industry.

"Innovation" means creating opportunities through development and making up for short board resources, such as leveraging media technology innovation to promote online video websites and APP construction, and building a long-term and effective new media exchange platform for Macao's interaction with other regional film industry, so as to persevere in cultivating local creative and performing talents.

"Cooperation" refers to cooperation with other regions with the help of policy advantages, location advantages and language advantages. For Macao's film industry, it is better to consider two ways mainly to promote common production: investing in high-level film production, establishing production teams from other regions to come to Macao for shooting and production, encouraging local producers to go abroad and introducing excellent film talents from other regions to Macao.

From the perspective of the overall industrial structure, the film industry will be a vital part of the process of industrial diversification in Macao and a momentous factor in constructing the cultural identity of Guangdong, Hong Kong and Macao Bay area and helping its cultural construction. Therefore, Macao's film industry needs to be based on its unique resource advantages to build a virtuous circle of industry helping cultural construction and creating industrial benefits with culture.

5 Conclusion

After returning to the embrace of the motherland, Macao's film industry has developed rapidly with the support of various policies of the state, G-H-M Great Bay area and the SAR government. The integration of various advantageous resources makes the film business, a cultural industry with high intelligence, high investment and high yield, have stronger penetration and vitality.

From the system angle of view, using the "diamond theory" and the "new diamond modal" developed on this basis to analyze Macao's film industry and development can fully understand and make rational use of advantageous resources, cherish and seize the historical opportunities brought by the construction of Guangdong, Hong Kong and Macao Bay area, and make full use of Macao's capital resources, market demand and its unique networking advantages, seize development opportunities, remedy the shortcomings of resources make full use of policy, location and language advantages, and vigorously carry out cooperation with other regions, so as to better promote the leap forward development of Macao's film industry.

Acknowledgment. This paper is the phased achievement of the National Social Science Foundation youth project "Research on the Development Model Innovation of China's online Drama Industry under the Background of Media Integration", (No. 17cxw032).

References

1. Williamson, J.: Historical dictionary of the Hong Kong SAR and the Macao SAR. Ref. Rev. **21**(7), 63–64 (2007)
2. Elliot, E.A., Xiao, Y., Wilson, E.: A multicultural blend: metaphors, cognitive social capital and multiculturalism. Int. Mark. Rev. **32**(02), 200–218 (2015)
3. Čeginskas, V.L.A.: Multicultural belonging, individuals across cultures, languages and places. University of Turku, Turku (2016)
4. Svetlicinii, A.: The limits for the global proliferation of competition law: the case of Macao SAR. Soc. Sci. Electron. Publ. **2**, 31–39 (2014)
5. Sun, J.: Cultural challenges in the Guangdong-Hong Kong-Macao Greater Bay Area: a clue by the hongkong film before and after the return. J. Guangzhou Univ. (Soc. Sci. Edn.) **17** (12), 47–52 (2018). (in Chinese)
6. Ho, C.H.: Challenges and innovations of the movie-television industry in the Guangdong-Hong Kong-Macao Greater Bay Area. In: The 6th International Conference on Social Sciences and Economic Development (ICSSED 2021) (2021)
7. Vasilkov, Y., Gushina, L.: Analysis of the effectiveness and efficiency of management systems based on system analysis methodology. Int. J. Qual. Res. **8**(3), 347–356 (2014)
8. Barman, D., Chowdhury, N.: A method of movie business prediction using back-propagation neural network. Int. J. Inf. Technol. Comput. Sci. **4**(11), 67–73 (2012)
9. Porter, M.E.: The Competitiveness Advantage of Nations, pp. 23–25. The Free Press, New York (1990)

10. Cuiling, Y.: Analysis of competitive advantages in enterprise intellectual property based on diamond-model. J. Jiangsu Polytech. Univ. (Soc. Sci. Edn.) **10**(02), 37–40 (2009). (in Chinese)
11. Ndou, P., Obi, A.: An analysis of the competitiveness of the South African Citrus industry using the constant market share and Porter's diamond model approaches. Int. J. Agricult. Manage. **2**(03), 160–169 (2013)
12. Zhao, H., et al.: Analysis of environmental protection industrial cluster in Yixing based on diamond model. Industrial Electronics & Applications. IEEE (2015)
13. Dunning, J.H.: Internationalizing Porter's diamond. Manag. Int. Rev. **33**(2), 7–15 (1993)
14. Zhao, D.: Evolution and Application of Diamond Theory. China Renmin University Press, Beijing, 110–152 (2005). (in Chinese)
15. Mingjie, R.: The "new diamond model" of industrial competitiveness. Soc. Sci. **04**, 68–73 (2006). (in Chinese)
16. Several opinions of the state administration of radio, film and television on accelerating the development of the film industry [EB/OL]. (2020–08–03) (in Chinese). http://www.china.com.cn/policy/txt/2004-02/20/content_5501559.htm
17. Barman, D., Chowdhury, N.: Estimation of possible profit/ loss of a new movie using "natural grouping" of movie genres. Int. J. Inf. Eng. Electron. Bus. **5**(04), 24–33 (2013)
18. Barman, D., Singha, R.K., Chowdhury, N.: Prediction of possible business of a newly launched film using ordinal values of film-genres. Int. J. Intell. Syst. App. **5**(6), 53–60 (2013). https://doi.org/10.5815/ijisa.2013.06.07
19. Alguliyev, R.M., Mahmudova, R.S.: Information culture formation as the most promising direction of individual`s general culture. Int. J. Modern Educ. Comput. Sci. **7**(3), 54–61 (2015). https://doi.org/10.5815/ijmecs.2015.03.08
20. Jouhki, J., Browne, R.: Macau as a gambling city: observations through visual anthropology. In: The Second Asia Pacific Conference on Gambling & Commercial Gaming Research (2013)
21. Yuan, T.: Overview of Macao film development in the past 20 years. Contemp. Film **4**, 99 (2018). (in Chinese)
22. Wu, R.: The 20th anniversary of Macao's return to the motherland, the Great Bay area of Guangdong, Hong Kong and Macao has brought new development to Macao's film and television industry[EB/OL].(2020–08–03) (in Chinese). http://ent.cri.cn/20191217/40a1f3e5–42b9-e3c4-e0bb-bf78c1954ea5.html
23. Wang, S., Ye, J.: Research on the development of cinemas in Macao. Art. Educ. **1**, 126 (2020). (in Chinese)
24. China report.com.cn. Statistics of average movie ticket prices, average box office and number of viewers per indoor screen in Macao, China from 2013 to 2017[EB/OL]. (2020–08–03) (in Chinese). http://data.chinabaogao.com/chuanmei/2019/12304H31R019.html
25. China report. Statistics of total box office revenue of all films released in Macao, China from 2013 to 2017[EB/OL]. (2020–08–03) (in Chinese). http://data.chinabaogao.com/chuanmei/2019/12304H0312019.html
26. Hongyan, G.: Independent Brazilian film industry. Contempor. Film **11**, 100 (2015). (in Chinese)

Empirical Research on Influence Mechanism of COVID-19 on University Students' Decision to Study Abroad

Lixia Wu(✉), Jianting Zheng, and Lixia Zhang

Wuhan Technology and Business University, Wuhan, Hubei, China

Abstract. Based on the cognitive theory and the model of "stimulation, cognition, response", this paper studies the influence mechanism of COVID-19 epidemic crisis, the image perception of the receiving country and the willingness to study abroad from the perspective of "cognition security emotion" three-dimensional structure. Using SPSS and AMOS software to conduct structural equation modeling analysis, the results show that COVID-19 crisis will have a significant negative effect on the potential to study abroad. The epidemic crisis has a notable impact on the perception of foreign image. The perception of security image strongly affects the decision to study abroad. The negative impact of epidemic situation on emotional image perception and cognitive image perception is not significant. It is expected that this study can help Chinese college students make reasonable decisions about studying abroad.

Keywords: COVID-19 · Study abroad · Influence mechanism

1 Introduction

The purpose of studying abroad is generally understood by the public to improve the existing educational background, learn practical language, master professional skills, expand international vision, break through limited space, seek a better future and experience the new course of life [1, 2]. The choice of destination country involves many different factors, and all aspects need to be considered [3–5].

According to IIE's 2020 Open Door Report, in 2019–2020 academic year, there are 372,532 Chinese students studying in the United States, accounting for 34.6% of the total number of international students in American colleges and universities. China is continuously the largest source of foreign students in Japan, Australia, the United States, New Zealand, the United Kingdom and Germany [6].

In May 2020, QS released the report on how COVID-19 affects global students, which shows that 96% of the Chinese students interviewed said that the epidemic would not interrupt the study abroad program [7, 8]. After all, studying abroad (whether long-term or short-term) is a rare opportunity for college students [9–12]. However, the report also points out that although the total number of students studying in the United States has continued to increase in the past 15 years, its growth rate has decreased from 29.9% in 2009–2010 to 0.8% in 2019–2020 academic years. Since the outbreak of COVID-19, in order to control the epidemic, countries all over the world

Z. Hu et al. (Eds.): AIMEE 2021, LNDECT 107, pp. 255–267, 2022.
https://doi.org/10.1007/978-3-030-92537-6_24

have issued policies to restrict studying abroad. Many college students who want to go abroad must postpone or cancel their plans, and college students also have a certain sense of crisis and anxiety about studying abroad [13].

Based on the outbreak of COVID-19 in 2020, this paper analyzes the impact of university students' perception of crisis and other factors on the willingness to go abroad from the perspective of image perception of the receiving countries abroad, and tries to find out the relationship between the elements, so as to provide valuable reference for the students studying abroad after the outbreak [14]. It also provides reference and suggestions for the economic development of study abroad service industry.

2 Model Construction and Research Hypothesis of Study Abroad Decision-Making

There are many decision-making factors affecting college students' willingness to study abroad. The image perception of the country and region studying abroad (including the overall image, safety image, etc.) and the risk perception under the epidemic background (containing crisis intensity, crisis duration, and so on.) will have various effects on the willingness to study in other contries [15–18]. To explore the impact of COVID-19 epidemic on the decision-making of studying abroad, it is needed to analyze it on the basis of crisis theory [19].

Crisis theory is a comprehensive theory involving psychology, sociology, management, economics and natural science, which can be widely used in the research of various decision-making problems in different fields and industries. Cognitive theory studies human perception of social situation and the process of information processing. People usually form the feeling of a social situation into a specific form. No matter how complex the social situation is, they can get some certain cognition. At the same time, the cognition of the environment will affect human response to social conditions. The most important thing to study abroad is the destination country. It is necessary to study the connotation of destination image from the subjective level.

The research shows that the target terrain image is composed of three elements: cognition, emotion and behavior. Cognition refers to the understanding of the knowledge and belief of the destination; Emotion implies to the feeling of destination; Behavior, also known as action elements, reflects the behavior characteristics of destination selection.

Based on previous studies and the COVID-19 incident, this paper constructs a theoretical model of studying abroad in Fig. 1, where, the COVID-19 incident is a "stimulus" factor; in the context of the epidemic, college students' perception of the image of the country studying abroad is "cognition"; in the context of the new crown epidemic situation, whether the image perception of the studying countries will affect the students' willingness to study abroad is "behavior intention", or simply, "behavior".

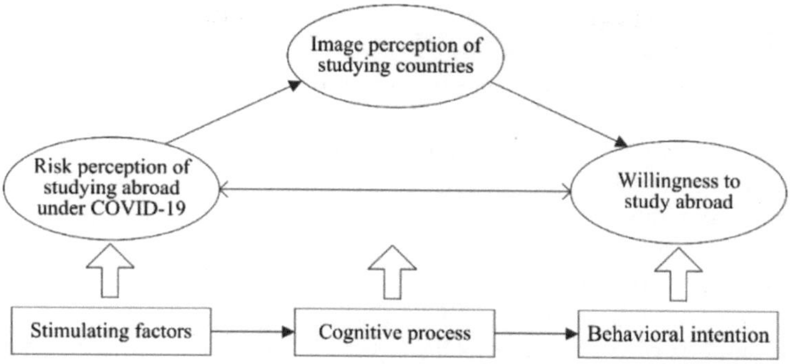

Fig. 1. Construction of study abroad decision-making theoretical model

2.1 Construction of Empirical Model

It has been proved to be a reasonable and effective method to use the means of empirical analysis and build a theoretical model to analyze various elements and their impact on the decision-making of studying abroad [20–22]. This paper constructs the theoretical model of Fig. 1 to explore the relationship between epidemic crisis, image perception of places to study abroad and willingness to go abroad. It can be seen from the model that the epidemic crisis will not only affect their willingness to go abroad directly, but also indirectly influence their willingness to go abroad through their perception of foreign image, as shown in Fig. 2.

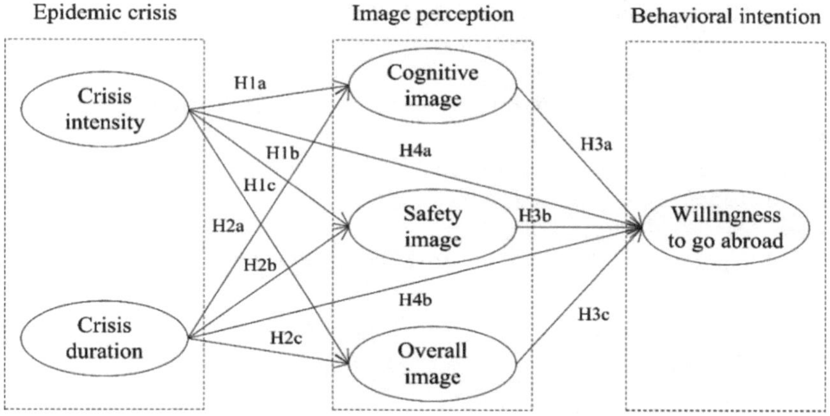

Fig. 2. Model construction of influencing factors of willingness to go abroad

2.2 Proposal of Research Hypothesis

On the base of construction of the research model and relevant theoretical research, the following three assumptions are put forward:

(1) Hypothesis on the relationship between epidemic crisis and image perception of places to study abroad.

H_1: the intensity of epidemic crisis has a significant impact on the image perception of overseas students;

H_1a: the intensity of epidemic crisis negatively affects the cognitive image of students studying abroad;

H_1b: the intensity of epidemic crisis negatively influences the safety image of the place of study abroad;

H_1c: the intensity of epidemic crisis negatively affects the overall image of the place of study abroad.

H_2: the duration of epidemic crisis has a significant impact on the image perception of overseas students;

H_2a: the duration of the epidemic crisis negatively acts on the cognitive image of the place of study abroad;

H_2b: the duration of the epidemic crisis negatively affects the safety image of the place of study abroad;

H_2c: the duration of epidemic crisis has a negative impact on the emotional image of students studying abroad.

(2) Hypothesis of the relationship between image perception and willingness to study abroad.

H_3: the image of studying abroad has a significant impact on the willingness to study abroad;

H_3a: the cognitive image of the place to study abroad has a positive impact on the willingness to study abroad;

H_3b: the safety image of the place to study abroad has a positive impact on the willingness to study abroad;

H_3c: the overall image of the place to study abroad has a positive act on the willingness to study abroad.

(3) Hypothesis on the relationship between epidemic crisis and willingness to study abroad.

H_4: the epidemic crisis has a significant impact on the willingness to study abroad;

H_4a: the intensity of epidemic crisis negatively affects the willingness to study abroad;

H_4b: the duration of the epidemic crisis negatively influences the willingness to study abroad.

2.3 Scale Design

According to the above model construction and research assumptions, the image perception scale in Table 1 is designed.

Table 1. COVID-19's image perception scale for studying abroad

Variables	Items
Crisis Strength (CS)	CS_1: CS of COVID-19 will affect my judge on WA
	CS_2: Warning of COVID-19 will affect my judge on WA
	CS_3: Range of COVID-19 will affect my judge on WA
	CS_4: Death toll of COVID-19 will affect my judge on WA
Crisis Duration (CD)	CD_1: CD of COVID-19 will affect my judge on WA
	CD_2: Different stage of COVID-19 will affect my judge on WA
Cognitive Image (CI)	CI_1: I think the environment of foreign universities is beautiful
	CI_2: I think foreign universities have higher professional heat
	CI_3: I think it is rather easy to find a job after studying abroad
	CI_4: I think the return salary is high after studying abroad
Safety Image (SI)	SI_1: I think public health abroad is safer
	SI_2: I think the social security in foreign countries is good
	SI_3: I think foreigners have a higher sense of crisis management
	SI_4: I think foreign countries can better control the epidemic
Overall Image (OI)	OI_1: After COVID-19 the impression of foreign cities deepened
	OI_2: After COVID-19, the overseas environment beautified
	OI_3: After COVID-19, their brand image has been strengthened
	OI_4: After COVID-19, their reputation has been improved
Willingness to Abroad (WA)	WA_1: I am eager to study abroad
	WA_2: I may consider studying abroad in the near future
	WA_3: I may consider studying abroad in the future

3 Empirical Research on Influencing Factors of Decision-Making

3.1 Questionnaire Design and Data Collection

The first part is the demographic survey, including gender, age, occupation and education level; The second part is the perception of epidemic crisis, which mainly has two dimensions: epidemic intensity and epidemic duration, a total of 6 questions; The third part is the image perception of studying abroad, which mainly has three dimensions: cognitive image, safety image and overall image, with a total of 12 questions; The fourth part is the willingness to study abroad, one dimension and three questions.

The questionnaire used LIKERT 5 scale and multiple-choice questions. It helps to use scientific methods for statistics. The relationship between multiple variables and their indicators can be understood by constructing the path coefficient structural equation. A total of 300 questionnaires were distributed and 282 questionnaires were recovered, of which 255 were valid documents, and the effective recovery rate was 90.4%.

3.2 Reliability and Validity Test

After sorting and analyzing the data collected from the questionnaire, firstly, the reliability and validity were analyzed by using the statistical software SPSS. The results are as follows:

(1) Reliability Analysis
The reliability test is carried out from six dimensions: crisis intensity, crisis duration, cognitive image, safety image, overall image and willingness to study abroad. The overall alpha value of the scale is 0.913 (>0.9), indicating that the overall reliability is very good. In the measurement of the dimension of the scale, the Clone Bach Alpha values of four variables are above 0.8, and only two variables have Clone Bach Alpha values below 0.8, but also above 0.75. Overall, the Clone Bach Alpha values of all dimensions are bigger than 0.7, with good reliability, indicating that the questionnaire has high consistency, reliability and stability, and can be used in this study. The overall reliability and dimensional reliability of the questionnaire are shown in Tables 2 and 3 respectively.

Table 2. Overall reliability analysis of the questionnaire

Clone Bach Alpha	Clone Bach alpha based on standardized terms	Number of items
0.913	0.927	21

Table 3. Reliability analysis of questionnaire variables

Variables	Number of items	Clone Bach Alpha
Crisis strength	4	0.856
Crisis duration	2	0.782
Cognitive image	4	0.853
Safety image	4	0.802
Overall image	4	0.856
Willingness to abroad	3	0.864

(2) Validity Analysis
In Table 4, KMO value is 0.914 (>0.9), Bartlett's spherical test P = 0.000 (<0.005) reaches a significant level, and the approximate chi-square value is 2870.133. Since the discussion of crisis duration only involves two issues, the KMO value must be 0.5 at this time. The KMO values of the other five dimensions were greater than 0.75, P = 0.000 (<0.005), achieving a significant level. It shows that the validity of this questionnaire is good, and the data obtained are very suitable for factor analysis. The validity analysis of questionnaire dimensions is shown in Table 5.

Table 4. KMO test

Kaiser Meyer Olkin measure of sampling adequacy		0.925
Bartlett's Spherical inspection	Approximate chi-square	2870.133
	freedom	2
	conspicuousness	0.000

Table 5. Dimension validity analysis of questionnaire

Variables	Items	KMO	Bartlett's Spherical inspection value	Remarkable
Crisis strength	4	0.824	457.236	0.000
Crisis duration	2	0.5	109.647	0.000
Cognitive image	4	0.779	276.172	0.000
Safety image	4	0.776	295.510	0.000
Overall image	4	0.825	397.864	0.000
Willingness to abroad	3	0.665	275.122	0.000

3.3 Confirmatory Factor Analysis

Confirmatory factor analysis is a research method to predict whether the measured factors and items are consistent with the researchers.

Table 6. Matching index values of structural equation model

Adaptation index	Recommended value	Fitting results	Fitting degree
χ^2/df	<3.0	2.118	Reasonable
RMSEA	<0.08	0.067	Reasonable
AGFI	>0.8	0.913	Good
GFI	>0.8	0.885	Good
CFI	>0.9	0.952	Reasonable
TFI	>0.9	0.938	Reasonable
IFI	>0.9	0.943	Reasonable

The questionnaire design of this paper adopts the maturity scale, takes the COVID-19 epidemic crisis as the background, appropriately modifies the specific research object, sets the structural equation model for hypothesis research analysis and model fitness value test.

(1) Construct Validity
Construct validity is also called devise validity. Its function is to check whether the corresponding relationship between the factor and the measurement item

Table 7. Factor load

	Route		Estimate	AVE	Composite reliability
CS_1	< –	Crisis strength	0.833		
CS_2	<–	Crisis strength	0.758	0.6615	0.8653
CS_3	<—	Crisis strength	0.782		
CS_4	<—	Crisis strength	0.786		
CD_1	<—	Crisis duration	0.819	0.5647	0.7804
CD_2	<—	Crisis duration	0.842		
CI_1	<—	cognitive image	0.831		
CI_2	< —	cognitive image	0.734	0.5483	0.8642
CI_3	<—	cognitive image	0.644		
CI_4	<—	cognitive image	0.758		
SI_1	<—	Safety image	0.769		
SI_2	<—	Safety image	0.728	0.5642	0.8234
SI_3	<—	Safety image	0.778		
SI_4	<—	Safety image	0.741		
OI_1	<—	Overall image	0.752		
OI_2	<—	Overall image	0.786	0.6394	0.8712
OI_3	<—	Overall image	0.821		
OI_4	<—	Overall image	0.846		
WA_1	<—	Willingness to abroad	0.921		
WA_2	<—	Willingness to abroad	0.741	0.6872	0.8682
WA_3	<—	Willingness to abroad	0.767		

meets the expectation. If the expectation is met, it indicates that its structural validity is acceptable, see Table 6 for details.
According to the overall fitting coefficient, χ^2/DF is 2.228 (<3), the value of RMSEA is 0.067 and the value of AGFI is 0.913; the value of GFI is 0.885; CFI, TFI and IFI are all greater than 0.9. In terms of index fitting degree, the model fitting degree is reasonable.

(2) Aggregate validity.
Aggregation validity can be analyzed by AVE and CR. If the AVE value of each factor is bigger than 0.5 and the CR value is greater than 0.7, it indicates that a good aggregation validity has been got. At the same time, the factor load coefficient corresponding to each measurement item is generally required to be greater than 0.7. The specific indexes are shown in Table 7.
It can be seen from Table 7 that the scale of this study is represented by 6 factors. Among the AVE values corresponding to these factors, the minimum is 0.5483 and the maximum is 0.6872, which are averagely significantly higher than the standard of 0.5; Moreover, the combined reliability CR value is 0.7804–0.8712, which is significantly higher than the standard of 0.7. Therefore, this study shows that the scale has excellent aggregate validity. Taken together, it can be proved that the data of this research scale has excellent aggregate validity.

4 Structural Equation Model Analysis

4.1 Construction of Structural Equation Model

In this paper, SPSS AMOS structural equation modeling (SEM) software is used for multivariable analysis.

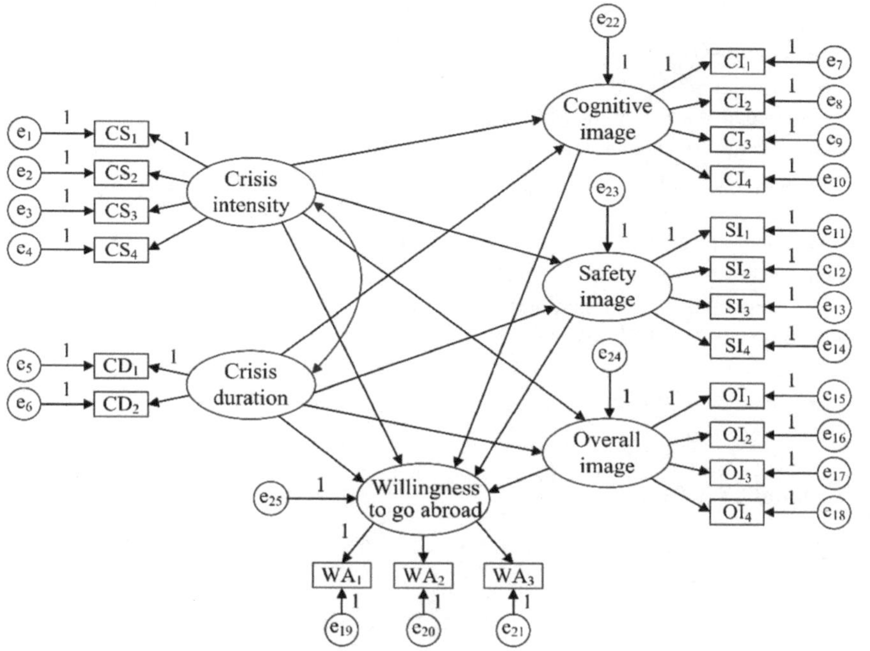

Fig. 3. Structural equation model

Structural equation model SEM is a multivariate data analysis method, including measurement relationship and influence relationship, as shown in Fig. 3. It can not only measure the relationship between the internal structure and correlation of various factors, but also measure the influence relationship between multiple independent variables and multiple dependent variables. It can be used to study the influence relationship between multiple latent variables. In Fig. 3, e1, e2,..., e25 in separate circles represent potential variables, while the squares represent observed variables. The structural equation model mainly represents the measurement relationship and influence relationship. According to the results of reliability and validity analysis, the questionnaire designed in this paper meets the standards and has ideal reliability and effectiveness.

4.2 Hypothesis Test and Analysis

Through the structural equation model and AMOS software, the standardized path coefficients of each potential variable can be calculated, as shown in Table 8. According to the operation results of path coefficient estimation, the test results shown in Table 9 can be obtained.

Table 8. Standardized path coefficient of each latent variable

Variable relationship	Estimate	S.E	C.R	P	Remarkable
Crisis strength → Cognitive image	–2.054	.660	–2.884	0.004	Yes
Crisis strength → Safety image	–2.506	.710	–3.020	0.003	Yes
Crisis strength → Overall image	–2.188	.618	–2.953	0.003	Yes
Crisis duration → Cognitive image	2.976	.699	4.116	***	Yes
Crisis duration → Safety image	2.479	.753	4.035	***	Yes
Crisis duration → Overall image	3.055	.722	4.049	***	Yes
cognitive image → Willingness to abroad	0.921	.474	1.944	0.052	Yes
Safety image → Willingness to abroad	–.500	.635	–.788	0.031	Yes
Overall image → Willingness to abroad	–.572	.290	1.973	0.011	Yes
Crisis strength → Willingness to abroad	–1.977	.525	–3.765	***	Yes
Crisis duration → Willingness to abroad	2.880	.637	4.519	***	Yes

4.3 Model Test Results

Through the above research, the following four model tests and hypothesis verification results are obtained.

(1) For hypothesis H_1, the intensity of epidemic crisis has a significant impact on the image of places to study abroad, which has been verified. As shown in Table 8, the P value is significant at the level of 0.05. Therefore, this paper believes that during the epidemic period, the intensity of the epidemic crisis has a significant negative impact on the image of studying abroad.

(2) For hypothesis H_2, the duration of epidemic crisis has a significant act on the image of studying abroad, which has been verified. As can be seen from Table 8, the value of P is significant at the level of 0.01. That means, the longer the epidemic lasts, the lower the perception of the safety image of the place of study abroad;

(3) For hypothesis H_3, the image perception of studying abroad has a significant impact on the willingness to study abroad. H_3a and H_{3C} are supported. Table 8 shows that they are significant at the level of 0.05. In the image perception of studying abroad, the impact of safety image on the willingness to study abroad is negative. Safety demand is the basic demand of the people. In that sense, under the influence of COVID-19, college students have higher requirements for the safety image of studying abroad.

(4) For hypothesis H_4, the epidemic crisis has a significant impact on the willingness to study abroad, which has been verified. As shown in Table 8, the P value is

significant at the level of 0.001. Then, the stronger the crisis perception of college students, the more they will increase their fear of the image of studying abroad, greatly affect their confidence, and their willingness to study abroad will decline.

Table 9. Hypothesis test results

Order number	Hypothesis	Test results
$H1_a$	The intensity of epidemic crisis negatively affects the cognitive image of places studying abroad	Support
$H1_b$	The intensity of the epidemic crisis negatively affects the safety image of places studying abroad	Support
$H1_c$	The intensity of the epidemic crisis has a negative impact on the overall image of places studying abroad	Support
H2a	The duration of the epidemic crisis negatively affects the cognitive image of students studying abroad	Support
$H2_b$	The duration of the epidemic crisis negatively affects the safety image of places studying abroad	Support
H2c	The duration of the epidemic crisis negatively affects the emotional image of students studying abroad	Support
H3a	The cognitive image of studying abroad has a positive impact on the willingness to study abroad	Support
$H3_b$	The safety image of studying abroad has a positive impact on the willingness to study abroad	Support
$H3_c$	The overall image of studying abroad has a positive impact on the willingness to study abroad	Support
$H4_a$	The intensity of the epidemic crisis negatively affects the willingness to study abroad	Support
$H4_b$	The duration of the epidemic crisis negatively affects the willingness to study abroad	Support

5 Conclusion

Through the research and summary of relevant theories, based on the COVID-19 epidemic background, the paper constructs the research model of epidemic crisis, the image of studying abroad and the willingness to study abroad, verifies and analyzes it on the basis of investigation and research, and draws the following three conclusions:

(1) The epidemic crisis has a significant impact on the image perception of places studying abroad. The paper divides the epidemic crisis into two dimensions: crisis intensity and crisis duration. The greater the intensity of the crisis and the longer the duration of the crisis, the stronger the perception of the epidemic crisis is and the more obvious the perception of the image of studying abroad should be. The transparent broadcast of the outbreak of COVID-19 made it possible for college students to clearly understand the situation of the development of the epidemic,

and the less likely it was to cause panic, the smaller the impact on image perception of studying abroad.

(2) The epidemic crisis may have a significant impact on college students' willingness to study abroad. COVID-19 is a sudden public crisis. In the outbreak stage, the willingness of university students to study abroad is relatively low. However, at different stages of the epidemic, governments of separate countries adopt different preventive measures to rebuild confidence and willing to go to the destination. At that time, the willingness to study abroad will increase correspondingly.

(3) The image perception of studying abroad has a certain impact on college students' willingness to study abroad. The assumption of the safety image of the place of studying abroad on the willingness of college students to go abroad is established. Generally, college students will yearn for a safe place. During the epidemic crisis, college students have low perception of the safety image of the place to study abroad, so they will not choose to study abroad. On the contrary, if college students have a better impression of the place where they study abroad and feel that the place is safer, they are more likely to choose to go abroad to study in that country.

Acknowledgment. This paper is supported by three projects as follows: (1) Research on technological innovation mechanism of strategic emerging industries in East Lake National Independent Innovation Demonstration Zone (2019ADC039); (2) Comparative study on scientific research efficiency of colleges and universities in Hubei Province from the perspective of "double first class" (20G079); (3) Study on spatial differentiation and growth effect of logistics industry in Central China (D2014005).

References

1. Gilliland, I., et al.: Building cultural sensitivity and interprofessional collaboration through a study abroad experience. J. Nurs. Educ. **55**(1), 45–48 (2016)
2. Sagyndykova, Z., Gazdiyeva B., Zhakupova, A.: Developing intercultural competency: study abroad experience. In: 8th International Conference on Education and New Learning Technologies Edulearn 2016, pp. 8280–8284 (2016)
3. Talburt, S.: What's the subject of study abroad? race, gender, and living culture. Modern Lang. J. **83**(2), 163–175 (1999)
4. Salisbury, M.H., et al.: Going global: understanding the choice process of the intent to study abroad. Res. High. Educ. **50**(2), 119–143 (2009)
5. Carlson, J.S., Widaman, K.F.: The effects of study abroad during college on attitudes toward other cultures. Int. J. Intercult. Relat. **12**(1), 1–17 (1988). https://doi.org/10.1016/0147-1767(88)90003-X
6. Wen, H.: Increasingly diversified study abroad in China in the post epidemic era - CCG released the blue book of the development report on study abroad in China (2020–2021). China's Foreign Trade **3**, 78–80 (2021). (in Chinese)
7. Hussain, M.W., Mirza, T., Hassan, M.M.: Impact of COVID-19 pandemic on the human behavior. Int. J. Educ. Manage. Eng. (IJEME) **10**(5), 35–61 (2020)

8. Hassan, A.H.M., Qasem, A.A.M., Abdalla, W.F.M., Elhassan, O.H.: Visualization & prediction of COVID-19 future outbreak by using machine learning. Int. J. Inf. Technol. Comput. Sci. (IJITCS) **13**(3), 16–32 (2021)
9. Anderson, P.H., et al.: Short-term study abroad and intercultural sensitivity: a pilot study. Int. J. Intercult. Relat. **30**(4), 457–469 (2006)
10. Walters, C., Charles, J., Bingham, S.: Impact of short-term study abroad experiences on transformative learning: a comparison of programs at 6 weeks. J. Transform. Educ. **15**(2), 103–121 (2017)
11. O'Connell, R., Ayllon, M.: Student perceptions of global knowledge and skills acquired during a five-week study abroad program. In: 2016 IEEE Frontiers In Education Conference (FIE) (2016)
12. Dwyer, M.M.: More is better: the impact of study abroad program duration. Frontiers **10**, 151–163 (2004)
13. Qi, Q.: Overseas students under the epidemic: hesitation and choice, responsibility and affection. Study Abroad **05**, 32–35 (2020)
14. Hassan, M.M., Mirza, T., Hussain, M.W.: A critical review by teachers on the online teaching-learning during the COVID-19. Int. J. Educ. Manage. Eng. (IJEME) **10**(5), 17–27 (2020)
15. Chen, Y., Yue, C.: Trends and influencing factors of college students studying abroad in China: 2005–2017. Educ. Acad. Monthly. **05**, 40-45+111 (2020). (in Chinese)
16. Li, W.: Analysis of influencing factors and promotion mechanism of college students' willingness to study abroad-taking Guizhou University of Finance and economics as an example. Indust. Sci. Technol. Forum **19**(13), 97–98 (2020). (in Chinese)
17. Li, W.: Analysis on influencing factors of college students' willingness to study abroad. Think Tank Era. **35**, 251–254 (2019). (in Chinese)
18. Wang, Z., Crawford, I.: Factors motivating destination decisions of Chinese study abroad students. Int. J. Educ. Manag. **35**(2), 408–425 (2020)
19. Yongchang, W., Fangyu, C.: Empirical study on influencing factors of college students' willingness to study abroad. High. Educ. Rev. **5**(02), 204–219 (2017). (in Chinese)
20. Zijian, M., Rui, W.: Empirical study on the influencing factors of college students studying abroad. Manage. Observ. **04**, 145–147 (2017). (in Chinese)
21. Yongchang, W., Fangyu, C.: Empirical study on influencing factors of college students' willingness to study abroad. High. Educ. Rev. **2**, 204–219 (2017). (in Chinese)
22. Emily, F.S.: Review of Crisis intervention: Building resilience in troubled times. J. Crisis Interven. Suicide Prevent. **41**(3), 237–238 (2020)

Importance of Air Pollutants in Controlling the Epidemic During the City Lockdown for COVID-19 in China: Analysis Based on Environmental Monitoring

Kai Liu(✉)

Business School, Sichuan University, Chengdu 610064, China

Abstract. The novel coronavirus pneumonia, namely COVID-19, has become a global public health problem. To control the spread of COVID-19, various studies have been conducted to explore important factors affecting the transmission of SARS-CoV-2. Industrial production contributes to the study of these two paths to improve air quality. Starting from the air pollutants themselves as a risk factor of respiratory tract infections, based on the perspective of environmental protection, this article aims to explore the new path of the relationship between ambient air pollutants and the new coronavirus infection, as well as the additional effects of the closure of the city. The research is based on the daily new confirmed cases, air pollution concentration and meteorological variables in 304 cities in my country from January 23, 2020 to March 5, 2020. The generalized additive model is used to study the relationship between six air pollutants and COVID-19 infection association. This study highlights that there is a significant relationship between air pollution and COVID-19 infection, which can partly to explain the positive impact on the city lockdown in order to control the epidemic when the epidemic breaks for, and explore and improve the path mechanism of the current research.

Keywords: Sustainable development · Health monitoring · COVID-19 · Generalized additive model

1 Instruction

In the context of the rapid outbreak of the COVID-19 epidemic, to control the spread of COVID-19, various studies have been conducted to explore important factors affecting the transmission of SARS-CoV-2. Several early studies have demonstrated that human-to-human contact could increase the risk of COVID-19 infection [1]. Besides, population mobility has a significant effect on the COVID-19 epidemic. In addition, a recent study has shown an association of ambient temperature with the infection of COVID-19. How to meet local authority's needs, improve efficiency, and reduce costs are important issues that cloud service providers must deal with [2]. However, the impact of short-term exposure to air pollution lacks careful consideration.

Previous studies have suggested that ambient air pollutants are risk factors for respiratory infection by carrying microorganisms to make pathogens more invasive to

Z. Hu et al. (Eds.): AIMEE 2021, LNDECT 107, pp. 268–276, 2022.
https://doi.org/10.1007/978-3-030-92537-6_25

humans and affecting body's immunity to make people more susceptible to pathogens [3–5]. Since COVID-19 is a respiratory disease and SARS-CoV-2 could remain viable in aerosols for hours, it is interesting to investigate the effect of air pollution on COVID-19 infection [6]. To provide useful implications for the control and prevention of this novel coronavirus disease, we aimed to explore the relationship between concentrations of six air pollutants and daily confirmed COVID-19 cases in 304 cities in China.

2 Literature Review and Hypothesis

Scholars have carried out extensive research in the field of COVID-19, mainly focusing on pathology, epidemiology, and sociology. The spread of the new type of coronary pneumonia through respiratory droplets and contact has become a pandemic worldwide.

The lockdown of the city also has the environmental effect of improving air quality and promoting public health benefits. Under the state of urban blockade, restrictions on private car travel and public transportation, and the suspension of most commercial activities and industrial production help to improve air quality. After the closure of the city, the average monthly air quality index of Wuhan was 59.7, which was 33.9% lower than before the closure (January 23, 2020) and 47.5% lower than the period from 2015 to 2019 (113.6). Compared with before the closure of the city, fine particulate matter (PM) has dropped by 36.9% and remains the main pollutant at 2.5. Nitrogen dioxide (NO2) decreased the most, about 53.3%, and ozone (O3) increased by 116.6%. Previous studies have shown that air pollutants are a risk factor of respiratory infections. Air pollutants make pathogens more invasive to humans by carrying microorganisms, and affect human immunity, making people more susceptible to pathogens. As research has pointed out, viruses are completely dependent on their hosts to replicate and survive. With the development of the past few decades, China is suffering from severe air pollution, causing significant economic losses and health consequences. With the expectation of controlling air pollution, the Chinese government issued and implemented the "Air Pollution Prevention and Control Action Plan" in September 2013 [7, 8].

Based on the situation of the epidemic as a public health intervention measure, this article specifically carried out the extension of the Spring Festival holiday, the suspension of urban bus, subway, ferry, and long-distance passenger transportation, the temporary closure of exit channels at airports and railway stations, the closure of schools and entertainment venues, and the prohibition of public gatherings, etc.. The closure of the city has the direct effect of effectively restricting and controlling the spread of the COVID-19 epidemic by restricting the movement tpwards people, isolating patients and close contacts, etc.; at the same time, the closure of the city restricts private car travel and public transportation, halting the most commercial activities and Industrial production helps to improve air quality, reduces the concentration of air pollutants, helps reduce air pollutants' invasion of respiratory tract infections and human body immunity, has the environmental protection effect of improving air quality and promotes public health benefits.

3 Research Methodology

According to statistics from the National Health Commission, as of March 5, 2020, a total of 79,968 confirmed cases of COVID-19 have been found across the country. Due to the limitation of weather data and air pollution data, we will focus our analysis on 304 (4 municipalities directly under the central government, 302 prefecture-level cities) cities in China. The cities we studied to cover 86% of confirmed cases. From January 23, 2020 to March 1, 2020, the newly confirmed cases of each city are obtained from the National Health Commission of China. We will use January 23, 2020 as the starting point for the research phase, and reduce the number of cases of Wuhan The potential impact on imported cases of each city. According to the report on the National Health Commission, starting from March 5 in other regions of the country except Hubei, the number of new cases will be basically cleared at the end of February. We chose March 5 as the end of the research phase.

Air pollution data were collected from an online platform (https://www.aqistudy.cn) monitoring and analyzing the air quality. Daily concentrations of six air pollutants were measured, including particles with diameters ≤ 2.5 μm (PM2.5), particles with diameters ≤ 10 μm (PM10), sulfur dioxide (SO2), carbon monoxide (CO), nitrogen dioxide (NO2), and ozone (O3). Meteorological data on daily mean temperature, relative humidity, air pressure, and wind speed during the study period were obtained from the National Meteorological Information Center (http://data.cma.cn) (Fig. 1).

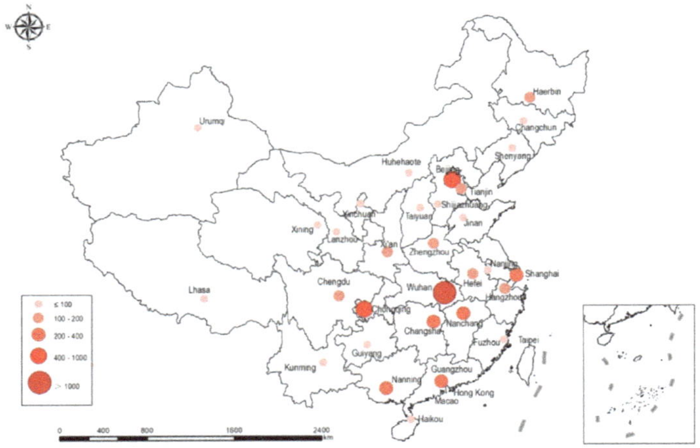

Fig. 1. Geographic patterns of COVID-19 confirmed case counts in 30 provincial capital cities of China as of March 5, 2020

3.1 Model Construction and Variable Selection

As previous studies have proved, the impact on air pollution is continuous. In addition, the National Health Commission of China reported that the incubation period of COVID-19 is 1 to 14 days. Therefore, it is a reasonable choice to use the moving average method to discuss the cumulative hysteresis effect of ambient air pollution. Therefore, in this study, we used the Generalized Additive Model (GAM) to estimate the association between the moving average concentration of air pollutants (Lag0–7, Lag0–14, Lag0–21) and the daily confirmed cases of COVID-19. Specifically, we studied the effects of six air pollutants on the reduction of collinearity in six separate models (ie, single pollution model), because some of these pollutants are highly correlated.

The basic model is defined as follows:

$$\log(y_{it}) = a + Zil + S(mte_{il}) + S(rhu_{il}) + S(wsp_{il}) + S(sho_{il}) + S(pre_{il}) + S(mai_{il}) + \log(y_{i,t-1}) + city_i + day_t + \varepsilon_{it} \quad (1)$$

Table 1. New cases confirmed daily, air pollution concentration, and descriptive statistics of meteorological variables for all cities and days

	Valid	Missing	Mean	Std. deviation	Minimum	Maximum
AQI	11856	0	70.55	45.784	13	500
PM2_5	11856	0	46.63	38.21	1	554
PM10	11856	0	63.31	50.507	3	781
SO2	11856	0	10.97	9.468	1	104
NO2	11856	0	18.38	10.671	2	95
O3	11856	0	77.78	21.577	4	189
CO	11856	0	0.853	0.4298	0.1	7.4
Altitude	11856	0	486.095	687.4177	–10.8	3393.5
Mean temperature	11856	0	4.959	8.9763	–33.8	26.9
Relative humidity	11856	0	72.32	300.821	6	32766
Wind speed	11856	0	2.148	1.2137	0	14.9
Sunshine hours	11856	0	5.114	3.8908	0	29.8
Precipitation	11856	0	10.03	917.48	0	99900
Mean air pressure	11856	0	96820.4	7879.474	66810	327660
Add	11856	0	6.73	142.674	0	13436

Zil represents the linear term of the (l 1) day moving average concentration of air pollutants (lag0 – l) in city i. Here, log(yit) represents the log-transformed COVID-19 count in city i reported on the i-th day (plus 1 to avoid taking the logarithm to 0) (Liu et al. 2020). a is the intercept. Zil represents the linear term of the (l + 1) day moving average concentration of air pollutants (lag + 1) in the i-th city. The possible confounding effects of meteorological factors during the same period are controlled, including average temperature (mte), relative humidity (rhu), wind speed (wsp), sunshine duration (sho), average air pressure (pre), and average surface temperature (mai). s is a smooth function of a certain meteorological factor (a thin plate spline function with a maximum degree of freedom of 3. log(yi,t − 1) represents the log-transformed COVID-19 count reported on day t-1 of city i to illustrate the potential serial correlation in our data [7]. In addition, we include urban fixed effects (cityi) to control the urban characteristics that change over time such as population size and density, and we also include daily fixed effects (dayt) to control unobservable factors such as regional blockades that affect all cities every day.

3.2 Method

All analyses in this study were performed in the R statistical software (version 4.0.2) by using the "mgcv (1.8–31)" software package. The statistical test is a two-sided test, and $p < 0.05$ is considered statistically significant. The effect estimate is shown as the percentage change (%) of daily COVID-19 confirmed cases as the concentration of air pollutants per unit increases (ie PM2.5, PM10, SO2, NO2, O3 for every 10 $\mu g/m^3$ increase or CO for every 1 mg increase/m^3).

4 Result

4.1 Descriptive Analysis

Table 1 shows the statistics of daily confirmed COVID-19 cases, new cases, air pollution concentration and meteorological variables. During the observation period, this study included 78913 cases of average daily concentrations of PM2.5, PM10, SO2, NO2, O3, and CO of 46.63 $\mu g/m^3$, 63.31 $\mu g/m^3$, 10.97 $\mu g/m^3$, 18.38 $\mu g/m^3$, 77.78 $\mu g/m^3$, 0.853 mg/m^3. The average daily temperature, relative humidity, wind speed, sunshine hours, precipitation, and air pressure is 4.959 °C, 72.32%, 2.148 m/s, 5.114 h, 10.03 mm, 96820.4 Pa.

4.2 The Relationship Between Air Pollution and Confirmed Cases of COVID-19

Table 2 shows the moving average hysteresis effects (Lag0–7, Lag0–14, Lag0–21) of different air pollutants on the single-pollution model COVID-19 confirmed cases of a day. When the weekly average air pollutant concentration increases by 1 $\mu g/m^3$ (or 1 mg/m^3), the relative risk (RelativeRisk, RR) and excess risk (ER) of COVID-19 are calculated by the following formulas:

$$RR = \exp(\beta),\ ER = (RR - 1) * 100\%.$$

Table 2. The number of new confirmed cases of COVID per 1 μg/m^3 (or 1 mg/m^3) increase in air pollutant concentration ER (%)

Air pollutants	Lag = 0–7	Lag = 0–14	Lag = 0–21
PM2.5 (μg/m^3)	0.0259	0.0283	0.046
PM10 (μg/m^3)	0.00804	0.0078	0.0129
SO2 (μg/m^3)	2.72**	2.80**	2.10**
CO (mg/m^3)	3.525**	4.377**	2.990**
NO2 (μg/m^3)	0.38	0.001**	0.0078**
O3 (μg/m^3)	−0.22*	−0.37**	−0.19*

*When the test level (two-sided) is 0.05, the model is significant; **When the test level (two-sided) is 0.01, the model is significant.

Based on the results of the previous study, four air pollutants with significant lag effects: PM2.5, SO2, NO2, and O3 were selected to model the impact of newly diagnosed COVID-19. The models all passed the parameter t test ($P < 0.05$), the results show that after controlling the lag effect of the disease itself, meteorological factors, and the trend and short-term fluctuations of the incidence of COVID-19, these air pollutants are risk factors of the incidence of COVID-19. The table shows the relevant statistics of the single pollutant model that passed the test (Table 3).

Table 3. Single pollutant model (Lag = 0–14) results

Model	Estimation	Standard deviation	P	RR	95% CI	95% CI	R square	QAIC
SO2	0.028	0.006	<0.001	1.028	1.016	1.034	0.909	277.698
CO	0.363	0.114	0.002	1.044	1.15	1.797	0.902	277.822
NO2	0.009	0.002	<0.001	1.009	1.005	1.014	0.911	277.99
O3	−0.004	0.001	<0.001	0.996	0.994	0.998	0.91	279.494

4.3 Multi-pollutant Model

For the selected single air pollutant variables that has a significant impact on the number of newly confirmed COVID cases, a multi-pollutant model is established separately, and the model is tested for the significance of the parameters, and the models that have not passed the test are eliminated, and the QIAC is optimized Criteria to get the optimal multi-pollutant model (Table 4).

Table 4. Comparison of multi-pollution factor models (Lag = 0–14)

Model	R square	QAIC
SO2 + CO*	0.909	279.475
NO2 + CO	0.908	279.041
SO2 + NO2	0.912	279.107
SO2 + O3	0.924	284.54
O3 + CO*	0.912	281.658
O3 + NO2*	0.914	282.103
SO2 + CO + O3	0.924	286.513
SO2 + CO + NO2	0.908	280.345
SO2 + NO2 + O3	0.921	285.569
NO2 + CO + O3	0.912	283.195
SO2 + CO + O3 + NO2	0.924	287.294

Table 5 shows the estimation results of the optimal multi-pollutant model. It can be seen that after controlling the long-term trend, short-term fluctuations of COVID and the mixed influence of meteorological factors, for every increase in the weekly average concentration of SO2 and CO by 1 $\mu g/m^3$ (or 1 mg/m^3), the weekly incidence of COVID will increase by 2.5 respectively. % And 2.6%.

Among them, after SO2 is included in CO, the weekly For every increase in the average concentration of 1 $\mu g/m^3$, the weekly incidence of COVID ER will decrease from 2.8% to 2.45%, and after the inclusion of SO2, for every increase in the weekly average concentration of 1 mg/m^3, the increase in the weekly incidence of CVOVID will decrease from 4.378% to 2.610%.

Table 5. Optimal multi-pollutant SO2 + CO two-factor model results

Model	Estimation	Standard deviation	P	RR	95%CI	95%CI
Interception	5.151	0.134	<0.001	172.604	132.736	224.447
SO2	0.024	0.006	<0.001	1.025	1.025	1.03
CO	0.232	0.115	0.046	1.026	1.261	1.58

5 Summary and Conclusion

This article uses a generalized additive model to explore the relationship between ambient air pollutants and the number of new COVID-19 cases. We found that PM2.5, PM10, SO2, and CO were significantly positively correlated with confirmed cases of COVID-19, while O3 was negatively correlated with the number of confirmed cases per day. These findings supports the original findings that air pollution is an important factor in COVID-19 infection.

As proven by previous literature, air pollution is also closely related to respiratory infections caused by other microorganisms. Therefore, we compared our main findings with previous studies to find similarities and differences.

As evidenced by previous literature, air pollution is also very closely related to respiratory infections caused by other microorganisms [8]. Therefore, we compared the main findings with previous studies to find the similarities and differences. Some scholars have adopted case crossover designed, and short-term exposure to higher PM2.5 will be linked to more medical care for acute lower respiratory tract infections. An important correlated distributed hysteresis nonlinear model was also found to evaluate atmospheric particulate matter (PM2.5 and PM10) and hospitalizations for respiratory diseases. A time series analysis conducted in Thailand found that PM10, SO2, CO, NO2, and O3 were significantly associated with increased respiratory risk in hospitalization [9]. A literature review shows that exposure to SO2, CO and NO2 is harmful to our health and increases the risk of respiratory diseases [10]. In general, all six air pollutants may be risk factors of respiratory infections. However, since we have observed a negative correlation between O3 and COVID-19, our results are different from previous studies. case. The bactericidal properties of O3 may be a possible reason [11, 12], whether this is due to the sterilization and disinfection effects of ozone itself. Further research is needed to determine the biological mechanism behind this phenomenon.

Our research has certain significance of the control and prevention of COVID-9. First of all, the government and the public should pay more attention to areas with high concentrations of SO2, CO and NO2, as these areas may suffer more severe COVID-19 pneumonia. In other words, reducing air pollutants may be a useful way to control COVID-19 infection. In addition, it is worth not that, therefore, O3 is negatively correlated with confirmed cases of COVID-19, and further laboratory studies are needed to clarify its underlying mechanism. Our results show that there is a significant relationship between air pollution and COVID-19 infection [13], which can partly to explain the positive impact on the city lockdown in order to control the epidemic during of the outbreak, and provide more flexible analytical framework that decomposes the important components of the time series, incorporates the prior information, and captures the evolving nature of model parameters [14].

This article still has several limitations that might be addressed in future research. First of all, we only focus on the relationship between air pollutants and confirmed cases of COVID-19, and do not focus on the causal relationship between air pollution and COVID-19 infection. Secondly, by observing and selecting Chinese cities, our research results lack global representation. Future research needs to overcome these limitations.

References

1. Berendt, R.F., Dorsey, E.L., Hearn, H.J.: Virucidal properties of light and SO2 II. Effect of a low gas concentration on aerosolized virus. Proc. Soc. Exp. Biol. Med. **138**(3), 1005–1008 (1971)

2. Saxena, S., Khan, M.Z., Singh, R.: Green computing: an era of energy saving computing of cloud resources. Int. J. Math. Sci. Comput. **7**(2), 42–48 (2021)
3. Chen, N., Zhou, M., Dong, X., et al.: Epidemiological and clinical characteristics of 99 cases of 2019 novel coronavirus pneumonia in Wuhan, China: a descriptive study. Lancet **395**(10), 507–513 (2020)
4. Chen, T.M., Kuschner, W.G., Gokhale, J., et al.: Outdoor air pollution: nitrogen dioxide, sulfur dioxide, and carbon monoxide health effects. Am. J. Med. Sci. **333**(4), 249–256 (2007)
5. Horne, B.D., Joy, E.A., Hofmann, M.G., et al.: Short-term elevation of fine particulate matter air pollution and acute lower respiratory infection. Am. J. Respir. Crit. Care Med. **198**(6), 759–766 (2018)
6. Lin, H., Tao, J., Kan, H., et al.: Ambient particulate matter air pollution associated with acute respiratory distress syndrome in Guangzhou, China. J. Eposure Sci. Environ. Epidemiol. **28** (4), 392–399 (2018)
7. Liu, K., Hou, X., Ren, Z., et al.: Climate factors and the East Asian summer monsoon may drive large outbreaks of dengue in China. Environ. Res. **6**(4), 183–197 (2020)
8. Lu, H., Stratton, C.W., Tang, Y.: Outbreak of pneumonia of unknown etiology in Wuhan, China: the mystery and the miracle. J. Med. Virol. **92**(4), 401–402 (2020)
9. Loeppky, J.A., Cagle, A.S., Sherriff, M., et al.: A local initiative for mobile monitoring to measure residential wood smoke concentration and distribution. Air Qual. Atmos. Health **6** (3), 641–653 (2013)
10. Wm, A., Hla, B., Hkb, C.: Short-term air pollution exposure and emergency department visits for amyotrophic lateral sclerosis: a time-stratified case-crossover analysis - ScienceDirect. Environ. Int. **12**(3), 467–475 (2019)
11. Zhu, Y., Xie, J., Huang, F., et al.: Association between short-term exposure to air pollution and COVID-19 infection: evidence from China. Sci. Total Environ. **6**(4), 710–727 (2020)
12. Yang, Z., Zeng, Z., Wang, K., et al.: Modified SEIR and AI prediction of the epidemics trend of COVID-19 in China under public health interventions. J. Thorac. Dis. **12**(3), 152–165 (2020)
13. Sahed, A., Mekidiche, M., Kahoui, H.: Forecasting natural gas prices using nonlinear autoregressive neural network. Int. J. Math. Sci. Comput. **6**(5), 37–46 (2020)
14. Hassan, M.M., Mirza, T.: Using time series forecasting for analysis of GDP growth in India. Int. J. Educ. Manage. Eng. **11**(3), 40–49 (2021)

Process and Intelligent Automation for the Parts Production from Heat-Resistant Alloys Based on the Morphological Approach

Rafael J. Sukhorukov[1], Lubov V. Gavrilina[1], Marin A. Pecheykina[2], and D. L. Rakov[1(✉)]

[1] Mechanical Engineering Research Institute of the Russian Academy of Sciences (IMASH RAN), Moscow, Russia

[2] Moscow Power Engineering Institute (National Research University), Moscow, Russia

Abstract. The article discussed the results of theoretical and experimental studies aimed at creating production intelligent technologies for forming parts from heat-resistant alloys by rolling under isothermal conditions. A methodological approach to the development of technology and equipment for forming axisymmetric parts made of heat-resistant alloys is given. During the research, mathematical and physical modeling of the process of rolling out parts was carried out. The possibility of controlling the structure formation of the workpiece material in the process of rolling under superplastic conditions is shown by choosing the optimal modes of temperature, magnitude and rate of deformation. The morphological approach was used for the analysis and selection of technological processes. On the basis of the proposed approach, an assessment of technologies, equipment and control systems for the manufacture of blanks of long hollow shafts of gas turbine engines from heat-resistant nickel and titanium alloys was carried out. Key objective is to find a solution space with the potential to fulfil the top-level technological processes requirements. Usage of cluster analysis, set theory, set of rules allows to identify the clusters of innovative technologies combining high performance potential with robustness regarding requirement changes. These investigations verify the weighty potential of the proposed approach compared to present methods.

Keywords: Process and intelligent automation · Heat-resistant alloys · Isothermal conditions solutions space · Morphological matrix · Forming axisymmetric parts

1 Introduction

The industrial concept “Industry 4.0” is a global, complex, multi-level organizational and technical system [1–3]. Special attention is paid to the intellectualization of production processes. Currently, in mechanical engineering, much attention is paid to the development of intelligent technologies for forming parts from heat-resistant alloys by rolling under isothermal conditions. In particular, the mathematical modelling of the

Z. Hu et al. (Eds.): AIMEE 2021, LNDECT 107, pp. 277–286, 2022.
https://doi.org/10.1007/978-3-030-92537-6_26

deformation-thermal process is carried out. This process ensures the formation of a homogeneous ultra-fine-grained structure in workpieces made of heat-resistant alloys.

One of the criteria for choosing a technology is to obtain materials with a given structure [4, 5]. In particular, the following were:

- methods for calculating kinetic, energy-force, thermomechanical parameters of equipment for rolling axisymmetric parts under isothermal conditions;
- finite-elementary modelling of the rolling process, the stress-strain state of the workpiece material and the tool, the effect of temperature on the rigidity of the tool [6, 7];
- methods of designing control systems for automated technological equipment;
- development and creation of systems for intelligent control of production technologies;
- for forming parts made of heat-resistant alloys and "control" of the microstructure of the material [8].

2 Intellectual Automated Control System

The automated control system (ACS) provides the movement of carriages, rollers, rotation and feeding of rollers, rotation of spindles, setting and maintaining the temperature in the heating furnace, temperature control of bearing supports, load adjustment on the rollers, control of the hydroelectric power station and pinole movement mechanisms, control of the manipulator of the vertical and horizontal drive of the heating furnace lid. All operations are displayed in real time in the form of mnemonic diagrams, graphs and numerical values on the operator's display.

The Intelligent synthesis of ACSs plays an important role in industrial automation [9]. The ACS allows performing operational programming of the technological process with automatic or manual control, calculating the deviation of the executive bodies based on the current information about the actual load on the rollers, the temperature of the heating chamber with the necessary adjustments made to the control technological program. The automated control system provides automatic synchronization of the rotation speeds of the spindles and rollers, the shaping of the part according to a given program when displaying on the screen the progress of all technological operations and the position of the working bodies of the mill (the display of the operator and the technologist).

The automated control system is made in the form of a multi-level distributed control system:

- the upper level - controllers that control the course of the technological process as a whole and provide a human-machine interface.
- middle level - controllers that perform online calculations of the parameters of the rotation speed of the rollers and pinole spindles, as well as synchronization of their operation, adjustment of the speed of movement of the rolling heads depending on changes in the loads on the rollers.

- the lower level - PLC controllers that control the rotation speeds of pinole spindles and rollers, as well as controllers that control the movement of carriages with rolling heads.

The software implementation of data exchange is performed via the Modbus RTV data exchange protocol.

3 Analysis and Synthesis of Technological Processes

One of the features of the first stages of creating new technologies is the need to study a large number of options. The more options are analyzed, the better the research and the greater the confidence in achieving the goal. Morphological approaches can be used for this purpose [10, 11]. These methods are suitable for a wide range of tasks in solving problems [12].

Morphological methods are based on a combinatorial approach. The procedure of morphological analysis allows you to purposefully, systematically lay a huge number of options in the morphological sets of technological solutions.

First, a morphological matrix is formed from known solutions, projects, and patents [13]. One or more reference solutions are entered into the morphological matrix, a number of criteria are selected, which are assigned expert scores (Fig. 1).

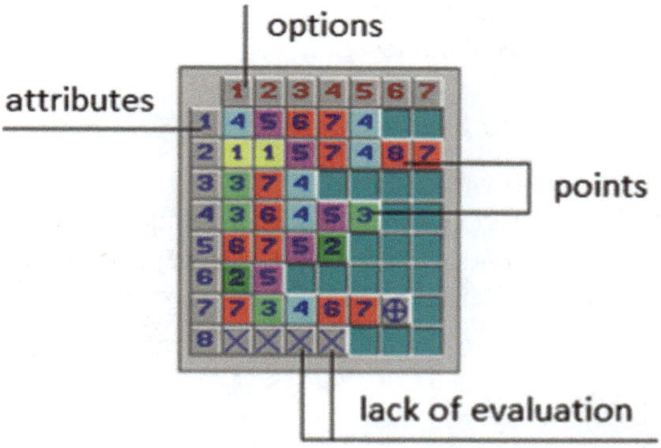

Fig. 1. A fragment of the criteria table

In the future, a certain number of alternatives are generated and a morphological set of solutions is formed, which is divided into clusters (Fig. 2). First, the solutions of the cluster in which the reference variant is located are investigated. In the future, other clusters are being developed. Technical solutions are compared with the reference version, and the possibility of improving it is found out.

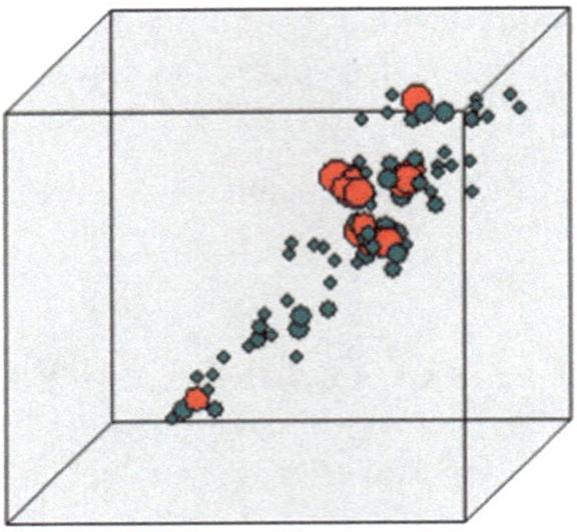

Fig. 2. Morphological set of solutions

To evaluate the variants with the help of expert assessments, a matrix was compiled [14]. Any set of elements of all features (one from each row of the morphological table) is a possible variant of the manufacturing technology of parts. The power of the morphological set is equal to 1728 variants.

The following attributes were selected:

- Technological process of preparation of the workpiece material;
- The initial material of the workpiece;
- The structure of the workpiece material;
- Temperature regime of the workpiece;
- Strain rate;
- The process of manufacturing the workpiece.

In the future, each alternative was given a score for each criterion. 480 variants were generated and 120 were selected for further clustering. (Fig. 3) All variants were calculated for a measure of similarity. Out of 120 generated and selected variants, 18 clusters containing similar technical solutions were formed (Fig. 4).

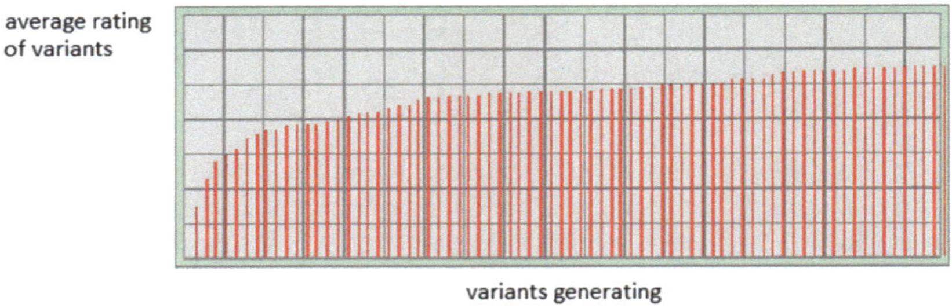

Fig. 3. Increase in the average score of the selected options when generating them

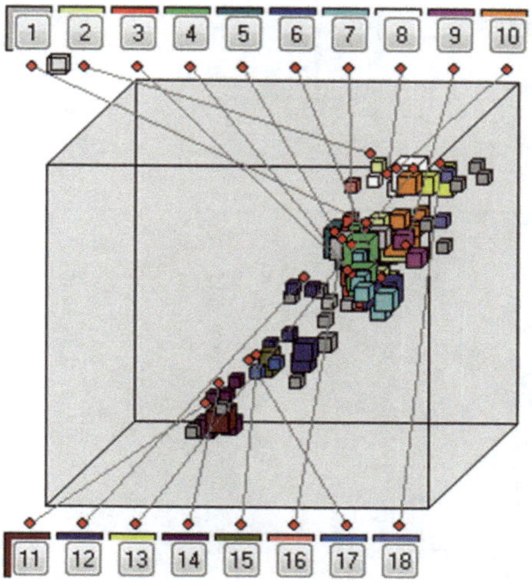

Fig. 4. The location of clusters in the morphological field of solutions

4 Automation of the Technological Process

The experimental mill is designed for the production of experimental samples of parts made of titanium and nickel-based alloys, conducting research, testing and testing the modes of technological processes for manufacturing axisymmetric parts such as disks, rings, shells made of various heat-resistant alloys under conditions of superplastic deformation in manual and automated modes (Figs. 5, 6 and 7) [15, 16].

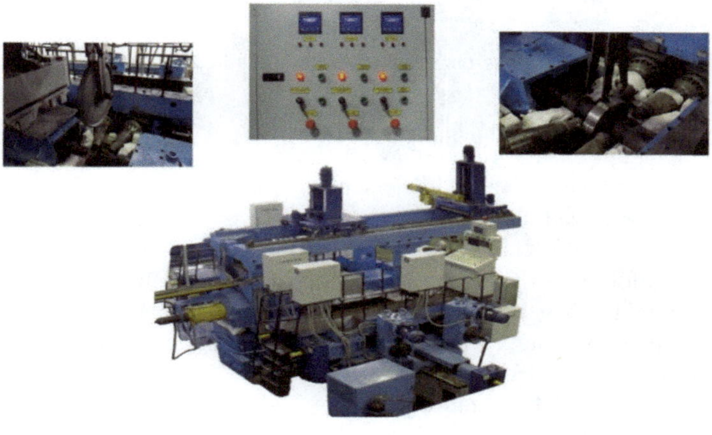

Fig. 5. General view of the SRZHD-800 mill and automated technological processes

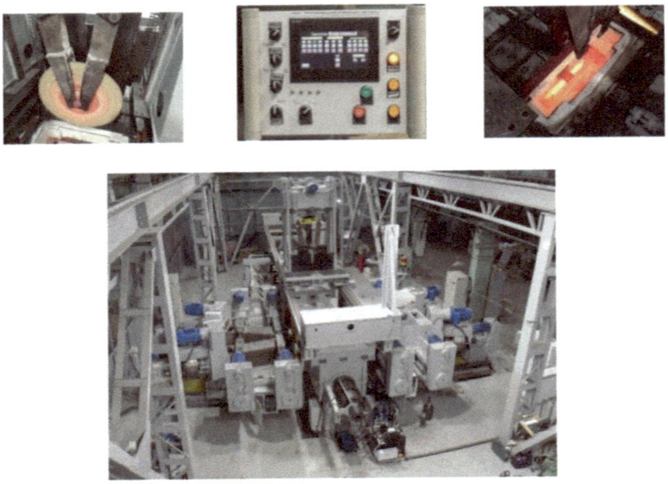

Fig. 6. Automated technological line

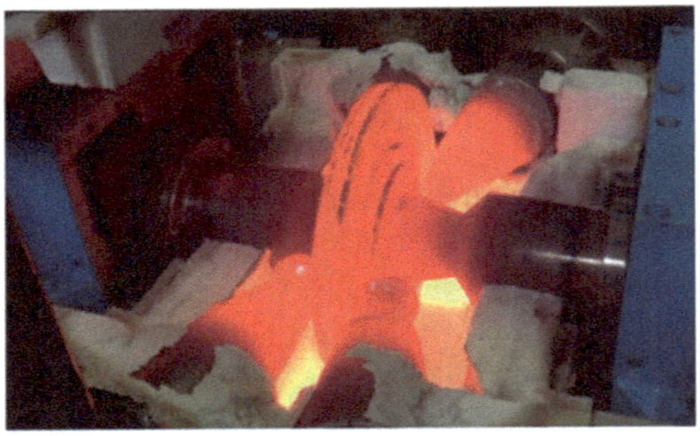

Fig. 7. Rolling out the workpiece in the heating furnace

Advantages of the created technology:

- ensuring optimal efforts for the implementation of isothermal rolling of disks made of titanium and nickel alloys;
- improvement of the mechanical properties of the material of the parts due to a high degree of uniformity of the structure and subsequent heat treatment;
- increase in productivity by 2–3 times due to the implementation of a single cycle of the rolling process using a working furnace and a furnace for preheating and heat treatment of the workpiece;
- increasing the material utilization factor by 3–5 times due to reducing the labor intensity of machining the workpiece;
- improving the reliability of operation through the use of separate hydraulic stations for the main components (carriages, rollers, pinoles);

- using an effective pinole cooling system that provides clamping and rotation of the workpiece, as well as cooling in the bearing supports of tool rollers;
- presence of a system for determining whether the rollers touch the workpiece by moving the pinoles. The hydraulic circuit allows you to automatically perform the operation of its compression in addition to the exact positioning of the workpiece;
- providing monitoring of the rolling parameters and the state of the main nodes by a mode control system consisting of 14 sensors.

The control system provides simultaneous operation of power mechanisms for the implementation of the stages of the technological process: rolling, transporting disks, controlling furnaces and storage devices, heating and cooling of workpieces in preheating furnaces and in regulated cooling furnaces.

- roller and spindle control station allow the joint operation of the rolling rollers and the spindles supporting the product along a given technological chain, observing the temperature and power modes of rolling;
- two carriage control stations ensure that the carriage is positioned at the specified angles and the synchronicity of carriage movement necessary for the technological process (Fig. 8);
- control station provides positioning of floating spindles, performs the process of preloading workpieces with a fixed force, as well as preloading rollers for rolling discs;
- manipulator control station provides loading and unloading of the product to/from the heating, processing, cooling zone and provides information to the computer of the central console.

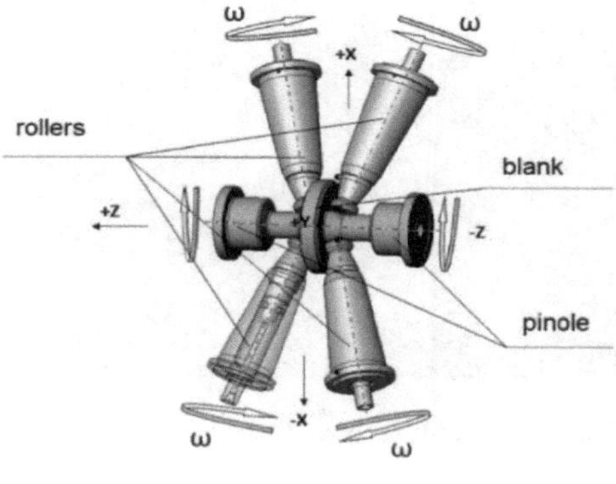

Fig. 8. Rolling out scheme

The control system provides control and automatic maintenance of the following parameters.

- temperature of the workpiece at all stages, from the beginning of its installation in the furnace and until its removal with an accuracy of ± 5 °C in the range from 200 °C to 1140 °C;
- execution time of each operation and the full technological cycle;
- values of displacements, tool feed (with an accuracy of ± 0.5%);
- speed of rotation of the workpiece (with an accuracy of ± 1%) in the range from 0.05 to 4 rpm;
- rotation speeds of each of the 4 tool rollers (with an accuracy of ± 0.1%) in the range from 0.1 to 8 rpm;
- clamping forces of the workpiece in pinoles up to 400 kN (with an accuracy of ± 1%).
- In addition, the control system ensures that the rotation speeds of the rollers and the workpiece are matched within the specified limits (the mismatch is not more than ± 1%).

The synthesized technologies are a modern alternative to those currently used in the manufacture of shafts not only at enterprises in Russia, but also in foreign companies (Fig. 9). The main qualitative indicators of the project are a significant increase in the utilization rate of expensive metal to the level of 0.5–0.6, the quality of the product due to an increase in mechanical properties by 20–30%, a reduction in labor intensity, the power of deforming equipment and the metal capacity of the tool.

Fig. 9. Rolled discs Ø800 mm–Ø 850 mm made of VT9 alloy

5 Conclusion

The proposed approach provides clarification and ordering of the task. At the same time, the degree of validity of the decisions made increases and a variety of technology options are expanded, among which a choice is made.

The conclusions can be drawn as follows:

- The proposed approach allows us to qualitatively consider a large number of alternative technical solutions and, taking into account the limitations, choose the most rational solutions located near the reference vehicles.
- On the basis of the proposed approach, an assessment of technologies, equipment and control systems for the manufacture of blanks of long hollow shafts of gas turbine engines from heat-resistant nickel and titanium alloys was carried out. The total number of possible technological solutions in the morphological matrix is 1728 variants. The studied options have high ratings, which allows us to conclude about the high technical and economic indicators.

The main quality indicators of the technology and system are a significant increase in the metal utilization rate to the level of 0.5–0.6. The quality of the product also increases due to an increase in mechanical properties by 20–30%. Also, the labor intensity and power of the deforming equipment is reduced.

References

1. Basl, J.: Companies on the way to Industry 4.0 and their readiness. J. Syst. Integr. **9**(3), 3–6 (2018)
2. Horst, D., Duvoisin, C., Almeida, R.: Additive manufacturing at Industry 40: a review. Int. J. Eng. Technol. **8**(8), 3–8 (2018)
3. Wang, L., Wang, G.: Big data in cyber-physical systems, digital manufacturing and industry 4.0. Int. J. Eng. Manufact. **6**(4), 1–8 (2016)
4. Valiev, R., Enikeev, N., Murashkin, M.: Superstrength of ultrafine-grained aluminum alloys produced by severe plastic deformation. Dokl. Ross. Akad. Nauk **432**(6), 757–760 (2010)
5. Utyashev, F., Raab, G.: Deformation Methods for Obtaining and Processing Ultrafine-Grained and Nanostructured Materials. Gilem, Ufa (2013)
6. Huang, D.W., Lambert, D., Semiatin, S.: Computer Simulation of Microstructure Evolution during Hot Forging of Waspaloy and Nickel Alloy 718. Scientific Forming Technologies Corporation, Columbus (2013)
7. Yeom, J., Lee, C., Kima, J., Park, N.: Finite-element analysis of microstructure evolution in the cogging of an Alloy 718. Mater. Sci. Eng. **449–451**, 722–726 (2007)
8. Valiev, R., Enikeev, N.A., Murashkin, M.: Using intensive plastic deformations for manufacturing bulk nanostructure metallic materials. Mech. Solids **47**(4), 463–474 (2012)
9. Zwicky, F.: The Morphological Method of Analysis and Construction Courant Anniversary. Intersciences Publish, New York (1948)
10. Zwicky, F.: Discovery, Invention, Research-Through the Morphological Approach. The Macmillan Company, Toronto (1969)
11. Phuyal, S., Bista, D., Izykowski, J., Bista, R.: Design and implementation of cost-efficient SCADA system for industrial automation. Int. J. Eng. Manufact. **10**(2), 15–28 (2020)
12. Shi, Y.: Problem solving, computer technology, and students motivation in learning mathematics. Int. J. Educ. Manage. Eng. **1**(1), 1–5 (2011)
13. Klimenko, B., Rakov, D.: Analysis and synthesis of innovative engineering solutions and technologies based on advanced morphological approach. Adv. Intell. Syst. Comput. **658**, 274–283 (2018)

14. Bardenhagen, A., Pecheykina, M., Rakov, D.: Advanced morphological approach for knowledge-based engineering (KBE) in aerospace. Adv. Artif. Syst. Med. Educ. II **902**, 495–505 (2018)
15. Rakov, D., Sukhorukov, R., Gavrilina, L.: Analysis and evaluation of technologies and equipment for the manufacture of long hollow shafts of gas turbine engines from heat-resistant nickel and titanium alloys on the basis of morphological approach. Probl. Mashinostr. Avtomatiz **4**, 136–143 (2015)
16. Rakov, D., Sukhorukov, R., Pecheykina, M.: Choosing and evaluating fabrication processes by means of a computer-aided innovation support system based on the morphological approach. J. Mach. Manuf. Reliab. **48**(2), 173–178 (2019)

Research on Organization and Optimization of Inbound Picking Area in Logistics Distribution Center

Taiwei Yang, Hanhua Zhu, Mengjian Wang(✉), Jie Wu, and Liangyan Zheng

School of Transportation and Logistics Engineering, Wuhan University of Technology, Wuhan 430070, China

Abstract. Logistics distribution center plays an increasingly prominent role in the whole supply chain. In reality, however, the operation level of logistics distribution center has not reached the ideal grade and not got an expected effect. Focusing on this problem, based on the analysis of the working characteristics and current situation of the logistics distribution center of a beverage company, this paper puts forward the countermeasures of establishing the logistics distribution center. The workflow model of distribution center system is established by using Flexsim simulation software; through simulation analysis, the problems and deficiencies of the distribution center are found; When selecting the transportation equipment of the logistics distribution center, considering the characteristics of the distribution center, a special attention of the research is paid on the optimization of the overall benefit. A targeted improvement plan is proposed and imitated, and the simulation results are compared so as to verify the feasibility of the improvement scheme.

Keywords: Distribution center · Flexsim · Program optimization · Modeling and simulation

1 Introduction

Logistics has developed to the present, and its integrity, system, and comprehensiveness have been paid more and more attention by more and more logistics companies, meeting consumer needs with the smallest cost. The most primitive goal of this logistics is manifested in reality as follows: The logistics cost is consumed a lot in the warehousing and transportation of goods.

The distribution center is a very intensive logistics transfer station. It is engaged in the equipment (collection, storage, processing, sorting, picking, and distribution) of the goods and the organization of the delivery (transportation) to the users with high efficiency. Realize logistics nodes and organizations that support logistics activities such as production and sales within a certain area. In the distribution center, the logistics cost of the entire enterprise chain can be greatly reduced by rationally exerting various functions of the distribution center.

Z. Hu et al. (Eds.): AIMEE 2021, LNDECT 107, pp. 287–299, 2022.
https://doi.org/10.1007/978-3-030-92537-6_27

The operation flow of the distribution center is determined according to the type and function of the distribution center, the general operation flow as shown in Fig. 1.

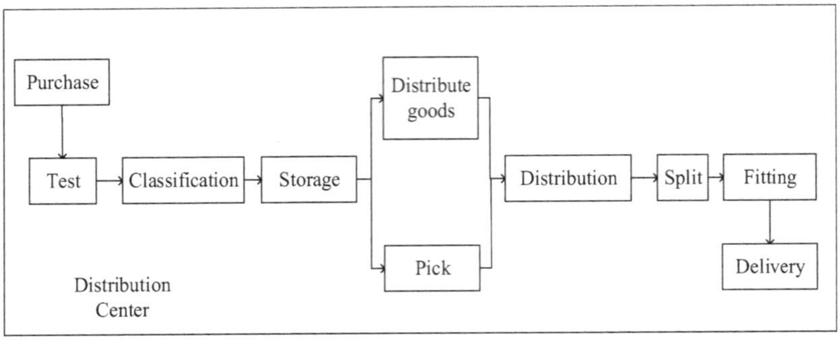

Fig. 1. Distribution center workflow

Young HoonLee et al. studied the operational planning of distribution centers under a single procurement constraint [1]. Kang Jeongyoon et al. used simulation, element matching and response surface methods to optimize the design parameters of the automatic distribution center model to optimize the design of the distribution center [2]. Paweł Droździel et al. analyzed and studied the existing postal network based on graph theory to optimize the logistics network [3]. Jardas Mladen et al. took the delivery of goods in Rijeka as an example, with the goal of reducing transportation costs and reducing the traffic burden on the urban transportation network, using transportation problem solving methods to optimize the delivery of goods. According to the basic functions of the warehouse [4]. Jin xiang et al. can be divided into four types of problems: receiving, storage, picking and transportation, explaining the current planning models and methods that can be used for warehouse operations and future research opportunities [5]. Isao Tsushima et al. applied the Petri net model to find an efficient picking operation method in the automated warehouse system [6]. Saleh Alyahya et al. studied the inventory management system of the interactive method of automatic storage and retrieval mechanism [7]. Gao Changyong et al. analyzed in detail the reasons for the blockage of the sorting line and gave corresponding solutions [8]. There is also a research foundation for the optimization and scheduling of transportation vehicles. Mohammad Ali H et al. proposes a collision-free path planning algorithm based on the generation of random paths between two points [9]. Mbida Mohamed introduced a hybrid architecture model using neural network and barcode reader 1D/2D vision technology in intelligent warehouse management [10]. Rohit Nagar et al. studied the problem of effective workflow scheduling using genetic algorithms in cloud computing [11]. Abhishek S. Rao studied an optimization method to improve the queuing model of the prepaid automatic counter [12]. Dai Jinshan et al. applied the improved grey prediction method and used the Monte Carlo method to simulate the demanding amount for next three cycles. It makes the setting and optimization of safety stock more accurate and effective data support [13]. Wang Yong et al. Questioned the existence of the warehouse center in Huang pi District and optimized it [14].

Qi Zhu conducted a demand analysis on the actual operation process and functions in logistics distribution, and designed a logistics distribution information system [15].

Zhibin Wu et al. studied a multi-attribute group decision (MAGDM) framework to promote the assessment of lean practices in logistics distribution centers [16]. Yuan Ruiping et al. analyzed the order picking process of an e-commerce logistics distribution center based on logistics AGV, and proposed two operating modes: synchronous picking with multiple picking stations and asynchronous picking with multiple picking stations [17]. He Yi elaborated on the significance of research on the optimization of order picking paths in logistics distribution centers [18]. Kim Jonghoon et al. conducted research on productivity growth and cost-effectiveness based on the introduction of automated guided vehicles [19]. Xi Wu et al. used the association drawing method to design the best layout of the distribution center, draw the final layout, rationally arrange the location of each functional area, shorten the walking path, and improve the operational efficiency of the distribution center [20]. Zhang Tanjun et al. analyzed the sorting operation requirements of the distribution center of the chain supermarket, and looked for the problems in the existing distribution center [21]. Jie Jin uses the availability analysis method to analyze the reliability of the logistics system's good receiving and storage process [22].

The order is the lifeline of the logistics distribution center. If there is no order, the logistics distribution center loses its meaning. Once you have mastered the order, you can understand the characteristics of the distribution center. Through the analysis of Entry and Quantity (EQ), we can understand the distribution of the order quantity of a single order, which can determine the order processing, the planning of the picking system, the delivery method and the planning of the delivery area. The EQ distribution type is shown in Fig. 2.

Through the analysis of Item and Quantity (IQ), we can know the distribution of various product shipments. It is helpful to analyze the importance of the product and the transportation situation. The IQ distribution type is shown in Fig. 3.

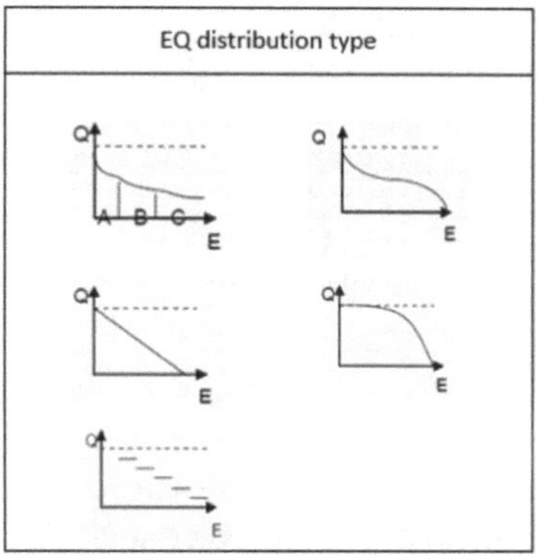

Fig. 2. EQ distribution type

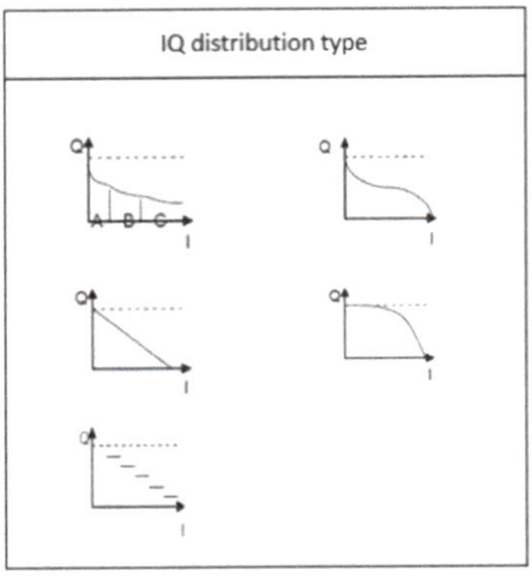

Fig. 3. IQ distribution type

This paper uses Flexsim to conduct a modeling and simulation study on a beverage distribution center, which can not only observe the operation of the logistics distribution center system in real time, but also calculate the data and results of each entity in the model. Through the analysis of the data results, the problem of the distribution center is obtained and the targeted improvement plan is proposed. At the same time, Flexsim is further used to verify the improved plan.

2 Distribution Center Organization and Structure

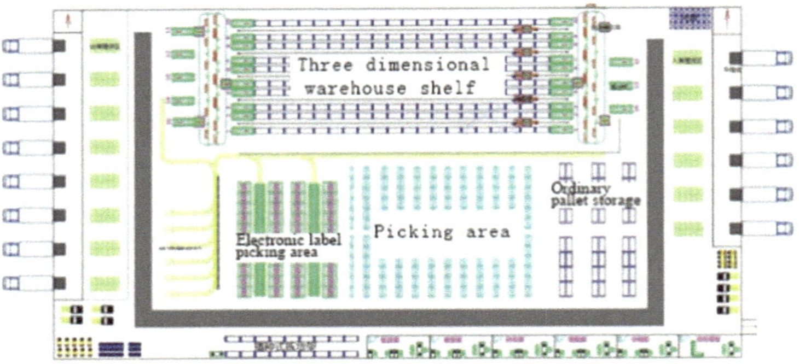

Fig. 4. The layout of distribution center

The beverage distribution center is responsible for collecting, sorting and distributing the products of a beverage manufacturer, preparing the goods according to the needs of various places, and delivering them with a satisfactory service level. The manufacturer first transports the goods to the distribution center, and the distribution center performs storage, picking, and distribution operations, and the goods are shipped to the places of demand in time. The operational functional area of the beverage company's distribution center is divided into three-dimensional warehouse shelf storage area, ordinary pallet storage area, picking area, electronic label picking area, etc. As shown in Fig. 4.

The equipment required for the shelf area of the three-dimensional warehouse includes 10 three-dimensional warehouse shelves, 3 Automated Guided Vehicle (AGV) automatic guided vehicles, and 5 stackers. The ordinary pallet storage area has 12 ordinary pallet shelves, and the picking area has 28 medium-sized shelves and electronic labels. There are 4 electronic label shelves in the picking area, and 2 sorting conveyor belts and 14 ordinary conveyor belts are needed to realize the transportation of goods between various functional areas. As shown in Table 1.

The beverage manufacturer has high requirements for the delivery cost and timeliness of the goods. In the actual operation process, it was found that the distribution center could not meet the needs of customers in time during the peak beverage demand period, thus reducing customer satisfaction and losing Some potential customers.

Table 1. Facilities and equipment of distribution center

Functional area and organizational structure		Major equipment	Equipment quantity	Functional area and equipment usage
Storage area	Three-dimensional warehouse area	Stereoscopic warehouse shelves	10	Make full use of warehouse space, high warehouse capacity utilization, and expand warehouse storage capacity
		AGV automatic guided vehicle	3	An unmanned transport vehicle equipped with electromagnetic or optical automatic guidance devices that can travel along a prescribed guidance path, with safety protection and various transfer functions
		Stacker	5	Special cranes that use forks to move and stack or unload unit cargo from high-rise shelves
Picking area	Ordinary pallet storage area	Ordinary pallet rack	12	Used to store whole pallets of goods. Pallet size: (1200 mm × 1000 mm)
	Picking area	Medium shelf	46	Used to store box cargo. Cargo box size: (600 mm × 500 mm)
	Electronic label picking area	Electronic label shelf	4	It is an electronic display device that can be placed on the shelf to replace the traditional paper price tag
Conveyor belt		Ordinary conveyor belt	5	Realize cargo transportation
		Sorting conveyor belt	2	Realize cargo transportation

3 Modeling and Simulation of Distribution Center

3.1 Distribution Center Model Establishment

In the three-dimensional warehouse area, pass the storage conveyor belt to the AGV trolley, the AGV trolley transports the goods to the stacker, and the stacker sends the goods to the three-dimensional warehouse shelf. In the ordinary storage area, the goods are transported to the designated location by manual or forklift. When leaving the warehouse, the stacker in the three-dimensional rack will send the goods to the outbound conveyor belt, and the outbound conveyor belt will send the goods to the sorting conveyor belt, which will be transported to the outbound position through the sorting conveyor belt. The ordinary racks will pick the goods by manual picking. To the sorting conveyor belt.

According to the actual layout of the distribution center, plan the area of each functional area and the distance between the regions, and establish the overall layout model of the distribution center. The overall layout of the model established in Flexsim is shown in Fig. 5.

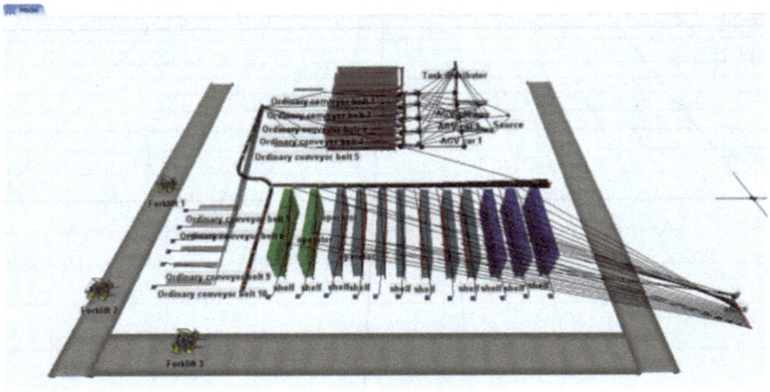

Fig. 5. The three dimensional model of distribution center overall planning

3.2 Distribution Center Parameter Settings

1) Set the flow arrival parameters of flow items in the Source entity, as shown in Table 2 is the timetable for one of the products in the distribution center when they arrive, double-click Source, and select "Arrivals Editors" in the arrival mode drop-down menu. And set the parameters, set other different types of goods one by one in the same way.

Table 2. The arrival parameters temporary solid flow

	Arrival time	Entry name	Type	Number
Arrive 1	0	Product	1	5
Arrive 2	10	Product	1	8
Arrive 3	33	Product	1	3
Arrive 4	55	Product	1	7

2) Set the shelf specifications and capacity of each storage area. Set the size and capacity of each shelf in the attribute window of each shelf. The specific parameters are shown in Table 3.

Table 3. Shelf specifications and storage capacity

Shelf name	Column (x)	Row (y)	Layer (z)	Storage capacity
Three dimensional warehouse shelf	10	42	8	Total 3360 storage spaces (unit: tray)
Ordinary pallet shelf	3	16	3	Total 288 storage spaces (unit: tray)
Medium-duty rack	23	12	3	There are 828 storage spaces in total, each of which can hold 3 boxes and 2484 boxes of goods in total
Electronic label shelf	4	48	2	A total of 384 storage spaces, each of which can hold 6 boxes, and a total of 2304 boxes of goods (unit: box)

3) Set the parameters of the conveyor belt and transportation equipment in each area, as shown in Table 4 and Table 5.

Table 4. The parameters of conveyor belt

Belt name	Number of conveyor belts	Belt length (m)	Conveyor speed (m/s)
Conveyor belt in storage area of three-dimensional warehouse	3	10	0.8
Conveyor belt in warehouse out area	5	10	0.8
Three dimensional warehouse sorting conveyor	1	85	1
Sorting conveyor in other areas	1	78	1

Table 5. The parameters of each transport equipment

Device name	Number	Maximum operating speed (m/s)
AGV	3	2
Stacker	5	3
Forklift	3	3
Operator	>10	1

3.3 Simulation Result Output and Analysis

This simulation has been running for 6 months. During the process of running the model, it can be seen that all functional areas of the distribution center can operate normally under normal conditions, but during peak periods, it can be seen that the AGV trolley cannot meet the needs and the goods will be in three-dimensional Pile up at the conveyor belt in the warehouse's warehousing area, as shown in Fig. 6.

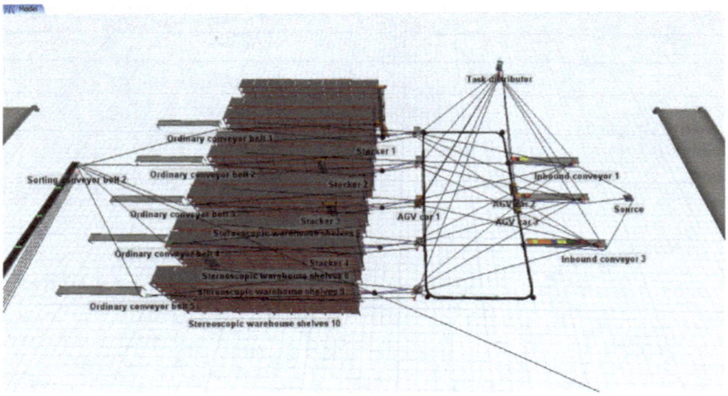

Fig. 6. Simulation effect

After the simulation is over, we use the statistical report generation function of the Flexsim software to perform statistics on the simulation data of the model. This simulation is for the idle time (idle) and the shortest stay time (stats-staytimemin) of the AGV trolley and the conveyor belt in the storage area. The longest stay time (stats-staytimemax), blocked time (blocked), conveying time (coveying) were counted. The specific data results are shown in Table 6.

Table 6. Simulation result data (unit: h)

Object	Idle	Stats_staytimemin	Stats_staytimemax	Blocked	Conveying
AGV car 1	337.2	6.3	10.4	890.8	3058.8
AGV car 2	401.5	6.7	12.5	759.0	3274.6
AGV car 3	344.5	6.8	15.3	833.4	3378.3
Inbound conveyor 1	358.8	7.6	14.2	654.2	2139.2
Inbound conveyor 2	362.4	6.4	15.6	665.6	2367.6
Inbound conveyor 3	379.5	6.0	15.2	655.3	2167.9

4 Proposal of Improvement Plan

4.1 Analysis of the "Bottleneck" of the Original Plan and Improvement

Through the simulation result data of idle time, shortest stay time, longest stay time, blocking time, and conveying time, it can be seen that the distribution center uses AGV trolleys in the storage area of the three-dimensional warehouse. During the peak period of beverage demand, AGV trolleys cannot meet the requirements. The warehousing function is required, and the peak period of beverages lasts for a long time. Therefore, the adverse effects of the AGV trolley cannot be ignored.

In response to the problems in the distribution center, an improved plan was proposed: replace the AGV trolley with the sorting conveyor belt, and the two sorting conveyor belts transport goods to both sides. This improves efficiency and makes full use of space, and the sorting conveyor belt is also It can realize the intelligent control of the transportation of goods by the trolley.

4.2 Modeling and Simulation of Improved Scheme

On the basis of the original model, the AGV trolley is partially replaced with the sorting conveyor belt, and the parameters of the conveyor belt are set at the same time. The parameters of the sorting conveyor belt are shown in Table 7, and the simulation model layout of the improved scheme is shown in Fig. 7.

Table 7. The parameters of the sorting conveyer belt in the storage area of the warehouse

Conveyor name	Number of conveyor belts	Conveyor belt length (m)	Conveyor speed (m/s)
Sorting conveyor belt	2	30	1

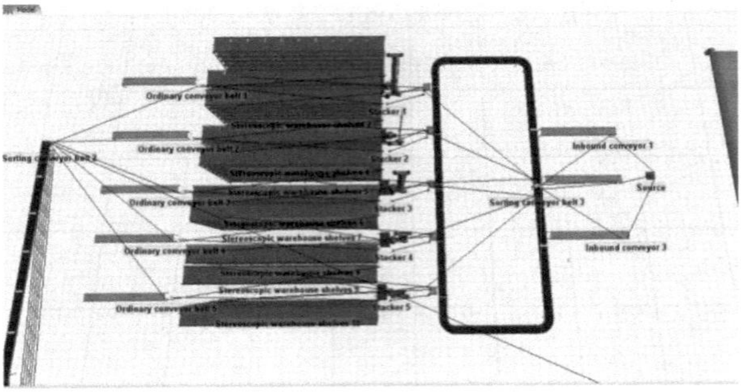

Fig. 7. The simulation model of improved scheme

4.3 Simulation Result Output and Comparative Analysis

Running the improved scheme for the same time, the accumulation of the conveyor belt area in the storage area of the three-dimensional warehouse has been significantly improved. The statistics of the idle time, the shortest stay time, the longest stay time, the blocking time, and the conveying time can also be obtained. The status of the sorting conveyor belt and the conveyor belt in the storage area after the improvement is shown in Table 8.

Table 8. The improved simulation result data (unit: h)

Object	Idle	Stats_staytimemin	Stats_staytimemax	Blocked	Conveying
Sorting conveyor in storage area 1	857.2	5.3	11.8	589.4	3061.2
Sorting conveyor in storage area 2	837.6	5.2	11.7	559.6	3066.0
Inbound conveyor 1	454.3	3.8	12.5	400.6	3337.3
Inbound conveyor 2	447.6	3.5	12.4	412.7	3394.1
Inbound conveyor 3	454.3	3.7	12.3	419.5	3329.7

By comparing the data in Table 6 and Table 8, it can be found that the longest residence time, blocking time, and conveying time of the improved plan warehousing conveyor belt are less than the original plan. The idle time, blocking time, and longest stay time of the conveyor belt in the storage area are counted. The specific data is shown in Table 9. The idle time of the conveyor belt has increased by 23.2%, the blocking time has been reduced by 37.6%, and the longest stay time has been reduced 19.87%. This shows that the improved plan has effectively alleviated the problem of accumulation in the storage area.

Table 9. Data comparison of the scheme (unit: h)

Program	Idle	Blocked	Stats_staytimemax
Original plan	1100.7	1975.1	15.6
Improvement plan	1356.3	1232.8	12.5

Figure 8 (a) shows the relationship between the accumulation of goods in the picking area of the original distribution center and the simulation run time. From the figure, we find that the accumulation of goods in the picking area of the warehouse increases with the number of simulations. The continuous increase indicates that the goods have been stranded in the storage area and proves that the layout of the storage area of the logistics distribution center is unreasonable. Figure 8 (b) shows the improved relationship between the accumulation of goods in the picking area of the warehouse and the simulation running time. From the figure, it can be seen that the accumulation of goods in the picking area of the warehouse is finally between 23 and 25. It fluctuates from time to time, and finally stabilizes. The point in the blue circle in the figure is the point where the accumulation of goods in the picking area of the warehouse reaches a stable point. At this time, the distribution center has been running at 990 h.

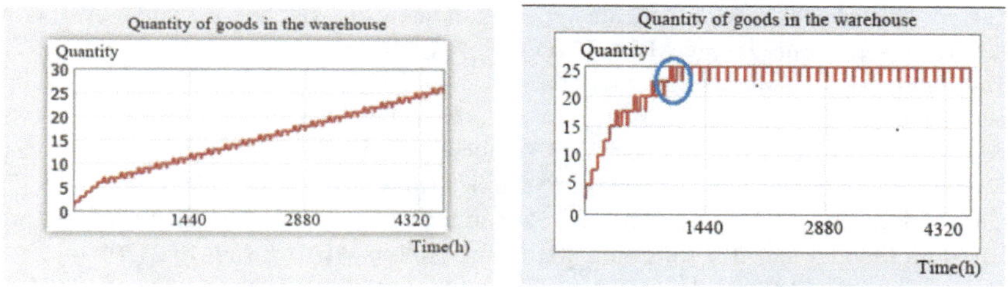

(a) (b)

Fig. 8. The relationship between the accumulation of goods in the picking area of the distribution enter and the simulation run time

5 Conclusion

In the case of this article, the efficiency of the AGV is lower than the efficiency of the sorting conveyor belt, which shows that the selection of transportation means should be combined with the characteristics and operating characteristics of the distribution center. It is can't blindly choosing intelligent equipment to improve the entire system efficiency.

This article mainly introduced the functional structure and working methods of a beverage distribution center. It uses Flexsim simulation software to conduct simulation research on the distribution center, and establishes warehouse shelf models, storage and delivery models, stackers, forklifts, and operations through Flexsim simulation software. member. Attribute setting and physical space connection of these models, unifying logistics data and planning logistics processes, setting up the operation methods and data output of each module. After running the model, wait until the result data, analyze the distribution center model through the actual simulation results and propose an optimized improvement plan. At the same time, the optimized plan is also simulated and verified, and the original plan and the improved plan are compared and analyzed through the simulation data. After comparison and analysis, it can be seen that the improved plan effectively improved the existing problems in the original plan and greatly improved the work efficiency of the distribution center.

References

1. Lee, Y.H., Kwon, S.G.: The hybrid planning algorithm for the distribution center operation using tabu search and decomposed optimization. Expert Syst. Appl. **37**(4), 3094–3103 (2009)
2. Kang, J.Y., Lee Hong, C.E.I.: A study for design optimization of an automated distribution center using the simulation and metamodel. J. Korea Soc. Simul. **15**(3), 103–114 (2006)
3. Drożdziel, P., Wińska, M., Madleňák, R., Szumski, P.: Optimization of the post logistics network and location of the local distribution center in selected area of the lublin province. Procedia Eng. **192**, 130–135 (2017)
4. Jardas, M., Krljan, T., Perić, H.A., Grubišić, N.: Distribution center logistics optimization model–city of Rijeka case study Pomorstvo **34**(1), 185–194 (2020)
5. Gu, J., Goetschalckx, M., McGinnis, L.F.: Research on warehouse operation: a comprehensive review. Eur. J. Oper. Res. **177**(1), 1–21 (2006)
6. Tsushima, I., Mitsumori, S., Komoda, N.: An efficient picking operation method in automatic warehouse system. IEEJ Trans. Electron. Inf. Syst. **107**(9), 843–849 (2008)
7. Alyahya, S., Wang, Q., Bennett, N.: Application and integration of an RFID-enabled warehousing management system – a feasibility study. J. Ind. Inf. Integr. **4**, 15–25 (2016)
8. Changyong, G., Jiandao, Z., Miao, C.: Research on the optimization of the sorting process based on the automated sorting line. Manufact. Autom. **41**(08), 129–131 (2019)
9. Eljinini, M.A.H., Tayyar, A.: Collision-free random paths between two points. Int. J. Intell. Syst. Appl. **12**(3), 27–34 (2020)
10. Mohamed, M.: Smart warehouse management using hybrid architecture of neural network with barcode reader 1D/2D vision technology. Int. J. Intell. Syst. Appl. **11**(11), 16–24 (2019)
11. Nagar, R., Gupta, D.K., Singh, R.M.: Time effective workflow scheduling using genetic algorithm in cloud computing. Int. J. Inf. Technol. Comput. Sci. **10**(1), 68–75 (2018)
12. Rao, A.S.: Improving the serviceability of a prepaid autorickshaw counter using queuing model: an optimization approach. Int. J. Inf. Technol. Comput. Sci. **9**(12), 19–27 (2017)
13. Jinshan, D., Yao, Z., Mengya, Z., Qingying, Z.: Optimization of stock basing on improved grey prediction model: a case study on garment supply chain. In: Proceeding of the 3rd International Conference on Computer Science, Engineering and Education Applications, Advances in Computer Science for Engineering and Education III, Ukraine, pp. 465–474 (2020)
14. Wang, Y., Zhang, P.L., Lu, Q., Semere, D.T., Li, G.: Research and application of storage location optimization in warehousing center of TPL enterprises. In: Proceeding of the 3rd International Conference on Computer Science, Engineering and Education Applications, Advances in Computer Science for Engineering and Education III, Ukraine, pp. 153–166 (2020)

15. Zhu, Q.: Design and analysis of information system of modern supermarket logistics distribution center. Int. Core J. Eng. **6**(8), 151–157 (2020)
16. Zhibin, W., Jiuping, X., Zeshui, X.: A multiple attribute group decision making framework for the evaluation of lean practices at logistics distribution centers. Ann. Oper. Res. **247**(2), 735–757 (2016)
17. Ruiping, Y., Huiling, W., Lirui, S., Juntao, L.: Research on task scheduling of "goods-to-person" order picking system based on logistics AGV. Oper. Res. Manage. **27**(10), 133–138 (2018)
18. Yi, H.: Optimization of order picking path in logistics distribution center based on genetic algorithm. Bus. Econ. Res. **21**, 110–111 (2016)
19. Jonghoon, K., Yeonjin, K., Hongcheol, L.: Simulation analysis of AGV introduction in the convenience store logistics distribution centers. J. Korean Soc. Ind. Acad. Technol. **17**(6), 61–69 (2016)
20. Xi, W., WenJi, S., Nan, L.: Data processing in optimization and design of regional layout based on pharmaceutical logistics distribution center. Appl. Mech. Mater. **3627**, 740–743 (2014)
21. Tanjun, Z., Wenlong, L.: EIQ analysis of the design of chain supermarket logistics distribution center. Logistics Technol. **39**(11), 10–12 (2016)
22. Jie, J., Wei, P.Y.: The reliability research of the good receipting and being put in storage process of X city's tobacco logistics distribution center. Appl. Mech. Mater. **2796**, 1665–1671 (2014)

Method to Prevent Hypothermia and Muscle Tremors in Burn Patients

Andrey Alekseev[1], Victor Vinogradov[2], Mihail Modestov[3], Nikolay Chilikin[1], and Alexander Shul'zhenko[3](✉)

[1] A.V. Vishnevsky National Medical Research Center of Surgery, Moscow, Russian Federation

[2] Federal State Institution A. I. Burnazyan Federal Medical and Biophysical Center, Moscow, Russian Federation

[3] Mechanical Engineering Research Institute of the Russian Academy of Sciences, Moscow, Russian Federation

Abstract. During the study, the reasons for the occurrence of unintentional hypothermia and muscle tremor in burn patients with lesions of a large surface area of the skin were clarified. It was determined that in addition to the direct effect of a burn - damage to the skin, causing high heat loss in the affected area, there are additional conditions that affect the growth of heat loss. This is the presence of water-soluble ointments applied to dressings and/or wound dressings put on the affected areas, and convective air movement. Various ways of possible compensation for heat losses were discussed. As the main means providing thermal compensation, a woven electric heater was chosen, which produced local heating of a burn patient, and this is a novelty. To calculate the technical characteristics of a woven electric heater, a method based on the calculation of heat loss in a burn patient was used. During the approbation of the method at A.V. Vishnevsky National Medical Research Center of Surgery, the results were obtained that showed stable thermal compensation of heat losses in burn patients during the entire time of the operation, which made it possible to prevent the occurrence of hypothermia and muscle tremor. This method can be used both for the treatment of burn patients in stationary conditions and during their evacuation. The breadth of applicability of the method became possible due to the creation of effective heating using a low-power, low-voltage woven electric heater capable of operating from stationary, on-board, and autonomous power sources.

Keywords: Burns · Burn patients · Woven electric heaters · Heat loss compensation · Ointment dressings and coatings

1 Introduction

Burns, unfortunately, are quite common in our life. "Burns are not just trauma, but a disease with many complications affecting to vary degrees all organs and systems of the body and accompanied by a polymorphic clinical picture, unpredictable treatment results and complications" [1].

Z. Hu et al. (Eds.): AIMEE 2021, LNDECT 107, pp. 300–308, 2022.
https://doi.org/10.1007/978-3-030-92537-6_28

2 The Main Causes of Hypothermia and Muscle Tremors in Burn Patients (Literature Review)

2.1 Heat Loss Due to Skin Lesions

Man is a rather powerful energy system. However, even within the framework of this system, in conditions of comfortable heat sensation, there is a lack of heat. Its value is up to 121 kJ [2].

The magnitude of the heat deficit in the human body is correlated with a person's general feeling of heat. If we assume that a healthy person feels comfortable in the operating room, then their heat deficit does not exceed the above value. In this case, additional heating with external heaters is not required.

In patients with deep and extensive thermal lesions, body parts with open wounds under the same conditions are in almost direct contact with the external environment. Since the temperature of the core of the human body and the temperature on the surface of the skin, even in the absence of burn lesions, differ by 1 ÷ 1.5 °C, and in the case of extensive burn lesions, this temperature difference increases due to the increased temperature of the core of the human body, this leads to the appearance of increased heat loss, which must be compensated by the thermoregulation system. The possibilities of the thermoregulation system are not limitless. The body can reduce the intensity of heat transfer by only 15–20% [1]. When a large surface of the skin is affected, at some point the thermoregulation system ceases to cope and provide compensation for heat losses, as a result, unintentional hypothermia appears. Preventing hypothermia is very important, as otherwise, it can lead to: arrhythmias; decreased immunity; hypokalemia; violation of the acid-base state; coagulopathy (platelet dysfunction); postoperative stress response and protein catabolism; change in mental status; impaired renal function; changes in the metabolism of drugs; the development of a syndrome of chills and muscle tremors.

Thus, the main cause of hypothermia in burn patients is excessive heat loss arising from damage to the skin.

2.2 Effect of Dressings/Coatings with Water-Based Ointments on the Occurrence and Development of Hypothermia

Most methods of treating burns use dressings impregnated with water-soluble ointments and/or wound dressings [3].

Methods of treating burns using ointments with a wide spectrum of therapeutic action are well-known [3–6].

The ointment is applied with a thin layer of 0.5 ÷ 3 mm on a dense cotton fabric, put on the affected area and the dressing is changed 1 ÷ 3 times a day for 3 ÷ 14 days in the method described in [4]. The choice of the thickness of the layer of ointment applied to the tissue (when using a bandage, the effectiveness of the ointment is sharply reduced), is conditioned by the greatest effectiveness of its impact on the wound. This occurs when the thickness of the ointment layer is from 0.5 mm to 3 mm.

However, the imposition of bandages and/or wound dressings on burns and donor wounds, which have a rather thick layer of water-soluble ointments, contributes to an even greater increase in heat loss, since heat transfer from a surface covered with water-soluble ointments is several times greater than heat transfer from the surface to dry air [7]. This leads to the creation of additional conditions for the emergence and development of unintentional hypothermia.

2.3 Influence of Convection on the Onset and Development of Hypothermia

In a person who is at rest, usually up to 90% of heat loss occurs in a convective-radiation way [8].

Most often, such heat loss occurs due to air flows of natural and/or forced origin, which significantly increase the evaporation of moisture in water-soluble ointments, which ultimately leads to a decrease in temperature on the surface of the human body [3].

2.4 The Main Causes of Muscle Tremors

The presence of hypothermia in itself is not a good phenomenon, and the occurrence of muscle tremors aggravates the situation of the burn patient.

The negative consequences of muscle tremors for patients include: increase in oxygen demand by 200 $\div$ 400%; difficulty in normal gas exchange due to hypertonicity of the respiratory muscles, which creates conditions for the aggravation of the existing tissue hypoxia and hypercapnia; release into the blood of biologically active substances: norepinephrine, acetylcholine and histamine, which determine the change in the geodynamics of the large and small circulation, which poses an additional risk for patients with reduced reserves; patient discomfort; vigorous tremors can cause displacement of the graft; interference with ECG registration and peripheral pulse oximetry.

At the same time, the occurrence of muscle tremors, according to B.R. Baxendale, et al. as well as D.Buggy and N.Hunges, is more influenced by the gradient between peripheral and central temperatures [9, 10].

3 Methods to Eliminate Hypothermia and Tremors

As shown in the previous section, excess heat loss is the main cause of emergence and development of hypothermia and muscle tremor. In order to eliminate the manifestation of hypothermia and the occurrence of tremors in the treatment of burns, it is necessary to apply methods of compensation for heat losses in patients [3]. There are a variety of methods. Let's consider the most used ones.

3.1 The Fluidised Bed

One of these methods is the use of fluidized beds not only for their main purpose, to reduce the pressure on the burned area of the body and ensure the drying of the eschar, but also to combat hypothermia, since heated air circulates in them [3, 11]. However, these devices can only be used in stationary conditions, and in stationary conditions they are not suitable for operations.

Their use is also complicated during the evacuation of burn patients. In addition, the warm air currents generated by these devices increase the evaporation of moisture from the ointments, which reduces the effectiveness of this treatment.

Studies have shown the following evaporation rate from 1 sq. m at a temperature of ~30 °C [12].

Table 1. The rate of evaporation

State of the environment	Calm	Mean motion
The ratio of the rate of evaporation	1,056	1,365

As can be seen from Table 1, even with a relatively small increase in the movement of the medium, the rate of moisture evaporation increases by more than ~1.3 times.

Moreover, the movement of air when using ventilators on the body surface in the area affected by the burn can cause the effect of muscle contraction, which contributes to the displacement of the transplanted grafts.

3.2 Total Heating

According to the studies, even reaching a heating temperature in a room equal to 36 ÷ 40° is not enough to compensate for heat loss in a burn patient; nevertheless, it is quite high [3]. Therefore, the use of sources of general heating, for example, heaters that heat the room with natural air convection, is inconvenient, for example, during operations, as it leads to unnecessary overheating of personnel. General heating is difficult to use when transporting a burn patient.

3.3 Warm Fluid Infusion Method

It is possible to use the warm fluid infusion method. However, the effect of preventing an increase in body temperature is not achieved immediately, but after a rather long time (from 1 to 3 h), and the tremor, although it decreases, persists [13].

3.4 Heating with Woven Electric Heaters

The most convenient heating method, devoid of most of the disadvantages listed above, is electrical heating using woven heaters.

Woven electric heaters have different physical properties. They use various electrically conductive and insulating materials, with their help it is possible to achieve the required levels of electrical parameters and generated heat fluxes.

The requirements for the observance of the temperature regime by these heaters for burn patients are rather stringent, since when implementing the contact heating method, they work directly on the surface of the human body. Therefore, they are most suitable for compensating for heat losses, at which it is necessary to ensure the temperature modes of operation of the heaters such that, on the one hand, do not cause additional thermal injuries, do not aggravate the existing ones, and on the other hand, ensure the transition of a person to a normal temperature regime.

When implementing the method, which allows ensuring the state of normothermia by compensating for heat losses using local elastic woven heaters, at the same time, constant temperature control was carried out in the burn patient. Temperature measurements were made in various ways using thermal sensors: rectally and/or orally, and/or on the body surface and on the surface of an elastic electric heater. The obtained temperature data were used to provide thermal control.

There is a wide variety of woven electric heaters available.

Woven heaters meet the set requirements to a greater extent, since at the same time they have high elasticity, strength, properties familiar to humans, as when wearing ordinary clothes, for example, air permeability, moisture permeability. Woven heaters, due to their high elasticity, are best suited for contact heating, which helps to reduce overhead and create a precise warm effect.

Woven heaters have a repetitive matrix structure (Fig. 1) that provides good operational reliability even under the condition of partial damage to individual fragments of the heater [14]. This is especially important when used in operating theaters where there may be cases of partial damage to the heaters.

In woven structures, the following can usually be used as a fuel element: metal threads; metalized threads; carbon filaments (with a high carbon content); threads with a layer of carbon applied to their surface; polymer threads with carbon fibers included in their structure, etc.

Based on these fuel elements, electric heaters are currently being produced, consisting of: only electrically conductive threads; both of electrically conductive and electrically insulating threads, which are among themselves in a variety of combinations; consisting of both electrically conductive threads with areas with variable conductivity, and electrical insulating threads, with their various combinations throughout the entire fabric, etc. [15, 16].

Such heaters also provide a high degree of electrical safety, as they operate from ultra-low voltages $\sim 6 \div 24$ V.

The use of low voltage levels of electrical power allows the use of woven heaters not only in hospitals, but also in vehicles, field hospitals, etc.

Fig. 1. Appearance of a woven electric heater

3.5 Calculation of Woven Electric Heaters

The electrical calculation of woven heaters and the thermal effect required from them was based on the assessment of heat loss occurring in patients considering the area of the burn injury, the level of the core temperature of the human body, and the presence of water-soluble ointments [17–22]. The formula for the estimated calculation of the required power of a woven electric heater was obtained in [23]

$$K \cdot \frac{U^2}{R \cdot S_3} = \frac{M}{S} \cdot \frac{T_{яд.} - T_{срв}}{T_{срв} - T_{окр.ср.}} \frac{S_1}{S_3} \tag{1}$$

U is the voltage of the power supply, R is the electrical resistance of the heater, K is the power loss factor of the heater, M is the human energy consumption, T_1 is the core temperature of the human body, T_2 is the weighted average temperature of the human body, T_3 is the ambient temperature, S is the surface area of human skin; S_1 is the surface area of the skin affected by the burn; S_3 is the surface area of a woven electric heater.

As can be seen from the formula (1), all the data obtained on the heat loss of the body are recalculated to determine the power parameters of the heater, for which the recalculation is performed through the surface area of the heater.

Using the values: the density of heat loss of a person in a normal state in the supine position, weighted average skin temperature, core temperature, ambient temperature, the area of the affected and unaffected skin surface (these are the known data), the parameters of the heater are determined: the electrical resistance of the heating element, its area, the types and number of threads used, the distance between them. Based on the obtained parameters of the heating part, the electrical and geometric parameters of the current supply part of the heater are calculated.

If additional conditions arise associated with changes in external thermal parameters, such as an increase in the heat transfer coefficient due to the use of dressings/coatings impregnated with water-containing ointments for treatment, an increase in convection effects, etc., then these processes must be considered through the corresponding coefficients, boundary conditions included in the system of equations formulated in the statement of the problem [24].

4 Practical Approbation

In A.V. Vishnevsky National Medical Research Center of Surgery, clinical testing of the efficiency of this technique was carried out.

4.1 Used Woven Electric Heater

To compensate for heat losses in order to prevent hypothermia and muscle tremors during graft transplant operations, we used electric woven heaters, such as those shown in Fig. 1.

In a woven electric heater, heat generation was provided with 99% carbon filaments, consisting of carbon.

The heater was calculated using formula (1).

The heater had a power of ~14 W.

An autonomous power supply was used as a source of electricity - a battery with a voltage of 12 V.

The heater was located under the body of the burn patient on the operating table.

4.2 Approbation Results

The effectiveness of the considered method for eliminating the phenomena of hypothermia and muscle tremors in the treatment of burn lesions is confirmed by the results obtained during the transplantation of autodermotransplants, which are presented in Figs. 2, 3.

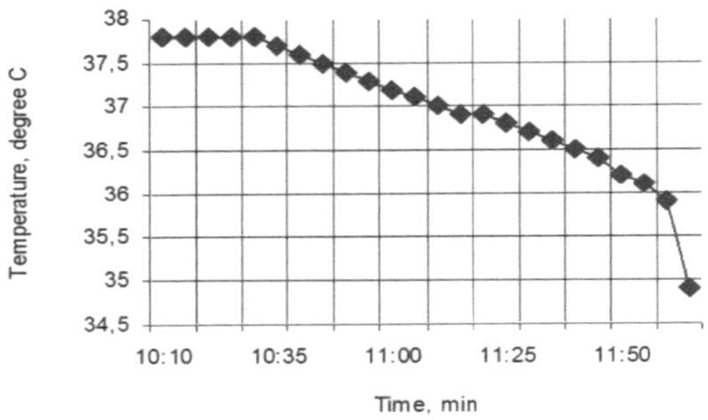

Fig. 2. Changes in rectal temperature over time in a burn patient with the woven heater turned off

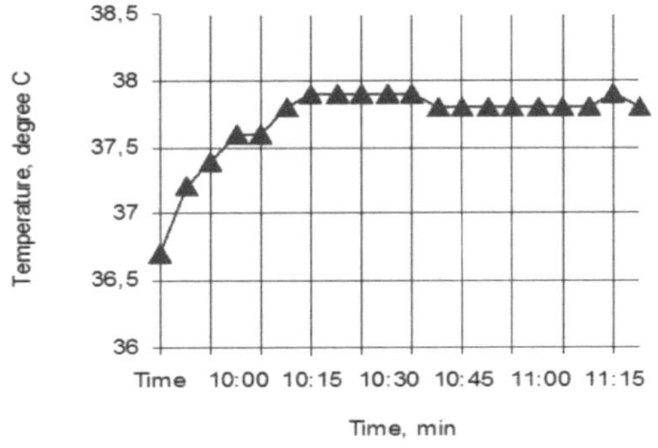

Fig. 3. Change of rectal temperature over time in a burn patient with a woven electric heater turned on

Figure 2 shows the change in temperature over time in a burn patient in the absence of a heater, where it is clearly seen how the temperature dropped in a burn patient in the absence of thermal compensation for heat loss. Already after ~1.5 h, the rectal temperature reached a level of 35 °C, which corresponds to the state of hypothermia. The occurrence of muscle tremors has been noted.

Figure 3 shows the change in temperature over time in a burn patient with a woven electric heater turned on.

In this case, the body temperature did not fall below 36.7 °C.

The use of woven electric heaters for heating made it possible to enter the mode of thermal compensation of heat losses [25]. Rectal temperature within 15 min was stable in the comfortable temperature range from 37.8 °C to 37.9 °C. Muscle tremors were absent.

5 Conclusions

The considered method using woven electric heaters provides several advantages:

1) Provide stable thermal compensation for heat losses arising in burn patients as a result of burns, not after an hour, as with the warm water infusion method, but after 15 min, which prevents unintentional hypothermia and muscle tremors.
2) Provide the possibility of contact heating with thermal compensation of heat losses of a burn patient, which allows significantly reducing the power of the heaters and using autonomous power supplies. In this case, the application of the method becomes possible not only in a stationary mode during operations, in the postoperative period, but also during the evacuation of burn patients.
3) Provide thermal compensation for heat losses without the use of additional devices that increase the evaporation of moisture from dressings impregnated with water-soluble ointments and / or wound dressings and prolong their service life.
4) Provide the ability to build a heating system with a non-invasive or semi-invasive nature of body temperature control.

References

1. Long, J.: Post-mortem appearances found afterburns. London Gaz. **25**, 743–750 (1840)
2. Afanasyeva, R.F.: Hyhienic Bases for Designing Clothes for Protection Against Cold. Legkaya Industriya, Moscow (1977)
3. Vinogradov, V.L.: Protocols of General Anesthesia and Regular Internal Audit as Components of the Safety of Anesthetic Provision for Patients with Thermal Lesions. A. V. Vishnevsky National Medical Research Center of Surgery, Russia (2003)
4. Potapenkova, E.O., Maximova, M.V.: Ointment for the treatment of burns, suppurative and infected wound of different etiology and method for treatment of burn, suppurative and infected wound. RU2003134532 (A). Accessed 20 May 2005
5. Bing, B.: Traditional Chinese medicine scalds ointment for treating burn and scald, and preparation method of traditional Chinese medicine scald ointment. CN110123902. Accessed 16 Aug (2019)

6. Jeschke, M.G., Kamolz, L.-P., Shahriar, S. (eds.): Burn Care and Treatment. A Practical Guide. Springer, Vienna (2015). https://doi.org/10.1111/bjd.12704
7. Vissler, E.: Human Temperature Control. Springer, Berlin (2018)
8. Markus, G.A., Morris, E.N.: Buildings, Climate and Energy Hydrometeoizdat. Pitman Publishing Limited, London (1985)
9. Baxendale, B.R., Mahajan, R.P., Crossley, A.W.: Anticholinergic premedication influences the incidence of postoperative shivering. BJA **72**, 291–294 (1994)
10. Buggy, D., Hunges, N.: Pre-emptive use of the space blanket reduces shivering after general anaesthesia. BJA **72**, 393–396 (1994)
11. Ryan, D.W.: The fluidised bed. Intensive Care Med. **21**(3), 270–276 (1995)
12. Cholpan, P.P.: Physics. Higher school (2003)
13. You Zhi-jian, X., Hong-xia, C.-M.: Infusion of warm fluid during abdominal surgery prevents hypothermia and postanaesthetic shivering. Int. J. Eng. Manufact **1**(5), 26–30 (2011)
14. Shul'zhenko, A.A., Kornev, V.N., Modestov, M.B.: Electric heating fabric, a heating element based on it and a means of connecting the heating element with a power source (options). RU2145984 (C1). Accessed 27 Feb 2000
15. Kazakov, M.E., et al.: Cloth from activated carbon fiber and method of its manufacture. RU2000360 (C1). Accessed 7 Sep 1993
16. Grinevich, I.A., Filippov, D.I., Tolochik, V.S., Shelemekh, A.V.: Electric heating fabric. RU2187907 (C1). Accessed 20 Aug 2002
17. Shul'zhenko, A.A.: Calculation of the thermal field of woven heating elements. J. Mach. Manuf. Reliab. **37**(2), 198–204 (2008)
18. Shul'zhenko, A.A., Modestov, M.B., Mnev, B.A.: Calculations for a woven electrical heater with a uniform temperature distribution on its surface. J. Mach. Manuf. Reliab. **40**(6), 573–578 (2011)
19. Shul'zhenko, A.A., Modestov, M.B., Mnev, B.A.: Woven heaters and their contacts. J. Mach. Manuf. Reliab. **39**(3), 295–300 (2010)
20. Fei, J., Ying, L., Fangjiong, C., Xiaodong, C., Shangkun, X.: Joint beam-power coordinative scheduling of neighboring sectors in cellular systems. Int. J. Wireless Microw. Technol. **1**(6), 10–18 (2011)
21. Chen, L., Ming, H., Xiao-Ming, L., Yuan, L.: Research on trustworthy distributed system. Int. J. Wireless Microw. Technol. **2**(3), 41–47 (2012)
22. Abebe, Y.M., Rao, P.M., Nak, M.G.: Load flow analysis of a power system network in the presence of uncertainty using complex affine arithmetic. Int. J. Eng. Manufact. **7**(5), 48–64 (2017)
23. Shul'zhenko, A.A., Larionov, I.Y., Modestov, M.B.: Requirement for heaters intended for warming burn patients. J. Mach. Manuf. Reliab. **36**(6), 575–579 (2007)
24. Samarskii, A.A., Vabishevich, P.N.: Computational Heat Transfer. Kn. Dom LIBEROKOM (2009)
25. Alekseyev, A.A., Vinogradov, V.L., Modestov, M.B., Modestova, O.B., Chuiko, T.V., Shul'zhenko, A.A.: Method of heat losses compensation in burn patients. RU2701170 (71) Accessed 25 Sep 2019

Current Situation and Countermeasures of Medical Materials Transformation and Expansion Under Major Public Health Emergencies

Lijing Du[1], Te Ke[1(✉)], Xiaohuan Li[1], and Shengtao Liu[2]

[1] School of Safety Science and Emergency Management, Wuhan University of Technology, Wuhan 430072, China
303317@whut.edu.cn

[2] Wuhan Building Material Industry Design and Research Institute Co., Ltd., Wuhan 430223, China

Abstract. Based on the scarcity, timeliness and irreplaceability of medical supplies under major public health emergencies, this article discusses the patterns of influences over the transformation and expansion of medical supplies on the evolution of the epidemic under the COVID-19 epidemic, the Decision-Making on the transformation and expansion of medical materials under major public health emergencies, and the research progress of the mobilization mechanism for the transformation and expansion of medical materials. Starting from the two scenarios of centralized and decentralized Decision-Making, this article summarizes and puts forward new issues that need further research, including analysis of different types of medical material needs, the Decision-Making mechanism for the transformation and expansion of medical materials under evolution of the epidemic and the government's transformation and expansion under coupling of multiple policies. Finally, several corresponding countermeasures and suggestions were given.

Keywords: Major public health emergencies · Transformation and expansion of medical materials · Mobilization mechanism

1 Introduction

Major public health emergencies refer to major infectious disease outbreaks, mass diseases of unknown causes, major food and occupational poisonings, and other events that seriously affect public health that occur suddenly, which cause or may cause serious damage to public health. The guarantee of medical materials directly affects the overall response to major public health emergencies. The transformation and expansion of medical materials is the key to ensure sustainable supply of medical materials. Based on the demand for medical materials from the public and medical institutions, the government has designed a series of mobilization mechanisms and guarantee policies that encourage, guide or require social enterprises to switch and expand production of medical materials in a timely and appropriately, and they are uniformly allocated or

Z. Hu et al. (Eds.): AIMEE 2021, LNDECT 107, pp. 309–318, 2022.
https://doi.org/10.1007/978-3-030-92537-6_29

sold by the government at a reasonable price. Medical supplies are a general term for the material information provided by society to health departments and the public for providing medical services under certain social and economic conditions. When major public health emergencies occur, medical materials often show the following characteristics: (1) Scarcity. (2) Timeliness; (3) Irreplaceability. Under the background of major public health emergencies, the transformation and expansion of medical materials has the following characteristics: (1) Temporary, enterprises need quickly respond to the transformation and expansion policy in the early stage of the epidemic and jointly bear the risk of plunging demands and overcapacity in the later period with the government; (2) Multi-subjectivity, the public, medical institutions, transformation and expansion enterprises and the government are the main participants in the transformation and expansion of medical materials, and they have different target requirements and mutual impact.

The scarcity, timeliness and irreplaceability of medical materials, as well as the temporary, and multi-subject nature of the transformation and expansion work, have brought many difficulties to the works in response to major public health emergencies. The following phenomena have appeared in the practice of the transformation and expansion of medical materials: (1) The production capacity was severely insufficient in the early stage of the epidemic; (2) Overcapacity and excessive waste was common in the later period; (3) The timing of policies implementation was ambiguous, which is impossible to accurately determine when policies should be carried out to meet the supply and demand relationship of medical supplies.

This article discusses the patterns of influences over the transformation and expansion of medical supplies on the evolution of the epidemic under the COVID-19 epidemic, the Decision-Making on the transformation and expansion of medical materials under major public health emergencies, and the research progress of the mobilization mechanism for the transformation and expansion of medical materials. Starting from the two scenarios of centralized and decentralized Decision-Making, this article summarizes and puts forward new issues that need further research, including analysis of different types of medical material needs, the Decision-Making mechanism for the transformation and expansion of medical materials under evolution of the epidemic and the government's transformation and expansion under coupling of multiple policies.

2 Transformation and Expansion of Medical Materials Under the COVID-19 Epidemic

The difficulties and deficiencies in transformation and expansion of medical materials under the COVID-19 epidemic are described as follows.

2.1 The Unclear Evolution Law of the Epidemic

The basis for the implementation of the Decision-Making work on the transformation and expansion of medical materials is the forecast of medical materials demand based on the evolution of the epidemic. However, as one of the most serious infectious diseases in a century, the source of infection of COVID-19 epidemic is unknown in a short period of time, the pathogenic mechanism of transmission is unknown, and the risk of virus mutation is uncertain, making it unpredictable in duration, intensity and scope of impact; the relationship between the evolution of the epidemic and the production capacity expansion of medical materials is unclear. The demand for medical materials is highly dynamic and uncertain, and it is difficult to accurately predict.

2.2 The Mismatch Between the Production Capacity and Demand of Medical Materials

Under the COVID-19 epidemic, the expansion of medical materials is facing the dilemma of double interruption of supply and demand. On the one hand, the demand for masks in a short period of time, the inventory of masks has been quickly consumed. The transformation and expansion of medical materials require a large amount of manpower, material and capital investments such as technological researches, upgrading of production equipment and personnel recruitment and training. It coincides with the Spring Festival period, facing labor shortages, insufficient supply of production raw materials, parts and high shortage of raw material supply cost. On the other hand, the demand for medical materials has surged and is time-sensitive. In the early stage of the epidemic, companies are required to have quick responses speed and be able to switch to expand the production of medical materials in a short period of time; in the later stage of the epidemic, the spread of the epidemic has slowed down and the demand for masks has plummeted.

2.3 Multivariate Policies Coupled Mutually

A reasonable mechanism for mobilizing medical materials for transformation and expansion is the guarantee for the balance between production capacity and demand for production expansion. The outbreak of the COVID-19 epidemic coincides with the Spring Festival. There is a shortage of labor, the cost of transportation and epidemic prevention and protection has risen, and the price of medical materials has skyrocketed. Some manufacturers and wholesalers have disrupted market order, such as driving up prices and hoarding goods, and set their own prices, which are far higher than usual prices.

3 The Government-Led Multi-subject Medical Material Transformation and Expansion Mechanism

The government sorted out the problems exposed by the transfer and expansion of medical materials in COVID-19 epidemic prevention and control, combined with the interaction between the government, enterprises and demanders (Fig. 1), extracted the following three scientific problems to be studied, in order to provide scientific assistance to major public health emergencies in the future.

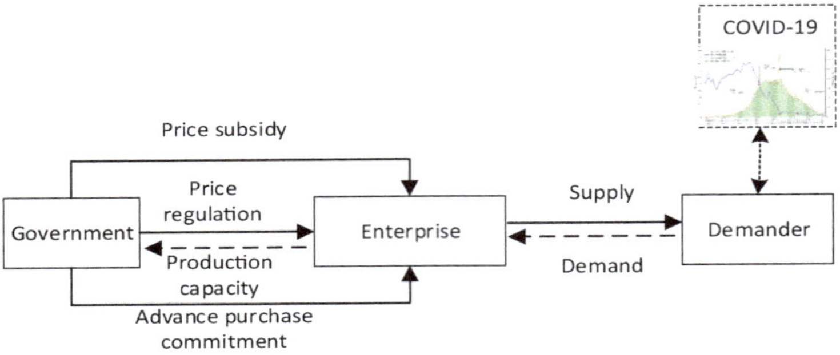

Fig. 1. Government-led multi-subject medical material transformation and expansion mechanism

As shown in Fig. 1, under major public health emergencies, the production capacity of enterprises to transfer and expand production will affect the supply of medical materials, further affecting the demand-side epidemic prevention and control efforts and the capacity and efficiency of medical institutions to receive and treat, which have effects on the evolution of the epidemic. The first scientific question is: in the prevention and control of the epidemic, what is the law of influence between the expansion of medical materials and the evolution of the epidemic? This is the basis for the design of the mobilization mechanism for the transformation and expansion of medical materials. Under the evolution of the epidemic, based on the dynamic changes in the supply costs of raw materials, labor, etc. and the demand for medical materials, the incentive effect of the government's production expansion mobilization mechanism and market adjustment mechanism should be integrated, considering the corporate social responsibility awareness and corporate image, and how to make reasonable medical decisions for material transformation to expand production capacity? This is the scientific basis for balancing the production and expansion of medical materials during the entire epidemic period, and is also an important second scientific issue that needs further study. In the government's mobilization mechanism for production transformation and expansion, push and pull policies and regulatory policies are mutually coupled. On this basis, how to design a medical materials mobilization mechanism? Achieving the balance between production capacity and demand for medical materials during the outbreak, stabilization, and recovery periods is the third scientific issue that requires in-depth study.

4 Key Issues in the Transformation and Expansion of Medical Materials

4.1 The Law of the Impact of the Transformation and Expansion of Medical Materials on the Evolution of the Epidemic

In the relevant research on COVID-19 epidemic, some studies have used, for example, the SEIR (Susceptible-Exposed-Infected-Removed) model or SIR (Susceptible-Infected-Removed) model to simulate the epidemic dynamics. The main goal is to estimate the spread of the epidemic, assess the impact of non-profit organizations on the evolution of the epidemic, or predict the spread of COVID-19, as shown in Table 1.

Table 1. The research status of the evolution law of the epidemic

Research Subject	Model	Model influencing factors	Author (Year)
COVID-19 Infections	Agent-based Infectious Disease Model	Demographics information, host mobility, social contact rates, and human communication possibilities in the workplace, schools, and families	Koo et al. (2020)
Epidemic spread	Room model of age structure	Physical distance limitation, isolation degree	Tuite et al. (2020)
COVID-19	SEIR model	Voluntary isolation, isolation of infected individuals, increased individual immunity, and inpatient treatment interventions	Nita et al. (2020)
Different intervention scenarios across China under COVID-19	Modeling framework based on the SEIR Model	City closure, travel bans, and social distance	Lai et al. (2020)
Hospital capacity Requirements	SIRM model	Population number and number of cases	Weissman et al. (2020)

In related studies on COVID-19, some studies utilized models such as the SEIR (Susceptible-Exposed-Infected-Removed) model [1] or the SIR (Susceptible-Infected-Removed) model [2] to simulate the infectious disease dynamics of epidemics. The main goals focus on estimating the spread rate [3], assessing the impact of non-profit organizations on the evolution of the epidemic, or predicting the spread of COVID-19 [4]. However, in these models, the impact of medical material transformation and expansion on the number of cases was not considered.

4.2 Decision of Medical Materials Under Major Public Health Emergency

The decision to convert medical supplies and expand production capacity under major public health emergencies means that in order to ensure the sustainable supply of medical supplies, under the government mobilization mechanism, through technological transformation, adding production lines (equipment), and increasing schedules, etc., in a timely manner Increase production capacity appropriately to meet the dynamic demand for medical supplies under the evolution of the epidemic (Du Lijing, 2020). The relevant literature on existing capacity Decision-Making is sorted out, as shown in Table 2.

Table 2. Current status of capacity decision research

Material	Decision	Method	Author (Year)
Product	Profit maximization	System dynamics model	Vlachos et al. (2007)
		Newsboy model	Zhang and Du (2009)
		Analytical model for predictive evolution	Biçer and Seifert (2017)
Innovative Products	Minimum cost	System dynamics Model	Narasimha and Roy (2006)
		Bass diffusion Model	Lukas et al. (2017)
Product	Reduce the loss	Production and profit function development model	Ho and Fang (2013)
Product	Profit maximization	Flexible Decision-Making model of production capacity	Zhang Zili et al. (2007)
		Single-cycle production capacity decision model	Li Chunfa et al. (2020)
	Minimum cost	Capacity expansion model	Ryan (2004)
		Two-stage random program	Shao (2020)
Electrical product	Optimum expected profit	Job cost method, M/M/1 decision model	Xie (2015, 2016)
High-quality products	Profit maximization	Game	Hsieh and Lai (2017)
Emergency products	Growth Optimization of Production Increments	System dynamics	Zhang Zili et al. (2009)

Aiming at enterprises with a certain surplus production capacity under sudden disasters [5, 6], the model introduces the cost of increasing production capacity, such as increasing scheduling [7], purchasing or renting equipment [8], hiring labor [9], emergency procurement [10], and the cost of emergency procurement due to shortage of raw materials [11].

4.3 Mobilization Mechanism for the Transformation and Expansion of Medical Materials

This section first summarizes the policy for the conversion and expansion of medical supplies under the COVID-19 epidemic; secondly, analyzes the current status of the application of price control policies and push-pull mobilization mechanisms in supply chain management, in order to obtain the conversion and expansion of medical supplies under major public health emergencies views and methods that can be used for reference.

Table 3. Policy on the transformation and expansion of medical materials under the COVID-19 epidemic

Medical materials		Policy	Author (Year)
Medical protection materials	Mask	Improve the production process, simplify the mask design, and choose an alternative supply chain	Shokrani et al. (2020)
	Protective clothing	Improve the production process, choose an alternative supply chain	Shokrani et al. (2020)
	Medical gloves, Alcohol hand sanitizer		Salmi et al. (2020)
		Issue guidelines for the development of standardized ingredients and formula products	Berardi et al. [30] (2020)
	Nasopharyngeal swab	Using 3D printing technology	Salmi et al. (2020)
Medical treatment device	Ventilator and its parts, etc		Iyengar et al. (2020) Manero et al. (2020)
		Activate the Defense Production Act (DPA),	Ranney et al. (2020) Greenspoon & Class (2020)

As shown in Table 3, the main policies and recommendations for the transformation and expansion of medical materials in the literature are: (1) The government requires enterprises to transform and expand production, and provide financial and convenient support policies [12, 13]. (2) Relaxation of production standards [14]; (3) Optimize production processes and make full use of existing alternative supply chains [15, 16]. (4) Announce production and testing standards for masks, protective clothing, etc., and provide guidelines for the development of standardized ingredients and formula products for products such as disinfectants to avoid a large number of substandard products on the market [17].

Price control means that the government uses effective means to regulate and control the price of materials and regulate the market to ensure the scientific and reasonable supply and distribution of medical materials. Push-type mobilization mechanism refers to the government directly granting certain production cost subsidies to promote the transformation and expansion of medical materials before or during the production expansion of medical materials. Pull mobilization mechanism means that the government sets a certain amount of purchase quantity or price for companies that can successfully transform and expand medical materials, so as to pull companies to transform and expand medical materials.

Judging from the current theoretical research status, how to introduce the interactive relationship between the transformation and expansion of medical materials and the evolution of the epidemic into the forecast of medical materials demand is a major factor in improving the accuracy of medical materials forecasting, it is also the basis for improving the scientific Decision-Making of medical materials to expand production capacity.

5 Suggestions and Conclusion

5.1 Evolution-Based Medical Supplies Prediction

Collect epidemiological data of the COVID-19 epidemic, establish an extended SEIR model [18, 19], draw an epidemiological curve of the cumulative production capacity of transformation and expansion and the number of new cases per day, and introduce the transformation and expansion mechanism and capacity into the SEIR model. Based on the questionnaire and actual demand data of medical materials, we analyze the demand behavior of medical materials during the COVID-19 pneumonia epidemic, and establish a reasonable model of the utility of medical materials. Based on relevant data such as the supply of medical materials, demand, and the number of people diagnosed, characterize the interaction between the evolution of the epidemic and production capacity, improve the existing infectious disease model.

5.2 Production Capacity Decision-Making Under Centralized and Decentralized Decisions

Research on the impact mechanism of the dynamic changes of supply and demand, corporate social responsibility and corporate image on the income of medical materials, and establish a model of the expected return of the company's medical materials. Under both centralized and decentralized Decision-Making scenarios, study the push-pull mechanism of production subsidies and bottom-of-the-line purchases in the mobilization mechanism, as well as the adjustment effect of the price control mechanism; Combining the demand utility function of medical materials under the evolution of the epidemic, and the profit model enterprises, clarify the optimal production capacity and range of medical materials at different stages of the epidemic, and promote the balanced production of medical materials throughout the epidemic [20].

5.3 Mobilization Mechanism Under the Coupling of Multiple Policies

Establish a sequential model between Demanders-Transforming and expanding Enterprises-Government in two scenarios: centralized Decision-Making and decentralized Decision-Making. At the same time, in this sequential model, the research is carried out in the order of the number of enterprises that have been transformed and expanded. Analyze the optimization design of production capacity and the mobilization mechanism for the transformation and expansion under the competition and cooperation of different transformation and expansion enterprises.

5.4 Conclusion

Under major public health emergencies, the demand for medical materials has exploded, and supply and demand have a double impact, it is difficult to match the demand for production capacity. Precise medical material demand forecasting is the basis for the implementation of the Decision-Making work on the production of medical materials. Under the government's policy of increasing production and supply, companies have made every effort to release their production capacity through the support policy, thereby increasing the supply of medical materials. A reasonable mechanism for mobilizing medical materials is the guarantee for the balance of production capacity and demand. It requires a comprehensive market mechanism and government mobilization mechanism. Starting from the two scenarios of centralized and decentralized Decision-Making, it analyzes the demand for different types of medical materials, the Decision-Making mechanism for production capacity expansion, and the government's mobilization mechanism under the coupling of multiple policies. The study of these issues will improve my country's medical material supply guarantee mechanism and further enhance the ability to respond to major public health emergencies.

Acknowledgment. This project is supported by Foundation of Social Science and Humanity, China Ministry of Education (Grant No. 20YJC630018), Hubei Provincial Natural Science Foundation (Grant No. 2020CFB162), Fundamental Research Funds for the Central Universities (WUT: 2020VI009).

References

1. Koo, J.R., et al.: Interventions to mitigate early spread of SARS-CoV-2 in Singapore: a modelling study. Lancet Infect. Dis. **20**(6), 678–688 (2020)
2. Tuite, A.R., Fisman, D.N., Greer, A.L.: Mathematical modelling of COVID-19 transmission and mitigation strategies in the population of Ontario, Canada. Can. Med. Assoc. J. **192**(19), 497–505 (2020)
3. Lai, S., et al.: Effect of non-pharmaceutical interventions to contain COVID-19 in China. Nature **585**(7825), 410–413 (2020)
4. Weissman, G.E., Crane-Droesch, A., Chivers, C., et al.: Locally informed simulation to predict hospital capacity needs during the COVID-19 pandemic. Ann. Intern. Med. **173**(1), 21–28 (2020)
5. Vlachos, D., Georgiadis, P., Iakovou, E.: A system dynamics model for dynamic capacity planning of remanufacturing in closed-loop supply chains. Comput. Oper. Res. **34**(2), 367–394 (2007)
6. Biçer, I., Seifert, R.W.: Optimal dynamic order scheduling under capacity constraints given demand-forecast evolution. Prod. Oper. Manag. **26**(12), 2266–2286 (2017)
7. Narasimha, B., Roy, K.R.: Capacity augmentation of a supply chain for a short lifecycle product: a system dynamics framework. Eur. J. Oper. Res. **179**(2), 334–351 (2006)
8. Lukas, E., Spengler, T.S., Kupfer, S., et al.: When and how much to invest? investment and capacity choice under product life cycle uncertainty. Eur. J. Oper. Res. **260**(3), 1105–1114 (2017)
9. Ryan, S.M.: Capacity expansion for random exponential demand growth with lead times. Manage. Sci. **50**(6), 740–748 (2004)

10. Shao, L.: Capacity Investment in a new production location. Decis. Sci. (2020). https://doi.org/10.1111/deci.12483
11. Hsieh, C.C., Lai, H.H.: Capacity allocation with differentiated product demands under dual sourcing. Int. J. Prod. Econ. **193**, 757–769 (2017)
12. Shokrani, A., Loukaides, E.G., Elias, E., et al.: Exploration of alternative supply chains and distributed manufacturing in response to COVID-19; a case study of medical face shields. Mater. Des. **192**, 10879 (2020)
13. Salmi, M., Akmal, J.S., Pei, E., et al.: 3D Printing in COVID-19: productivity estimation of the most promising open source solutions in emergency situations. Appl. Sci. **10**(11), 4004 (2020)
14. Berardi, A., Perinelli, D.R., Merchant, H.A., et al.: Hand sanitisers amid CoViD-19: a critical review of alcohol-based products on the market and formulation approaches to respond to increasing demand. Int. J. Pharm. **584**(119431), 1–14 (2020)
15. Iyengar, K., Bahl, S., et al.: Challenges and solutions in meeting up the urgent requirement of ventilators for COVID-19 patients. Diab. Metab. Synd. **14**, 499–501 (2020)
16. Manero, A., Smith, P., et al.: Leveraging 3D printing capacity in times of crisis: recommendations for COVID-19 distributed manufacturing for medical equipment rapid response. Int. J. Environ. Res. Public Health **17**(4634), 1–17 (2020)
17. Greenspoon, L.R., Class, R.D.: Defending the producers: examining product liability protection for compelled manufacturing under the defense production act. Def. Couns. J. **87**(3), 1–18 (2020)
18. Ahmed, H., Arfan, A., et al.: Visualization & Prediction of COVID-19 future outbreak by using machine learning. Int. J. Inf. Technol. Comput. Sci. (IJITCS) **13**(3), 16–32 (2021)
19. Joseph, I., Divine, O., et al.: Machine learning based on kernel function controlled gaussian process regression method for in-depth extrapolative analysis of Covid-19 daily cases drift rates. Int. J. Math. Sci. Comput. (IJMSC) **7**(2), 14–23 (2021)
20. Mirza, W., Tabasum, M., et al.: Impact of COVID-19 pandemic on the human behavior. Int. J. Educ. Manag. Eng. (IJEME) **10**(5), 35–61 (2022)

Research on Influencing Factors and Promotion Path of Rural Embeddedness Willingness of Young Talents from the Perspective of Rural Revitalization

Hui Liu(✉)

School of Logistics, Wuhan Technology and Business University, Wuhan 430065, China

Abstract. Youth is the backbone of national development and the new force of Rural Revitalization. The key to Rural Revitalization is the revitalization of young talents. Through the systematic review of related literature on the impact of factors such as public services, personnel policies and rural environment on young talents' rural rooting willingness in Rural Revitalization, based on the data survey of 950 young talents in Lichuan City, Enshi Prefecture and the omni-directional measure on the 20 selected rural rooting willingness of young talents-related indicators in Rural Revitalization, comprehensive analysis of rural revitalization of perspective the influence factors of rural roots intend to young talents, innovative put forward of the will of promoting rural youth talents grounding path: perfecting rural public service level; improving the level of rural government services; strengthening the publicity and implementation of the policy of rural revitalization for young talents; optimizing the rural environment; strengthening the degree of attention to agriculture-related majors in higher education. In order to guide the young talent training, to help the country keep youth talents, and promote rural revitalization, build a well-off society in an all-round way.

Keywords: Young talents · Rural Revitalization · Rooting willingness · Lichuan city

1 Introduction

The report of the 19th National Congress of the Communist Party of China emphasizes that if the youth are strong, the country will be strong; If youth prospers, the country prospers [1]. Under the background of a new era, young talents and Rural Revitalization have a high degree of internal fit [2]. Youth is the new force and backbone of the national economic and social development, and the source of vitality of Rural Revitalization. China's rural problems are prominent, and the most arduous task in completing the building of a moderately prosperous society in all respects lies in rural areas. The Party and the state have given great space for young talents to participate in Rural Revitalization from the policy level, but they are still faced with practical

Z. Hu et al. (Eds.): AIMEE 2021, LNDECT 107, pp. 319–328, 2022.
https://doi.org/10.1007/978-3-030-92537-6_30

difficulties such as the lack of willingness of young talents to take root in rural areas. Therefore, this study has certain theoretical and practical significance.

There are three standards for the age division of "youth" in China: first, 14–28 years old (Chinese Communist Youth League); Second, 15–34 years old (National Bureau of Statistics); The third is 14–35 years old (Medium and Long Term Youth Development Plan (2016–2025) [3, 4]. Based on the research practice, this paper defines the age of youth as 20–35 years old.

Many researches have been done on the affinity of young talents, but few on the rural rooting of young talents. Foreign studies mainly include: Ravenstein and Lee believe that young talents' rural rooting willingness starts in the economic field [5]. Weng et al. empirically analyzed the impact of environmental characteristics on talent rooting in industrial clusters [6]. Christopher R. Ettington et al. deem that leisure can improve life satisfaction [7]. Florida research shows that regional leisure living facilities are significantly positively correlated with talent grounding intention, with the correlation coefficient reaching 0.335 [8].The domestic research mainly include: Ying Jianguo believes that the improvement of rural talent training still needs the strong support of the government [9]. Hu Bei et al. also believe that the level of regional economy is the most fundamental guarantee to meet the demand for talents [10]. He Jie et al. Hold that personal life needs and family life security are the important reasons for rural talent retention[11]. Zhang Yuzhong points out that the willingness of high-level talents to take root does not come from rational considerations. Chinese people generally concern the affection for hometown, local people and family, and some emotional factors also deserve attention [12]. Chen Jie et al. believe that the weight of international talents in economic environment and living environment is higher than that in policy environment [13]. In the context of Rural Revitalization, it is an urgent topic to pay attention to the rooting willingness of young talents in rural areas.

2 The Current Situation of Rural Revitalization in the Absence of Young Talents

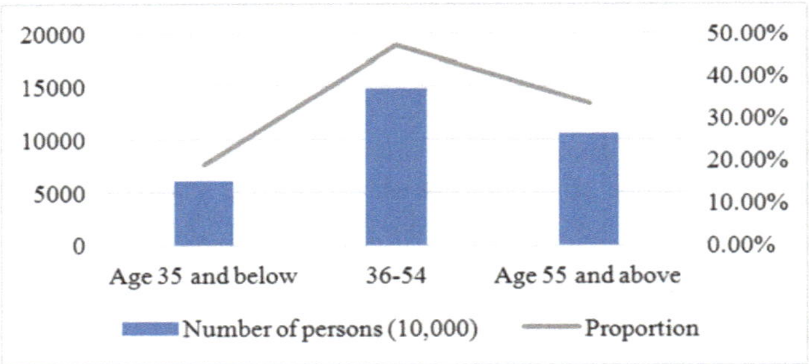

Fig. 1. Proportion of youth in agricultural production and operation in China

China's rural resident population has been declining for many years [14, 15], and the outflow of youth group is an important factor leading to the continuous decline of rural resident population and employment population. As shown in Fig. 1.

China's rural resident population has been declining for many years [14, 15], and the outflow of youth group is an important factor leading to the continuous decline of rural resident population and employment population. As shown in Fig. 1. The number of highly educated young talents participating in Rural Revitalization is small. According to the current survey results, most of them are low educated, As shown in Fig. 2, only about 11% of them have bachelor's degree and graduate degree.

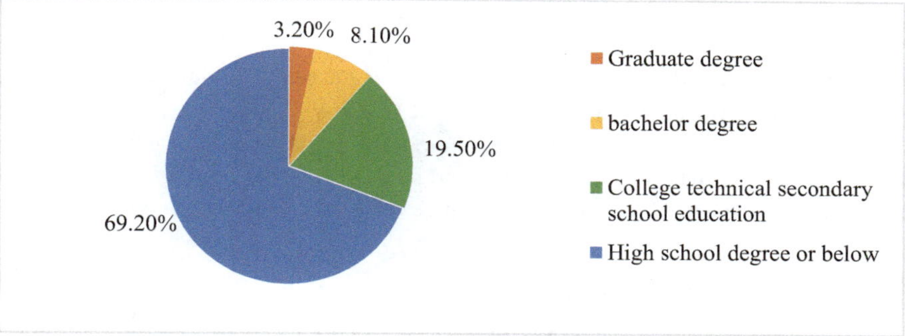

Fig. 2. Distribution of educational qualifications of young people participating in Rural Revitalization

There is no doubt that the effective implementation of rural revitalization urgently needs young talents to return to their hometowns [16–18]. Kashiwa Chiba found that farmers who took part in cultural skills training significantly improved agricultural productivity. Karl Polanyi believed that the stability of the European countryside contributed to the stability of the cities of that era. Yang Xijiyu's investigation on the values of Japanese rural revitalization believes that the key to rural revitalization lies in the initiative of human nature to return to the countryside [19]. Yu emphasizes the youth of the country revitalization is especially needed to reserve talents [20]. Ye Xingqing proposed that rural revitalization should accelerate the training of modern young farmers, leaders of new agricultural operation subjects and agricultural professional managers [21]. Wang Yahua et al. stated that the rural revitalization project requires a large number of young people to return to their hometowns to start their own businesses and promote more rural people to find jobs [22]. In the new round of rural vitalization strategy practice, how to attract and retain young people is the issue that governments at all levels should be concerned about [23]. Rural rooted willingness of young talents, that is, the willingness of talents to work and live in rural areas for a long time, has an important impact on rural revitalization.

3 Factor Analysis and Data Sources

3.1 Factor Analysis

In order to achieve a better analysis effect under the condition of less information loss, this study firstly uses factor analysis method to reduce dimensions, conducts factor analysis on the relevant variable indexes that affect the rural rooting willingness of young talents, and finds out a few mutually independent common factors that can replace many measurement indexes. The general form of the factor analysis mathematical model is as follows:

$$X_{p\times 1} = A_{p\times m} \times F_{m\times 1} + e_{p\times 1} \tag{1}$$

Whereinto, X is the p-dimensional random variable measured in the study, and each of its components represents an index. $F_{m\times 1} = (F_1, F_2, \cdots, F_m)^T$ is an unobservable random vector of m(m $\leq$ p) dimension, namely the common factor vector. The matrix A is the factor load matrix, and a_{ij} is the load of the i variable on the j common factor. The vector e is a special factor that cannot be explained by the first m common factors.

3.2 Data Source and Definition

Due to factors such as economic foundation, location conditions and population quality, the economic development of Lichuan City in Enshi Prefecture lags behind relatively, and the vast majority of young people leave home to seek opportunities [24]. With the in-depth promotion of the Rural Revitalization strategy, the problem of the lack of young talents in Lichuan City is becoming more and more prominent.

Table 1. Basic Information of the survey sample

Statistical indicator		Ratio/%	Statistical indicator		Ratio/%
Sex	Male	52	Home	In the city	19.8
	Female	48		Out of the city in the province	34.2
Age	Under 22	2.3		Out of province or state	46.0
	22-27	32.1	Graduation-city	In province	48.6
	28-35	42.7		Out of province	51.4
	Over 35	22.9	Income	Under ￥80,000	28.9
Background	College degree and below	18.2		￥80,000-$150,000	38.2
	Undergraduate course	59.0		￥160,000-$260,000	18.2
	Master's degree	17.8		￥260,000-$350,000	7.6
	doctor's degree	5.0		OVER￥350,000	7.1

From July to September in 2020, the author conducted a questionnaire survey in Lichuan City, Enshi Prefecture. The survey of the on-site visit lasted nearly 2 months, with youth as the survey object. A total of 1000 questionnaires were sent out and 950 were recovered. After eliminating 45 questionnaires with incomplete and invalid data, a total of 905 valid questionnaires were obtained with effective recovery rate of 90.5%. As shown in Table 1, from the perspective of age and education background, the survey sample has the characteristics of youth and high education background.

On the basis of summarizing previous research literature and the preliminary analysis of survey sample, it found that fourteen indicator variables, namely, talent subsidies, housing security, household registration policy, government efficiency, openness of government affairs, equality of government affairs, public transport, medical and health, basic education, rural culture, employment opportunities, economic level, public security and rural sanitation affect the willingness of young talents to take root in the countryside. This study classifies the evaluation results of talent rooting willingness. Specifically, it classifies "very willing" and "willing" into "willing", and assigns them "1", takes "very unwilling" and "unwilling" as the reference frame, and assigns them "0".

4 Variable Factor Analysis on Influencing Factors of Young Talents' Rural Rooting Willingness

In order to ensure the reliability of the analysis results and eliminate the influence of the original index value of the variable in dimension and order of magnitude, the standard deviation method was adopted to standardize the original data before the factor analysis. In order to express relevant indicator variables more intuitively, the evaluation indicators in Table 2 were set as $(X_2, X_3, X_4, \cdots, X_{20})$, Then, KMO test was conducted, and the results showed that KMO value was 0.866(greater than 0.7) and p value was 0.000(less than 0.05). The test results were very significant and there was correlation between surface data thus suitable for factor analysis.

In this study, the principal component method was adopted, the standard with eigenvalue greater than 1 was taken into account, and information loss and factor generality were considered. Six common factors were extracted, and the internal subload matrix was rotated by the orthogonal selection method with maximum variance. According to the analysis results, the load of the two variables(gender and economic development level) on each common factor was relatively small, so they were deleted. The common degree of the remaining variables is relatively high, most of which are above 0.7 (see Table 2).

According to the data results of factor rotation load matrix, the main factor F_1-the government efficiency, the openness of government affairs, and the fairness of government affairs are named as the government service factor. The main factor F_2-public transport, health care, basic education and employment opportunities, was named as the

Table 2. Factor rotation load Matrix

Evaluation indicator	F1	F_2	F_3	F_4	F_5	F_6	Degree of community
X2	.010	−.015	.109	.770	.104	.013	.616
X_3	−.010	−.054	−.040	.023	−.865	−.043	.756
X_4	.001	−.188	.141	.566	−.270	.303	.541
X_5	−.022	.005	.055	−.053	.832	.029	.700
X_6	.106	−.051	.103	.864	−.088	.003	.778
X_7	.376	.450	−.100	.079	−.028	.523	.634
X_8	.384	.480	.076	.051	.089	.546	.692
X_9	.088	−.004	.066	.082	.063	.870	.779
X_{10}	.828	.203	.290	.087	−.015	.127	.835
X_{11}	.843	.240	.267	.025	.005	.154	.863
X_{12}	.856	.219	.283	.003	−.019	.125	.876
X_{13}	.125	.734	.292	−.185	.054	.018	.677
X_{14}	.201	.716	.3771	−.107	−.004	.061	.706
X_{15}	.167	.785	.219	−.138	.043	.120	.727
X_{16}	.190	.282	.766	−.002	.028	.051	.706
X_{17}	.438	.605	−.072	.282	.023	−.008	.643
X_{19}	.270	.127	.818	.091	.046	.030	.769
X_{20}	.211	.210	.773	.038	.052	.106	.701

public service factor; Main factor F_3-rural culture, public safety, rural sanitation, named as rural environment factor; The main factor F_4, such as age, education background, income and other personal characteristics, was named as personal characteristics factor; The main factor, F_5, represented the relationship between the surveyed talents and the countryside, such as hometown and graduation city, was named the rural affective connection factor. Among the main factors F_6, housing security, household registration policy and talent subsidy are named as talent policy factors.

In order to obtain the common factor expression (factor score function), the regression method was used to calculate the factor rotation score. According to the weight of corresponding variance contribution ratio of each factor, the influence of each indicator on talent's rooting willingness could be judged, and the total score of each variable could be calculated (see Table 3).

In order to facilitate the next analysis, the common factor expression was obtained according to the factor scoring coefficient matrix, and the scoring values of each common factor were calculated respectively. The factor score comprehensive evaluation model is as follows:

Table 3. Rotation factor score sheet

Indicators	F1	F_2	F_3	F_4	F_5	F_6	Score	Sort
X2	−0.100	0.075	0.055	0.472	0.119	−0.093	0.0755	1
X_3	−0.020	0.022	0.025	−0.048	−0.003	0.002	0.0687	18
X_4	−0.110	−0.086	0.125	0.259	−0.568	0.0217	0.0251	16
X_5	0.009	−0.050	0.002	0.504	0.149	0.003	0.0608	4
X_6	0.006	0.087	−0.086	0.001	0.545	−0.120	0.0630	2
X_7	0.007	0.125	−0.095	0.007	−0.014	0.317	0.0426	8
X_8	0.013	0.134	−0.122	−0.070	0.042	0.331	0.0503	5
X_9	−0.120	−0.129	0.018	−0.031	0.033	0.715	0.0246	17
X_{10}	0.414	0.146	−0.025	−0.067	0.015	−0.078	0.0423	9
X_{11}	0.422	0.135	−0.049	−0.084	0.005	−0.055	0.0405	12
X_{12}	0.440	−0.154	−0.036	−0.022	−0.013	−0.076	0.0361	14
X_{13}	−0.160	0.359	0.030	0.004	−0.019	−0.070	0.0350	15
X_{14}	−0.140	0.322	0.069	0.003	−0.053	−0.054	0.0404	13
X_{15}	−0.150	0.389	−0.034	−0.004	−0.024	−0.003	0.0406	11
X_{16}	−0.110	−0.010	0.395	0.216	−0.023	−0.027	0.0413	10
X_{17}	0.119	0.327	−0.247	0.216	−.030	−0.194	0.0620	3
X_{19}	−0.030	−0.127	0.436	0.033	0.005	−0.054	0.0457	6
X_{20}	−0.100	0.065	0.405	0.010	−0.001	0.019	0.0443	7

$$
\begin{aligned}
F = {} & 0.0755X_2 - 0.0687X_3 + 0.0251X_4 + 0.0608X_5 + 0.0630X_6 + 0.0426X_7 \\
& + 0.0503X_8 + 0.0246X_9 + 0.0423X_{10} + 0.04051X_{11} + 0.0361X_{12} + 0.0246X_{13} \\
& + 0.0404X_{14} + 0.0406X_{15} + 0.0413X_{16} + 0.062X_{17} + 0.0457X_{19} + 0.0443X_{20}
\end{aligned}
$$

Based on the result of factor analysis, the following analysis can be made.

1) The first major factor F_1 is the common factor of government services. This factor comprehensively reflects the information of government efficiency, government affairs openness and government affairs fairness [25], indicating that the government's service consciousness, service ability and service level are the important influencing factors of talent's rural rooting willingness.
2) The second major factor F_2 is the common factor of rural public service. This factor comprehensively reflects the information of rural basic education, public transport, health care and employment opportunities and other public services. The more job opportunities there are in rural areas, the better the public transportation system, and the better the educational and medical resources, the stronger the willingness of young talents to stay in rural areas for long-term development.
3) The third major factor, F_3, is the common factor of rural environment. This factor comprehensively reflects the rural soft and hard environment information such as social security, rural culture, urban management and sanitation, etc. The richer the rural cultural entertainment activities, the more harmonious the rural relationship, and the stronger the willingness of young talents to root in the countryside.

4) The fourth main factor, F_4, is the main factor of personal characteristics. This factor comprehensively reflects the information of education background, income and age of young talents. The influence of age on the satisfaction of talents' environment is complex, but in general, education and income are positively correlated. The higher the income, the stronger the willingness of young talents to settle down in rural areas.
5) The fifth main factor, F_5, is the main factor of rural affective connection. This factor comprehensively reflects the special emotional connection between young talents and a village. It is not entirely rational for young talents to choose whether to stay in a rural area for a long time. Emotional factors also play an important role. The more emotional they are, the more willing they are to stay in the rural area.
6) The sixth major factor, F_6, is talent policy. This factor comprehensively reflects the information of housing security, talent subsidy and household registration policy, and the talent policy factor has a positive impact on the rural rooting of young talents.

5 Conclusion and Suggestions

This study conducted a survey on 950 young talents working in Lichuan City from July to September in 2020 to analyze the influencing factors of young talents' rural rooting willingness from the perspective of Rural Revitalization. According to the factor molecule, although the talent policy has a positive impact on the rural rooting willingness of young talents, the rural public service factors, government service factors and rural environment factors (rural culture, rural sanitation, public security) have a greater positive impact. In addition, the emotional connection between talents and the countryside will also significantly affect the talents' intention to be rooted in the countryside.

Combined with the result of factor analysis, it is suggested to innovate the path for young talents to root in the countryside from the following five aspects.

1) Continuously improve the level of public services in rural areas. In reality, it is often difficult to bring in talents but neglects the management and service after attracting talents. This requires rural managers to establish a public service input mechanism to retain young talents, improve rural education and medical care, public transportation, social security and other supporting facilities, so as to match the supply structure of public services with young talents.
2) Improve the level of rural government services. To retain talents, it is also necessary to continuously deepen the reform of township governments and improve the level of government services. When introducing public policies related to the immediate needs of talents, the government should take the initiative to disclose information, try to attract talents to participate in decision-making, timely and accurately convey policies to young talents in need, and make decision-making and services closer to the actual needs.
3) Enhance the publicity and implementation of the Rural Revitalization policy involving young talents. To strengthen the implementation and publicity of policies

will constantly improve the ideas and feelings of young talents rooted in the countryside. Relevant departments should implement incentive policies on the training of young talents, professional title evaluation, financial support, etc., do a good job in sorting out the policies, precise publicity, and active service.

4) Optimize the rural environment. Young talents not only focus on economic interests but more concerned about the working environment. To strengthen the "concentration" of public opinion that young talents are rooted in Rural Revitalization, and create a positive rural public opinion atmosphere, so the young talents can get good praise and respect.
5) Strengthen the importance of agriculture-related majors in higher education [26], deepen the reform of agriculture-related personnel education system. Combining with the construction requirements of "double first-class", further improve the comprehensive strength of agriculture-related colleges and universities, and expand the employment market of agriculture-related majors by using the rural revitalization policy.

Acknowledgment. This paper is one of the achievements of the Innovation and Entrepreneurship Project of Wuhan Technology and Business University (Project No. ST2020005).

References

1. The central committee of the communist youth league medium and longterm youth development plan (2016–2025), pp. 2–3. Beijing People's Publishing House, Beijing (2017)
2. Wang, Y., Gu, R.: Investigation and thinking on the youth talent training organized by the youth league for rural revitalization taking the west coast new area of Qingdao city as an example. Chin. Commun. Youth League **08**, 18–20 (2019)
3. Huang, Z.: Who is a youth research report on the definition of youth age. China Youth Res. **11**, 31–41 (2003)
4. Zhao, H.: Youth in the International perspective definition, attributes and problems. J. Youth Stud. **07**, 1–5 (2005)
5. Lee, E.: A theory of migration. Demography **10**(3), 47–57 (1966)
6. Weng, M.: HR environment and regional attraction an empirical study of industrial clusters. Aust. J. Manag. **6**(3), 121–131 (2010)
7. Christopher, R., Jordan, D., Donald, G., et al.: Trans. by Du Yongming. Leisure and Life Satisfaction, pp. 3–22. China Economics Press, Beijing (2009)
8. Florida, R.: The economic geography of talent. Ann. Assoc. Am. Geograph. **11**(4), 98–107 (2002)
9. Ying, J.: Research on the service of vocational education to local economic development. Mod. Voc. Educ. **12**, 30–33 (2018)
10. Bei, H., Weng, Q.: The impact of industrial cluster characteristics on talent embeddedness intention: an empirical study based on four industrial clusters in China. Ind. Eng. Manag. (05), 113–119 (2008)
11. He, J., Wang, H., Zheng, X.: University of science and technology talent flow will status and related factors analysis. Pop. Dev. **3**, 24–32 (2014)
12. Zhang, Y.: Research on the relationship between urban development and inclusion. J. Shanghai Inst. Adm. **01**, 85–95 (2010)

13. Chen, J., Liu, Z., Chen, M., et al.: An empirical study on the impact of talent environment perception on overseas highlevel talent mobility intention a case study of Guangdong province. Sci. Technol. Manag. Res. (01), 163–169 (2018)
14. Li, B.: Talent revitalization and its promotion path in rural revitalization: based on the internal logic between different talents and rural revitalization. Yunnan Social Sci. **04**, 137–143 (2020)
15. Zhu, J.: Review and research progress of population migration theory. Jiangsu City Plan. **07**, 40–44 (2008)
16. Hossain, D., Gorman, D., Lawrence, J., Burton, L.: Factors that Impact on urral and remote students' participation in higher education. Int. J. Mod. Educ. Comput. Sci. **4**(7), 1–8 (2012)
17. Anand, R., Saxena, S., Saxena, S.: E-learning and its impact on rural areas. Int. J. Mod. Educ. Comput. Sci. **4**(5), 46–52 (2012)
18. Liu, S., Chen, P.: Research on cultivation of ethnic minorities IT talents in nationalities universities. Int. J. Mod. Educ. Comput. Sci. **6**(2), 33–40 (2014)
19. Xi, Y.: The breakthrough of values in Japanese rural revitalization. Urban Plan. Int. **05**, 115–120 (2016)
20. Yu, J.: Rural revitalization requires orderly public participation. People's BBS **12**, 74–75 (2018)
21. Ye, X.: Outline of China's rural vitalization strategy in the new era. J. Reform **01**, 65–73 (2018)
22. Wang, Y., Su, Y.: Rural vitalization a new strategy for China's rural development. J. Central Soc. Coll. **06**, 49–55 (2017)
23. Zhang, X., Zhang, G.: Transformation of social principal contradiction, rural governance transformation and rural revitalization. J. Northwest A&F Univ. (Social Sci. Ed.) **03**, 63–71 (2018)
24. Lai, X., Luo, X., Xu, Z., Gan, J., Huang, S.: Research on the training mechanism of rural young talents under the background of rural revitalization strategy a case study of Western Guangdong. Think Tank Era (02), 161–163 (2019)
25. Siiman, L.A., Pedaste, M., Tõnisson, E., Sell, R., Jaakkola, T., Alimisis, D.: A review of interventions to recruit and retain ICT students. Int. J. Mod. Educ. Comput. Sci. **6**(3), 45–54 (2014)
26. Wichitra, Y.: Rural empowerment through education case study of a learning community telecentre in Indonesia. Int. J. Mod. Educ. Comput. Sci. **12**(4), 12–26 (2020)

Fractal Antenna Systems with Chiral Metamaterials Substrates for MIMO Systems

Veronika S. Beliaeva[1], Dmitriy S. Klyuev[2], Anatoly M. Neshcheret[2], Alexander A. Potapov[3,4](✉), and Yulia V. Sokolova[2]

[1] Higher School of Economics, Moscow, Russian Federation
[2] Povolzhskiy State University of Telecommunications and Informatics, Samara, Russian Federation
[3] V.A. Kotelnikov Institute of Radio Engineering and Electronics, Russian Academy of Sciences, Moscow, Russian Federation
potapov@cplire.ru
[4] JNU-IREE RAS Joint Laboratory of Information Technology and Fractal Processing of Signals, JiNan University, Guangzhou, China

Abstract. This article is devoted to the study of the possibilities of increasing the spectral efficiency in MIMO systems by using antennas with substrates of biisotropic and bianisotropic chiral metamaterials and various types of fractal emitters. Fractal structures in the form of a Sierpinski triangle, Koch and Gilbert curves, as well as a dipole triangular antenna of complex configuration were used. The method of designing a chiral metamaterial based on helices, which provides the implementation of the specified values of resonant frequencies, is shown. The spectral efficiency was calculated by using one of the variations of the Shannon formula, which includes a complete matrix of Z-parameters calculated using electrodynamic modeling methods based on integral equations. It is shown that the use of such antennas with the fractal geometry of the emitters located on chiral substrates allows to reduce the mutual influence between the emitters, and, in turn, to increase the spectral efficiency in several frequency ranges compared to traditional solutions.

Keywords: Fractal · Radio systems · Antenna · Chiral metamaterial · MIMO · Spectral efficiency

1 Introduction

Due to the availability of high-quality multimedia content, as well as an increase in channel capacity, it is obvious that at present the requirements for the quality of service of wireless networks in general are increasing, which in turn leads to their modernization of both software and hardware. An example of such an integrated approach is the MIMO (Massive MIMO) technology [1, 2], which has found wide application in modern wireless networks, including IEEE 802.11 ax (Wi-Fi 6) networks and 5G networks, which allows for a significant increase in spectral efficiency. This technology is used for multipath propagation of signals that occur, including in urban development conditions.

Z. Hu et al. (Eds.): AIMEE 2021, LNDECT 107, pp. 329–345, 2022.
https://doi.org/10.1007/978-3-030-92537-6_31

It was shown in [3, 4] that an increase in spectral efficiency, from the point of view of hardware modernization, makes it possible to ensure the use of antenna complexes in the design of which chiral metamaterials are used [5, 6]. In general, metamaterials are artificial composite materials that are a combination of two or more materials with radically different electrophysical parameters (for example, a dielectric with metal inserts included in it in a special way). Such metamaterials have electrophysical properties that are not typical for natural materials, and which, in turn, are due to the periodic structure of elements of a certain shape. Among metamaterials, of particular interest is the study of spectral metamaterials, in which conducting inclusions of a mirror-asymmetric shape are evenly distributed over a dielectric container.

The increase in spectral efficiency in MIMO systems by using such antenna complexes is achieved by reducing the mutual influence between the emitters by creating high-resistance surfaces using chiral metamaterials [7, 8].

In modern communication networks, in some cases, there may be situations when several radio transmitters have to work on one antenna, and in different frequency ranges. In turn, there is a need for multi-frequency antennas that can function in several frequency bands. One of the types of such multi-frequency antennas are antennas and antenna complexes with a fractal (self-similar) emitter geometry. In addition, since there is a large variety of fractal geometric structures, such antennas can be performed as part of exterior objects, which, in turn, will reduce their visibility, and can also ensure their placement on objects (buildings) where distortion of their architectural appearance is unacceptable.

In this regard, this article is devoted to the study of antenna complexes using chiral metamaterials and fractal geometry of emitters for MIMO systems. The emitters are fractal structures in the form of a Sierpinski triangle, Koch and Gilbert curves, as well as a dipole triangular antenna of a complex configuration - FSM, located on substrates of chiral metamaterials.

2 Literature Review

The analysis of modern literature shows that traditional approaches to the development of strip and microstrip antennas have almost reached their limit regarding the further increase in their electrical and mass-dimensional characteristics, and therefore, new solutions and approaches to improving their characteristics are currently being sought. One of these approaches is the using of artificial composite materials in the antenna design, which are called metamaterials [9–12].

Research results show that the using of metamaterials in antenna-feeder devices can significantly improve their characteristics [13], in particular, compensate for the reactivity of electrically small antennas [14], significantly increase directional properties [15], improve coordination, reduce the mutual influence between emitters in the antenna arrays of MIMO systems [16]. In [17], D. M. Pozar noted that the using of more complex metamaterials (bianisotropic and biisotropic media) makes it possible to improve the characteristics of antennas even better. Such metamaterials, for example chiral media, consist of a dielectric container with conducting elements of a mirror-asymmetric shape (helices, open rings, S-elements, etc.) are uniformly located. One of the important features of such media is the possibility of radiation redistribution (incident radiation transforms into azimuthal) [18], which can be used in creating highly efficient antennas for mobile radio networks' subscriber stations.

3 Methodology for Evaluating the Spectral Efficiency of MIMO Systems When Using Antenna Systems with a Fractal Radiator Located on a Substrate of Chiral Metamaterials

It was shown in [1, 3, 4] that the mutual influence of the radiating elements of the antenna system has a sufficiently strong influence on the spectral efficiency of MIMO systems. It is proposed to evaluate the spectral efficiency of the MIMO system taking into account the mutual influence of the emitters of the antenna system using the following formula [1]:

$$C = \log_2 \det\left(I + \frac{1}{D_Z} Z_{RX}^{-1} H Z_{TX} R Z_{TX}^H H^H (Z_{RX}^{-1})^H\right) \quad (1)$$

where D_N is the total noise power in each receiving antenna; Z_{RX}, Z_{TX} are the matrix of the impedances of the receiving and transmitting antenna systems, respectively; H is channel matrix; H^H is Hermitian conjugation operation of the channel matrix; R is the correlation matrix of the input signal.

It follows from this formula, which, generally speaking, is a variation of K. Shannon's formula [19] for the case of the MIMO system, that the mutual influence of emitters is described by the impedance matrices of the transmitting and receiving antenna systems forming the antenna complex. The general view of the impedance matrix of an antenna system having n emitters is written as follows:

$$Z = \begin{Bmatrix} z_{11} & z_{12} & \cdots & z_{1n} \\ z_{21} & z_{22} & \cdots & z_{2n} \\ \vdots & \vdots & \vdots & \vdots \\ z_{n1} & z_{2n} & \cdots & z_{nn} \end{Bmatrix} \quad (2)$$

In (2), the diagonal of the matrix has its own impedances of each of the inputs of the antenna system. Non-diagonal elements are calculated as the ratio of the voltage induced at the passive input of the antenna system radiator to the current at the active input of the antenna system radiator:

$$z_{ij} = \frac{\dot{U}_i}{\dot{I}_j} \quad (3)$$

where $\dot{U}_i$ is the complex tension on a passive radiator; $\dot{I}_j$ is the complex current on the active radiator.

To evaluate the spectral efficiency of the MIMO system, taking into account the mutual influence between the emitters, we will conditionally assume that there is no correlation of signals in the partial MIMO channels, and the same antenna systems are installed in the source and receiver. In this regard, the formula (1) is written as follows:

$$C = \log_2 \det\left(I + \frac{q}{N_{TX}} Z^{-1} H Z Z^H H^H \left(Z^{-1}\right)^H\right) \tag{4}$$

where q is signal strength; N_{TX} is number of partial channels.

It should be noted that in order to ensure the condition of equality of the energy characteristics of the MIMO system, it is necessary to normalize the impedance matrix as follows:

$$Z^{norm} = \frac{Z}{\sqrt{tr(ZZ^H)}} \tag{5}$$

where $tr(X)$ is the trace of the matrix X.

In this study, the MIMO 2×2 system was considered, in which several pairs of fractal antenna systems with substrates of chiral metamaterials were used (one receiving, the other, respectively, transmitting). In addition, in order to evaluate the gain in spectral efficiency, antenna systems with dielectric substrates were also used, the relative permittivity of which was assumed to be equal to 1 ($\varepsilon_c = 1$).

The following types of substrates made of chiral metamaterials were considered:

– Biisotropic substrates with chaotic orientation of chiral elements-spirals;
– Bianisotropic substrates with the same orientation of chiral elements-spirals.

These chiral elements (spirals) were placed in a homogeneous dielectric container with a relative permittivity of 1 ($\varepsilon_c = 1$). This is due to the fact that in practice, a material with a small dielectric constant is usually used, for example, expanded polystyrene.

An important issue is the determination of the resonant frequency (resonant frequencies) of the chiral metamaterial, near which these chiral properties manifest themselves. One of the approaches to determining the resonant frequency of a chiral metamaterial is the use of the Thomson formula:

$$f_0 = \frac{1}{2\pi\sqrt{LC}} \tag{6}$$

where L is inductance of chiral elements; C is the capacity of the elements.

Consider a single thin-wire metal spiral (Fig. 1). This spiral is made of a thin wire, the radius of which is r_H. The spiral consists of N_H turns located at a distance of h_H from each other, and its internal radius is R_H. The length of the spiral in the unfolded state (the length of the thin wire) is l_H. The height of the spiral is H_H, and the winding angle is a_H.

To calculate the inductance of the spiral, we use the formula for determining the inductance of the solenoid:

$$L_H = \mu_0 \mu_c \frac{S_H N_H^2}{H_H} = \mu_0 \mu_c \frac{\pi R_H^2 N_H^2}{H_H} \tag{7}$$

where μ_c is relative magnetic permeability of the medium; $S_H = \pi R_H^2$ is the area of the spiral turn.

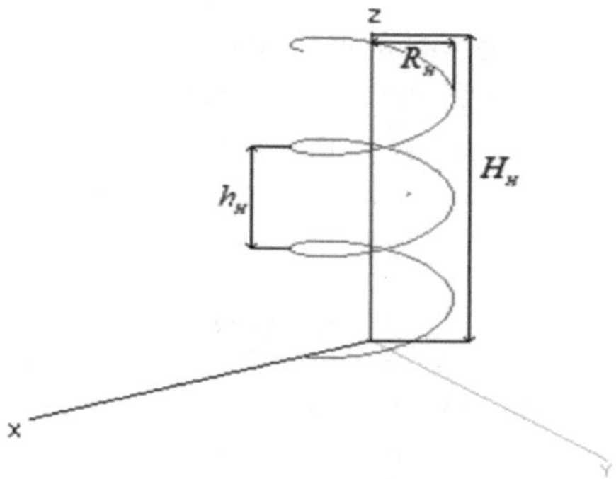

Fig. 1. Helix

The capacity of the spiral is determined by the main three components (8): firstly, the capacity of the wire C_{wire}, from which it is made, secondly, the inter-turn capacity C_{it} and, thirdly, in the presence of a group of spirals located close to each other, the inter-element capacity C_{elem}:

$$C_H = C_{wire} + C_{it} + C_{elem} \tag{8}$$

The capacity of the spiral wire is calculated in accordance with the expression for determining the capacity of a rectilinear conductor:

$$C_{wire} = \varepsilon_0 \varepsilon_c \frac{l_H}{18 \ln\left(\frac{2 l_H}{r_H}\right) - 1} 10^{-9} \tag{9}$$

where $l_H = 2\pi(R_H + 2r_H)N_H + H_H$ is the length of the spiral in the unfolded state.

The inter-turn capacity is defined as follows:

$$C_{it} = \varepsilon_0 \varepsilon_c \frac{S_{it}(N_H - 1)}{h_H} \tag{10}$$

where $S_{it} = \pi\left[(R_H + 2r_H)^2 - R_H^2\right]$ is the area of the ring formed by one turn of the wire.

The distance between adjacent turns h_H is determined by the following expression:

$$h_H = \frac{H_H}{N_H + 1} \tag{11}$$

Finally, the inter-element capacity is determined in accordance with the following expression:

$$C_{elem} = \varepsilon_0 \varepsilon_c k_H \frac{S_{elem}}{A_0} \tag{12}$$

where $S_{elem} = \frac{N_H(2R_H + 2r_H)}{\cos(a_H)} 2r_H$ is the area of space occupied by spiral elements; k_H is spatial coefficient; $a_H = \frac{\pi}{2(N_H + 1)}$ is the angle of the wire winding.

The spatial coefficient k_H, depends on the location and orientation of the spirals. For the case of the considered isotropic chiral substrate (when the spirals have a chaotic orientation and each spiral is surrounded by from 3 to 8 spirals), this coefficient was set equal to $k_H = 0.25$.

Substituting expressions (9), (10) and (12) into (8), we obtain an analytical expression for determining the total capacity of a chiral metamaterial based on spirals.

However, the Thomson formula (6) determines only the first resonant frequency of the chiral element. In fact, there may be much more such resonant frequencies. A rigorous electrodynamic analysis (the method of moments was used to calculate the current function) of a single spiral, as well as a chiral metamaterial as a whole, showed that in a certain frequency band, the resonant frequencies of the spiral are proportional to the difference of two close resonant frequencies, i.e. the proportionality coefficient is equal to $\Delta f \approx f_{0n} - f_{0n-1}$. In this case, the resonant frequencies are determined in accordance with the following expression:

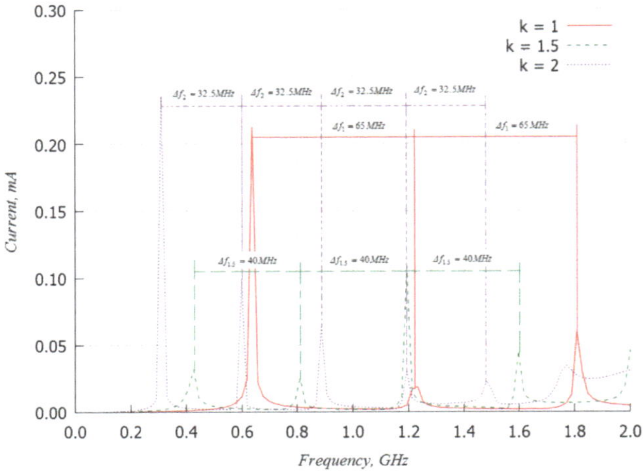

Fig. 2. Dependence of the current on the spiral on the frequency

$$f_{0n} \approx n\Delta f \tag{13}$$

If a certain error is allowed, a zero value can be selected as the frequency f_{0n-1}, which, in turn, allows us to determine the proportionality coefficient as $\Delta f \approx f_0$. It should be noted that the formula (13) is valid for cases when there is a simultaneous proportional change in the height H_H of the spiral and its radius R_H (Fig. 2), i.e. $f_{0n} \approx n\Delta f$, where $k(H_H, R_H), k \in (0, +\infty)$.

Figure 3 shows an electrodynamic model of an antenna system with a bianisotropic chiral substrate on which the emitter is located, which is a Sierpinsky dipole.

The physics of the fractal emitter (Sierpinsky dipole) is explained by the fact that when an EMF source is connected to the top of the antenna arm (triangle), an electric current begins to propagate along it in the direction of its base. After reaching the cluster, the size of which is comparable to the wavelength, electromagnetic waves are emitted. Similarly, at each frequency, the surface current is localized in the regions of clusters corresponding in size, and therefore, the antenna system as a whole exhibits the resonant properties of each such cluster. The number of clusters, in turn, is determined by the order of fractality. Figure 3 shows an emitter having a 4-order fractality, i.e. the division into triangles occurs in 4 iterations.

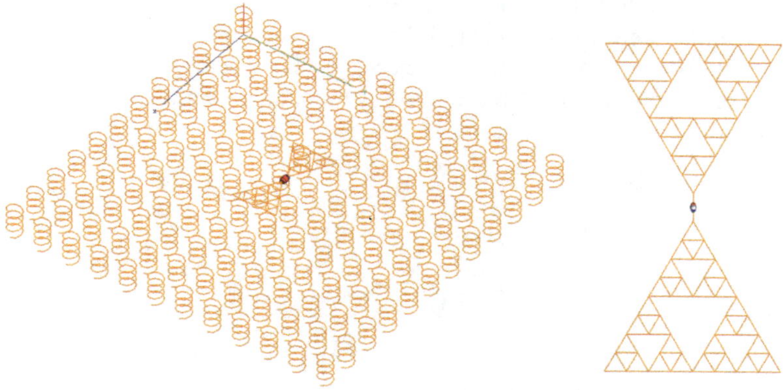

Fig. 3. Model of an antenna system with a Sierpinski dipole

Figure 4 shows the dependences of the reflection coefficients on the frequency of antenna systems with dielectric, bianisotropic and biisotropic chiral substrates (Fig. 4 a), including those metallized from the lower side (Fig. 4b), on which the Sierpinsky dipole (AS) is located.

As can be seen from Fig. 4, such speakers operate in two, rather strongly spaced, frequency ranges, and in the case of metallization, antenna systems with substrates made of chiral metamaterial are much better coordinated.

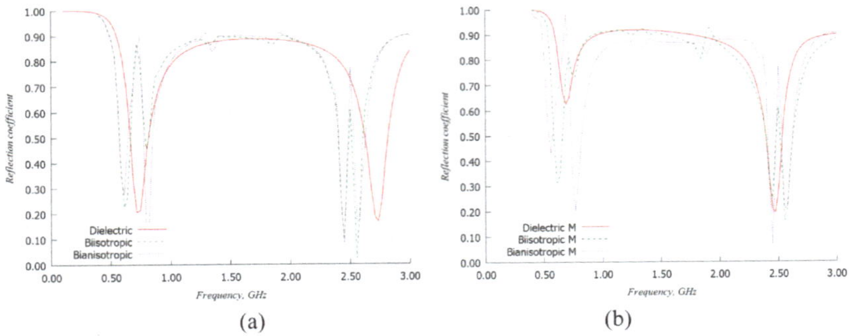

Fig. 4. Dependences of reflection coefficients on the AS frequency

Figure 5 shows the directional diagrams (DD) of AS with a bianisotropic chiral substrate made of right-helical spirals in the vertical (Fig. 5a) and azimuthal (Fig. 5b) planes. As can be seen from the data of the DD, at the lower frequency of 675 MHz, such an AS has azimuthal radiation of electromagnetic waves, i.e. at the resonance frequencies of a bianisotropic chiral substrate, azimuthal scattering of normally incident waves occurs.

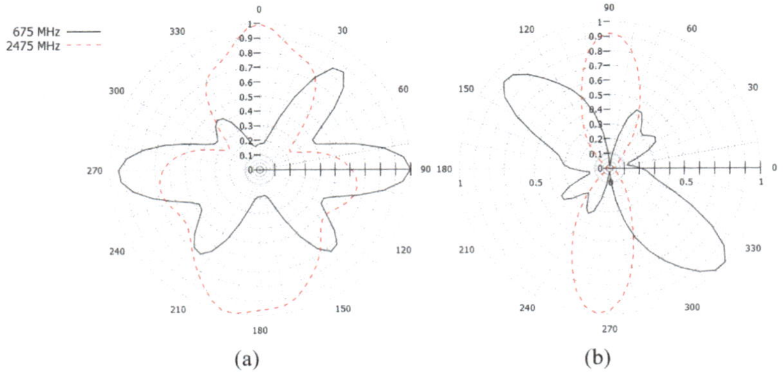

Fig. 5. Directional diagrams of AS

Fig. 6. Model of an antenna system with a FSM radiator

Figure 6 shows an electrodynamic model of an antenna system with an isotropic chiral substrate on which a fractal emitter of the FSM is located. The arms of FSM are triangles that are obtained as a result of performing a recursive procedure (the number of iterations in this case is 3) connecting the centers of mass of triangles and their angles (by analogy with drawing the medians of a triangle). It is thanks to this analogy-fractal splitting by medians-that this antenna got its name.

Figure 7 shows the dependences of the reflection coefficients on the frequency of antenna systems with dielectric, bianisotropic and biisotropic chiral substrates on which the FSM dipole (ASFSM) is located. As can be seen, such a FSM radiator as a whole can also work in two frequency ranges.

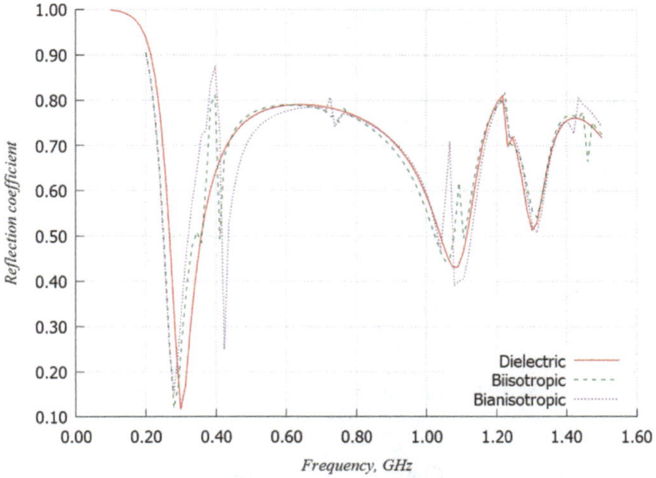

Fig. 7. Dependences of reflection coefficients on the frequency of ASFSM

Figure 8 shows the directional diagrams of ASFSM with an isotropic chiral substrate made of left-screw spirals in the vertical (Fig. 8a) and azimuthal (Fig. 8b) planes. As can be seen from the Fig. 8, the ASFSM has azimuthal radiation of electromagnetic waves already at two resonant frequencies.

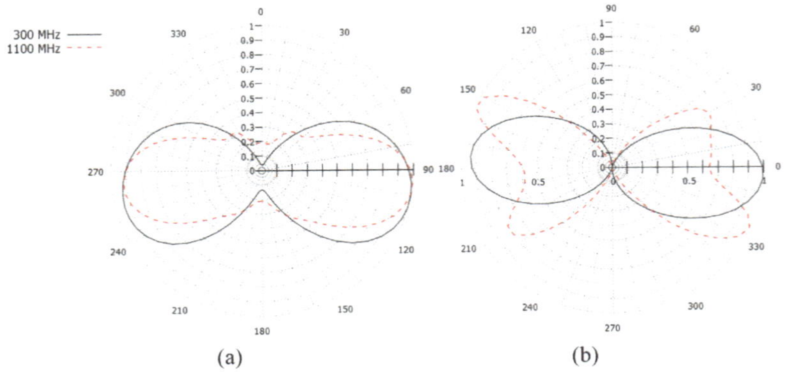

Fig. 8. Directional diagrams of ASFSM

Figure 9 shows an electrodynamic model of an antenna system with a bianisotropic chiral substrate on which a fractal radiator is located, which is a Hilbert curve (ASH). It should be noted that in order to equalize the potentials, which, in turn, will avoid the occurrence of anti-rollback electromagnetic waves, a shunt connection was used in the lower part of this antenna system.

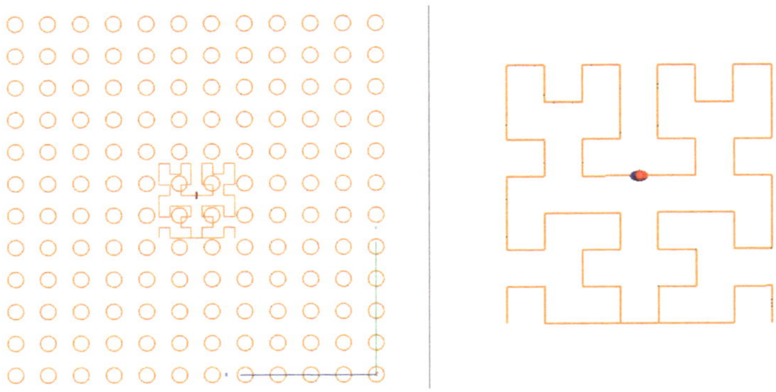

Fig. 9. Model of an antenna system with a Hilbert radiator

Figure 10 shows the dependences of the reflection coefficients on the frequency of antenna systems with dielectric, bianisotropic and biisotropic chiral substrates on which the emitter is located in the form of a fractal Hilbert curve (ASH). Figure 10 shows that the ASH has at least three resonances at a wave resistance equal to 50 Ω, however, with a different wave resistance of the feeder, it is possible to coordinate the antenna system in other frequency ranges.

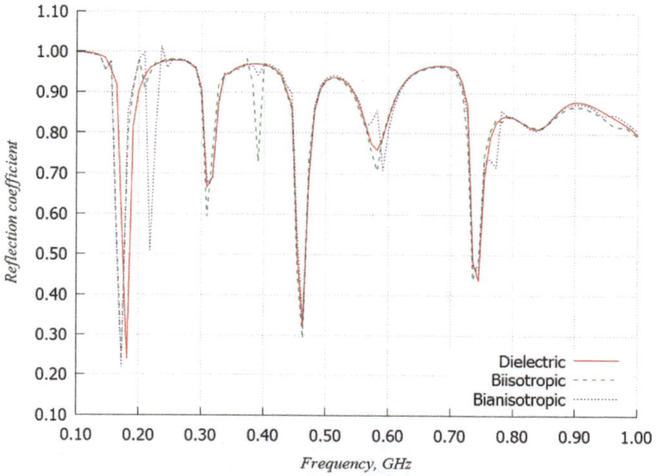

Fig. 10. Dependences of reflection coefficients on the frequency of ASH

Figure 11 shows the radiation patterns of an ASC with a bianisotropic chiral substrate made of left-screw spirals in the vertical (Fig. 11a) and azimuthal (Fig. 11b) planes calculated at resonant frequencies, in particular, at 310 and 460 MHz. At the considered resonant frequencies, as well as in the previous cases, the effect of azimuthal scattering takes place.

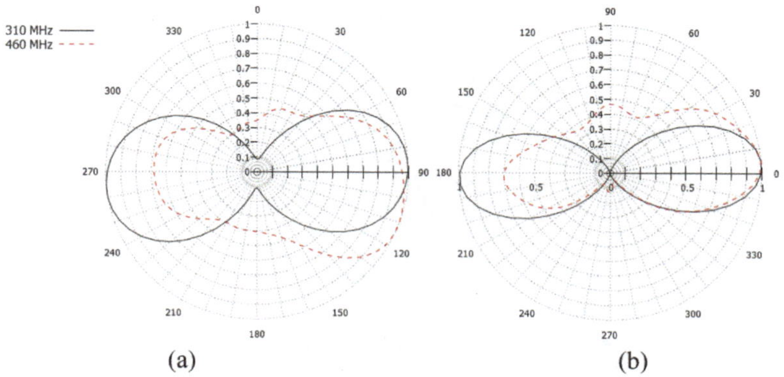

Fig. 11. Directional diagrams of ASH

Finally, Fig. 12 shows an electrodynamic model of an antenna system with an isotropic chiral substrate on which a fractal emitter is located, which is a Koch curve (ASK).

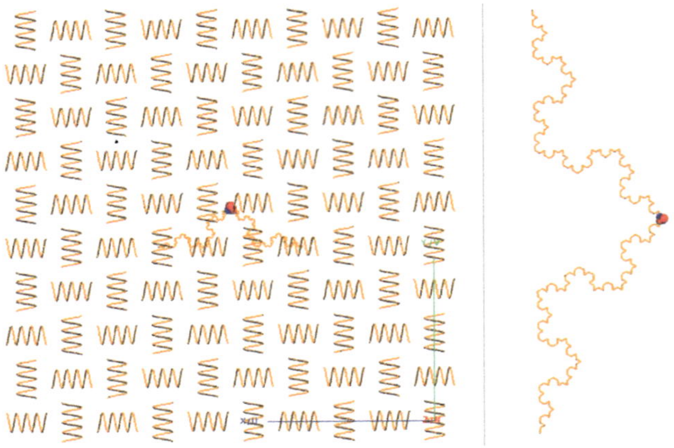

Fig. 12. Model of an antenna system with a Koch radiator

Figure 13 shows the dependences of the reflection coefficients on the frequency of antenna systems with dielectric, bianisotropic and biisotropic chiral substrates on which the emitter is located in the form of a fractal Koch curve (part of the Koch snowflake) (ASK). In this case, such an antenna system with a fractal Koch emitter clearly has 3 operating frequency ranges.

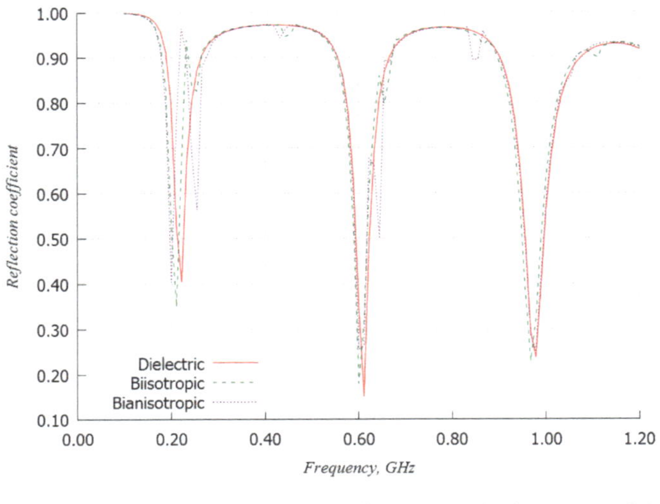

Fig. 13. Dependences of reflection coefficients on the frequency of ASK

Figure 14 shows the directivity diagrams of the ASK with a biisotropic chiral substrate made of left-screw spirals in the vertical (Fig. 14a) and azimuthal (Fig. 14b) planes calculated at resonant frequencies, in particular, at 220, 610 and 970 MHz. In all operating ranges, as can be seen from the directional diagrams shown in Fig. 14, this antenna with a biisotropic chiral substrate, on which the Koch fractal emitter is located, has an azimuthal direction of electromagnetic wave radiation.

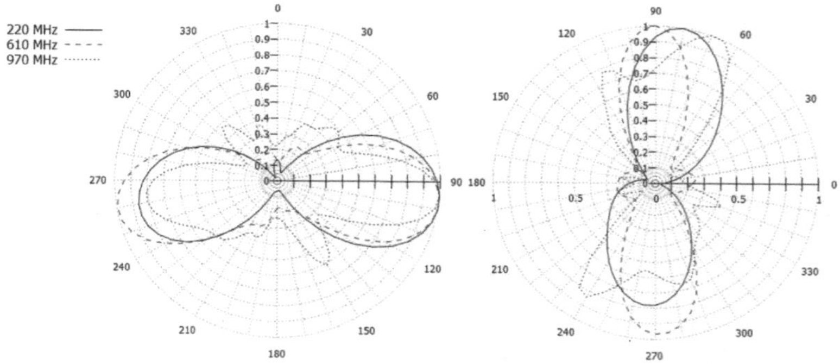

Fig. 14. Directional diagrams of ASK

4 Results and Discussion

To evaluate the spectral efficiency of MIMO systems, antenna complexes with substrates of chiral metamaterials and fractal geometry of emitters were used. On such chiral substrates of antenna systems, two fractal emitters of various types, considered above, were located parallel to each other at a fixed distance.

Figure 15 shows the dependences of spectral efficiency on frequency when using antenna systems with dielectric, biisotropic and bianisotropic substrates on which a pair of Sierpinsky dipoles is located, and Fig. 16 shows similar dependences with the only difference that such antenna systems are located on an infinitely thin and ideal conducting surface (base).

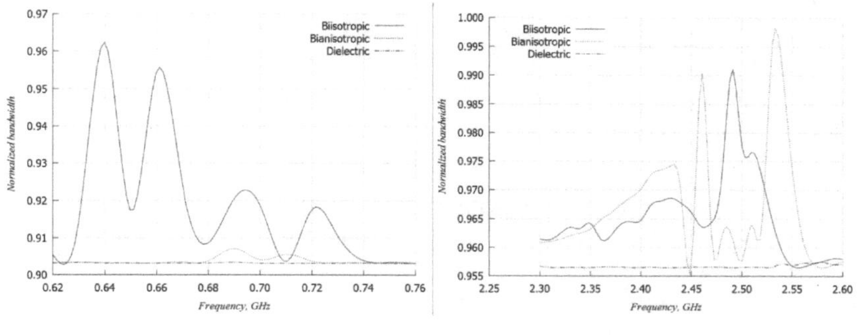

Fig. 15. The dependence of the spectral efficiency on the frequency when using AS with different substrates

As can be seen from the presented dependencies, antenna systems on spectral substrates on which Sierpinski dipoles are located can provide a gain in spectral efficiency compared to antenna systems on dielectric substrates, and in two frequency ranges. In the lower range, the greatest gain is provided by antenna systems on biisotropic chiral substrates, in the upper range both on biisotropic and bianisotropic.

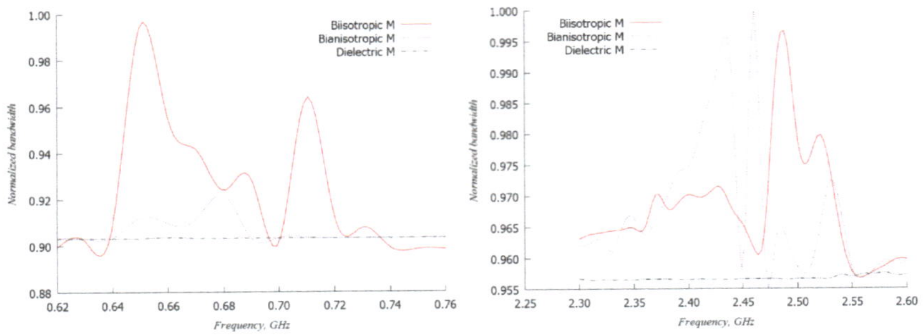

Fig. 16. The dependence of the spectral efficiency on the frequency when using AS with different substrates

Figure 17 shows the dependence of the spectral efficiency on the frequency when using antenna systems with dielectric, biisotropic and bianisotropic substrates, on which a pair of fractal emitters of the FSM is located.

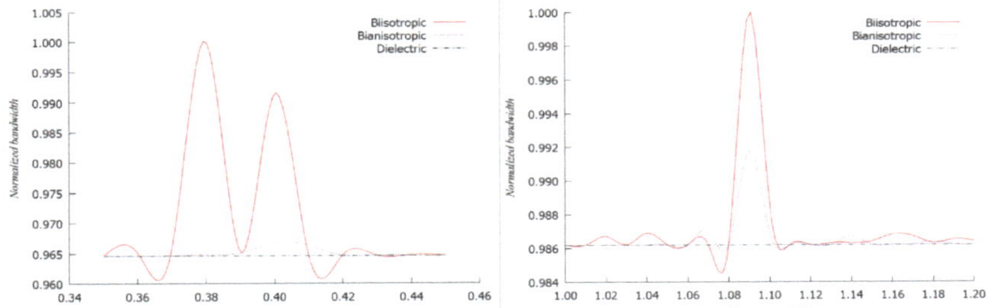

Fig. 17. The dependence of the spectral efficiency on the frequency when using ASFSM with different substrates

Similarly, such antenna systems with a FRM emitter on chiral substrates provide a gain in spectral efficiency in two frequency ranges. However, in comparison with antenna systems with a Sierpinski dipole, ASFSM data, firstly, have a smaller frequency band in which there is a gain, and, secondly, a much smaller gain in spectral efficiency.

Figure 18 shows the dependence of the spectral efficiency on the frequency when using antenna systems with dielectric, biisotropic and bianisotropic substrates, on which a pair of emitters is located, made in the form of fractal Hilbert curves (Fig. 18a) and Koch (Fig. 18b).

As can be seen from the presented graphs, such antenna systems with emitters in the form of fractal Hilbert and Koch curves generally allow to increase the spectral efficiency in three small operating frequency bands. However, in practice, large ranges are usually used in such a frequency domain, and therefore, such solutions of antenna systems are very specific.

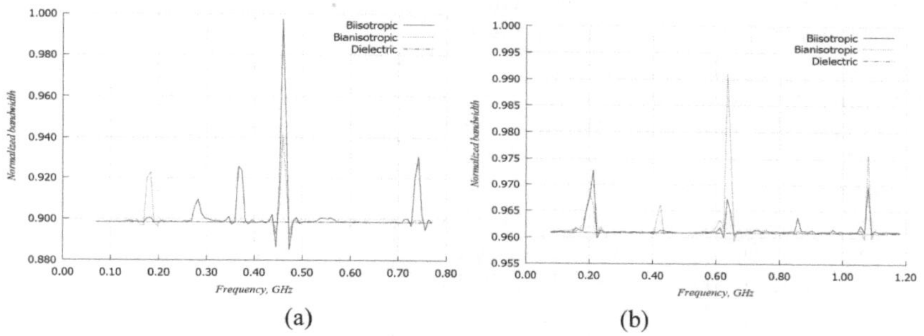

Fig. 18. The dependence of the spectral efficiency on the frequency when using ASH (a) and ASK (b) with different substrates

As already noted, in antennas with substrates made of chiral metamaterials, at resonant frequencies, the surface of the substrate becomes highly resistive, and therefore there is a greater attenuation of surface waves, which leads to a decrease in the mutual influence of emitters (increasing the isolation between them), which in turn contributes to an increase in spectral efficiency.

5 Conclusions

Thus, the study of antenna complexes using chiral metamaterials and fractal geometry of emitters for MIMO systems makes it possible to increase their spectral efficiency in comparison with antenna complexes on dielectric substrates. It should be noted that such a solution is very specific, since it has a relatively small operating frequency band.

The obtained results can be used in the development of a new generation of antennas used in modern communication and telecommunications systems, including medical monitoring systems and also, in radar photonics [20, 21]. Further research is related to the development of mathematical and electrodynamic models of other types of antennas and antenna arrays based on chiral metamaterials to improve their characteristics.

Acknowledgments. Author A.M. Neshcheret thanks the Foundation for the Promotion of Innovations for the support provided in the framework of the implementation of research work under the contract No. 16259GU/2021.

References

1. Gershman, A.B., Sidiropoulos, N.D.: Space-Time Processing for MIMO Communication. John Wiley & Sons, Chichester (2006)
2. Telatar, I.E.: Capacity of multi-antenna Gaussian channels. Eur. Trans. Telecom. **10**, 585–595 (1999)
3. Klyuev, D.S., et al.: Antennas based on biisotropic and bianisotropic chiral metamaterials for MIMO system SNC-URSI radio science meeting. USNC/URSI Proc. **2020**, 51–52 (2020)
4. Klyuev, D.S., Neshcheret, A.M., Osipov, O.V., Potapov, A.A., Sokolova, J.V.: Microstrip and fractal antennas based on chiral metamaterials in MIMO systems. In: Skiadas, C.H., Dimotikalis, Y. (eds.) 12th Chaotic Modeling and Simulation International Conference, pp. 295–306. Springer International Publishing, Cham (2020). https://doi.org/10.1007/978-3-030-39515-5_23
5. Klyuev, D.S., Neshcheret, A.M., Osipov, O.V., Potapov, A.A., Sokolova, J.V.: The method of singular integral equations in the theory of microstrip antennas based on chiral metamaterials. In: Skiadas, C.H., Dimotikalis, Y. (eds.) CHAOS 2019. SPC, pp. 267–294. Springer, Cham (2019). https://doi.org/10.1007/978-3-030-39515-5_22
6. Lindell, I.V., Sihvola, A.H., Tretyakov, S.A.: Electromagnetic-waves in Chiral and Bi-isotropic Media. Artech House, Boston (1994)
7. Buzov, A.L., et al.: Calculation of characteristics of planar antenna arrays with substrates made of chiral metamaterials taking into account the dispersion of macroscopic parameters. In: 15th European Conference on Antennas and Propagation EuCAP (2021)
8. Hirvonen, M., Sten, J.C.-E.: Power and Q of a horizontal dipole over a met-amaterial coated conducting surface. IEEE Trans. Antennas Propag. **56**(3), 684–690 (2008)
9. Kaur, J.: Dual band high directivity microstrip patch antenna rotatedstepped-impedance array loaded with CSRRs for WLAN applications. Int. J. Wirel. Microw. Technol. **6**(4), 1–11 (2016). https://doi.org/10.5815/ijwmt.2016.04.01
10. Kaur, P., Aggarwal, S.K., De, A.: Design and investigation of circularly polarized RMPA with chiral metamaterial cover. Int. J. Wirel. Microw. Technol. **6**(3), 61–70 (2016). https://doi.org/10.5815/ijwmt.2016.03.07
11. Shareef, A.N., Shaalan, A.B.: Fractal Peano antenna covered by two layers of modified ring resonator. Int. J. Wirel. Microw. Technol. **5**(2), 1–11 (2015). https://doi.org/10.5815/ijwmt.2015.02.01
12. Ali A Saleh, Abdulkareem S Abdullah. A Novel Design of Patch Antenna Loaded with Complementary Split-Ring Resonator and L- Shape Slot for (WiMAX/WLAN) Applications [J]. I.J. Wireless and Microwave Technologies (IJMECS), 2014, 4(3): 16–25
13. Engheta, N., Pelet, P.: Modes in chirowaveguides. Opt. Lett. **14**(11), 593–595 (1989)
14. Toscano, A., Vegni, L.: Evaluation of the resonant frequencies and bandwidth in microstrip antennas with a chiral grounded slab. Int. J. Electron. **81**(6), 671–676 (1996)
15. Zebiri, C., Benabdelaziz, F., Lashab, M.: Bianisotropic superstrate effect on rectangular microstrip patch antenna parameters. META 2012 (2012)
16. Zebiri, C., Lashab, M., Benabdelaziz, F.: Asymmetrical effects of bi-anisotropic substrate-superstrate sandwich structure on patch resonator. Prog. Electromagn. Res. B **49**, 319–337 (2013). https://doi.org/10.2528/PIERB13012115

17. Zebiri, C., et al.: Gyro-chirality effect of bianisotropic substrate on the operational of rectangular microstrip patch antenna. Int. J. Appl. Electromag. Mech **51**, 249–260 (2016)
18. Abramov, V.Y., Klyuev, D.S., Neshcheret, A.M., et al.: Input impedance of a microstrip antenna with a chiral substrate based on left-handed spirals[J]. The Journal of Engineering **19**, 6218–6221 (2019)
19. Sibille, A., Oestges, C., Zanella, A.: MIMO from Theory to Implementation. Elsvier, UK (2011)
20. Potapov, A.A., Wu, H., Shan, X.: Fractality of Wave Fields and Processes in Radar and Control. South China University of Technology Press, Guangzhou (2020)
21. Feng, T., et al.: Manipulating light scattering by nanoparticles with magnetoelectric coupling. Phys. Rev. B **102**(20), 205428 (2020). https://doi.org/10.1103/PhysRevB.102.205428

Experimental Study of the Positioning System in the Centralized Wi-Fi Network

Elena Kokoreva(✉), Anatoliy Kostyukovich, Ksenia Shurygina, and Ilya Doshchinsky

Siberian State University of Telecommunications and Information Sciences, Novosibirsk 630102, Russian Federation

Abstract. The task before the developers was to create a geolocation system using the Russian-produced equipment, so that the positioning error would not exceed 5 m. A pilot area for conducting the experiments was defined and a test Wi-Fi network was designed and built for the local positioning system implementation as the parts of the project. The signal strength level at the input of the receiving device was chosen as a physical parameter, which must be measured to calculate the coordinated of the target object. Then a combined positioning algorithm, based on the closest access point, spatial patterns differentiation and trilateration mechanisms, was developed. More than a thousand measurements were carried out in the test network, and during that process the main problems of the location system were identified and the ways to solve them were proposed.

Keywords: Positioning · Location · RSSI · Model · Wi-Fi · Trilateration

1 Introduction

Geolocation systems are found in various sectors of economy, science, technology, and everyday life in the modern world. Traditionally, global and local positioning are distinguished, and each of these approaches has its own areas of application. Operation of global navigation systems such as GPS, GLONASS and Galileo is based on satellite equipment, which is characterized by long signal travel path and its high frequency. These principles of satellite signal propagation explain the limited use of global navigation systems inside buildings due to the influence of multipath and significant signal power losses when passing through partitions and ceilings [1–3].

Location systems are in demand in consumer services, advertising and trade, education, medicine, cultural and educational activities, logistics and security companies.

Systems based on Wi-Fi networks available in any state or commercial enterprise is widespread.

Such solutions based on Wi-Fi networks, RFID tags, iBeacons, etc. are offered by many foreign companies in the market, such as Cisco, Huawei, Aruba, Ubitel, etc.[4–7].

Z. Hu et al. (Eds.): AIMEE 2021, LNDECT 107, pp. 346–357, 2022.
https://doi.org/10.1007/978-3-030-92537-6_32

The main disadvantages of these proposals are, firstly, that geolocation service offered by these companies is an additional and very expensive option; and secondly, geolocation solutions provided by foreign companies are unacceptable for a number of public institutions due to non-compliance with information security.

Thus, the authors are to develop and study the local positioning system based on the Russian equipment from the Eltex company, as well as to test the modes of its operation and to achieve the accuracy of determining the object's coordinates up to 5 m.

It is necessary to measure the received signal strength indicator (RSSI) to implement the positioning system. RSSI is measured in dBm by the subscriber device and sent to location server [8, 9]. To determine the mobile object's coordinates, the authors use a combined algorithm based on the following mechanisms:

- closest access point (AP): the coordinates of the object are equated to the cordinates of the AP with the strongest received signal;
- comparison with samples: RSSI is evaluated in every point in the premises in advance and written into database with the point's known coordinates, and in the process of location every incoming RSSI value is compared to those existing in the database, assuming that matching RSSI values are located in the same points;
- trilateration: the method based on finding three (or more) spheres' intersection, where the spheres' radii are the distances from the APs to the receiving device.

The mechanisms in question are further discussed in [10, 11].

2 Experimental Wi-Fi Network for Conducting the Tests

2.1 Building an Experimental Wi-Fi Network

The pilot zone of the location system based on a centralized Wi-Fi network is located on one of the educational institution's buildings three floors, from 2nd to 4th. On each floor, there are five to six APs connected to the university's local network with Internet access. The distance between the APs on each floor was chosen on the basis of the optimal throughput provision of wireless access to the corporate network for the employees of all premises in the defined experimental zone.

The block diagram of the centralized network for the researched system is presented in Fig. 1.

The block diagram contains the main devices, monitoring and control protocols that support the operability of the network and the positioning system. The software to support location procedures is implemented on the Positioning Server, which in this architecture interacts with SoftWLC controller to manage the wireless LAN.

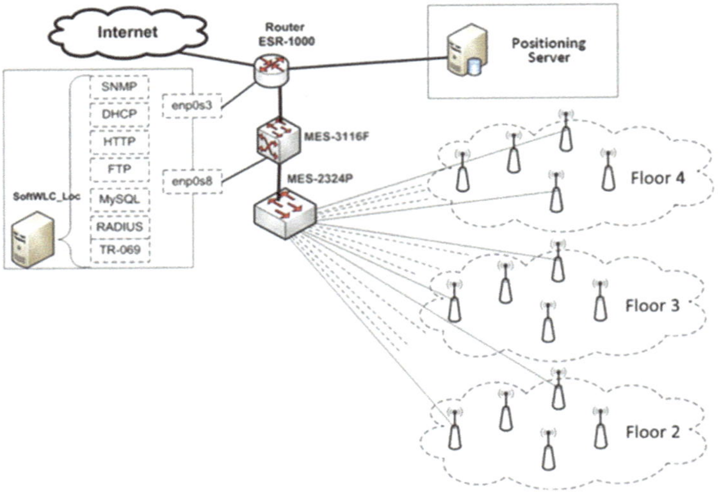

Fig. 1. Block diagram of the Wi-Fi network for the positioning system implementation

The transmission medium between the active network devices is wired. All nodes except subscriber devices are connected by UTP 5 cable and use Ethernet technology for transmission. For more flexible radio coverage planning during field studies, all access points are connected to MES-2324P switch with 24 PoE ports and configured to operate in "Cluster" mode, which ensures centralized management and automatic configuration of their parameters.

DHCP server on the controller distributes IP-addresses to access points and network users. The controller also acts as a router between networks. All packets routed to the external network via the enp0s3 interface are subjected to Network Address Translation (NAT). Authentication of Wi-Fi network users is performed by the SoftWLC controller through the web-portal.

2.2 Frequency-Territorial Planning of the Test Wi-Fi Network

In order to visualize the radio space coverage areas inside the premises with complex structure, as well as composite walls and ceilings materials, the heat maps were simulated in Matlab environment. Comparison of these heat maps with the experimental data obtained in the course of field measurements in the real network allowed to perform frequency-territorial planning of the test network.

Wi-Fi heat map is a graph of Wi-Fi network's signal strength distribution inside the room. It allows one to assess the coverage area provided by the AP, see the unreached zones and, based on the comparison of model maps with the results of measurements, perform more efficient APs' placement and distribution of frequency channels. It is recommended to provide an RSSI level of at least –65 dBm with a signal-to-noise ratio of at least 2 dB for each room.

The program created by Salaheddin Hosseinzadeh, Multi Wall (COST231) Signal Propagation Models was selected from the publicly available software tools for wireless network planning [12].

The examples of AP's radio coverage maps located on the 4th floor of the experimental area are shown in Figs. 2 and 3.

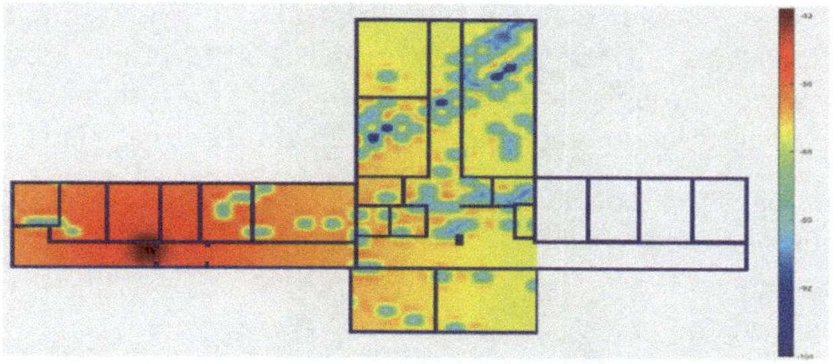

Fig. 2. Heat map for AP 1, the original location

The program uses the following parameters: the frequency at which the access point is operating, the transmission power, the losses in the transmission antenna.

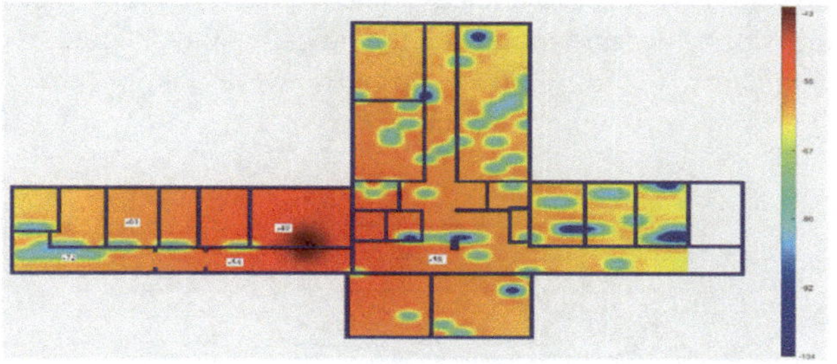

Fig. 3. Heat map for AP 1, the corrected location and measurements results

In the course of its work, this program analyzes the layout of the room, calculates where the partitions and walls are located, inspects the position of the AP, using the free space and multi-wall signal propagation models (COST-231), and then generates heat maps images.

The coverage area found in the field study according to the results of measurements is an exact match with the results of modeling reflected on the heat maps (Fig. 3).

Due to building the heat maps it was possible to find places with the lowest signal level in the premises, make measurements in these problem areas, analyze the differences between the model maps and real maps, take these differences into account in order to subsequently move access points to those places where more even radio coverage is provided.

3 Choosing the Indoor Radio Wave Propagation Models

3.1 Indoor Propagation Models Comparative Characteristic

More than a thousand measurements of RSSI have been conducted at sampling points in the experimental area.

A formula for determining the power loss in the transmission of the signal indoors must be used to determine the distance from the sampling point to the access point by the measured RSSI value. There are many models that take into account the propagation of radio waves through the partitions and ceilings. Of these, the following six models were selected through empirical studies for further analysis [13–16]:

- free space propagation model;
- linear attenuation model;
- one-slope model;
- multi-wall model;
- the model recommended by International Telecommunication Union (recommendation ITU-R P.1238–8);
- the model developed by Lucent Technologies Bell Laboratories company for IEEE 802.11 networks.

Figure 4 shows the results of signal power loss depending on the distance calculations using the above models of radio wave propagation and their comparison with RSSI values measured at these distances. The coordinates of APs, the transmitter output power, the transmitter and receiver antenna gains, the channel frequency and the power loss coefficients for different conditions were taken into account in the calculations.

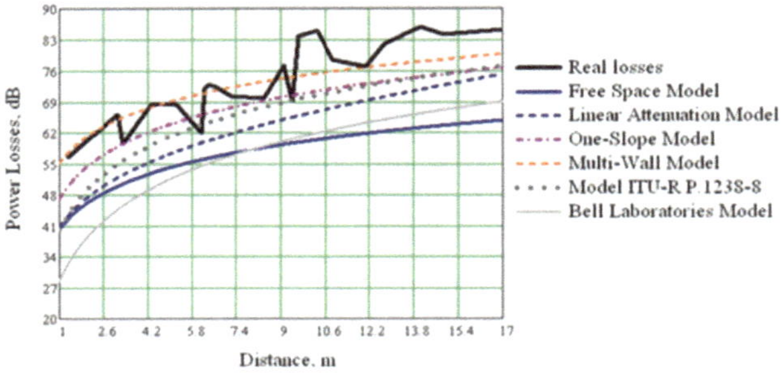

Fig. 4. The comparison of the calculated values with the measured power losses of the signal from one of test AP in the sampling points

Analyzing the dependencies shown in the figure, it can be concluded that the measured values curve corresponds to the one-slope model curve at the points separated from the access point by one partition, and coincides with the multi-wall model curve at the points separated from the access point by two or more partitions. Additionally, it comes close to ITU model calculation in some cases. It must be noted that at a large distance, about 12–17 m from the access point, the effect of multipath, interference and other obstructing factors is amplified, creating an additional loss of about 5 dB.

The most suitable models for the trilateration positioning system are one-slope model, multi-wall model and the model recommended by the International Telecommunication Union. Other models are excluded from further consideration.

Moreover, the analysis of the data obtained as a calculation and measurement result shows that it is necessary to apply not one selected model of radio-wave propagation inside the premises for positioning purposes, but a set of models, applying each of them depending on the room's configuration and the relative locations of the access point and the object.

3.2 Experimental Verification of Trilateration Positioning Method with Different Indoor Propagation Models

Experimental results have shown that the trilateration mechanism requires an especially careful placement of APs and accounting the material and thickness of walls, partitions, furniture, doors and other objects that can be in the path of signal propagation.

Let's consider several sampling points on different floors of the building and apply the trilateration method to the parameters measured by several devices (Fig. 5, 6, 7 and 8).

Example 1. The object being located is in fact at coordinates (10.6, 13.8, 5.0) on the second floor and receives signals from six APs with some of them located on the same floor and others one floor apart. The location will be determined via three APs located on the same floor with the object. The distance to each of them does not exceed 20 m and RSSI is not lower than –65 dBm. The location results are shown in Fig. 5.

Different symbols in the figure indicate the results of calculating the coordinates using three different signal propagation models. The blue circle around the sampling point defines the specified calculation accuracy of 5 m.

The figure shows that there are no permanent obstacles in the path when the signal from AP №2–4 passes to the object, the signal from AP №2–1 passes through two plasterboard walls and the signal from AP №2–5 passes through three similar walls.

It can be seen that the results of object's coordinates calculation by three models listed above fall into the zone of the sampling point outlined by the blue circle, which indicates sufficient accuracy for the project.

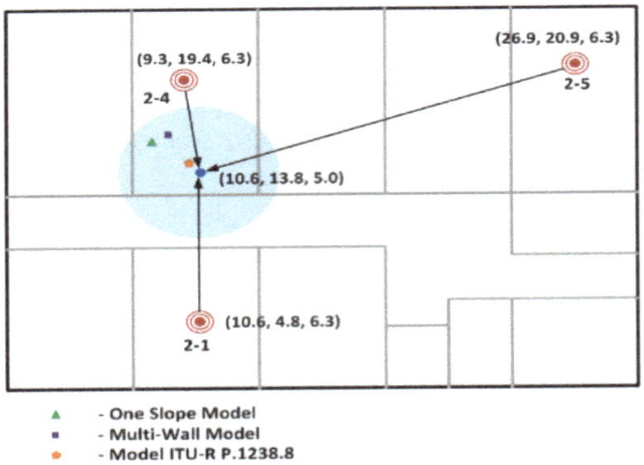

Fig. 5. Determining the object's location in the sampling point (10.6, 13.8, 5.0) when all access points are on the same floor as the object

Example 2. The target object is located at coordinates (8.6, 19.8, 5.0) on the 2nd floor. Here, a signal is received from the AP on the same floor separated by 5 walls from the target and signals from one AP on the floor below (no walls but one ceiling) and one on the floor above (one wall and one ceiling).

It can be seen in Fig. 6 that none of the methods gave acceptable results in this situation, because the signal level at the reception is too weak for accurate positioning due to the large distance from the object to APs and the quantity of partitions and ceilings.

Example 3. The sampling point has coordinates (4.4, 13.4, 8.6) and is located on the third floor. The wall placement between the receiving and transmitting antennas and the positioning results are shown in Fig. 7. The object receives signals from two APs located in the same horizontal plane and from one AP from the lower floor. The figure shows that the one-slope model does not work when calculating the characteristics of a system where the AP and the target are located on different floors, while ITU-R P.1238–8 and COST 231 models show acceptable results.

Example 4. The object is located at the sampling point with coordinates (9.6, 19.9, 12.2) on the fourth floor and receives a signal from three access points located in the same horizontal plane.

As seen in Fig. 8, the passing of the signal is obstructed by 0, 2 and 3 plasterboard partitions, respectively.

To summarize briefly, when the target object was placed in the same horizontal plane with the access points and received a sufficiently strong signal, almost all of the studied indoor radio wave propagation models gave a satisfactory result.

Thus, it can be concluded that in each specific situation it is necessary to select the suitable mechanism and methodology for determining the location of the object.

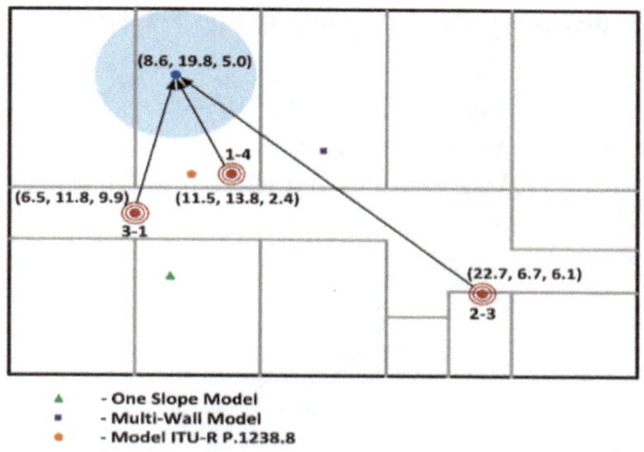

Fig. 6. Determining the object's location in the sampling point (8.6, 19.8, 5.0) when all APs are on different floors

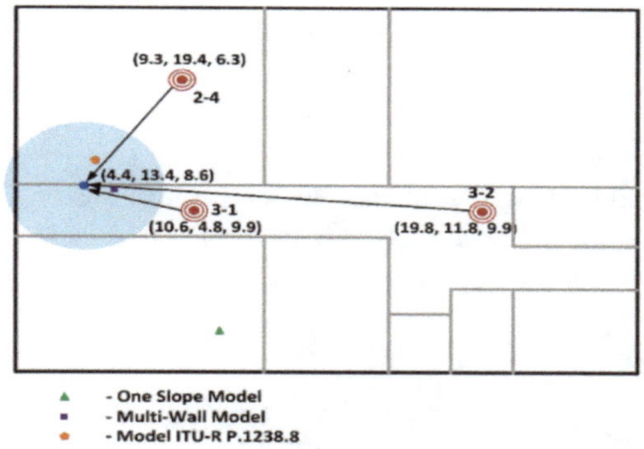

Fig. 7. Determining the object's location in the sampling point (4.4, 13.4, 8.6)

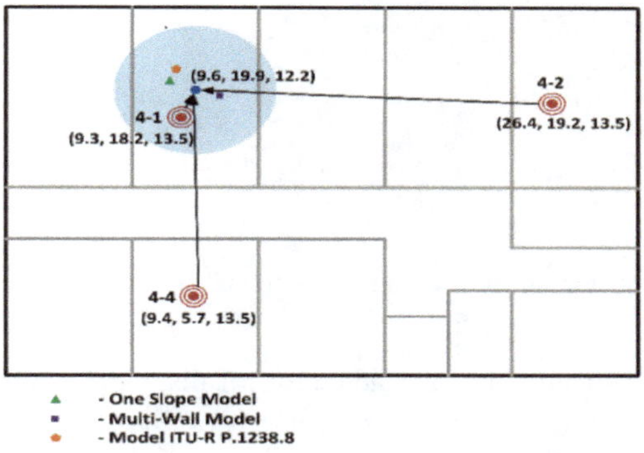

Fig. 8. Determining the object's location in the sampling point (9.6, 19.9, 12.2)

4 Results of Processing the Experimental Data

To assess the measurement error, the actual coordinates of the sampling points were compared with their coordinates obtained in the process of trilateration calculations. The results of the experimental data statistical processing are presented as histograms in Fig. 9, 10 and 11 for three radio wave propagation models.

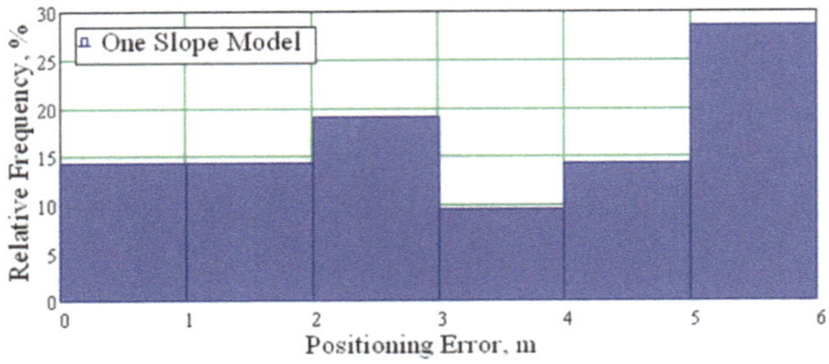

Fig. 9. The histogram of the one-slope model positioning error distribution

The one-slope model (Fig. 9) is suitable for positioning only in some cases, for example, when there is one wall between the transmitter and the receiver and the distance between them is small. In most cases, the measurement error exceeds the permissible value of 5 m.

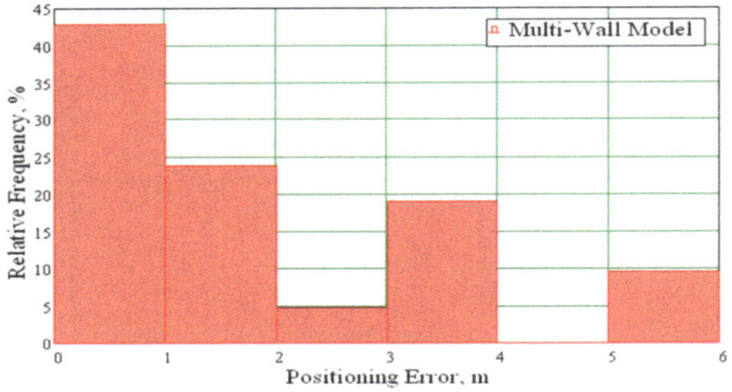

Fig. 10. The histogram of the multi-wall model positioning error distribution

Multi-wall model is suitable for calculation of distances from access point to the client in most cases, since the measurement error does not exceed 2 m in 70% of experiments, and exceeds the acceptable values in less than 10% of experiments.

The model recommended by International Telecommunication Union is the most questionable for positioning purposes, as it shows worse positioning error results than the multi-wall model, but also has a lower percentage of failed experiments (those with the error of more than 5 m) compared to it.

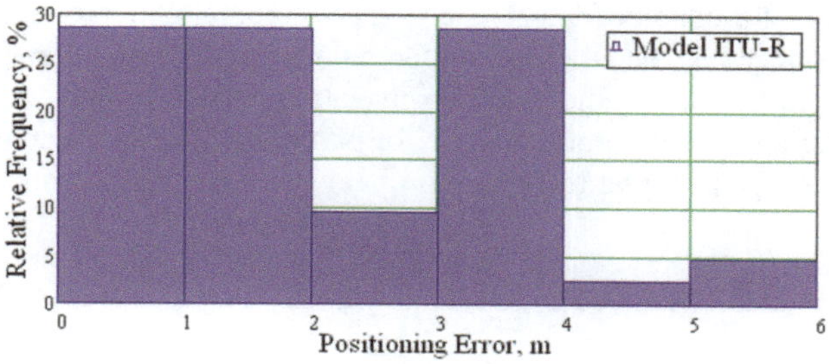

Fig. 11. The histogram of the ITU-R P.1238–8 model positioning error distribution

Therefore, comparing the experimental evaluations of the error in determining the object's location using three radio wave propagation models shows that, in most cases, the most suitable is multi-wall model, although the ITU-R P.1238–8 model cannot be completely discarded.

5 Conclusions

The results presented in this article can convincingly prove the necessity of a flexible approach to determining the distances between the receiving and transmitting antennas and further calculation of the target point's coordinates.

The data in Fig. 5, 6, 7, 8, 9, 10 and 11 show that it is virtually impossible to apply the same model for different signal propagation conditions on the path between the transmitting and receiving antennas.

For example, the one-slope model is sufficiently accurate to determine the distance when the signal passes through one wall, but is not suitable at all for the cases where the transmitter and receiver are located on different floors. In practice, there is rarely a situation where all access points are on the same floor with a subscriber and separated from him by exactly one wall, so the application of this model is limited to specific conditions.

The multi-wall model COST 231 covers situations where the signal goes through one or more partitions and/or through one or more floors – though unfortunately the test network did not allow to considering the cases where the signal goes through several floors. In case of line-of-sight between receiver and transmitter, it behaves like the free space radio wave propagation model and gives large deviations from the values measured in the sampling points. In other cases, the model is usually more accurate than the other models discussed here. Sometimes the model needs the specification of power loss coefficients when the signal passes through the partitions and obstacles made of different materials and at different angles.

The model recommended by International Telecommunication Union ITU-R P.1238–8 usually gives an acceptable result, except the situations when access points are located at a sufficiently large distance from the target object and separated from it by a large number of obstacles (the best result here is obtained using COST 231).

Hence, it can be concluded that a universal indoor radio propagation model to calculate the distance between the transmitter and receiver does not yet exist. Therefore it is necessary to select a radio wave propagation model basing on the particular situation - configuration of a room, quantity of partitions, obstacles and overlaps on the signal propagation path, signal level and, probably, some other aspects of the premise.

References

1. Bensky, A.: Wireless Positioning Technologies and Applications, 2nd edn. Artech House, Boston, London (2016)
2. Yang, G.: Principle of satellite navigation orbit and positioning. Int. J. Educ. Manage. Eng. **3** (2), 40–45 (2013)
3. Galileo. Open service. Service definition document [EB], (1.1) (2019). https://galileognss.eu/wp-content/uploads/2020/08/Galileo-OS-SDD_v1.1.pdf
4. Cisco: Cisco connected mobile experiences data sheet [EB] (2015). (in Russian). https://www.cisco.com/assets/global/RU/about/brochures/datasheet-c78-734648.pdf
5. Huawei technologies co. ltd. Huawei CloudCampus WLAN location technology white paper [EB] (2021). https://e.huawei.com/ru/material/networking/wlan/9dc74c8ef89f4bdab96382f6028ea78d
6. Ubitel: RLTS systems[EB]. https://ubitel.ru/services/rtls-systems/
7. Aruba: location services[EB]. https://www.arubanetworks.com/products/location-services/
8. Li, G., Geng, E., Ye, Z., Xu, Y., Yu, J.L.P.: Indoor positioning algorithm based on the improved RSSI distance model. Sensors **18**(9), 2820 (2018). https://doi.org/10.3390/s18092820
9. Bullmann, M., Fetzer, T., Ebner, F., Ebner, M., Deinzer, F., Grzegorzek, M.: Comparison of 2.4 GHz WiFi FTM- and RSSI-based indoor positioning methods in realistic scenarios. Sensors **20**(16), 4515 (2020). https://doi.org/10.3390/s20164515
10. Kokoreva, E., Kostyukovich, A., Doshchinsky, I., Shurygina, K.: A combined location method with indoor signal strength measurement. In: Proceedings of the 1st International Conference Problems of Informatics, Electronics and Radio Engineering PIERE-2020 in Novosibirsk. IEEE, pp. 281–286 (2020)
11. Kokoreva, E., Kostyukovich, A., Doshchinsky, I.: Analysis of the error in determining the location inside the logistics warehouse complexes. In: Popovic, Z., Manakov, A., Breskich, V. (eds.) VIII International Scientific Siberian Transport Forum: TransSiberia 2019, Volume 2, pp. 1086–1094. Springer International Publishing, Cham (2020). https://doi.org/10.1007/978-3-030-37919-3_106
12. Hosseinzadeh, S.: Multi wall (COST231) signal propagation [EB]. https://www.mathworks.com/MatLabcentral/fileexchange/61340-multi-wall-cost231-signal-propagation-models-python-code
13. Recommendation ITU-R P.525–4: Calculation of free-space attenuation. Electronic Publication, Geneva (2019)
14. Recommendation ITU-R P.1238–10: Propagation data and prediction methods for the planning of indoor radio communication systems and the radio local area networks in the frequency range 300 MHz to 450 GHz[S]. Electronic Publication, Geneva (2019)

15. Kumari, P., Prabha, I.S.: Signal Propagation analysis at 28 GHz and 73 GHz millimeter wave bands for next generation networks. Int. J. Image Graph. Signal Process. **11**, 19–27 (2019)
16. Kostyukovich, A.E., Doshchinsky, I.V.: Selection of a radio waves propagation model when determining the location. In: Proceedings of the Modern Telecommunications Problems. Novosibirsk: SibSUTIS, pp. 459–466 (2020). (in Russian)

Influence of the Structure of Coatings on the Nature of Deformation and the Type of the Indentation Diagram for Topocomposites on Ductile Substrates

Nikolay A. Voronin(✉)

Mechanical Engineering Research Institute of the RAS, 101990 Moscow, Russia
voroninn@imash.ru

Abstract. It is carried out a qualitative analysis of experimental indentation diagrams of topocomposites with coatings of titanium nitride and aluminum nitride, having substrates of aluminum alloy D16T. The reasons for obtaining the unloading curves of the indentation diagrams of the topocomposites under study – with a linear and nonlinear section of the end of the unloading curve characterizing the interfacial delamination along the coating-substrate interface are investigated. It is shown that the delamination mechanism in both topocomposites implements a linear dependence between the load and the depth of indentation. The high level of hysteresis losses of the aluminum nitride coating was established and explained. It is found that the nonlinear shape of the end of the unloading curves of the indentation diagrams of topocomposites with coatings of aluminum nitride is associated with solid-phase structural changes in the coating material. The connection between the diverse aspects of the material world, considered in this work on the example of a change in the structure of matter under the influence of mechanical forces, can serve as an example of the development of models for knowledge management.

Keywords: Instrumented indentation · Indentation diagram · Topocomposites · Ductility · Interfacial fracture · Hysteresis

1 Introduction

The instrumental indentation has now become one of the indispensable tools in scientific and engineering disciplines. In mechanical engineering, recent attention has been focused on contact sensing of functional surfaces of tribotechnical topocomposites [1–3]. Analysis both of the deformation and failure of coating-substrate systems is an important problem in the tribology of such systems. Tribotechnical topocomposites are a representative variant of layered systems. By definition, topocomposites are such surface-layered materials, the functional properties of which, in this case, the coefficient of friction and wear resistance (which in some cases directly correlates with the resource of the tribological object), depend not only on the contact geometry and

Z. Hu et al. (Eds.): AIMEE 2021, LNDECT 107, pp. 358–368, 2022.
https://doi.org/10.1007/978-3-030-92537-6_33

coating thickness, but also on the ratio of the mechanical characteristics of the components of the layered system (the ratio of hardness and elastic modulus) [3].

1.1 Literature Review

The results of the tests for pressing the diamond pyramid into the solid surface under study are presented in the form of indentation diagrams - the loading curve and the unloading curve. The nature of deformation, damage and destruction of the material in the surface volume of the test sample is shown on the loading and unloading curves in the form of specific local changes in the curves (angles of inclination, jumps, breaks, etc.) of the indentation diagram. Periodic displacements in the horizontal plane (pop-ins), marked on the loading curves and occurring during loading, are associated first with the appearance of radial and picture-frame cracks [4, 5], and then through-thickness cracks [1, 2, 5, 6]. The latter are explained by the tensile stresses in the coating, which are associated with the lowering of the edges of the print due to the significant plasticity of the substrate [4, 6, 7].

In several indentation diagrams, there is a noticeable typical linear or nonlinear section of the unloading curve with a sharply different slope, in relation to the slope of the unloading curve that is observed at the initial section of the unloading curve [1, 2, 7]. Such types of changes in the unloading curve indicate the presence of interfacial delamination between the coating and the substrate. Interfacial destruction occurs at the coating-substrate interface or near it on a material that is weaker in strength [4]. In experimental studies and numerical simulations of the processes of pressing spherical and pyramidal indentors into thin solid coatings, the effect of interfacial delamination is often noted, which is not accompanied by the destruction of the coating with the formation of surface cracks [8–10]. To quantify the adhesive strength of such coatings, an analysis of the modified part of the unloading curve associated with the elastic recovery of the delamination coating is used.

In the indentation diagrams obtained for solid coatings on elastic substrates with good adhesion of the coating to the substrate, the delamination of the coating from the substrate is shown as a clear linear section on the unloading curve [2, 6, 11]. The linear form of the unloading curve is associated with the elastic recovery of the exfoliated coating due to the loss of the coating bond with the substrate under the indenter tip. In this case, it is assumed that the coating remains attached at the edge of the contact and as the load is removed from the indenter, the coating straightens (unbends), like a flat elastic membrane. The quantitative assessment of the adhesive strength of the coating to the substrate in such topocomposites, carried out using the model of an elastic circular membrane with a fixed edge at the periphery and a loaded central force, is in good agreement with the results of the assessment of the adhesive strength by other experimental methods.

In the indentation diagrams obtained for hard coatings on ductile substrates, the delamination of the coating from the substrate appears as a highly nonlinear section on the unloading curve [9, 10]. This type of unloading curve characterizes the elastic interfacial delamination at the interface of the layered system not only under the top of the indenter, but also at a certain distance from the center of the load application, reaching up to 2–3 times the diameter of the contact print. The model of cohesive zone fracture is widely used to describe the origin and propagation of cracks in continuum mechanics [12].

Thus, today there are two types of indentation diagrams that differ greatly in the form of unloading curves, which are processed for the purpose of evaluating the interfacial strength using two models: the model of a round thin elastic membrane with a pinched edge and the model of cohesive zone failure. For diagrams of the introduction of layered systems with a linear section of the unloading curve, there is a good correlation between the understanding of the mechanism of interfacial stratification and the type of unloading curve. The model of a thin elastic membrane assumes a linear relationship between the load and the penetration depth, which is confirmed by the shape of the unloading curve of the experimental indentation diagram. For indentation diagrams with a steep nonlinear section of the unloading curve, there is no understanding of the relationship between the mechanism of interfacial delamination and the type of unloading curve. The nature and source of additional elastic deformation in the mechanism of coating delamination is not clear, which a priori represents the elastic recovery of the exfoliated coating, which implies a linear dependence of the penetration depth on the load.

In this work, is made an attempt to describe the mechanisms of deformation of topocomposites during instrumental indentation, which lead to the appearance of a nonlinear section of the unloading curve of the penetration indentation diagram. It is necessary to determine the contribution of the coating structure to the process of deformation of layered systems.

The solution is proposed to be sought by phenomenological analysis of the experimental indentation diagrams for topocomposites of two variants: the first one, which gives a unloading curve with a linear section, and the second one, with a nonlinear section, using known information about the behavior of materials of topocomposite components under deformation conditions, including a pyramidal indenter.

2 Materials and Methods

In this paper, experimental studies were carried out on two model samples of topocomposites, for which the indentation diagrams were obtained by instrumental indentation in their surface with a diamond pyramid. Both model samples were made of D16T aluminum alloy. A titanium nitride (TiN) coating was applied to the flat surface of one sample, and an aluminum nitride (AlN) coating was applied to the surface of the second sample. The technology for obtaining coatings is magnetron sputtering at the MIR-2 installation by sputtering pure metal (aluminum or titanium) in a mixture of gases – argon and nitrogen. The thickness of the coatings was $\sim$5 microns. The titanium nitride coating had a dense, microfine columnar structure with crystallites of 80–200 nm in size. Electronographic analysis revealed that the coating consists of titanium nitride with a FCC lattice and corresponds to the stoichiometric composition. The aluminum nitride coating had a wurtzite-type crystal lattice of hexagonal syngony and also corresponded to the stoichiometric composition. The coating had a fine columnar structure and consisted of crystallites with a size of 100–150 nm. Electronographic studies have shown the presence of a texture in the coating with the orientation of the crystallites with a plane (0001) parallel to the surface of the substrate.

The measurement of the hardness (H) and elastic modulus (E) of the topocomposite components was carried out based on the results of the analysis of the indentation diagrams obtained on the NanoScan4D nanohardometer. The three-sided diamond pyramid of Berkovich served as an indenter. From 3 to 5 indentation diagrams were used for the analysis. The type of indents was studied using a 3D optical microscope Sneox (SENSOFAR). The substrate had nanohardness 0.98 GPa and elastic modulus 93 GPa. The titanium nitride (TiN) coating had a nanohardness of 18 GPa and an elastic modulus of 210 GPa. The aluminum nitride (AlN) coating had a nanohardness of 36 GPa and an elastic modulus of 320 GPa. Experimental studies of topocomposites included single indentation tests and repeated double loading and unloading cycles.

3 Results and Discussion

3.1 Analysis of Experimental Indentation Diagrams

Typical experimental indentation diagrams for the two topocomposites under study are shown in Fig. 1. For the convenience of further comparison of the obtained diagrams, the indenter penetration depths were chosen to be close in size.

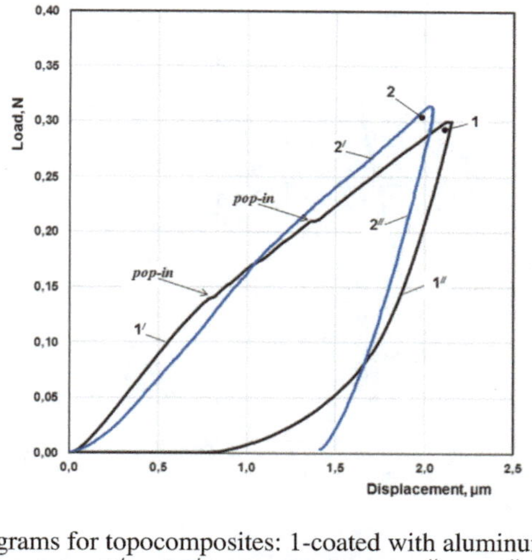

Fig. 1. Indentation diagrams for topocomposites: 1-coated with aluminum nitride, 2-coated with titanium nitride. Loading curves-1′ and 2′; unloading curves-1″ and 2″

Comparing the areas of the indentation diagrams (1 and 2, Fig. 1) it can be assumed that the work spent on the elastic-plastic deformation of the studied topocomposites is approximately the same. There is a slight difference in the loading curves, although the difference in the values of the hardness and modulus of elasticity of the coating materials is significantly different. The aluminum nitride coating material is 1.5 times more elastic and 2.0 times more rigid than the titanium nitride coating. However, according to the E/H parameters of the coatings, compared with the E/H parameter of the base material and characterizing the size of the plastic zone in the substrate, the coating materials have fairly close values. It can be expected that the elastic-plastic deformation of the selected layered systems will also be close, which allows us to hope for a reduction in the number of parameters significant for the deformation process. A much larger difference is observed both in the areas under the unloading curves (curves $1^{//}$ and $2^{//}$ in Fig. 1), and in the nature of these curves. It is known that the unloading curves in typical indentation diagrams of topocomposites and homogeneous solids describe the elastic deformation of a solid or layered body when the load is removed. For a topocomposite coated with titanium nitride, the shape of the unloading curve corresponds to a typical curve of elastic deformation of a layered body. And the view of the unloading curve is identical for a number of experimental indentation diagrams with the same coverage, but at lower final indentation loads (Fig. 2, a). For a topocomposite coated with aluminum nitride in a wide range of final indentation loads (Fig. 2, b) the unloading curve exhibits a strong non-linearity of the unloading curve with an increase in the value of the final load of the introduction. The observed nonlinearity begins with loads in the region of 0.15 N and increases with the increase in the value of the final load of the introduction. The strong curvature of the unloading curve in the lower part of the unloading curve for a top-composite coated with aluminum nitride is explained by the presence of damage to the coating in the form of [2, 12].

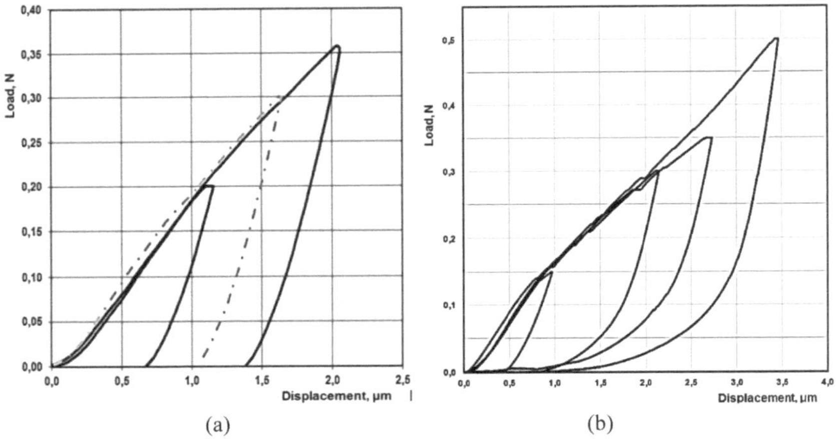

Fig. 2. Indentation diagrams for a topocomposite with a) titanium nitride coating and b) aluminum nitride coating at different values of the final load on the indenter

Returning to the analysis of the combined diagrams of the introduction of the studied topocomposites (see Fig. 1) it can be noted that if the difference in the angles of inclination of the unloading curves in the initial part of the unloading curves of the studied topocomposites can be explained by the difference in the elastic modulus of the coatings, then the very different curvature of the lower part of the unloading curves is associated with the presence of interfacial delamination in the system of aluminum nitride coating – D16T alloy substrate, which is probably due to the lower adhesive strength of this topocomposite from the topocomposite coated with titanium nitride.

A more careful analysis of the loading curves of the studied topocomposites shows the presence of the following characteristic features of the curves of the indentation diagrams. The feature is the presence of "irregularities" of individual sections of the loading curve of the topocomposite coated with aluminum nitride. Short-term periodic sharp changes in the depth of penetration under constant load, called "pop-ins" in the literature, correspond to various mechanisms – the formation of cracks on the surface, dislocation movement, and sliding along the grain boundary. As a rule, when the jumps have a certain, well-observed length, which is often emphasized in the name of such elements of the curves – "sliding pop-ins", this implies the presence of destruction of the coatings, in the form of transverse cracks (ring-like and picture-frame). For layered systems with substrates made of pliable materials, visually detectable pop-ins are most often associated with plastic flow at the coating – base interface or sliding along the boundaries of textured grains [1, 6]. It is precisely these pop-ins that we observe in the diagrams of the introduction of topocomposites coated with aluminum nitride (see Fig. 1 and Fig. 2, b). For topocomposites coated with titanium nitride, no such areas are observed on the loading curve.

Taking into account that in the studied topocomposites the base material is the same, the thickness of the coating and the penetration diagrams for the areas under the loading curves are approximately the same, we will try to find out the reason for the difference in the results of deformation and damage to these layered systems, as well as the difference in the form of the unloading curves of the indentation diagrams.

From the comparison of the results of the analysis of the two indentation diagrams for the studied topocomposites, it can be assumed that in the case of using titanium nitride as a coating at almost the same indentation depths, the presence of the same substrate material and almost the same energy consumption for elastic-plastic deformation, the only reason for the difference in the types of unloading curves is a more significant level of adhesive bond in this topocomposite compared to a topocomposite coated with aluminum nitride. Is it true? Let's try to take into account the degree of adhesive interaction in the topocomposite coated with titanium nitride in the comparative analysis of the indentation diagrams of the studied topocomposites.

3.2 Analysis of Indentation Diagrams with Repeated Loading and Unloading

We will bring the topocomposite coated with titanium nitride to a state where there will be a situation with a violation of the interfacial connection at the interface. Exceeding the final load value may lead to transverse cracks in the coating already during loading and will not allow for an objective comparison of the indentation diagrams with the manifestation of coating delamination. Therefore, we will conduct experiments with the studied topocomposites, without going beyond the accepted loads by conducting repeated loading and unloading cycles. To do this, in the first loading cycle with unloading, we will perform incomplete unloading (up to a load value of about 3% of the maximum loading value) and again perform two repeated deformation cycles with unloading.

Figure 3 shows the indentation diagrams with repeated loads for the studied TiN-coated topocomposite. Repeated loading of the topocomposite leads to delamination (see Fig. 3, insert, loading curve $2^{/}$, unloading curve $2^{//}$). The diagram demonstrates the well-known in the literature [2, 4, 7] type of the indentation diagram of a layered system with delamination, with a linear view of the curve at the end of unloading. To quantify the parameters of delamination in such layered systems, a model of an elastic circular membrane is used, pinched at the edge and loaded in the center on the axis of symmetry by a force. Such loading is characterized by a linear relationship between the load and the deflection.

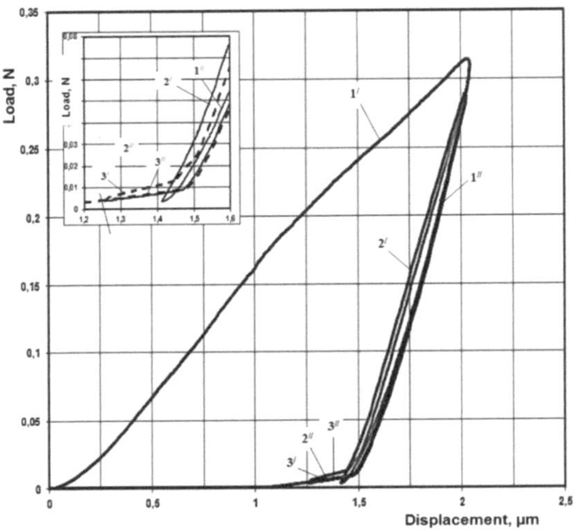

Fig. 3. Repeated indentation diagrams for a topocomposite coated with titanium nitride: $1^{/}$, $2^{/}$, $3^{/}$ – loading curves, $1^{//}$, $2^{//}$, $3^{//}$ – unloading curves for the first, second, and third deformation cycles

A similar linear dependence is observed in the topocomposite with a titanium nitride coating studied in this work also in the third cycle of deformation (see Fig. 3, curve $3^{//}$). The unloading curves of the second and third deformation cycles in most of their curves practically coincide with the unloading curve of the first deformation cycle (see Fig. 3, curves $2^{//}$, $3^{//}$ and $1^{//}$). The loading curves of the second and third deformation cycles, realizing the elastic deformation of the layered system, are located at an insignificant distance from the unloading curve of the first indentation cycle.

This behavior of the topocomposite coated with titanium nitride indicates a good preservation of the structure and composition of the components of the topocomposite (we can only talk about a slight plastic deformation under repeated loads) and, first of all, the coating material, when changing the direction of deformation. An additional confirmation of the conclusion about the good stability of the structure of the topocomposite in the process of deformation is also the behavior of the exfoliated coating, which can be judged by the close coincidence of the linear sections of the curves of unloading and restoration (straightening) of the coating as an elastic membrane.

For a topocomposite coated with aluminum nitride (see Fig. 4), the indentation diagram with repeated deformation cycles differs from the indentation diagram for a topocomposite coated with titanium nitride, both on the first cycle and on subsequent ones. The second deformation cycle shows the loading curve far away from the unloading curve of the first deformation cycle. However, the unloading curve of the second deformation cycle coincides along the entire length with the unloading curve of the first deformation cycle. A characteristic feature of these hysteresis loops is the significant width of the loops, compared to the hysteresis loops of topocomposite coated with titanium nitride. The last, third, deformation cycle in this experiment was brought to an end, that is, to complete unloading.

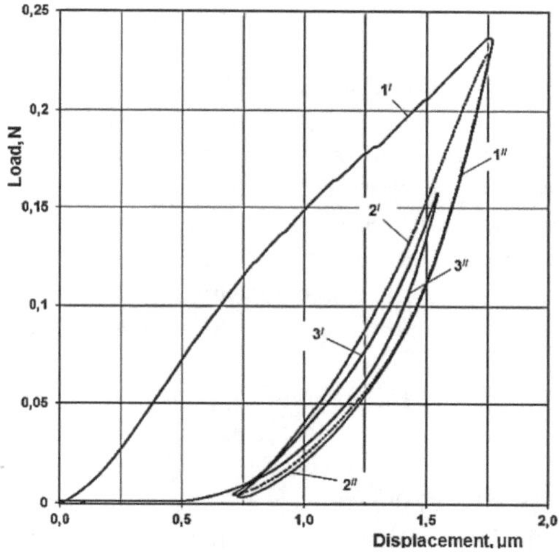

Fig. 4. Repeated indentation diagrams for a topocomposite coated with aluminum nitride: 1′, 2′, 3′ – loading curves and 1″, 2″, 3″ – unloading curves for the first, second, and third deformation cycles

The appearance of the hysteresis loops, their change during subsequent repeated deformation cycles, and the width of the loops in the topocomposite coated with aluminum nitride indicate significant structural rearrangements in the topocomposite material during each indentation cycle. The width of the hysteresis loops indicates the amount of energy spent on the structural adjustment of the topocomposite material. A comparison of the indentation diagrams of the studied topocomposites with repeated deformation cycles shows that the structural adjustment does not occur in the base material of the topocomposites, since the same material, aluminum alloy, is used as the base in the studied topocomposites, and hysteresis losses in the topocomposite coated with titanium nitride are practically absent (compare the width of the hysteresis loops in Fig. 3 and Fig. 4). Therefore, it can be concluded that the energy loss on the hysteresis in the topocomposite coated with aluminum nitride is due to structural or phase changes in the coating material.

3.3 Summary

An analysis of the literature [13–15] on the structure formation in covalent crystals (diamond and diamond-like phases BN, SiC, AlN) at the stage of plastic deformation suggests that the aluminum nitride coating in the topocomposite studied in this work underwent structural changes during the indentation period. In [13], the phase transformations during nanoindentation into the (0001) surface of monocrystalline AlN with a wurtzite structure were investigated using molecular dynamics simulation. With the indentation depth increasing, the following effects are observed consecutively: surface rupture; nucleation and growth of a graphite-like phase; the formation and flow of the amorphous phase; generation of a tetragonal intermediate structure. It was found that the amorphization of AlN leads to load drops during nanoindentation and the appearance of a noticeable nonlinear section on the unloading curve. The paper [14] summarizes the results of studies of the evolution of structural transformations under high-pressure deformation of compact materials based on a group of covalent substances: diamond, silicon carbide, aluminum nitrides, and boron with diamond-like wurtzite-type lattices. To date, it has been established that in the crystals of ceramic materials, sliding is realized at room temperature under various loading conditions. This is typical for substances with both the ionic type of chemical bond and the ionic-covalent and covalent types. Typical structural transformations of ceramic materials are plastic fragmentation of crystals due to the appearance of intracrystalline interfaces and slippage (displacement) without breaking the continuity along the intragrain boundaries of the section.

From the results of the work [15] on the comparison of the surface relief inside the Vickers pyramid print of films with the hexagonal and cubic surface structures, it follows that for both structures the mechanism of plastic deformation is realized by sliding the columnar elements of the structure parallel to the applied load. The presence and magnitude of plastic deformation in the coatings of the studied topocomposites can be judged by the type and size of the areas of the hysteresis loops. The hysteresis and width of the hysteresis loops revealed in repeated deformation diagrams of topocomposites coated with aluminum nitride (see Fig. 4) indicate polymorphic solid-phase structural transformations of the coating material.

The titanium nitride coatings in the topocomposite under study are a columnar structure of crystallites of equiaxed cubic syngony shape. During indentation, these crystallites and the grain structure of titanium nitride also undergo solid-phase structural transformations in the form of sliding along grain boundaries. But this deformation takes place only in the first indentation cycle, locally, in the coating area directly adjacent to the indenter, and during subsequent loading cycles, plastic deformation practically does not occur. This is confirmed by the absence of pop-ins, kinks or discontinuities on the loading curve of the indentation diagram (see Fig. 2, a) and narrow-width hysteresis loops (see Fig. 3).

Significant energy expenses for solid-phase structural transformations in the aluminum nitride coating, those were spent at the stage of introducing the indenter into the topocomposite, create a significant reserve of elastic energy in the coating (in the form of compression stresses), which is realized during the unloading process. These stresses, combined with the tensile stresses from the bending of the coating when the indenter is embedded in the topocomposite, lead to the appearance of a unloading curve with a highly nonlinear section at the bottom of the unloading curve of the indentation diagram. The area of the figure under the nonlinear part of the unloading curve, which characterizes the energy realized for the dissipation of energy from stresses in the coating, will also include the energy spent on elastic recovery of the delamination coating. The mechanism that implements the delamination of the coating is similar to the mechanism for restoring a round elastic diaphragm with an edge fixed at the periphery and curved under the action of a centrally located load.

Returning to the task set in this paper, we can note the following. The energy costs for the intergranular sliding of the fiber structure of the coating, the tangential elastic-plastic sliding of the material at the coating-substrate interface, the reduction of the adhesive bond at the interface outside the contact area of the indenter with the coating, the interfacial destruction of the contact and the elastic recovery of the exfoliated coating form a unloading curve with a significant deviation from the linear form of the unloading curve, that is typical for the case of energy costs only for overcoming the adhesive interaction forces between the coating and the substrate under the indenter. However, to assess the adhesive strength of layered systems with a pliable substrate, according to the author, the elastic thin diaphragm model is quite applicable, if we take into account the difference in work spent on elastoplastic deformation of a layered system and relaxation of internal stresses during unloading, as shown in the works [16, 17].

4 Conclusion

It is carried out a phenomenological analysis of experimental indentation diagrams of topocomposites with coatings of titanium nitride and aluminum nitride, having substrates of the same material – aluminum alloy D16T. A different type of unloading curves of the indentation diagrams of the studied topocomposites is established - with a linear and nonlinear section of the end of the unloading curve, which characterize the interfacial delamination along the coating-substrate interface. It is shown that:

1) The non-linear end of the unloading curve is due to solid-phase structural changes in the coating of aluminum nitride during instrumental indentation.
2) The delamination mechanism in both topocomposites implements a linear relationship between the load and the depth of penetration.
3) The level of solid-phase structural changes in the coatings can be estimated from the value of the hysteresis losses (the width of the hysteresis loops).
4) The level of hysteresis losses of the coating material is an essential parameter affecting the nature of deformation and the shape of the unloading curve of the indentation diagram and should be taken into account when analyzing the performance of layered systems.
5) It is possible to estimate the adhesive strength of hard coatings of an aluminum nitride coating to a substrate made of a plastic material using the energy approach and the elastic recovery model of a thin elastic circular membrane.

Acknowledgment. I thank the researcher Kravchuk K. S. for obtaining experimental implementation diagrams and images of indents.

References

1. Bhushan, B.: Depth-sensing nanoindentation measurement techniques and applications. Microsyst. Technol. **23**, 1595–1649 (2017)
2. Chen, J.: Indentation-based methods to assess fracture toughness for thin coatings. J. Phys. D Appl. Phys. **45**(203001), 1–14 (2012)
3. Voronin, N.A.: Actual problems of creation of topocomposite for tribotechnical using. Izvestiy Samarskogo Nauchnogo Tsentra Ros. Akad. Nauk. **13**, 4(3), 695–698, (2011) (in Russian)
4. Chen, J., Bull, S.J.: Indentation fracture and toughness assessment for thin optical coatings on glass. J. Phys. D Appl. Phys. **40**(18), 5401–5417 (2007)
5. Fu, K.: Analysis on cracking in hard thin films on a soft substrate under Berkovich indentation. Vacuum **112**, 29–32 (2015)
6. Fu, K., et al.: Analysis on multiple ring-like cracks in thin amorphous carbon film on soft substrate under nanoindentation. J. Phys. D Appl. Phys. **46**(505314), 1–10 (2013)
7. Fu, K., Tang, Y., Chang, L.: Toughness assessment and fracture mechanism of brittle thin films under nano-indentation. In: Fracture Mechanics - Properties, Patterns and Behaviour (2016). http://www.intechopen.com/books/fracture-mechanicsproperties-patterns-and-behaviours. Chapter 6
8. Chen, J., et al.: Experimental and modelling techniques for assessing the adhesion of very thin coatings on glass. J. Phys. D Appl. Phys. **42**(214003), 1–11 (2009)
9. Lu, M., Xie, H., Huang, H.: Characterization of interfacial adhesion of thin film/substrate systems using indentation-induced delamination: a focused review. Key Eng. Mater. **533**, 201–222 (2013)
10. Abdul-Baqi, A., Van der Giessen, E.: Delamination of a strong film from a ductile substrate during indentation unloading. J. Mater. Res. **16**(5), 1396–1407 (2001)
11. Lu, M., Huang, H.: Interfacial energy release rates of SiN/GaAs film/substrate systems determined using a cyclic loading dual-indentation method. Thin Solid Films **589**, 822–830 (2015)
12. Ruiz, G., Pandolfi, A., Ortiz, M.: Three-dimensional cohesive modeling of dynamic mixed-mode fracture. Int. J. Numer. Meth. Eng. **52**(12), 97–120 (2001)
13. Luo, X., Zhang, Z., et al.: Atomistic simulation of amorphization during AlN nanoindentation. Ceram. Int. **47**(11), 15968–15978 (2021)
14. Oleinyk, G.S.: Structural mechanisms of plastic deformation of ceramics. Electron Microsc. Strength Mater. **20**, 3–30 (2014). (in Russian)
15. Shtanskii, D.V., Kulinich, S.L., Levashov, E.L., Moore, J.J.: Structure and physical-mechanical properties of nanostructured thin films. Solid State Phys. **45**(6), 1177–1184 (2003). (in Russian)
16. Voronin, N.A.: Effect of substrate material compliance on the character of topocomposite damage under instrumental indentation. J. Mach. Manuf. Reliab. **49**(10), 862–869 (2020)
17. Zhao, J., Zhang, L.: Structural damage identification based on the modal data change. Int. J. Eng. Manuf. **4**, 59–66 (2012)

Review of Technical Solutions for Parallel Mechanisms with Flexible Links (Cable-Driven Robots (CDR))

Vyacheslav A. Borisov(✉), Vladimir S. Ramzhaev, Sergey V. Levin, and Natalia L. Kovaleva

Mechanical Engineering Research Institute of the Russian Academy of Sciences (IMASH RAN), Moscow 101000, Russia

Abstract. Parallel link mechanisms with flexible links (cable-driven robots) offer the following advantages: low link inertia, low noise, mobility, low material consumption and others. Cable-driven robots are finding more and more applications, examples of which are described in this article. After reviewing the examples, the possible classification criteria will be outlined and a classification for three- and four-coordinate flexible link manipulators will be developed. The S symbol indicates that the basic structural scheme is a mechanism that performs Schoen flies motion and the number of attachment points n on the output link and satisfy the following relations on the base i: $3 \leq n \leq 4$ and $3 \leq i \leq 4$, and the number of arrangement layouts of axes of the input links m and the combination of flexible links satisfy the following relations: $1 \leq m \leq 3$ and $1 \leq j \leq 3$ respectively. The article concludes with a summary of the work results and an outline of future research.

Keywords: Cable-driven robots · Tensioning technology · Motion simulation · Control technology · Development trend

1 Instruction

It is generally accepted that the concept of a parallel structure with flexible links first emerged in the USA in 1989 and was called Cable controlled crane or Robot crane, which were later developed into the Cable Robot. Today, the solutions based on the Cable Robot system are used in various applications.

One of the greatest advantages of the flexible links is the ability to mount the drive at any distance from the output link of the mechanism without any restrictions. This fact allows to mechanisms for solving non-trivial technical challenges as well as greatly reduces the weight-dimensional, static and dynamic loads in the developed solution. The disadvantage of this type of mechanism is the limited scope of application.

Z. Hu et al. (Eds.): AIMEE 2021, LNDECT 107, pp. 369–378, 2022.
https://doi.org/10.1007/978-3-030-92537-6_34

2 Review of Cable-Driven Robots (Literature Review)

The specific response of flexible links to the external load can be used to store mechanical energy. Such a solution has been implemented by NASA engineers in the Super Ball Bot (Fig. 1). A robot called Super Ball can be dropped from high altitude onto the surface of any planet, planet satellite or other space bodies. The potential energy of the drop height will be used by this robot to move like a ball (bounce without stopping). Flexible links in this structure act as a suspension or transmission and partly as a mechanical energy accumulator. The devices for collecting data about the planet's surface are mounted on the robot's rods. NASA is planning to use these robots for exploration missions to the satellites of Saturn and Titan [1].

Fig. 1. Pilot samples of the Super Ball Bot

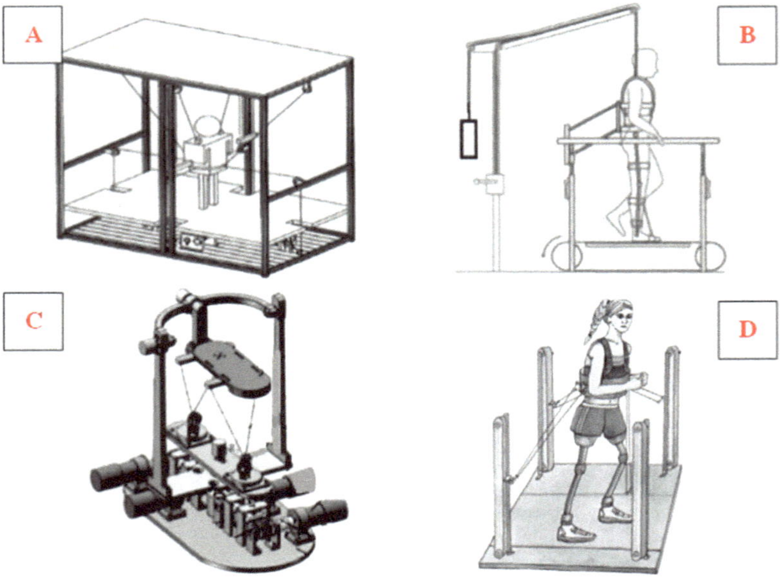

Fig. 2. Manipulators with flexible links for rehabilitation: a, d) hipjoint; b) multifunctional with feedback; c) ankle joint

In medical technology, the solutions have been proposed for the application of parallel mechanisms with flexible links in gravity balance devices (Fig. 2, a) for rehabilitation of the hip joint by means of six flexible links that ensure four degrees of freedom allowing successful rehabilitation of patients when combined with changes in motion parameters [2]. A rehabilitation robot designed in Germany is equipped with a feedback system and has two types of flexible links (Fig. 2, b): the first type flexible links are connected with the patient torso to control the rehabilitation, while the second type links monitor the results (feedback) [3]. The robot can be used not only for rehabilitation purposes, but also in wellness principles. The robot can be used for special training, e.g. to reduce overweight of patients who are prone to corpulence. In the next work, a cable-driven ankle joint structure was proposed for rehabilitation purposes [4]. This solution (Fig. 2, c) provides a more precise motion control that is achieved due to a special design of the manipulator. The rehabilitation task can also be solved on the basis of a mechanism with flexible links implementing Schoenflies motion with using the patient torso as an output link (Fig. 2, d) [5, 6].

With the increasing number of technical solutions based on the use of manipulation mechanisms with flexible links, the need for special design tools for such solutions is also growing [7]. In 2018, it was reported that a computer-aided system for the development of cable-driven robots has been developed [8]. The new system is called CARDSFlow (Fig. 3). Autodesk Fusion 360 is usedas a CAD solution, Gazebo Sim is used for simulation, the analysis and calculation of flexible links (cables) is performed in the CASPR software environment. This system was jointly developed by a team of the University of Hong Kong and the Roboy team. An important feature of this project is the open source code that allows the software to be upgraded and improved after the purchase of a license.

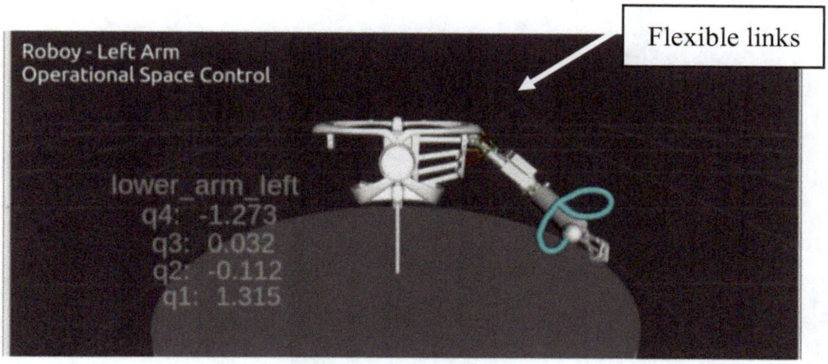

Fig. 3. Example of left hand motion simulation based on CARDSFlow system

3 Theory and Method: Creation Classification of CDR

Given the above, it can be concluded that the mechanisms with flexible links have three key application areas: positioning of objects in a limited space (cranes, spreaders, motion simulators), positioning of objects over long distances (logistics robots, research robots), and hybrid solutions (Fig. 4, a), in which a robotic arm can be installed e.g. on the platform driven by flexible links. Assemblies and mechanisms of the parallel manipulator with flexible links can be both cable and belt-driven. A combination of belts and cables is also possible. The belt can have pulleys of different diameters (Fig. 4, b), pulleys of same diameter and pulleys located on different axes (Fig. 4, c) [9]. Cable mechanisms can be classified into those where one end of the cable is attached to another link e.g. an output link, and those where a cable with "endless" length is used (Fig. 4, d) [9]. Some belt-driven solutions can also have different types of connection with other links of the mechanism, e.g. well-known Hain mechanisms and similar ones can be implemented [10].

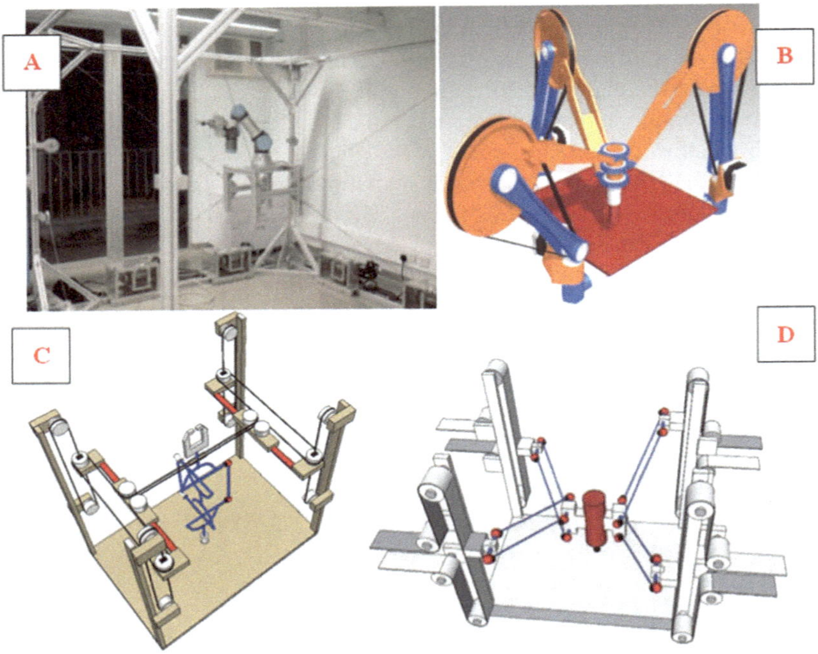

Fig. 4. Example of a structural solution for mechanisms with flexible links

Both Figs. 5 and 6 show the schemes of the selected groups of parallel mechanisms with flexible links.

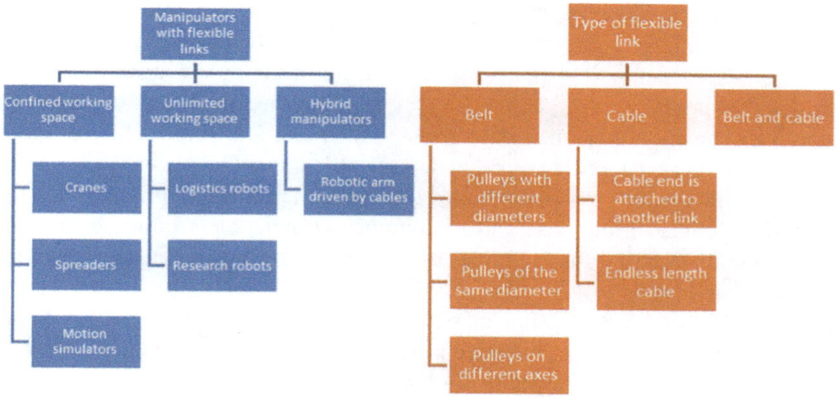

Fig. 5. Groups of manipulators with flexible links classified by application and type of link

The first basic stage in the study of parallel manipulator mechanisms is their classification. The analysis of motions, which are necessary for the intended use of the technical solution under study, shows that the most popular are mechanisms with four degrees of freedom that can be supplemented by a robotic chassis or by using the mechanism as an attachment, which is already implemented in studies of the tripod for agricultural machinery [11].

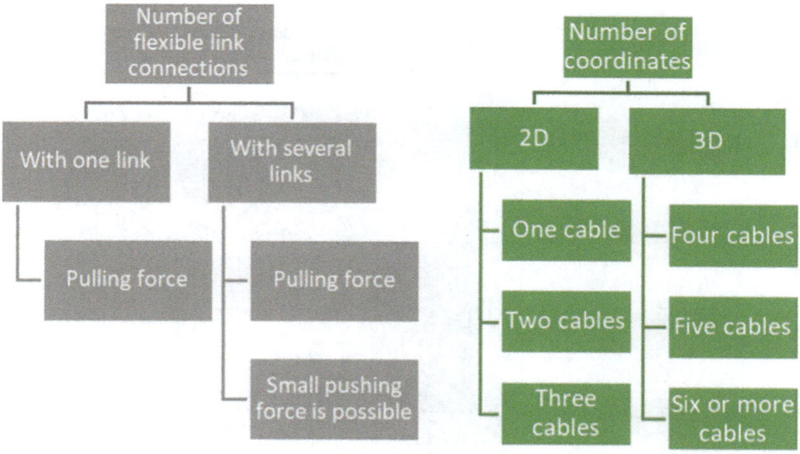

Fig. 6. Groups of manipulators with flexible links classified by the number of non-flexible links adjacent to the flexible link and by the number of coordinates

In the next work, classifications and calculation methods, in which the criteria of classification are the number of attachment points at the base of the output link and various variants their relations are considered, have been proposed [12–18]. In the most general form there are L(n, m) possible versions of structural schemes of mechanisms. The L symbol indicates that there are L-coordinate structural schemes, where the number of attachment points n on the output link and on the base m satisfy the following relations: $3 \leq n \leq 6$ and $3 \leq m \leq 6$ and the number of drive kinematic chains is six. The classification is based on a combinatorial method where all possible combinations of parameters are considered. Basically, the method is a simple numerical model. The following two parameters are selected as classification criteria: the number of attachment points on the base and the number of attachment points on the output link. The combination of different values of these parameters will determine the type of mechanism.

Let's see Table 1, where the first classification criteria is the combination of number of attachment points on the base and the guide arrangement layout, while the second is the combination of attachment points on the output link and the type of flexible link. Different versions of these combinations will be considered. Let's present possible versions of structural schemes of mechanisms in the most general form S(n, m, i, j). The S symbol indicates that the basic structural scheme is a mechanism that performs Schoen flies motion and the number of attachment points n on the output link and satisfy the following relations on the base i: $3 \leq n \leq 4$ and $3 \leq i \leq 4$, and the number of arrangement layouts of axes of the input links m and the combination of flexible links satisfy the following relations: $1 \leq m \leq 3$ and $1 \leq j \leq 3$ respectively.

Table 1. Criteria of mechanism classification

Parameters		Number of attachment points on the base and arrangement of the input link axes			
		3	4		
Number of attachment points on the input link and type of flexible link	3	3130			
		3131			
		3132			
		3133			
	4		4140	4240	4340
			4141	4241	4341
			4142	4242	4342
			4143	4243	4343
			4144	4244	4344

4 An Example of Using Classification

Let's consider all the rows in Table 1, in which the structural groups of mechanisms are categorized. The first row contains a group of structural schemes of mechanisms S (3130); S(3131); S(3132); S(3133). Structural schemes in this group have three points on the base (the first digit of the code), the arrangement layout of parts with perpendicular and vertically arranged guides on the base (the second digit of the code is scheme 1), the number of attachment points on the input link is three (the third digit of the code), the type of flexible link (0 — only belts, 1 — one cable and two belts, 2 — two cables and one belt, 3 — three cables and zero belts). Based on the proposed classification, a totally four main versions of structures can be identified, depending on the number of attachment points on the base and the output link, the arrangement of guides and the number of cables and belts used for flexible links (Table 2).

Table 2. Basic structure versions for a manipulator mechanism with four degrees of freedom and flexible links

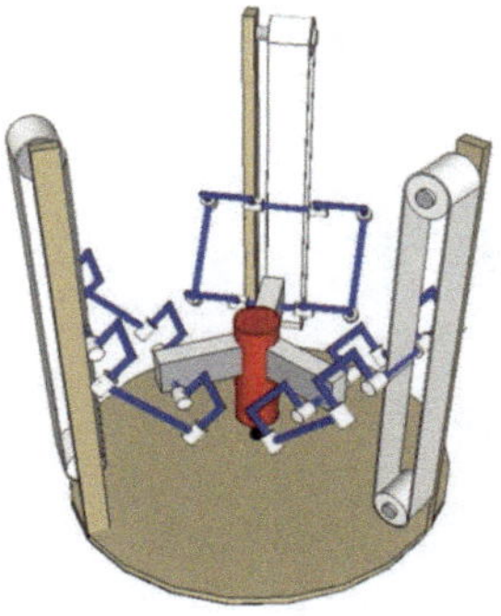 3130	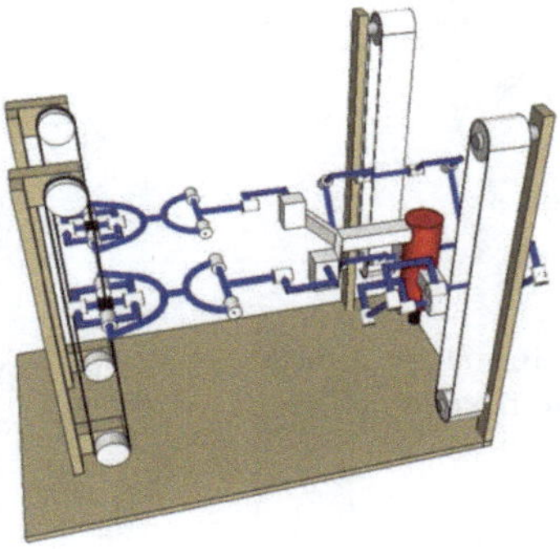 4142
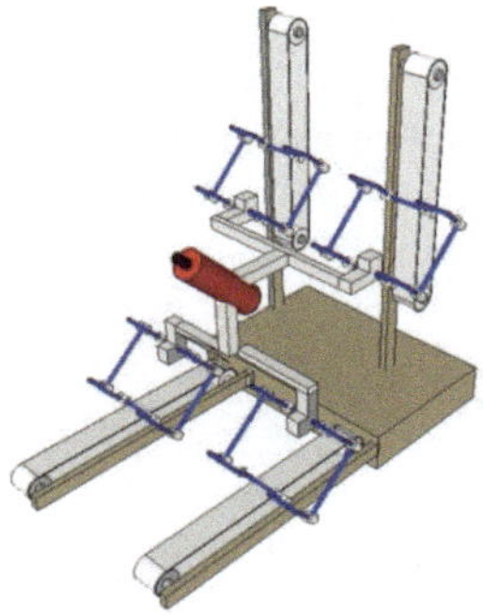 4240	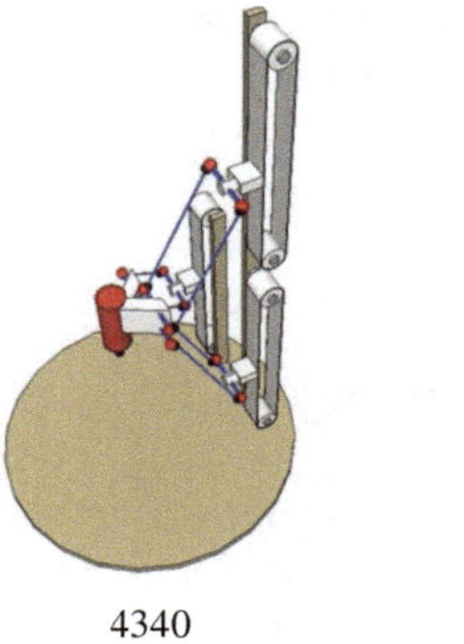 4340

It should be noted that in cases when only belts or only cables are used as input links, it is possible that special positions of the mechanism may exist. In order to avoid this possibility, patent-protected solutions are available for these schemes that have been proven to be effective. Another aspect of the use of flexible links is the specific dynamic behavior of the links. For this reason, it is important to take into account the properties of flexible links in the parametric structural design. The use of flexible links allows the development of structural solutions e.g. with variable external dimensions. The mechanism shown in Fig. 7, a can be built in telescopic version (Fig. 7, b). In this case, the accepted classification (Table 1) allows to write down an identification code S (4140) for the mechanism with the telescopic flexible links [19].

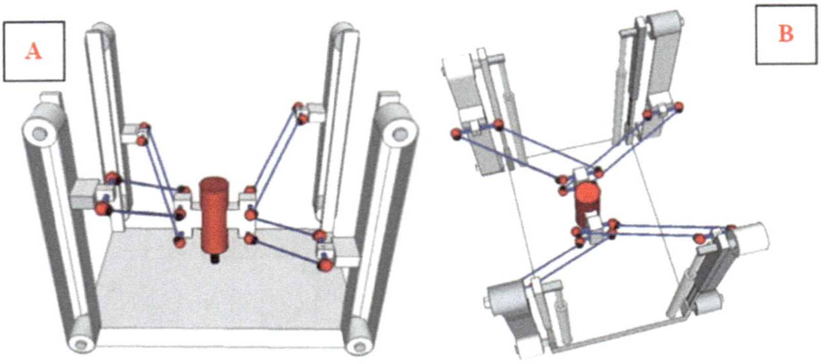

Fig. 7. Spatial mechanisms realizing the Schoenflies motion: a) with flexible links; b) with telescopic flexible links

Furthermore, the current Russian-language literature, e.g. the literature on the application of parallel manipulators in machine tools sector, provides a simple classification of mechanisms according to the number of coordinates (axes): 3-coordinate, 4-coordinate, 5-coordinate and 6-coordinate respectively. This classification cannot fully reflect the variety of solutions that have been developed.

The above examples and the ever-increasing amount of research and development show the growing attention to parallel mechanics with flexible links in all fields of advanced science and technology [16, 17, 20–22].

5 Summary and Conclusion

CDR are finding more and more applications, examples of which are described in this article. Described classification criteria are and a classification for three- and four-coordinate flexible link manipulators are developed. The S symbol indicates that the basic structural scheme is a mechanism that performs Schoen flies motion and the number of attachment points n on the output link and satisfy the following relations on the base i: $3 \leq n \leq 4$ and $3 \leq i \leq 4$, and the number of arrangement layouts of axes of the input links m and the combination of flexible links satisfy the following relations: $1 \leq m \leq 3$ and $1 \leq j \leq 3$ respectively.

Research on the parallel cable-driven mechanisms will require solving a number of practical tasks related to positioning accuracy, optimum constructions type, stiffness calculations and architecture of the robot control system. In turn, these tasks set new research directions: new types of materials, development of mathematical models of durability and tensile strength of a cable. A special place among the future research will be taken by the consideration of hybrid structures, in which the deficiencies in cable accuracy can be compensated by accompanying structural solutions.

References

1. Kim, K., Agogino, A.K., Agogino, A.M.: Rolling Locomotion of Cable-Driven Soft Spherical Tensegrity Robots. Soft Robotics, Mary Ann Liebert, Inc. (2020)
2. Weidong, W., et al.: Analysis of the rope-driven parallel rehabilitation robot. Mech. Design Manuf. Eng. **41**(11), 31–35 (2012)
3. Colombo, G., et al.: Treadmill training of paraplegic patients using a robotic orthosis. J. Rehabil. Res. Dev. **37**(6), 693 (2020)
4. Runtian, Y., Yuefa, F., Sheng, G.: Design and motion performance analysis of rope-driven ankle joint rehabilitation mechanism. Robot **37**(1), 53–62 (2015)
5. Glazunov, V., Borisov, V.A.: Development of parallel-structure mechanisms with four degrees of freedom and four kinematic chains. J. Mach. Manuf. Reliab. **46**, 417–425 (2017)
6. Olaronke, I., Oluwaseun, O., Rhoda, I.: State of the art: a study of human-robot interaction in healthcare. Int. J. Inf. Eng. Electron. Bus. **9**(3), 43–55 (2017)
7. Deng, Y., Bai, L., Long, Z., et al.: Research on cable-driven robots. In: Advanced Control, Automation and Artificial Intelligence, pp. 41–47 (2018)
8. Online Resource. www.roboy.org
9. Borisov, V.A.: Chapter 10: Development and research of parallel mechanisms with flexible links. In: Mechanisms of Advanced Robotic Systems. Technosphere, Moscow (2020)
10. Artobolevsky, I.I.: Mechanisms in the Modern Technology. Nauka, Moscow (1979)
11. Gerasun, V.M., et al.: Determination of service area for mobile manipulator-tripod. Mach. Build. Eng. Educ. **3**, 2–8 (2013)
12. Fomin, F.S., Antonov, A.V., Glazunov, V.A.: Forward kinematic analysis of a rotary hexapod. In: Venture, G., Solis, J., Takeda, Y., Konno, A. (eds.) ROMANSY 2020. CICMS, vol. 601, pp. 486–494. Springer, Cham (2021). https://doi.org/10.1007/978-3-030-58380-4_58
13. Filippov, G.S., Glazunov, V.A., Terekhova, A.N., Lastochkin, A.B., Chernetsov, R.A., Gavrilina, L.V.: 3-DOF spherical parallel mechanism. In: Zhengbing, H., Petoukhov, S., He, M. (eds.) Advances in Artificial Systems for Medicine and Education III, pp. 334–344. Springer, Cham (2020). https://doi.org/10.1007/978-3-030-39162-1_31
14. Fomin, A., Petelin, D., Glazunov, V.: Calculation of drive mechanism for novel rotary hexapod with single motor. In: Zawiślak, S., Rysiński, J. (eds.) Engineer of the XXI Century. MMS, vol. 70, pp. 17–28. Springer, Cham (2020). https://doi.org/10.1007/978-3-030-13321-4_2
15. Fomin, A., Ivanov, W., Glazunov, V.: Design and analysis of a mechanism for spherical surface processing. In: Kuo, C.-H., Lin, P.-C., Essomba, T., Chen, G.-C. (eds.) ISRM 2019. MMS, vol. 78, pp. 39–50. Springer, Cham (2019). https://doi.org/10.1007/978-3-030-30036-4_4
16. Antonov, A., et al.: Dynamics of a new parallel structure mechanism with motors mounted on the base outside the working area. In: Ronzhin, A., Shishlakov, V. (eds.) Proceedings of 14th International Conference on Electromechanics and Robotics "Zavalishin's Readings": ER(ZR) 2019, Kursk, Russia, 17–20 April 2019, pp. 183–195. Springer, Singapore (2020). https://doi.org/10.1007/978-981-13-9267-2_16
17. Glazunov, V.A., et al.: Velocity analysis of a spherical parallel robot. J. Phys. Conf. Ser. **1260**(11), 112012 (2019)

18. Rashoyan, G.V., et al.: Analysis and synthesis of parallel structure mechanism without singularities. Mechanical Science and Technology Update. IOP Conf. Series. J. Phys. Conf. Ser. **1260**, 1–8 (2019)
19. Utility Model Patent of Russian Federation No. 196059, MPK B25J 1/00. Spatial Mechanism with Four Degrees of Freedom. Glazunov, V.A., Borisov, V.A. Applicant and Patent Holder: Research Institute for Machine Science named after A.A. Blagonravov. - № 2019136996; Application. 19.11.19; publ. 14.02.20, bul. no. 5
20. Rashoyan, G.V., et al.: Analysis and synthesis of parallel structure mechanism without singularities. Mechanical Science and Technology Update. IOP Conf. Ser. J. Phys. Conf. Ser. **1260**, 1–8 (2019)
21. Polishchuk, M., Opashnianskyi, M., Suyazov, N.: Walking mobile robot of arbitrary orientation. Int. J. Eng. Manuf. **8**(3), 1–11 (2018)
22. Medina-Catzin, J.L., et al.: Body gestures recognition system to control a service robot. Int. J. Inf. Technol. Comput. Sci. **9**(9), 69–76 (2017)

Simulations of the Stretching of Two Models of Filamentous Actin at a Constant Velocity Using Method of Molecular Dynamics

Anna V. Glyakina[1,2(✉)], Ilya V. Likhachev[1,2], Nikolay K. Balabaev[1], and Oxana V. Galzitskaya[2,3(✉)]

[1] Institute of Mathematical Problems of Biology, Russian Academy of Sciences, Keldysh Institute of Applied Mathematics of Russian Academy of Sciences, Pushchino, Moscow Region 142290, Russia

[2] Institute of Protein Research, Russian Academy of Sciences, 4, Institutskaya Street, Pushchino, Moscow Region 142290, Russia
ogalzit@vega.protres.ru

[3] Institute of Theoretical and Experimental Biophysics, Russian Academy of Sciences, Pushchino, Moscow Region 142290, Russia

Abstract. Filamentous actin is one of the main components of the eukaryotic cellular cytoskeleton. In the cell, it performs such functions as cell motility, cell reshaping, cytokinesis, exo- and endocytosis, redistribution of surface receptors, and others. Since filamentous actin is a muscle tissue protein it also plays an important role in muscle contraction together with proteins myosin and titin. Therefore, the study of mechanical properties of filamentous actin is an important and urgent task in the fields of molecular biology and biophysics. In this work, the stretching of two models of filamentous actin was simulated at three constant velocities (0.1, 0.05 and 0.01 Å/ps) using the method of molecular dynamics. The first model was taken from the Protein Data Bank and the second model was built. From the analysis of the trajectories obtained, it follows that the second model of filamentous actin turned out to be more mechanically stable in comparison with the first model.

Keywords: Filamentous actin · Molecular dynamics · Mechanical stretching

1 Introduction

The actin protein belongs to the class of ATPases and is one of the most abundant proteins in eukaryotic cells. It has been shown that analogs of the actin protein are also present in the cells of bacteria and archaea [1–5].

Actin occurs naturally in two forms: monomeric globular (G-actin) and filamentous (F-actin). The transition of actin from one form to another is reversible and depends on the presence of ions in the solution. In the absence of salt, actin exists in monomeric form (G-actin). In the presence of a salt, actin monomers polymerize and form filaments (F-actin).

The monomeric form of actin is well studied and is a protein with a molecular weight of 42 kDa, consisting of 375 amino acid residues. The linear dimensions of

Z. Hu et al. (Eds.): AIMEE 2021, LNDECT 107, pp. 379–385, 2022.
https://doi.org/10.1007/978-3-030-92537-6_35

monomeric actin are approximately 5.5 nm × 5.5 nm × 3.5 nm (see Fig. 1). Monomeric actin consists of two domains: large and small. Each of these domains, in turn, consists of two subdomains. The small domain includes subdomains 1 (residues 1–32, 70–144, and 338–372) and 2 (residues 33–69), and the large domain includes subdomains 3 (residues 145–180 and 270–337) and 4 (residues 181–269) [6]. Large and small domains are separated by a cleft, that allows them to rotate relative to each other by a small angle.

In turn, the structure of filamentous actin has a strong polymorphism; therefore, it has not yet been possible to obtain its crystalline form. In this regard, the structure of filamentous actin has not been studied well enough. Although back in the 1960s, on the basis of the experimental data available at that time [6], it was suggested that filamentous actin has a double helix structure [7, 8]. Until now, this structure is the basis of all research and is included in all textbooks on molecular biology [9].

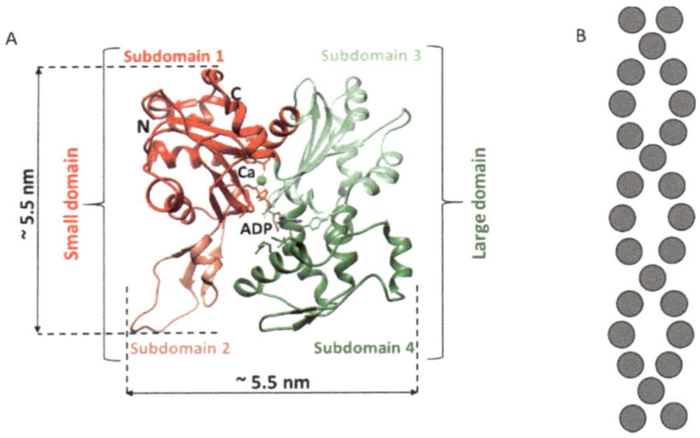

Fig. 1. A. Structure of monomeric globular actin (pdb 2zwh). B. The double helical model of filamentous actin.

As mentioned above, it has not yet been possible to crystallize filamentous actin and, accordingly, its X-ray structure is not presented in the Protein Data Bank (PDB). Although there are intensive studies of filamentous actin using cryogenic electron microscopy (cryo-EM) [10–14]. There are more than 50 models of filamentous actin in the UniProtKB database (https://www.uniprot.org), record number P68135, gene ACTA1, wild rabbit species (*Oryctolagus cuniculus*).

To date, a lot of scientific evidences (proteolysis, mass spectrometric analysis, electron microscopy, S-S bond introduction) has been accumulated that does not agree with the generally accepted double helix structure of filamentous actin [15, 16]. This means that there is also a misunderstanding of many of the key cellular processes involved in filamentous actin. For a detailed understanding of the mechanism of actin polymerization and the interaction of actin with myosin and various regulatory proteins, it is necessary to know the correct structure of the actin filament at the molecular level. The correct model will allow research and development of effective pharmacological drugs for the treatment of diseases associated with defects in actin molecules (nemaline myopathy, actin myopathy and other types of myopathies). In the paper [16], a new model of filamentous actin was proposed. Since the functions of filamentous actin are mainly related to its mechanical properties, they must be studied in detail.

In this paper, we study the mechanical stability of two models of filamentous actin. For this, molecular dynamics simulations of the stretching of these models were carried out.

2 Methods

2.1 Models of Filamentous Actin

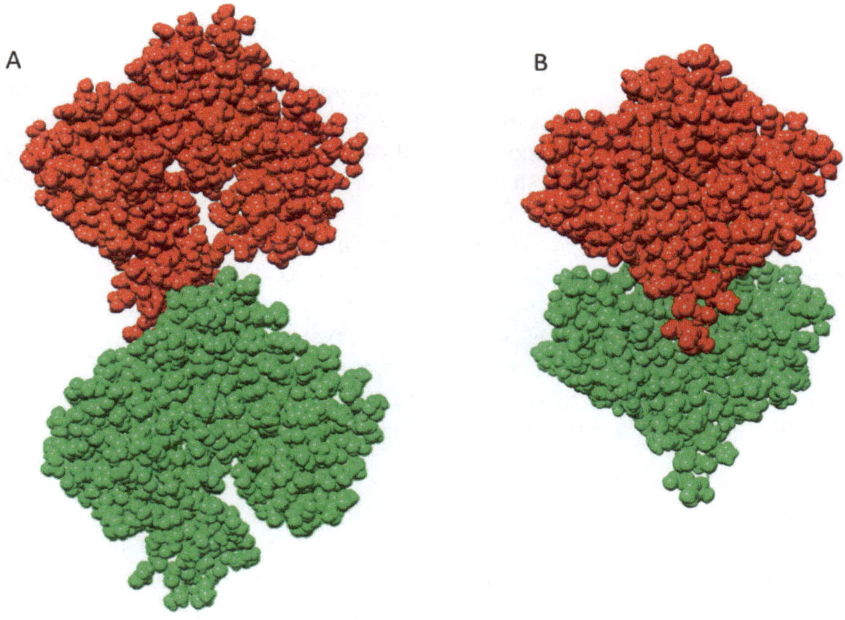

Fig. 2. A. Structure of filamentous actin from the Protein Data Bank (pdb 3j8k, chains E and G). B. Constructed structure of filamentous actin with overlapping structural units.

A search in the uniprot database (https://www.uniprot.org) showed that most of the resolved monomeric (globular) and filamentous actin structures are derived from the rabbit actin structure. Cα RMSD calculated between monomers in the five structures of filamentous actin with different resolutions (from 4.6 to 70 Å) and number of monomers (from 8 to 16) does not exceed 2 Å. It turns out that the resolution of the structures is different, but the Cα RMSD between the monomers in the structures are small. The same can be said about monomeric actin; a comparison of the Cα RMSD distribution of aligned 11781 pairs of rabbit actin structures showed that these values do not exceed 3 Å [17]. Thus, a model of filamentous actin from rabbit muscles was taken as the object of the study. The stretching of two filamentous actin models was simulated (see Fig. 2). The first structure was taken from the Protein Data Bank (pdb file is 3j8k, chains E and G) (Fig. 2 A). This structure was obtained by electron microscopy method with a resolution of 12 Å. The second structure was constructed by us (Fig. 2 B). The monomer structure from the Protein Data Bank (pdb file is 2zwh) was taken as a building unit. This monomeric structure of actin was determined by X-ray diffraction analysis with a resolution of 3.3 Å.

2.2 Method of Molecular Dynamics Simulations

The simulation was carried out using the PUMA-CUDA program [18, 19]. The AMBER99 force field [20] for the interactions of atoms and the TIP3P model of water [21] were used. A constant temperature was maintained using a collision thermostat [22, 23]. The mean collision frequency of atoms with virtual particles was 10 ps^{-1}, and the masses of virtual particles were 1 atomic mass unit.

The relaxation of the two models of filamentous actin was first carried out in vacuum (during 5 ns for the first model and 15 ns for the second model). Then water was added and relaxation continued for another 3 ns. After relaxation, computational experiments were performed to stretch two models of filamentous actin at a constant velocity.

The modeling was carried out as follows: in each model, one actin monomer remained stationary, and a constant velocity was applied to the center of mass of the other monomer. The simulation was carried out at three constant stretching velocities: 0.10, 0.05, and 0.01 Å/ps. For each stretching velocity, 8 independent computational experiments were performed. Thus, 48 trajectories were obtained (24 trajectories for each model of filamentous actin). The cloud storage was used to store a large amount of information obtained in molecular dynamics simulations [24, 25].

3 Results

The maximal forces arising from the simulations were determined (see Fig. 3). The faster the stretching velocity, the larger the force peak was observed. Thus, at the maximum stretching velocity (0.1 Å/ps), forces of ~6000 pN for the first model and ~8000 for the second model were observed. At medium stretching velocity (0.05 Å/ps) forces of ~4000 pN were observed for the first model and ~6000 for the second model. At the lowest stretching velocity (0.01 Å/ps) forces of ~2000 pN for the first model and ~4000 pN for the second model were observed.

From these data, it can be concluded that the second model of filamentous actin (constructed model) is more mechanically stable compared to the first model (model from the Protein Data Bank), since in the case of the second model, a higher force is required to separate the monomers apart from each other. It is also worth noting that the force arising from the separation of monomers in filamentous actin models is much greater than the force required to stretch small single domain proteins (completely α-helical or proteins in which both α-helices and β-sheets are present). When small proteins are stretched to their termini at constant velocity of 0.1 Å/ps, the resulting force does not exceed ~1200 pN for three-helical proteins [18] and ~1700 pN for immunoglobulin-binding domain of protein L and protein G with mixed α/β structure [26, 27]. It should be noted that for small three-helical proteins and for proteins L and G there is no large increase in force with an increase in the stretching velocity, the force increases only by 100–200 pN. In the case of filamentous actin, the force increases by 2000 pN.

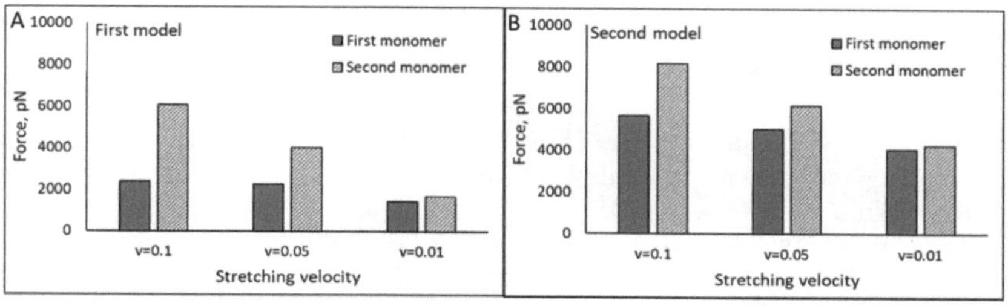

Fig. 3. The maximum force that occurs when two models of filamentous actin are stretched at a constant velocity. A. Structure of filamentous actin from the Protein Data Bank (pdb 3j8k, chains E and G). B. Constructed structure of filamentous actin with overlapping structural units.

It also follows from the analysis of the trajectories that the force reaction of the monomer, to the center of mass of which a constant velocity is applied, is higher than that of the monomer, which remains stationary at high and medium stretching velocities (0.10 and 0.05 Å/ps, respectively). This difference is more significant for the first model of filamentous actin compared to the second. At the lowest stretching velocity of 0.01 Å/ps, the force reaction of both actin monomers (the monomer that remains stationary and the monomer to which a constant velocity is applied) becomes approximately the same for two models of filamentous actin (see Fig. 3).

4 Conclusion

To date, there is no correct model of filamentous actin yet. It is very important to understand how the structures and conformation of monomeric actin confer the specific properties of filamentous actin. Filamentous actin is involved in many protein-protein interactions, which allows it to play an important role in cell motility and maintaining their shape. Therefore, it is very necessary to obtain the correct model of filamentous actin for the correct description of the interaction of actin with other proteins. Since filamentous actin is also a muscle protein, it is important to know its mechanical properties.

In this work, the mechanical properties of two models of filamentous actin were studied. The first structure was common model of filamentous actin which included in all textbooks on molecular biology. The second structure was constructed by us.

Molecular dynamics simulation of the stretching of two models of filamentous actin at three constant velocities (0.1, 0.05 and 0.01 Å/ps) was done. From the analysis of the data, it follows that the second model is more mechanically stable than the first model. The difference in the forces required to separate the monomers from each other for the first and second is ~2000 pN. This once again emphasizes that the model of filamentous actin taken as a basis requires correction.

Acknowledgment. This project is supported by the Russian Science Foundation for A.V.G. (19-74-10051).

References

1. Popp, D., et al.: Molecular structure of the ParM polymer and the mechanism leading to its nucleotide-driven dynamic instability. EMBO J. **27**, 570–579 (2008). https://doi.org/10.1038/sj.emboj.7601978
2. Galkin, V.E., Orlova, A., Rivera, C., Mullins, R.D., Egelman, E.H.: Structural polymorphism of the ParM filament and dynamic instability. Structure **17**, 1253–1264 (2009). https://doi.org/10.1016/j.str.2009.07.008
3. Popp, D., et al.: Polymeric structures and dynamic properties of the bacterial actin AlfA. J. Mol. Biol. **397**, 1031–1041 (2010). https://doi.org/10.1016/j.jmb.2010.02.010
4. Izoré, T., Duman, R., Kureisaite-Ciziene, D., Löwe, J.: Crenactin from Pyrobaculum calidifontis is closely related to actin in structure and forms steep helical filaments. FEBS Lett. **588**, 776–782 (2014). https://doi.org/10.1016/j.febslet.2014.01.029
5. Braun, T., Orlova, A., Valegård, K., Lindås, A.-C., Schröder, G.F., Egelman, E.H.: Archaeal actin from a hyperthermophile forms a single-stranded filament. Proc. Natl. Acad. Sci. U. S. A. **112**, 9340–9345 (2015). https://doi.org/10.1073/pnas.1509069112
6. Selby, C.C., Bear, R.S.: The structure of actin-rich filaments of muscles according to x-ray diffraction. J. Biophys. Biochem. Cytol. **2**, 71–85 (1956). https://doi.org/10.1083/jcb.2.1.71
7. Hanson, J., Lowy, J.: The structure of F-actin and of actin filaments isolated from muscle. J. Molecul. Biol. **1963**(6), 46-IN5 (1963). https://doi.org/10.1016/S0022-2836(63)80081-9
8. Holmes, K.C., Popp, D., Gebhard, W., Kabsch, W.: Atomic model of the actin filament. Nature **347**, 44–49 (1990). https://doi.org/10.1038/347044a0
9. Jegou, A., Romet-Lemonne, G.: The many implications of actin filament helicity. Semin. Cell. Dev. Biol. **102**, 65–72 (2020). https://doi.org/10.1016/j.semcdb.2019.10.018
10. Chen, L.F., Winkler, H., Reedy, M.K., Reedy, M.C., Taylor, K.A.: Molecular modeling of averaged rigor crossbridges from tomograms of insect flight muscle. J. Struct. Biol. **138**, 92–104 (2002). https://doi.org/10.1016/s1047-8477(02)00013-8
11. Galkin, V.E., Orlova, A., Vos, M.R., Schröder, G.F., Egelman, E.H.: Near-atomic resolution for one state of F-actin. Structure **2015**(23), 173–182 (2015). https://doi.org/10.1016/j.str.2014.11.006
12. Gurel, P.S., Kim, L.Y., Ruijgrok, P.V., Omabegho, T., Bryant, Z., Alushin, G.M.: Cryo-EM structures reveal specialization at the myosin VI-actin interface and a mechanism of force sensitivity. Elife **6**, e31125 (2017). https://doi.org/10.7554/eLife.31125
13. Murakami, K., et al.: Structural basis for actin assembly, activation of ATP hydrolysis, and delayed phosphate release. Cell **143**, 275–287 (2010). https://doi.org/10.1016/j.cell.2010.09.034

14. Wu, S., et al.: Electron tomography of cryofixed, isometrically contracting insect flight muscle reveals novel actin-myosin interactions. PLoS One **2010**(5), e12643 (2010). https://doi.org/10.1371/journal.pone.0012643
15. Durer, Z.A.O., Diraviyam, K., Sept, D., Kudryashov, D.S., Reisler, E.: F-actin structure destabilization and DNase-I binding loop fluctuations. J. Mol. Biol. **395**, 544–557 (2010). https://doi.org/10.1016/j.jmb.2009.11.001
16. Glyakina, A.V., et al.: New model for stacking monomers in filamentous actin from skeletal muscles of oryctolagus cuniculus. Int. J. Mol. Sci. **21**, E8319 (2020). https://doi.org/10.3390/ijms21218319
17. Glyakina, A.V., Galzitskaya, O.V.: Bioinformatics analysis of actin molecules: why quantity does not translate into quality? Front Genet. **11**, 617763 (2020). https://doi.org/10.3389/fgene.2020.617763
18. Glyakina, A.V., Likhachev, I.V., Balabaev, N.K., Galzitskaya, O.V.: Right- and left-handed three-helix proteins. II. Similarity and differences in mechanical unfolding of proteins. Proteins **82**, 90–102 (2014). https://doi.org/10.1002/prot.24373
19. Likhachev, I.V., Balabaev, N.K., Galzitskaya, O.V.: Elastic and Non-elastic properties of cadherin ectodomain: comparison with mechanical system. Adv. Comput. Sci. Eng. Educ. **II**, 555–566 (2020)
20. Wang, J., Cieplak, P., Kollman, P.A.: How well does a restrained electrostatic potential (RESP) model perform in calculating conformational energies of organic and biological molecules? J. Comput. Chem. **21**, 1049–1074 (2000). https://doi.org/10.1002/1096-987X(200009)21:12%3c1049::AID-JCC3%3e3.0.CO;2-F
21. Jorgensen, W.L., Chandrasekhar, J., Madura, J.D., Impey, R.W., Klein, M.L.: Comparison of simple potential functions for simulating liquid water. J. Chem. Phys. **79**, 926–935 (1983). https://doi.org/10.1063/1.445869
22. Lemak, A.S., Balabaev, N.K.: A comparison between collisional dynamics and Brownian dynamics. Molecul. Simulat. **15**, 223–231 (1995). https://doi.org/10.1080/08927029508022336
23. Lemak, A.S., Balabaev, N.K.: Molecular dynamics simulation of a polymer chain in solution by collisional dynamics method. J. Comput. Chem. **17**, 1685–1695 (1996). https://doi.org/10.1002/(SICI)1096-987X(19961130)17:15%3c1685::AID-JCC1%3e3.0.CO;2-L
24. Althagafy, E., Qureshi, M.R.J.: Novel cloud architecture to decrease problems related to big data. Int. J. Comput. Netw. Inf. Secur. **9**(2), 53–60 (2017)
25. Akomolafe, O.P., Abodunrin, M.O.: A hybrid cryptographic model for data storage in mobile cloud computing. Int. J. Comput. Netw. Inf. Secur. **9**(6), 53–60 (2017)
26. Glyakina, A.V., Balabaev, N.K., Galzitskaya, O.V.: Two-, three-, and four-state events occur in the mechanical unfolding of small protein L using molecular dynamics simulation. Protein Pept. Lett. **17**, 92–103 (2010). https://doi.org/10.2174/092986610789909449
27. Glyakina, A.V., Balabaev, N.K., Galzitskaya, O.V.: Multiple unfolding intermediates obtained by molecular dynamic simulations under stretching for immunoglobulin-binding domain of protein G. Open Biochem. J. **3**, 66–77 (2009). https://doi.org/10.2174/1874091X00903010066

Intelligent Remote Health Monitoring for Rehabilitation After COVID Disease

Lev I. Evelson[1(✉)], Boris V. Zingerman[2], Olga S. Kremenetskaya[3], and Nikita E. Shklovskiy-Kordi[4]

[1] Innovation Scientific Center of Information and Remote Technologies, 241050 Bryansk, Russian Federation
[2] TelePat, 196787 Moscow, Russian Federation
[3] Center for Theoretical Problems of Physicochemical Pharmacology, 119991 Moscow, Russian Federation
[4] National Medical Research Center for Hematology, 125167 Moscow, Russian Federation

Abstract. The paper is devoted to utilization of Artificial Intelligence in the remote health monitoring system intended for rehabilitation of people having complications after COVID disease. Described system had been created on a base of earlier developed telemedicine platform. It includes subsystem of intelligent agents tracking health statement and providing a patient with relevant information. The main part of the rehabilitation system is medical physical training directed against typical negative consequences of the COVID disease. Remote health monitoring allows involve into the rehabilitation many people living in various towns and countryside localities far from medical centers. The pilot project was implemented in several regions of Russia. The results described in this article corroborate high effectiveness of the proposed approach. Ways of the perspective development of the proposed system are connected with further development of the intelligent agents and periodical analysis of the personal medical data to reveal the individual health tendencies as well as with analysis of collected and depersonalized medical data to detect important general regularities. It is noted that proposed approach and created systems are corresponding to the perspective patient-centered concept of medical care.

Keywords: Intelligent agents · Rehabilitation after COVID · Remote health monitoring · Artificial intelligence

1 Introduction

Now remote health monitoring (RHM) is very quickly growing segment of telemedicine. Many scientific articles are devoted to various aspects of RHM [1–5]. The COVID-19 pandemic has caused many challenges and more accelerated development and application of such medical service. During the pandemic an opportunity to reduce number of face-to-face contacts is very important. After the COVID disease many people have various health problems. Two new problems can be noted. They are

Z. Hu et al. (Eds.): AIMEE 2021, LNDECT 107, pp. 386–393, 2022.
https://doi.org/10.1007/978-3-030-92537-6_36

necessity of rehabilitation after COVID for concrete persons and research of general regularity in complications of the COVID disease. The rehabilitation method including specific medical physical training had been developed. It must to be combined with permanent health monitoring and keeping some instructions concerning the life style during rehabilitation. In countries like Russia for which great distances are typical and there is shortage of doctors and other medical specialists in many small localities proposed rehabilitation method can be implemented only with RHM. However, remote monitoring also needs many physicians to observe the data transferring from the patients. Artificial Intelligence (AI) can help here. It can take a significant part of routine doctor's work. AI can automatically analyze the data flow and reveal situations when personal attention of a doctor is really necessary. AI can answer to typical patient questions and provide a patient with general information relevant for the situation and individual peculiarities. So, RHM with AI can do taking part in the rehabilitation course possible in all localities.

The telemedicine platform "Medsenger.AI" [6] had been developed earlier. It includes specific medical messenger which has important advantages as compare with usual messengers (for example like WhatsApp) as applied to RHM. The subsystem of intelligent agents (IA) automatically analyzes the data flows transferring from a patient to the attending doctor and detects alarms. IA also provides a patient with relevant information from the digital Patient Library preliminary completed by checked and ordered knowledge taken from reliable sources. Several RHM systems for various categories of patients (cancer patients, people after organ transplantation and others) had been created and the pilot implementation projects had taken place in different regions of Russia [7, 8].

This article is devoted to the new RHM system created on a base of Medsenger.AI and intended to rehabilitation after the COVID disease. The main purposes are to describe the system, used approach and methods and to present some fist results of the pilot running. Some problems and perspectives of further development of the system are displayed and discussed too.

2 Methodology

In contradistinction to many research papers devoted to RHM, the methodology presented in this article concerns first of all not technological but psychological and managerial aspects of RHM with AI elements. It reflects the methodology of the "Medsenger.AI". The main principle is "not injure to a patient and not injure to a doctor". It means that medical care and rehabilitation results for a patient and life and work of a physician involved in RHM must be at least not worse than without RHM. The COVID REHAB is proposed, first of all, for inhabitants of localities where involvement into the rehabilitation course without RHM is too difficult.

The main methods used in the presented solutions are corresponding to patient-centered concept of medical care. The main principles are: RHM for permanent remote health monitoring, permanent analysis of the data flow to reveal alarms and dangerous individual tendencies, unloading a doctor from routine part of his work by the way of entrusting it to AI. COVID REHAB uses IoT to measure objective parameters of health. However, it is not principal in the proposed methodology. Important role belongs also to analysis of subjective information generated by a patient. The methods of joint analysis of objective and subjective health parameters are used. Usually the objective parameters are quantitative (for example, blood pressure and the subjective ones are qualitative (for example, presence of psychological stress). However, the objective qualitative (for example, presence of syncope) and subjective quantitative (for example, number of days in the last month when the person had a headache) can also take place. Often there are intermediate cases when the boundaries between subjective and objective (for example, presence of a weak pain), quantitative and qualitative (for example, in estimation of sport activity) types of the parameters are relative.

In order to analyze the heterogeneous data, various methods can be used depending on the solved problem. A combination of the different types of methods should be implemented. In can include AI methods based on Expert Systems (ES) or Machine Learning (ML), traditional mathematical methods for analysis of quantitative data (in particular, analysis of temporal data), methods of multicriteria optimization, etc. The general base of the methodology is the system approach. The data describe a health system statement and it's change in the abstract multidimensional space.

The following principles and methods can be noted.

1) Asynchronous method of organization of interaction between a patient and a doctor: a patient asks when the question arises, and the doctor answers when it is convenient for him but necessarily within the agreed time (working day, 12 h, 24 h).
2) The Personal Electronic Medical Card (PEMC) is used. The whole information generated in RHM is ordered according with the PEMC structure and saved in the PEMC. The telemedicine platform includes own PEMC [9, 10] but there is also an opportunity to integrate the RHM system with PEMC used in some medical institution.
3) Collection of the data describing a patient health statement is done by measuring some quantitative parameters (for example, blood pressure, pulse, etc.) with wearable medical devices and by filling up various patient questionnaires in which the patient describes the qualitative parameters of the health statement. The qualitative parameters include characteristics of the psychological statement.
4) A great attention is paid to visualization of information sent from a patient. Some special tools had been developed: color diagrams reflecting a patient condition according with noted symptoms ("heat map"), graphs displaying dynamics of the quantitative parameters together with qualitative ones and with important current events (for example, taking drugs). The common time axis is used in the graphs.

Approximate example of the united picture including simultaneously the graphs, heat maps and events is presented in the Fig. 1.

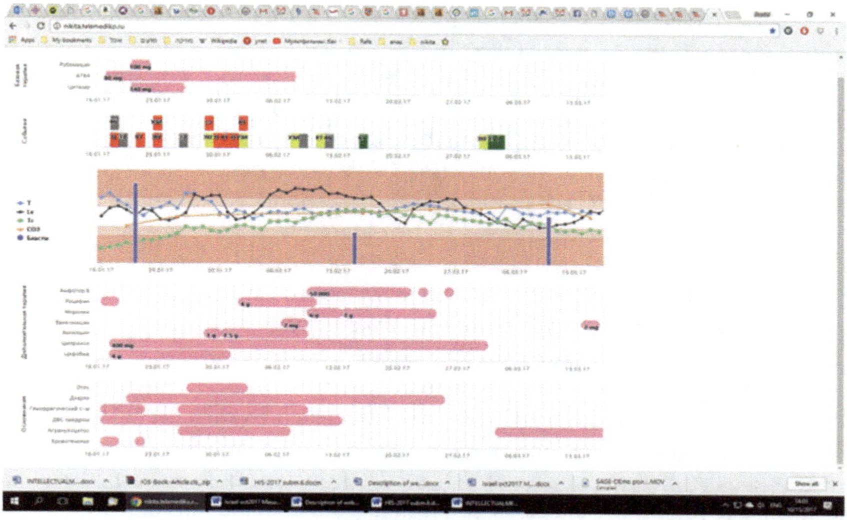

Fig. 1. An example of visualization utilizing the "Heat Map" together with the graphs

5) AI is used to classify automatically the situations: a) urgent actions are necessary; b) urgent actions are not necessary but personal attention of the attending doctor is needed; c) urgent actions are not necessary and personal attention of the attending doctor is not needed. In the case c) AI initiates standard response including recommendations for the patient and some useful information relevant to the situation. AI methods are realized by subsystem of intelligent agents (IA). The simplest method here is to store some list of quantitative and qualitative parameters with standard marks of the deviations values indicating the alarm situation. If any (at least one) parameter has such deviation the IA turns on the reaction procedure including sending information to the attending doctor and emergency service. For example, if the quantitative parameter "temperature" is more 40 °C or the qualitative parameter "consciousness" is cloud that the alarm situation takes place. More delicate methods should be used to detect situations which are not so alarm but demand quick personal attention of the physician. Here the combinations of parameters (syndromes) are estimated. In more simple cases the lists of syndromes can be utilized but more often it's better to use ES including Knowledge Base developed according with the medical standards and individual peculiarities.
6) The platform Medsenger.AI includes virtual Patient Library and Doctor Library. Information concerning the life conditions of various categories of patients under the COVID-19 quarantine situation and after postponed COVID disease had been added recently. There are, in particular, many important recommendations such as: how to control breath under the COVID-19, prophylaxis peculiarities for patients with different chronic diseases, psychological aspects during pandemic, etc.
7) IA are embedded in the patient's personal communication channel with his attending doctor. IA implement current analysis of information transferred from a patient to detect alarm situations when personal attention of the attending doctor or

emergency service is necessary saving the doctor's time recourse in other cases. Patient questions at first are analyzed by IA and typical answer can be generated if the question is typical too. IA of different levels can be used. The simplest analysis can be done according with clear algorithm and list of norm intervals for the quantitative health parameters. In this case detection of alarms can be done separately for every parameter. In order to receive automatically information from a patient in a natural language (besides of the questionnaires filled up by a patient) conversational IA can be used [11, 12]. IA of more high level of "intelligence" can analyze quantitative and qualitative parameters together utilizing, in particular, fuzzy sets method to transform patient fuzzy conversational information into the qualitative parameters description and to make decisions [13, 14].

8) Decision making support systems (DMSS) are used in form of interaction between IA. In more simple level it can be done according with a clear algorithm. In more high levels the AI methods developed for complicated IA systems [15] can be used. Besides of quantitative parameters measurement and qualitative parameters verbal description the medical images (MRI, CT, US, histopathological, etc.) should be taken into account. The methods of the images automatic analysis are developed and used in practice [16, 17]. However, joint automatic analysis of all above types of data is not so easy problem for DMSS used within the RHM system. Therefore, now the images analysis is done in COVID REHAB separately by the doctor of corresponding profile.

3 Results of the Pilot Running Project

The RHM system COVID REHAB created on a base of the platform Medsenger.AI was running from October 2020 till April 2021 in 6 Russian medical institutions situated in different towns from Moscow to Altai. 2032 patients took the remote rehabilitation course after the COVID disease. Many patients don't live in the towns where the medical institutions are situated but live in other localities including countryside. The total number of those localities is 472 but approximately a half of patients' number live in the same towns where the medical institutions are situated. 52% of the patient's number had attendant chronic diseases.

The main results of the rehabilitation were following.

- 34% of the patients number have stopped to feel weakness;
- breath difficulties have been eliminated in 20%;
- cough has disappeared in 15% of the patients number;
- vertigo has been chucked out in 13% of the patients number.
- 89% of the patients' number has noted improvement of the health.

Most of all patients and doctors involved to the RHM project estimated the course as very good, i.e. the objective results were corresponding to subjective marks.

So, the results of the pilot utilization of the proposed system were obviously good for the patients and for the doctors.

4 Discussion

Results of the pilot exploitation of COVID REHAB display that it is an effective and convenient tool for rehabilitation after COVID disease. RHM allows involve many people living far from medical centers into rehabilitation courses. It can save health and reduce risk of various heavy complications after the COVID disease. Further development of the COVID REHAB system should be done by two main ways. The telemedicine platform Medsenger.AI should continue to be developed in direction of the IA subsystem functionality expansion. It should perfect health monitoring in all RHM systems created on a base of the platform including the COVID REHAB. Wide exploitation of the COVID REHAB should give valuable experience which should be utilized in the further development.

The analytical subsystem should be added to analyze the data collected in RHM. Analysis of the personal medical data should reveal individual health tendencies and analysis of the depersonalized data should detect important general regularity. Methods of analysis of depersonalized data can be taken from a set of known methods depending on formulated problem and discussion about it is outside of this article. Analysis of the temporal personal medical data is corresponding to more specific class of problems. Now the methods presented in [18] are seen as appropriate ones. They probably can also be adapted to be used in DMSS for patients [19]. It is corresponding well to the general patient-centered concept which is included into foundation of the Medsenger. AI and COVID REHAB. More abstract methods like as JSM [20] also can be used in some RHM systems for a perspective. The method based on fuzzy clustering [21] should be also useful there.

It can be noted that RHM opens wide opportunities of research and taking into account individual peculiarities including the life style, psychological statement, etc. It is very important and it can improve medical care significantly. However, it is not so well combined with existing now public health system which is devoted not to a patient but to standards developed in investigation of big people groups. Probably the system should be changed and RHM systems should be considered as an engine of such transformation.

5 Conclusion

The combination of RHM with AI gives an effective and convenient tool to solve many important and actual problems of medical care. The telemedicine platform Medsenger. AI implementing such combination allows creates various intelligent RHM systems for different concrete problems. COVID pandemic has become a catalyst for acceleration of RHM and simultaneously brought many new challenges to modern public health. The COVID REHAB is a valuable mean to reduce complications and rehabilitate health after COVID disease. The pilot running has displayed that the COVID REHAB can solve the rehabilitation problems successfully.

AI doesn't replace a doctor in Medsenger.AI and in COVID REHAB but it helps a doctor unloading him from a great part of his routine work. IA utilizing the AI methods work as intelligent mediators between a patient and his attending doctor.

Further development of the proposed approach, telemedicine platform and the COVID REHAB system is associated with an expansion of AI functions connected with current analysis of the data flows and with development of analytical subsystem implementing temporal analysis of the personal medical data to reveal individual health tendencies and analysis of collected and depersonalized data to detect important general regularity. So, the RHM system would help to save and rehabilitate people health and to generate important research results for medical care.

The telemedicine platform Medsenger.AI and RHM systems created on a base of this platform like as COVID REHAB are corresponding well to a new patient-centered concept of medical care. Existing now in all countries public health system is founded on the standards developed as results of statistical investigations of big people groups. It is not suitable for the patient-centered concept. However, a patient who heeds medical help wants to be healthy. He is not so interested to keep the general standards. Probably the public health system should be changed and the RHM with AI components should be an engine and catalyst of such transformation.

Acknowledgments. This work was supported by the Ministry of Science and Higher Education of the Russian Federation (AAAA-A18-118012390247-0), by RFBR grant (project № 19-07-01235).

References

1. Topol, E.: Digital medicine: empowering both patients and clinicians. The Lancet **388** (10046), 740–741 (2016)
2. Lakmini, P., Naaem, R., Keshav, D.: Remote patient monitoring. J. Ambient. Intell. Humaniz. Comput. **10**, 57–76 (2019)
3. Moreno, H.B.R., Ramírez, M.R., Hurtado, C., Lobato, B.Y.M.: IoT in medical context: applications, diagnostics, and health care. In: Chen, Y.-W., Zimmermann, A., Howlett, R.J., Jain, L.C. (eds.) Innovation in Medicine and Healthcare Systems, and Multimedia. SIST, vol. 145, pp. 253–259. Springer, Singapore (2019). https://doi.org/10.1007/978-981-13-8566-7_25
4. Tsukiyama, T.: In-home health monitoring system for solitary elderly. Procedia Comput. Sci. **63**, 229–235 (2015)
5. Parthiban, N., Esterman, A., Mahajan, R., et al.: Remote monitoring of implantable cardioverter defibrillators: a systematic review and meta-analysis of clinical outcomes. Am. Coll. Cardiol. **65**, 2591–2600 (2015)
6. Zingerman, B., Vorobyev, A., Shklovsky-Kordy, N.: About telemedicine: patient-doctor. Phys. Inf. Technol. **1**, 61–79 (2017). (in Russian)
7. Shklovskiy-Kordi, N., Borodin, R., Zingerman, B., Shifrin, M., Kremenetskaya, O., Vorobiev, A.: Web-service medical messenger - intelligent algorithm of remote counseling. HIS LNCS **11148**, 193–197 (2018)
8. Evelson, L.I., et al.: Telemedicine monitoring with artificial intelligence elements. In: Hu, Z., Petoukhov, S., He, M. (eds.) AIMEE 2020. AISC, vol. 1315, pp. 212–221. Springer, Cham (2021). https://doi.org/10.1007/978-3-030-67133-4_20
9. Zingerman, B., Shklovsky-Kordi, N.: Electronic medical record and the principles of its organization. Phys. Inf. Technol. **2**, 37–57 (2013). (in Russian)

10. Zingerman, B., Shklovsky-Kordy, N., Karp, V., Vorobyov, A.: Integrated electronic medical card: tasks and problems. Phys. Inf. Technol. **1**, 24–34 (2015)
11. Galitsky, B.: Adjusting chatbot conversation to user personality and mood. In: Galitsky, B. (ed.) Artificial Intelligence for Customer Relationship Management: Solving Customer Problems, pp. 93–127. Springer, Cham (2021). https://doi.org/10.1007/978-3-030-61641-0_3
12. Riksa Meidy Karim, S.: Optimizing parameters of automatic speech segmentation into syllable units. Int. J. Intell. Syst. Appl. **11**(5), 9–17 (2019)
13. Sun, L., Dong, H., Liu, A.: Aggregation functions considering criteria interrelationships in fuzzy multi-criteria decision making. State-of-the-Art. IEEE Access **6**, 68104–68136 (2018)
14. Evelson, L., Dubovoy, I., Borisova, E.: Improving screening methods to determine belonging to the risk group for the consequences of alcohol effect (with information technology and mathematical modeling). Phys. Inf. Technol. **3**, 6–17 (2018). (in Russian)
15. Rosenfeld, A., Zuckerman, I., Segal-Halevi, E., Drein, O., Kraus, S.: NegoChat-A: a chat-based negotiation agent with bounded rationality. Auton. Agent. Multi-Agent Syst. **30**(1), 60–81 (2016)
16. Hu, Z., Dychka, I., Sulema, Y., Valchuk, Y., Shkurat, O.: Method of medical images similarity estimation based on feature analysis. Int. J. Intell. Syst. Appl. **10**(5), 14–22 (2018)
17. Xu, J., Luo, X., Wang, G., et al.: A deep convolutional neural network for segmenting and classifying epithelial and stromal regions in histopathological images. Neurocomputing **191**, 214–223 (2016)
18. Shahar, Y., Moskovitch, R., Klimov, D.: Mining medical knowledge from time-oriented clinical data using existing clinical knowledge. Harefuah **152**, 279–308 (2013)
19. Peleg, M., Shahar, Y., Quaglini, S., et al.: MobiGuide: a personalized and patient-centric decision-support system and its evaluation in the atrial fibrillation and gestational diabetes domains. User Model. User-Adap. Inter. **27**(2), 159–213 (2017)
20. Finn, V.K.: Distributive lattices of inductive JSM procedures. Autom. Document. Math. Linguist. **48**(6), 265–295 (2014)
21. Hu, Z., Bodyanskiy, Y., Tyshchenko, O., Samitova, V.: Fuzzy clastering data given in the ordinal scale. Int. J. Intell. Syst. Appl. **9**(1), 67–74 (2017)

Advances in Educational Approaches

Influencing Factors of Network Public Opinion Violence in Colleges and Universities Based on Rational Behavior Theory

Lixia Wu(✉), Huanmeng Cui, and Chenyu Li

Wuhan Technology and Business University, Wuhan, Hubei, China

Abstract. It is the popularity of the Internet and the anonymity and virtuality of the network that lead to the frequent occurrence of network public opinion violence in colleges and universities. As a unique form of public opinion, network public opinion violence in colleges and universities not only endangers the personal safety of college students, but also affects the development of universities. According to the characteristics of college students, it has become a very necessary research issue to strengthen the effective guidance of college network public opinion, promote the healthy development of campus network culture and maintain social stability. Based on the rational behavior theory, this paper constructs a theoretical research model based on the attitude and subjective norms of the network environment - the intention of network public opinion violence - the actual behavior of network public opinion force, reveals the factors affecting network public opinion violence, and puts forward relevant countermeasures and suggestions for the governance of public opinion injury by network in colleges and universities.

Keywords: Rational behavior theory · Network public opinion violence · Influencing factors of public opinion

1 Introduction

The 44th statistical report on China's Internet development pointed out that the number of Internet users is quite large, reaching 854 million, an increase of 25.98 million over last year. As an important part of Internet users, college students are willing to express their views and opinions on campus and social hot issues that have attracted extensive attention on the Internet platform, forward and discuss them, and form network public opinion [1]. It must be noted that network public opinion and its problems have brought a variety of impacts on public life, and network violence is one of the most prominent contents [2–5].

Under the guidance of theory, choosing appropriate analysis tools to analyze the causes and influencing factors of network public opinion violence is necessary and of great significance for social harmony and tranquility, especially for college students.

Z. Hu et al. (Eds.): AIMEE 2021, LNDECT 107, pp. 397–408, 2022.
https://doi.org/10.1007/978-3-030-92537-6_37

2 Research Status and Literature Induction

From the perspective of foreign research on cyber violence, by studying the foreign literature with the keywords of "cyberbulling", "online shading" and "cyber violence", it can be found that the research results of various countries on cyber violence are rich and profound [6]. Martínez-Monteagudo María Carmen and others found that victims of online violence increased the possibility of suicidal thoughts through a questionnaire survey of college students [7]; Khine Aye Thazin and other scholars conducted a one-year questionnaire survey on college students in Myanmar and found that victims of online bullying were less likely to discuss their experiences with others, and the victims not only suffered academic obstacles, but also contracted bad living habits [8]; Jia Ping. Chua, Angeline Meiyan. Liew believe that "cyber violence cannot prevent the occurrence of deviant behavior" [9]; From the sight of "silence spiral", scholar Zhang Kai analyzes the formation mechanism of network violence from internal and external causes, and considers that the internal cause is the conformity psychology of social media users, while the external cause is due to the anonymity, low threshold and rapid information dissemination of social media [10]; Li Yan summarized the violent effects of online discourse as the moral encirclement and suppression of others by violent language, anonymous cover of violence, easy infection of online force, violence "killing invisible", and the perpetrator seizing the right to speak [11]; From the angle of innovative social governance, Qiu Rui discussed the importance of Internet public opinion research and the basic strategies of public opinion guidance [12]; Xu Caiqi incorporated cyber violence into the criminal law and further improved the judicial interpretation by learning from the relevant legislative experience of foreign countries on cyber violence [13]; Wang Jiayu pointed out that the causes of Internet public opinion violence mainly include external environment, social conditions, emotional factors, behavior mobilization and network name itself [14]; Jiao Jieqing deemed that the occurrence of online public opinion violence in colleges and universities is closely related to factors such as network environment, psychological characteristics and social cognition, and stressed that we must pay attention to the governance of online public opinion violence [15]; Liu Yanfang and Qiu ting combined the network public opinion violence with the ternary theory to explore the causes of the formation of online public opinion force, and explore the formation mechanism of network public opinion violence from multiple perspectives [16]; Guan Yale expounds the connotation of network public opinion violence, analyzes its formation mechanism and communication environment and puts forward suggestions for governance of network public opinion violence [17]. Hui Xiaowen indicated that there are still problems in the network public opinion monitoring system under the condition of big data, and relevant measures should be taken to improve it to guide the society to develop in a positive and healthy direction [18].

Generally speaking, foreign theoretical research results generally start from the technical level and social level, and the research on the governance of cyber violence mostly stays at the legal stage. Special attention is paid to the impact of network violence on teenagers, and mostly application of the methods of quantitative analysis and empirical analysis. Compared with western countries, domestic scholars' research on network violence started late and the research is not deep enough [19, 20]. It mainly makes a more in-depth research from the aspects of performance, causes, hazards and governance countermeasures, but the research method is single, mostly qualitative research, and less quantitative research and empirical research. In any case, these research results play a role in standardizing the network environment and promoting the healthy growth of college students [21–23].

3 Research Hypothesis and Study Design

An American scholar, Fishbein, put forward the rational behavior theory, which holds that individual behavior can be inferred from human behavior intention, and attitude and subjective norms affect individual behavior intention. The assumption is that people will consider many factors before their behavior, and finally get the meaning and consequences after their action. Network media has the characteristics of virtuality and concealment. In such a virtual space, netizens do not have to worry about taking any responsibility and they have no worries, while express their views and opinions freely and confidently. Some netizens believe that what they say and do in the virtual environment will not have any impact on the parties in real life. They will even arrange dissatisfied remarks about someone or an event and phenomenon at will to slander others. Communication on the Internet is often easy to resonate, and some dissatisfaction will be amplified and strengthened, thus forming extreme views and behaviors.

3.1 Research Hypothesis

The analysis shows that the factors affecting the intention of network public opinion violence mainly include attitude, viewpoint, subjective will and code of conduct. The network environment will provide certain conditions for the intention of network public opinion violence. Accordingly, the theoretical model shown in Fig. 1 is proposed in this paper. In the figure, subjective norm refers to the social pressure that individuals feel about whether to take a specific behavior; Attitude is a stable psychological tendency held by individuals towards specific objects (people, ideas, emotions or events, etc.). This psychological tendency contains the individual's subjective evaluation and the resulting behavioral tendency; Point of view refers to the position or starting point when observing things, and the view on things or problems.

Based on the theoretical model, this study takes attitude and subjective norms as research variables, the network environment as regulatory variables, the intention of network public opinion violence as intermediate variables, and the actual behavior of network public opinion violence as outcome variables. The following assumptions are made:

H_1: there is a significant positive correlation between the intention of network public opinion violence and attitude;

H_2: there exist a significant positive correlation between the intention of network public opinion violence and subjective norms;

H_3: a significant positive correlation between the intention of network public opinion violence and the actual behavior of network public opinion violence is presented;

H_4: the positive influence of network environment on the correlation between attitude and view, the intention of network public opinion violence and the actual behavior of network public opinion violence;

H_5: the network environment positively affects the correlation between subjective norms and the intention of network public opinion violence and the actual behavior of network violence.

Fig. 1. Theoretical model of the impact of network public opinion violence

3.2 Study Design

The questionnaire is designed and distributed to the social media platform through a tool named questionnaire star. In order to ensure the quality of the questionnaire, an option jump is set in the first question of the questionnaire, i.e., users who have not seen network violence can directly end the answer and set their questionnaire as an invalid one; The questionnaire variable adopts Likert five level scale (see Table 1), and scores are used to reflect the intensity of attitude, which are recorded as 5 points, 4 points, 3 points, 2 points and 1 point respectively, correspond to very agree, agree, general, disagree and very disagree.

4 Analysis of Empirical Results

Due to regional restrictions, this questionnaire is mainly distributed online. The "questionnaire star" is used to establish an online questionnaire, and the distributed questionnaire is linked to QQ, Wechat and other platforms to fill in. 224 questionnaires were filled in and 224 were recovered, of which 218 were valid, and the effective recovery rate was 97%.

4.1 Reliability and Validity Analysis

According to the reliability test of the five dimensions of attitude and opinion, subjective norms, network environment, intention of network public opinion violence and actual behavior of network public opinion force by SPSS 24, the overall reliability of the scale is 0.950 (>0.7), and the confidence level coefficients of each variable are more than 0.7. The credibility is good, indicating that the questionnaire has good consistency, reliability and stability, which can be used in this study, shown in Tables 2 and 3.

Table 1. Questionnaire items

Variables	Items
VA (View and attitude)	VA_1: Netizens' attacks and boycotts are the embodiment of a sense of justice for the sake of social stability
	VA_2: Internet users' attacks and boycotts are to highlight the sense of existence and hope to be recognized
	VA_3: Internet users' attacks and boycotts are impulsive and affected by their own emotions
SN (Subject norms)	SN_1: Netizens attack and resist because of the influence of conformity psychology
	SN_2: Netizens' attacks and boycotts are due to the influence of the impetuous social atmosphere
	SN_3: Netizens' attacks and boycotts are caused by the one-sided and subjective guidance of the news
WE (Website environment)	WE_1: The attack and resistance of netizens are due to the lack of comprehensive network management and moral constraints
	WE_2: In an environment where everyone is the media, netizens are more likely to abuse others when they see negative news
	WE_3: In the context of imperfect network laws and regulations, netizens are easy to abuse others when they see negative news
	WE_4: In an anonymous environment, netizens are more likely to abuse others when they see negative news
BI (Behavioral intention)	BI_1: I may forward unconfirmed information in the future
	BI_2: I may use uncivilized language to evaluate cyber violence in the future
	BI_3: I may have compassion for the objects of online violence in the future
AB (Actual behavior)	AB_1: On the Internet, they will comment on the parties in insulting language
	AB_2: Make or disseminate pictures that insult or vilify the parties on the Internet
	AB_3: Make videos on the Internet that insult and vilify the parties
	AB_4: Publish or forward the telephone number, address and other personal privacy information of the parties on the Internet

In Table 4, the KMO value is 0.892, between 0.8 and 0.9, indicating that it is very suitable for factor analysis, and the approximate chi-square of Bartlett's spherical test is 1767.294, P = 0.000 (<0.005) reaches a significant or conspicuousness level, which means that the data is also suitable for factor analysis. It can be seen from Table 5 that there are 17 items in the five dimensions in the questionnaire, and the KMO value is between 0.683–0.782, which is greater than 0.6, and P = 0.000 implies that the data is adapted for factor analysis.

Table 2. Overall reliability analysis of the questionnaire

Clone Bach Alpha	Clone Bach alpha based on standardized terms	Number of items
0.950	0.950	15

Table 3. Reliability analysis of questionnaire variables

Variables	Number of items	Clone Bach Alpha
View and attitude	3	0.785
Subject norms	3	0.946
Website environment	4	0.956
Network public opinion violence behavioral intention	3	0.898
Network public opinion violence actual behavior	4	0.974

Table 4. Overall validity analysis of the questionnaire

Kaiser Meyer Olkin measure of sampling adequacy		0.892
Bartlett's Spherical test	Approximate chi-square	1767.294
	Freedom	105
	Conspicuousness	0.000

Table 5. Validity analysis of each dimension of the questionnaire

Variables	Items	KMO	Bartlett's Spherical test	Conspicuousness
View and attitude	3	0.683	89.301	0.000
Subject norms	3	0.771	278.848	0.000
Website environment	4	0.765	320.297	0.000
Network public opinion violence behavioral intention	3	0.695	220.836	0.000
Network public opinion violence actual behavior	4	0.782	407.425	0.000

Generally, when the factor variance is greater than 0.4, it means that the practice is consistent with the theory. As can be seen from Table 6, the common factor variance indicators of attitudes and views are not high, but they are all greater than 0.5, so the indicators basically tend to be consistent; The common factor variance of network public opinion violence intention BI1 and Bi2 is bigger than 0.8; The common factor variance of Bi3 is low, but still above 0.4, indicating that the item can reflect relevant variables; The common factor variance of the three dimensions of subjective norms, network environment and actual action of network public opinion violence is large, all of which are greater than 0.7, which can better explain the consistency of data.

Table 6. Common factor variance

Variables	Items	Numerical extraction
VA	VA_1	0.525
	VA_2	0.550
	VA_3	0.517
SN	SN_1	0.743
	SN_2	0.805
	SN_3	0.823
WE	WE_1	0.810
	WE_2	0.825
	WE_3	0.810
	WE_4	0.820
BI	BI_1	0.851
	BI_2	0.863
	BI_3	0.519
AB	AB_1	0.874
	AB_2	0.885
	AB_3	0.898
	AB_4	0.875

4.2 Correlation Analysis Among Dimensions

According to the data in Table 7, the $P < 0.001$ of the five dimensions indicates that the significance exists, and $R > 0$, and there is a positive correlation between them. The R value of VA is 0.663, which indicates that there is a significant positive correlation between the intention of network public opinion violence and attitude; The R value of SN is 0.547, which means that there is a significant medium positive correlation between the intention of network public opinion violence and subjective norms; The R value of WE is 0.470, indicating that there is a significant medium positive correlation between the intention of network public opinion violence and the network environment; The R value of BI is 0.849, indicating that there is a significant positive correlation between the actual behavior of network public opinion violence and the intention of network public opinion violence.

Table 7. Correlation between dimensions

	"r" or "P"	VA	SN	WE	BI	AB
VA	r	1	0.641*	0.500**	0.663**	0.701**
	P		0.000	0.000	0.000	0.000
SN	r	0.641**	1	0.764**	0.547**	0.505**
	p	0.000		0.000	0.000	0.000
WE	r	0.500**	0.764**	1	0.470**	0.431**
	p	0.000	0.000		0.000	0.000
BI	r	0.663**	0.547**	0.470**	1	0.849**
	p	0.000	0.000	0.000		0.000
AB	r	0.701**	0.505**	0.431**	0.849**	1
	p	0.000	0.000	0.000	0.000	

Note: In the second column, "r" refers to the Pearson correlation coefficient; "P" is the conspicuousness (two tailed)

4.3 Regression Coefficient Test

As can be seen from Table 8, the regression coefficients of the five hypotheses are: beta > 0, and the conspicuousness or significance is $P < 0.001$. Therefore, there is a significant positive correlation between the intention of online public opinion violence and attitudes, subjective norms and online environment.

Table 8. Regression coefficient

Hypothesis	Non standardized coefficient		Standardization coefficient		
	B	Standard deviation	Beta	t	Conspicuousness
H_1	0.702	0.080	0.663	8.774	0.000
H_2	0.551	0.085	0.547	6.472	0.000
H_3	0.508	0.097	0.470	5.268	0.000
H_4	0.967	0.061	0.849	15.908	0.000
H_5	0.865	0.045	0.634	7.287	0.000

There is a significant positive correlation between the intention of network public opinion violence and the actual behavior of network public opinion violence. This result is the same as the previous conclusion, which further verifies the validity of the previous hypothesis.

5 Conclusions and Recommendations

5.1 Research Conclusion

This paper draws the following three conclusions.

(1) There is a significant positive correlation between the intention of network public opinion violence and attitude.
Attitudes and opinions have a great impact on netizens themselves. Nowadays, the popularity of mobile phones has led to the network being closely related to almost everyone. The Three Outlooks of college students are not yet fully mature, so they are easily affected by the attitudes and views of others, and encounter network public opinion events. For example, some netizens lack the ability of self-emotion management, like to express their views willfully, and even use inflammatory and harmful remarks to fierce attack the parties of public opinion. On the other hand, because some people inevitably encounter difficulties in life and need to vent their depressed emotions, they may use the anonymity of netizens to make verbal attacks on other people's behavior on the Internet, which leads to the generation of entrapment public opinion violence.

(2) There is a significant medium positive correlation between the intention of network public opinion violence and subjective norms.
Netizens often show a herd mentality and follow others' remarks, which virtually make network public perspective evolve into netizens' public opinion violence. The impetuous social atmosphere and traditional violent thinking are one of the factors affecting behavior intention. The survey presents that under the influence of social atmosphere, college students generally have a certain psychology of "hating the rich" and "hating officials". The phenomenon of "flaunting wealth" on the net is prevalent. Many college students exaggerate "showing off their wealth" for vanity, which further intensifies the contradiction between the rich and the poor. The behavior intention is also affected by the one-sided and subjective publicity of the news media (including We Media). Since many students and Internet users lack the ability of independent thinking and judgment, the information transmitted by the media is likely to guide the thoughts of netizens to produce certain or even wrong tendencies. If the media people are ill intentioned, do everything they can to win the attention of netizens, deliberately strengthen the dark side of the event or the parties, and add personal views to deliberately exaggerate, in this way, it is inevitable that netizens, including the college students, will be misled and make wrong judgments.

(3) The network environment will affect the attitudes and subjective norms of Internet users, and there is a correlation between the Internet public opinion and the intention of violence.
In today's Internet globalization, Internet users can communicate without regional restrictions with the help of computers and mobile phones. In the network, everyone enjoys the right to freedom of speech. Because of the anonymity and virtuality of the network, Internet users can express their opinions freely on the network, which may lead to the occurrence of network public opinion violence.

College students get information and participate in discussions at the first time, but they are often full of offensive remarks, which will more convey their feelings. The characteristic of anonymity makes these bad critics more rampant and evaluate all kinds of events and characters at will, which results in the intention of online public opinion violence. The content of network media reports is also an important factor in the network environment. As a key to transmitting information, if the media blindly pursue to obtain "heat", release unconfirmed news content, stimulate the discussion of Internet users, and then produce the intention of network public opinion violence. The main reason for behavior intention is the lack of a sound network comprehensive management system and moral constraints. The characteristics of the network itself will lead to random comments by netizens. Coupled with the imperfect network system and the lack of moral constraints, netizens will use insulting and aggressive words on the network, which will aggravate the intention of network public opinion violence.

5.2 Policy Recommendations

Basing on the above research and conclusions, the following policy suggestions are put forward.

(1) Improving college students' moral quality
Guide college students to establish a correct attitude and control their behavior intention. If it is fully capable to improve the moral quality of college students, so that Internet users will first think rationally from the perspective of law when solving problems, so as to alleviate college students' emotions. Next, college students will not use uncivilized words in comment and communication, and learn rational thinking methods to create a harmonious network environment.
(2) Perfecting the comprehensive network management system
The media should be encouraged to produce and publicize themes and contents full of positive energy, such as publicizing and broadcasting the real stories of "moving China". At present, the legal definition of network public opinion violence is relatively vague, ignoring the legal restrictions on speech attack, slander, human flesh search and other acts to a certain extent, and objectively condoning the occurrence of such acts. For these reasons, the law should give a clear definition of network public opinion violence and enhance the strength and efficiency of regulation. In addition, it is necessary to strengthen content check and tracking, and review the information of media before release. Due to the increasing number of media in China, it seems to be available for the media to cooperate with the platform to reinforce the audit, implementation and inspection.
(3) Developing the network real name system
The anonymity and virtuality of the network will inevitably make the publishers of information publish bad content because of the loss of self-discipline and morality, resulting in the violence of network public opinion. The establishment of real name system can be regarded as an effective method to control this situation. Although China is implementing the real name system, the effect is not obvious yet. At present, it is impossible to register and run all websites with the real name in a real

sense. The implementation of the real name system is absolutely dependent on huge technical support, which still needs the continuous efforts of the majority of scientific and technical personnel in China.

Acknowledgment. This paper is supported by three projects as follows: (1) Research on technological innovation mechanism of strategic emerging industries in East Lake National Independent Innovation Demonstration Zone (2019ADC039); (2) Comparative study on scientific research efficiency of colleges and universities in Hubei Province from the perspective of "double first class" (20G079); (3) Study on spatial differentiation and growth effect of logistics industry in Central China (D2014005).

References

1. Al-garadi, M.A., Varathan, K.D., Ravana, S.D.: Cybercrime detection in online communications: the experimental case of cyberbullying detection in the Twitter network. Comput. Hum. Behav. **63**, 433–443 (2016)
2. Zhang, Y., Wu, C., Liu, F.: Exploration of attitude change theory in online public opinion guidance. E3S Web Conf. **253**, 3018 (2021)
3. Wegge, D., Vandebosch, H., Eggermont, S., Pabian, S.: Popularity through online harm: the longitudinal associations between cyberbullying and sociometric status in early adolescence. J. Early Adolesc. **36**(1), 86–107 (2016)
4. Shaw, C.D., Hadwen, T., Cardoso, K., et al.: Longitudinal impact of the cyber friendly schools program on adolescents' cyberbullying behavior. Aggress. Behav. **42**(2), 166–180 (2016)
5. Feng, X., Ma, J., Jiang, G.: Social trust, rational behavior, and government microblogs' dissemination: an empirical analysis based on text mining. J. Inf. **038**(009), 954–965 (2019). (in Chinese)
6. Khweiled, R., Jazzar, M., Eleyan, D.: Cybercrimes during COVID-19 pandemic. Int. J. Inf. Eng. Electron. Bus. **13**(2), 1–10 (2021)
7. Martínez-Monteagudo, M.C., Delgado, B., Díaz-Herrero, Á., et al.: Relationship between suicidal thinking, anxiety, depression and stress in university students who are victims of cyberbullying. Psychiat. Res. **286**, 112856 (2020)
8. Thazin, K.A., Mon, S.Y., Ye, H.Z., et al.: Assessing risk factors and impact of cyberbullying victimization among university students in Myanmar: a cross-sectional study. PLoS ONE **15** (1), 68–72 (2020)
9. Chua, J.P., Liew, A.M., Wong, K.H., Yeo, P.J.: The digital scarlet letter: online shaming in Singapore. Bench Bar Minnesota **12**, 56–60 (2018)
10. Zhang, K.: Formation mechanism of cyber violence from the perspective of "silent spiral." Young Report. **32**, 101–102 (2018). (in Chinese)
11. Li, Y.: Violence effect of network discourse - to interpret the generation of network violence. Contemp. Commun. **05**, 26–28 (2014). (in Chinese)
12. Qiu, R.: Research on Internet public opinion and public opinion guidance under the background of innovative social governance. J. Ocean Univ. China **06**, 88–91 (2016). (in Chinese)
13. Xu, C.: On criminal law regulation of cyber violence. Appl. Law **03**, 102–108 (2016). (in Chinese)

14. Wang, J.: On the causes of Internet public opinion violence. News Res. Guide **8**(08), 59–181 (2017) (in Chinese)
15. Jiao, J.: Research on Internet public opinion violence in colleges and universities under the background of Internet. News Front **12**, 122–123 (2017). (in Chinese)
16. Liu, Y., Qiu, T.: Research on the formation mechanism of Internet public opinion violence. Media Observ. **07**, 30–35 (2018). (in Chinese)
17. Guan, Y.: Research on the formation mechanism and governance of Internet public opinion violence. Mod. Bus. Indust. **40**(07), 140–141 (2019). (in Chinese)
18. Hui, X., Liu, B., Li, J.: Research on the monitoring system of government online public opinion under the background of big data. Econ. Res. Guide **30**, 164–165 (2018). (in Chinese)
19. Hanxiang, W., Mingjun, X., Weimin, L., Zhihua, N.: An approach to micro-blog sentiment intensity computing based on public opinion corpus. Int. J. Wirel. Microw. Technol. **2**(5), 53–61 (2012)
20. Xin, M., Wu, H., Li, W., Niu, Z.: A public opinion classification algorithm based on micro-blog text sentiment intensity: design and implementation. Int. J. Comput. Netw. Inf. Secur. **3** (3), 48–54 (2011)
21. Chen, X.G., Duan, S., Wang, L.: Research on trend prediction and evaluation of network public opinion. Concurr. Comput. Pract. Exp. **29**(24), 4212 (2017)
22. Xiao, R., Li, J., Chen, T.: Modeling and intelligent optimization of social collective behavior with online public opinion synchronization. Int. J. Mach. Learn. Cybernet. **10**(8), 1979–1996 (2019)
23. Hryshchuk, R., Molodetska, K., Syerov, Y.: Method of improving the information security of virtual communities in social networking services. CEUR Workshop Proc. **2392**, 23–41 (2019)

Training Mode and Evaluation Method of Entrepreneurial Talents in Higher Vocational Education

Guowei Han(✉)

Changjiang Polytechnic, Wuhan, China

Abstract. Mass entrepreneurship and innovation is a major strategy proposed by the state based on the needs of transformation and development and domestic innovation potential, which aims to optimize the innovation and entrepreneurship environment, stimulate people's infinite wisdom and creativity, give those who are capable and want to start a business and innovation the opportunity to display their talents, and realize self-reliance and brilliance by innovation. Under the social trend of mass entrepreneurship and innovation, the cultivation of entrepreneurial talents has become an important task of higher education. Focusing on the training mode of entrepreneurial talents in higher vocational colleges, this paper clarifies the concept and direction of talent training, analyzes the form of teaching organization and curriculum design, discusses the mode of educational management, designs evaluation methods and mathematical models, and comprehensively evaluates teachers and students from different evaluation subjects and evaluation contents, in order to provide reference and ideas for the improvement of entrepreneurial talent training in higher vocational colleges.

Keywords: Higher vocational education · Entrepreneurial talents · Evaluation method

1 Introduction

In the era of digital economy, continuous improvement and application of new technologies including big data, cloud computing, Internet of Things, blockchain, artificial intelligence, and 5G communication have facilitated modern industry into a more intelligent, effective, rapid and flexible smart-industry stage [1]. China is currently in an active period of innovation and entrepreneurship, and innovation and entrepreneurship are the driving force for a new-round technological progress and industrial transformation [2]. Traditional industries that rely on information technology for continuous transformation and upgrading urgently need entrepreneurial talents who are brave to innovate and daring to reform. Entrepreneurial talents training has become an important approach for higher vocational colleges to adapt to industry changes, support industry development, and respond to the call of the "mass entrepreneurship and innovation" era [3]. Consequently, a scientific and effective talents training mode play a vital role in enhancing the quality of entrepreneurial talents training [4].

Z. Hu et al. (Eds.): AIMEE 2021, LNDECT 107, pp. 409–421, 2022.
https://doi.org/10.1007/978-3-030-92537-6_38

In terms of competency requirements, different from other types of talents, entrepreneurial talents must possess entrepreneurial knowledge, entrepreneurial awareness, innovative spirit, entrepreneurial ability and social responsibility, making them high-level professionals with outstanding professionalism in great need of innovative spirit and practical ability [5, 6]. The talents training mode refers to the theoretical model and operating style of the talents training process composed of several elements, designed by the training subject for specific talents training goals under the guidance of certain educational concepts and the guarantee of a certain training system and characterized by systematicness, purposiveness and imitability.

The talents training mode doesn't equal talent training. It is only a subsystem in the talents training system, and involves eight elements including talents training concept, major setting mode, curriculum setting mode, teaching institutional system, organization forms of teaching, teaching management mode, hidden curriculum form, and teaching evaluation methods [7]. Entrepreneurial talents training mode for higher vocational education refers to the paradigm of talents training process designed and constructed on the basis of two key attributes as "higher vocational education" and "entrepreneurship-oriented" [8]. For this specific research object, five aspects including talents training concepts, curriculum settings, and teaching organization forms, teaching system and management mode, and teaching evaluation are the key elements for the research on the entrepreneurial talents training mode for higher vocational education [9, 10]. The five elements are interrelated and mutually supportive, for talents training concepts point out the design direction, the curriculum setting sets up the carrier of education, the teaching organization method provides education methods, the teaching system and management mode provides guarantee, and the teaching evaluation facilitates improvement through feedback [11–14].

2 Entrepreneurial Talents Training Concepts for Higher Vocational Education

Talents training concepts refer to the ideal pursuit and educational concept formed by the training subject after rational thinking about the related issues of talents training, and the basis for the design of the talents training mode. Entrepreneurial talents training concepts for higher vocational education includes two parts: vocational skills training concept and entrepreneurship education concept.

2.1 Vocational Skills Training Concept

Entrepreneurial talents training for higher vocational education should be based on vocational skills training. The growth of high-skilled talents is a gradual process and usually requires a relatively long training period [15]. Therefore, designers of entrepreneurial talents training modes for higher vocational education should have the awareness that skills learning cannot be accomplished instantly, and quick success is hard to achieve in skill training. As a matter of fact, the role of higher vocational education in the growing process of vocational skills is fundamental. By facing up to the basic position of higher vocational education in vocational skills training, the "rash advance" phenomenon that may occur in higher vocational skills training can be avoided to a certain extent, as well as the resource waste incurred by unrealistic vocational skills training orientation and low efficiency of the implementation of corresponding talents training modes.

2.2 Entrepreneurship Education Concept

The knowledge of entrepreneurship education is based on the knowledge of entrepreneurship. Entrepreneurship, based on a professional background, contains three stages including pre-employment learning, orientation and starting position-based entrepreneurship, establishing a new enterprise and becoming self-employment. Self-employment, as an advanced stage of entrepreneurship, is not feasible to everyone, but the first two stages of entrepreneurship can be experienced by anyone with entrepreneurial motivation. The growing of entrepreneurship requires continuous efforts of entrepreneurs, and receiving entrepreneurship education can help to improve the ability of innovation and entrepreneurship, but it is not a necessary condition for entrepreneurship. In particular, self-employment with greater achievements is not directly promoted by entrepreneurship education. Therefore, the goal of entrepreneurship education should not be self-employment or even considerable entrepreneurial performance that only a few people can achieve. Instead, more attention should be paid to the first two stages of entrepreneurship, making the basic vocational skills, innovative thinking, and entrepreneurial awareness training and inspiration required by entrepreneurship as the training objectives, thus avoiding elitism and utilitarianism in entrepreneurial education (Fig. 1).

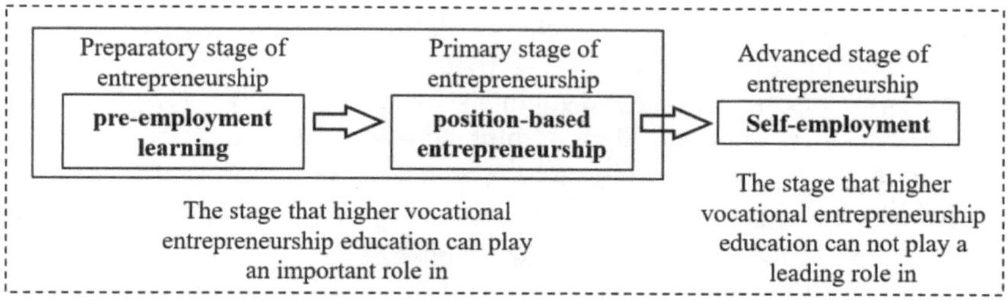

Fig. 1. Stage of entrepreneurship division

In summary, high-level skills and successful entrepreneurship, acquired and achieved through a long-term career and entrepreneurial career, are definitely not to be facilitated by three-year higher vocational education which is fundamentals for vocational skills training and entrepreneurial ability training. Entrepreneurial talents training for higher vocational education is not entrepreneurship-oriented, nor aimed at making students bosses who can successfully start a business, or high-level vocational skills training. It should be positioned at "access" level, with basic skills, innovative thinking ability, and training and inspiration of entrepreneurial awareness as the main tasks by integrating entrepreneurship education into professional education, and training entrepreneurial talents for basic positions and the initial stage of entrepreneurial activities [16–18]. Consequently, elitism and utilitarianism in talents training can be avoided and education can finally be student-oriented and universally suitable.

3 Organization of Entrepreneurial Talents Training for Higher Vocational Education

3.1 Entrepreneurial Talents Training System Organization

The teaching organization form is the concrete embodiment of the talents training mode in the course teaching activities, and depicts the way that teachers systematically arrange the teaching time, teaching site, teaching resources, and teacher-student activities in the teaching activities. Different teaching organization forms will exert differentiated impact on the knowledge acquisition, skill improvement and quality training of the students. There are two keys in the entrepreneurial talents training for higher vocational education. One is the cultivation of basic skills, and the other is the cultivation of innovative thinking, independent thinking and problem-solving abilities indispensable for entrepreneurship.

The traditional ways of teaching organization, which is teacher-centered and highly depends on teachers' lectures, students' listening, in-class learning, and after-class exercises, can't tackle the two key issues in the entrepreneurial talents training for higher vocational education. Only by giving students' initiative to full play with students being the center and applying the teaching organization forms of problem-oriented, task-oriented inquiry, action-oriented, flipped classroom with the real projects of the enterprise as the carrier can students' vocational ability and innovation and entrepreneurship ability be better cultivated [19, 20]. Inquiry-based teaching is problem-oriented.

In the process of enlightening and guiding students to actively solve problems, their abilities in innovative thinking, logical thinking, observation and the ability to use various learning tools can be fully exercised, which are very useful for innovation and entrepreneurship training. Therefore, it should become a teaching form often used in the process of entrepreneurial talents training for higher vocational education. Action-oriented teaching form, with simulation or real learning tasks derived from actual work as the carrier, by means of six-step teaching method as "information, planning, decision-making, implementation, examination and evaluation" [21], guides students or student teams to voluntarily completes the learning and application of vocational knowledge in the learning tasks completion process and to facilitate skills enhancement and work experience accumulation in thinking, innovation and practice, which can exert strong impact on the cultivation of vocational skills and innovation and entrepreneurship. By reverting the traditional "in-class lectures, after-class homework" teaching sequence, flipped classroom teaching form transfers more meaningful and in-depth teaching activities such as problem exploration, practical operation, evaluation feedback and reflection into the class while the large amount of knowledge teaching [22], especially teaching of simple knowledge stay outside the class, which can better develop students' initiatives, better utilize their after-class learning hours, improve their in-class efficiency for high-quality entrepreneurial talents training in higher vocational colleges.

The teaching organization forms above effectively use both in-class time and extra-curricular time regarding teaching time, so that students can use extra-curricular time to conduct independent learning and exploration under the careful arrangement of teachers, creating necessary conditions for more efficient in-class teaching activities. In terms of teaching venues, it extends from conventional teaching venues such as classrooms and training rooms to unconventional teaching venues such as libraries and dormitories. Moreover, teaching resources can be more diversified and information-based that teaching resources composed of multiple elements including text, pictures, video, and audio, material object, scenarios and tasks can be utilized in the form of paper material or digital materials including textbooks, task lists, micro-classes, PPT, video open courses. Regarding teacher-student activities, "teacher-centered" has been changed to "student-centered", "cramming education" to "doing it on your own." The arrangement of teaching organization in all aspects is changing towards what is conducive to entrepreneurial talents training for higher vocational education.

3.2 Curriculum Setting for Entrepreneurial Talents Training for Higher Vocational Education

Curriculum setting is the core component of entrepreneurial talents training mode for higher vocational education, and a synthesis of curriculum outlook, curriculum structure, curriculum content, and curriculum objectives [2].

3.2.1 Curriculum Outlook

Curriculum concept is the value pursuit of curriculum setting. The curriculum setting for entrepreneurial talents training in higher vocational colleges should hold "adhering to the development path of socialist education with Chinese characteristics, and cultivating socialist builders and successors with all-around development of moral, intellectual, physical, aesthetics and labor education" as the highest pursuit. As the basic unit for morality cultivation in higher vocational colleges, curriculum should provide an education platform for the cultivation of the five qualities of moral, intellectual, physical, aesthetics and labor education in its structure and content, and must provide carriers not only for the cultivation of vocational skills, innovative thinking, entrepreneurial awareness, and innovative spirit, but also for the cultivation and improvement of students' outlook on life, values, ideals and beliefs, as well as physical fitness, psychological qualities, aesthetic judgement, and labor values.

3.2.2 Curriculum Structure

The curriculum structure is a comprehensive combination of curriculum types, curriculum categories, curriculum linking, class hours and credit settings. As to the curriculum structure for entrepreneurial talents training in higher vocational colleges, five types of courses should be set up in accordance with talents training concept and curriculum outlook. The first type refers to public basic courses including courses on ideological and political, laws and regulations, PE, and aesthetic education. The second refers to professional basic courses that teach basic knowledge. The third involves the professional core courses for vital job ability training, and the fourth refers to professional practice courses aimed to cultivate comprehensive vocational skills and experience the workplace atmosphere and entrepreneurial procedures. The fifth refers to professional development courses in order to improve students' humanistic quality and aesthetic taste, and expand professional knowledge and competence. Each course type contains several specific courses. The organization of public basic courses should be based on socialism core values. The categories of professional basic courses and professional core courses should be jointly determined by schools and enterprises based on industry research and job analysis after clarifying the composition of important positions in the industry, the requirements of core positions, and the frontier dynamics of industry development.

The setting of professional practice courses should rely on in-depth school-enterprise cooperation, rich simulation and practical training projects, well-functioning training bases in and outside the school, with a series of courses such as step-by-step project training, entrepreneurial experience and entrepreneurial incubation included [23]. Professional development courses should be expanded around the center of higher vocational entrepreneurial talents training, and involve courses that are conducive to raising awareness, expanding knowledge, and improving competency. For several similar majors, professional groups can be organized to achieve coordinated development. The majors in the professional groups can share professional group platform courses and public basic courses, while professional basic courses, professional core courses, professional practice courses and professional development courses can be adjusted according to the characteristics of the major (Fig. 2).

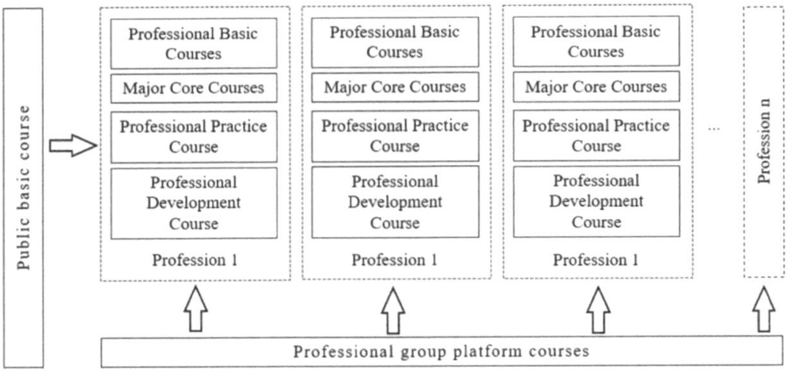

Fig. 2. Professional group curriculum system

Curriculum linkage in entrepreneurial talents training for higher vocational education should be determined on the basis of comprehensive consideration of career and entrepreneurial growth pattern, ability training pattern, students' learning characteristics, curriculum content and functions, by combining talents training ideas including "platform + module", "graded progression", "integration of theory and practice", and "learning, imitating, and doing". For example, according to the idea of "platform + module", professional basic courses can be firstly confirmed, and a professional core course for an entry-level position in accordance with the needs of entrepreneurship will be selected, and serve as the center for expansion and deepening. Then the relevant courses and the central courses are opened simultaneously to form a course module for ability training, and each course module contains multiple types of courses. Several course modules designed according to the above ideas, if organically integrated according to a certain progressive relationship, will constitute the entire "platform + module" curriculum system for the major. The design ideas for other curriculum linkages, with their own emphasis, must follow the law of career and entrepreneurial growth and ability training. The design of class hours and credits is based on the importance of the course, the amount of content and the actual situation of the teaching arrangement. In terms of class hours and credits, professional core courses and professional practice courses outnumber public basic courses, professional basic courses and professional development courses.

3.2.3 Curriculum Content

Curriculum content is the soil for the cultivation of knowledge and skills. Whether this soil is fertile is directly related to the selection and ordering of the curriculum content. The selection of curriculum content for entrepreneurial talents training in higher vocational colleges should keep pace with the times, match occupations, integrate theory and practice, keep proper amount and highlight practice. In particular, the content of professional courses should reflect the cutting-edge dynamics of the industry and the latest trends in the development of technical skills, and should be selected based on the feasibility to cultivate the abilities required by important positions. It should be summed up from actual work and can be used for work. Moreover, based on the basic status of vocational education, the theoretical content should be appropriate and the practical part sufficient, while the selected course content needs to be sequenced before it can be used in continuous teaching activities. The key procedure of entrepreneurial talents training in higher vocational colleges is to guide students to complete continuous practical learning tasks and entrepreneurial experience learning tasks. The ordering of these learning tasks should be based on the work process and entrepreneurial process. As a result, the curriculum content can be appropriately sequenced based on systematic analysis of work tasks, work process and entrepreneurial process on the premise of in-depth industry and enterprise research, extensive interviews on seniors and successful entrepreneurs.

3.2.4 Curriculum Objectives

From macro to micro dimensions, curriculum objectives can be divided into three levels: curriculum system teaching objectives, curriculum teaching objectives and classroom teaching objectives, which are closely related to talents training concepts, curriculum concept, job requirements, and students' learning characteristics. According to the talents training concepts and the curriculum concept proposed above, the teaching objectives of curriculum system for entrepreneurial talents training in higher vocational colleges can be summarized as: "through systematic curriculum teaching, while the students' socialist core values are strengthened, and their outlook on life and values well established, students should be proficient in the basic vocational knowledge and skills of key positions in a certain industry, familiar with the development status of the industry, understand the development trend of the industry, technological innovation and business model innovation direction, and possess the awareness of innovation and entrepreneurship, innovative thinking ability and basic entrepreneurship knowledge, thus becoming entrepreneurial and skilled talents for the comprehensive development of morality, intelligence, physical education, art, and labor". The curriculum teaching goal is to detail the teaching goals of the curriculum system, and should be determined according to the specific positions requirements or professional standards corresponding to the curriculum under the leadership of the curriculum system teaching goals. Classroom teaching objectives specify the curriculum teaching objectives, usually involve a goal to be achieved in the teaching of a certain knowledge or skill point, serve as the basis for teachers to implement specific classroom teaching behaviors, and should be based on the curriculum teaching objectives and students' learning characteristics (Fig. 3).

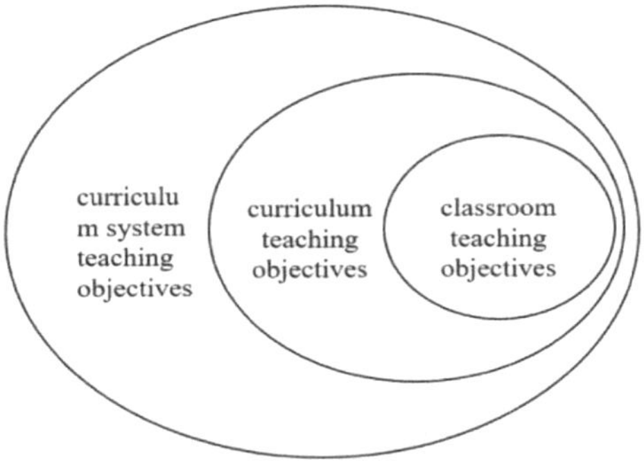

Fig. 3. Three levels of teaching objectives

4 Management and Evaluation of Entrepreneurial Talent Training in Higher Vocational Colleges

4.1 Management Mode of Entrepreneurial Talent Training in Higher Vocational Colleges

Teaching system is the general name of various rules, regulations, norms, methods and schemes closely related to talent training. It is the guarantee for the efficient and standardized implementation of talent training mode and the realization of predetermined talent training objectives. Teaching management mode is a means and method to manage the whole process of talent training.

The training and management objects of entrepreneurial talents in higher vocational colleges include people, matters and objects.

The management of "people" includes the introduction, training, management and assessment system of full-time teachers, the employment, management and assessment system of part-time teachers and entrepreneurial mentors in enterprises, the selection methods of students' entrepreneurial partners, the evaluation methods of students' entrepreneurial quality, students' code of conduct, etc.

The management system of "matters" includes the diagnosis and improvement system of entrepreneurial talent training quality, classroom teaching norms, classroom teaching supervision system, regular teaching and research meeting system, entrepreneurial performance and credit exchange system, assessment scheme, management methods of students' entrepreneurial projects, management methods of vocational skills competition, school enterprise cooperation mechanism, etc.

The management of "objects" includes the management system of classroom and training room, and the management measures of entrepreneurship incubation base and practice training base. The establishment of management mode needs to be based on the concept of entrepreneurial talent training in higher vocational colleges, combined with the characteristics of employment and entrepreneurship, the characteristics of enterprise operation, the mode of entrepreneurship incubation and other factors, and formulated with the joint participation of schools and enterprises. The teaching system of entrepreneurial talent training in higher vocational colleges has great room for improvement. It is necessary and urgent to establish a perfect teaching system.

4.2 Teaching Evaluation of Entrepreneurial Talent Training in Higher Vocational Colleges

Teaching evaluation involves many aspects. As an important part of talent training mode, it mainly involves two contents: teaching evaluation and learning evaluation, as shown in Fig. 4.

Teaching evaluation is to evaluate teachers' teaching. At present, it is mainly carried out through five ways: Teachers' self-evaluation, peer evaluation, students' evaluation, supervision evaluation and enterprise evaluation. The evaluation of teachers' ability and level mainly depends on whether they have comprehensive knowledge, solid skills and innovative thinking. It also depends on whether they can effectively organize practical teaching, teach students according to their aptitude, and realize the collaborative education between schools and enterprises.

Learning evaluation is to evaluate students' learning ability and effect, mainly by teachers and enterprise experts. The focus is on students' fundamental knowledge, basic skills, innovative thinking, entrepreneurial consciousness, and their performance in practical learning and training.

The cultivation of entrepreneurial talents in higher vocational colleges needs much more effective practical training and collaborative education between schools and enterprises. Much attention should be paid to the cultivation of basic knowledge, comprehensive skills, innovative thinking and entrepreneurial consciousness, which requires targeted evaluation to judge the effectiveness, and then put forward further improvement plans.

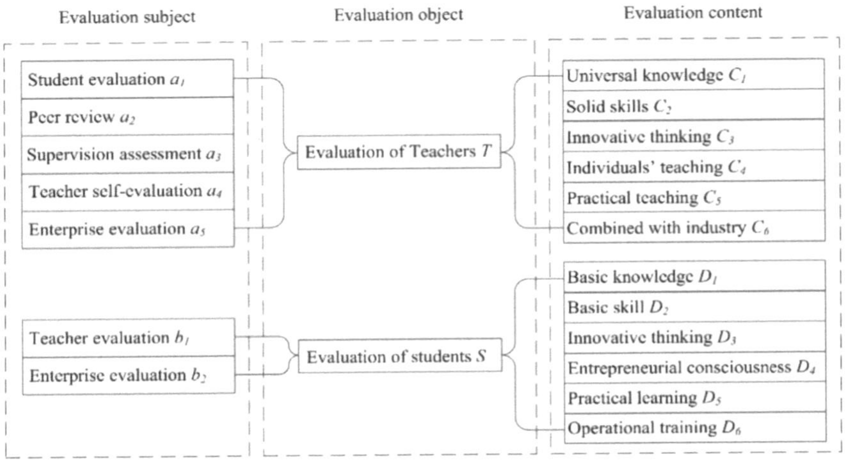

Fig. 4. Evaluation of teaching and learning

4.3 Mathematical Model of Teaching Evaluation

Teaching evaluation is divided into two parts: teacher evaluation and student evaluation. According to different evaluation objects, the mathematical models of teacher evaluation and student evaluation are established respectively, described as follows.

4.3.1 Mathematical Model of Teacher Evaluation

According to the evaluation method and content shown in Fig. 4, the mathematical model of teacher evaluation can be established.

Let $T = (t_1, t_2, \ldots, t_y)$, be the vector of n teachers who are to be evaluated;

$A = (a_1, a_2, \ldots, a_5)$, which represents the vector composed of a group of judges who evaluate teachers in a specific round, and its scoring weight is set as Φ, and,

$$\Phi = (\varphi_1, \varphi_2, \ldots, \varphi_5),\ \varphi_i > 0,\ i = 1, 2, \ldots, 5,\ \sum\nolimits_{i=1}^{5} \varphi_i = 1.$$

Besides, $C = (c_1, c_2, \ldots, c_6)$ represents the dimension vector of teacher evaluation, that is, the index vector, and its weight is expressed as ω, then,

$$\omega = (\omega_1, \omega_2, \ldots, \omega_6),\ \omega_i > 0,\ i = 1, 2, \ldots, 6,\ \sum\nolimits_{i=1}^{6} \omega_i = 1.$$

According to the requirements of the teaching evaluation system and the rules of teacher evaluation, each of them needs to organize P rounds of comprehensive evaluation for teachers according to the same weight value.

In a specific round of evaluation process, five judges A evaluate the evaluation matrix of six indicators C of n teachers T.

$$R = (R_1, R_2, \ldots, R_n),$$

Where, $R_x = A_{xj}\psi^T \cdot (C_{xk}\omega^T)^T$, $x = 1, 2, \ldots, n$; j = 1, 2, …, 5; k = 1, 2, …, 6.

After several rounds of evaluation, the final evaluation matrix for n teachers is:

$$\overline{R} = \left(\overline{R_1}, \overline{R_2}, \ldots, \overline{R_n}\right),$$

Where, $\overline{R_i} = \frac{\sum_{k=1}^{p} R_{ik}}{p}, (i = 1, 2, \ldots, n;\ k = 1, 2, \ldots, p).$

4.3.2 Mathematical Model of Student Evaluation

Let $S = (s_1, s_2, \ldots, s_x)$ be the vector of m students to be evaluated,

$B = (b_1, b_2)$ refers to a vector composed of a group of judges who evaluate and score students, and the scoring weight is shown as Y, then,

$$\Gamma = (\gamma_1, \gamma_2),\ \gamma_i > 0,\ \gamma_1 + \gamma_2 = 1;$$

$D = (d_1, d_2, \ldots, d_6)$ means the dimension vector of student evaluation, that is, the index vector, and its weight is $\varDelta$, then,

$$\Delta = (\sigma_1, \sigma_2, \ldots, \sigma_6),\ \sigma_i > 0,\ i = 1, 2, \ldots, 6,\ \sum_{i=1}^{6} \sigma_i = 1.$$

According to the requirements of the teaching evaluation system, it is necessary to organize q groups of judges composed of teachers of different subjects and experts of different enterprises, and each group will conduct systematic and comprehensive evaluation on students according to the same weight value.

The evaluation matrix of five judges B on six indicators D of m students S is as follows:

$$E = (E_1, E_2, \ldots, E_m),$$

Where, $E_y = B_{yj}\Gamma^T \cdot (D_{yk}\varDelta^T)^T$, $y = 1, 2, \ldots, m$, j = 1, 2, k = 1, 2,…,6.

After the comprehensive evaluation of students by teachers of multiple subjects and experts from multiple enterprises, the final evaluation matrix for m students is as follows:

$$\overline{E} = \left(\overline{E_1}, \overline{E_2}, \ldots, \overline{E_m}\right),$$

Where, $\overline{E_j} = \frac{\sum_{k=1}^{q} E_{jk}}{q}, (j = 1, 2, \ldots, m; k = 1, 2, \ldots, q).$

The purpose of evaluation is to find weak links and correct them in time to enhance the quality of the teaching and learning system. Evaluation and improvement are cross cutting. The continuity of improvement requires that the evaluation process should be dynamic, the process evaluation and result evaluation be combined, and the interaction between evaluation and improvement be realized through teaching management.

5 Conclusion

The entrepreneurial talents training mode for higher vocational education with its own connotation and extension is a subsystem of the higher vocational talents training system and a paradigm related to the entrepreneurial talents training process. The correct refinery should be based on its concept and connotation starting from the analysis of the component elements of the talents training mode, and a targeted summary in combination with specific talents training objects.

The research on the entrepreneurial talents training modes for higher vocational education should not be limited to the summary and refinery of the empirical and imitable procedures formed in the long-term practice of a certain school or a certain major, but should also include research results that explore the concept, element systems, and design ideas with theoretical and general guiding significance on the entrepreneurial talents training modes for higher vocational education. The research in this article belongs to the latter.

Aiming at entrepreneurial talents training in higher vocational colleges, it has carried out a basic discussion around the component elements of the talents training mode in order to provide guidance for the design of entrepreneurial talents training modes in vocational colleges in accordance with their actual conditions.

References

1. Tam, M.: Outcomes-based approach to quality assessment and curriculum improvement in higher education. Qual. Assur. Educ. **22**(2), 158–168 (2014)
2. Xiaolong, H., Lin, S., Anni, M.: Research on the entrepreneurial talents training mode in colleges and universities based on competency model. Heilongjiang High. Educ. Res. **38** (07), 106–110 (2020). (in Chinese)
3. Ristovska, N., Stankovska, S.: Enhancing entrepreneurial skills of employees in the education process. UTMS J. Econ. **10**(1), 109–123 (2019)
4. Ellis, A.K.: Research on Educational Innovations. Routledge, London (2014)
5. Pastore, F., Zimmermann, K.F.: Contributions to school-to-work transitions: vocational training, skill mismatch and policy. Int. J. Manpow. **40**(8), 1361–1363 (2019)
6. Jianqiang, A.: Research on the entrepreneurial talents training mode in colleges and universities. Educ. Vocat. **3**, 32–33 (2015). (in Chinese)
7. Zefang, D.: Concept definition and element analysis of the talent training model in colleges and universities. Univ. Educ. Sci. **03**, 30–36 (2012). (in Chinese)
8. Tshai, K.Y., Ho, J.H., Yap, E.H., et al.: Outcome-based education–the assessment of programme educational objectives for an engineering undergraduate degree. Eng. Educ. **9** (1), 74–85 (2014)
9. Bansal, S., Gaffar, A., Dalrymple, O.: Building faculty expertise in outcome-based education curriculum design. In: 2015 IEEE Frontiers in Education Conference (FIE), pp. 1–8. IEEE (2015)
10. Barman, L., Bolander-Laksov, K., Silén, C.: Policy enacted–teachers' approaches to an outcome-based framework for course design. Teach. High. Educ. **19**(7), 735–746 (2014)
11. Haji, E.E., Azmani, A., Harzli, M.E.: Using AHP method for educational and vocational guidance. Int. J. Inf. Technol. Comput. Sci. **9**(1), 9–17 (2017)

12. Liu, S., Chen, P.: Research on fuzzy comprehensive evaluation in practice teaching assessment of computer majors. Int. J. Mod. Educ. Comput. Sci. **7**(11), 12–19 (2015)
13. Kotevsk, Z., Tasevska, I.: Evaluating the potentials of educational systems to advance implementing multimedia technologies. Int. J. Mod. Educ. Comput. Sci. **9**(1), 26–35 (2017)
14. Choi, S.J., Jeong, J.C., Kim, S.N.: Impact of vocational education and training on adult skills and employment: an applied multilevel analysis. Int. J. Educ. Dev. **66**, 129–138 (2019)
15. Tiancong, H.: Misunderstandings and model reconstruction of china's high-skilled talent training-from the perspective of high-skilled talents' growth. China High. Educ. Res. **07**, 100–105 (2017). (in Chinese)
16. Jianpeng, G.: Flipped classroom teaching mode: variation and unity. China High. Educ. Res. **06**, 8–14 (2019). (in Chinese)
17. Kyrpychenko, O., Pushchyna, I., Kichuk, Y., Shevchenko, N., Luchaninova, O., Koval, V.: Communicative competence development in teaching professional discourse in educational establishments. Int. J. Mod. Educ. Comput. Sci. **13**(4), 16–27 (2021)
18. De Matas, S., Keegan, B.P.: A case study on adult and workplace learning. Int. J. Educ. Manag. Eng. 10(1): 11–19 (2020)
19. Abubakari, M.S., Arifin, F., Hungilo, G.G.: Predicting students' academic performance in educational data mining based on deep learning using TensorFlow. Int. J. Educ. Manag. Eng. **10**(6), 27–33 (2020)
20. Sathe, M.T., Adamuthe, A.C.: Comparative study of supervised algorithms for prediction of students' performance. Int. J. Mod. Educ. Comput. Sci. **13**(1), 1–21 (2021)
21. Daliang, Z.: Optimize the curriculum system and strengthen curriculum construction. China Univ. Teach. **12**, 10–12 (2018). (in Chinese)
22. Zhiqun, Z.: About action-oriented teaching. Voc. Educ. Forum **20**, 1 (2008). (in Chinese)
23. Zhen, H.: The vocational school-enterprise cooperation: problems and policy analysis. China High. Educ. Res. **1**, 90–93 (2013). (in Chinese)

Analysis and Evaluation on Practical Teaching of Space Transformation Design of Old Buildings Based on Place Perception

Simeng Fan[1] and Xianhui Zeng[2](✉)

[1] Department of Environmental Art and Design, Hubei Institute of Fine Arts, Wuhan, China
[2] College of Art, Hubei University of Education, Wuhan, China

Abstract. With the development of urban construction, the requirements of old building transformation are improved, and the teaching demand of old building space transformation is also gradually strengthened. This paper analyzes the significance and function of the characteristic analysis of old building space transformation and architectural decoration patterns, discusses the teaching practice of old building transformation design based on place perception from the aspects of the activity organization of field investigation, "place perception" field teaching method and the arrangement of architectural decoration patterns, and makes a comprehensive evaluation of "place perception" field teaching method by using analytic hierarchy process, The improvement direction of practical teaching in the space transformation design of old buildings is clarified.

Keywords: Building reconstruction · Place perception · Practice teaching · Field investigation · Analytic hierarchy process

1 Introduction

The pace of urbanization is accelerating, and the development of urban construction has entered a new stage. In this context, incremental construction is transformed into stock construction, and the demand for the transformation of old buildings is gradually increasing, followed by the teaching demand for the transformation of old building space. Under such a background, the course of old building space transformation design has become a new professional restricted course for junior and senior students of the department of environmental art and design, which aims to cultivate and train students' ability to complete the old building space transformation design, understand the restrictive preconditions of the old building space, and master the basic theoretical knowledge, design specifications and relevant technical treatment means of space transformation design of old buildings.

This course adopts the method of practical teaching, which is often used in environmental design courses [1, 2]. It enables students to obtain a large number of first-hand materials and form a perceptual cognition through the perception of the environmental scene. It is an important learning means.

Z. Hu et al. (Eds.): AIMEE 2021, LNDECT 107, pp. 422–432, 2022.
https://doi.org/10.1007/978-3-030-92537-6_39

Based on the historical background and development characteristics of Wuhan, the old building space reconstruction design course focuses on understanding the old buildings and Li-Fen (a unique local name of alleyway) residential buildings in different historical periods, paying special attention to getting familiar with the architectural decorative components and patterns, so as to maintain its distinctive historical characteristics in the reconstruction of old buildings.

2 Analysis of Characteristics of Architectural Decoration Patterns in the Space Transformation of Old Buildings

2.1 Significance and Principles of Space Transformation of Old Buildings

"Repair the old as the old" is a principle followed in the restoration of cultural relics and ancient buildings in China. Maintaining the unique style of the old buildings while making them glow with new vitality is an important content for a city with cultural background to reflect its historical imprint [3–5]. It is a pity if history is separated in the process of building renovation, especially for those cities with "stories".

Architecture is called the container of people's social behavior. The research on the old architectural space should not only contain architectural components such as walls, plinths, door openings, windows, mountain flowers, roofs, but also include localized patterns and decorative components in architectural components. The style and materials of building components can reflect the scientific and technological level and technical details of construction engineering during the construction period, while the architectural spatial scale and streamline arrangement can reflect the social activity process and lifestyle during the construction period. A very important principle in the transformation and utilization of old buildings is the principle of cultural inheritance. The historical and cultural value of architecture is not only the core and key of the value of old buildings, but also the essential difference from other new buildings.

2.2 Function of Characteristic Analysis of Decorative Patterns of Old Buildings

Architectural decoration pattern is the most direct way to integrate local humanistic symbols and traditional cultural symbols into architecture. Architectural decorative patterns and components are divided into architectural structural components, non-structural components, pure decorative components and architectural decorative patterns.

Different historical periods will have their unique aesthetic characteristics and artistic styles. Architectural decoration patterns are an important factor to reflect the historical and cultural background of old buildings. By studying decorative components and patterns, the students are capable to grasp the architectural style more comprehensively and provide clearer design ideas for the transformation of old buildings.

As a city with a long cultural history, Wuhan has many old-fashioned buildings with distinctive historical characteristics. In many Li-Fen folk houses, such as

Dongting village, Jianghan village, Shanghai Village, Tongxingli and other places, there are a large number of architectural decoration component patterns, Including Shikumen mountain flowers, window edge lines, window frame columns, door decoration patterns, anti-theft net patterns, hall floor tiling, starting step columns of wooden stairs in the building, railing column decoration, etc. [6]. These are important contents that need to be updated on the basis of retention in the transformation of old buildings.

The research focus of old buildings is to explore their value. The three part values shown in Fig. 1 are: time value/age value, historical value, and reference value for modern architecture.

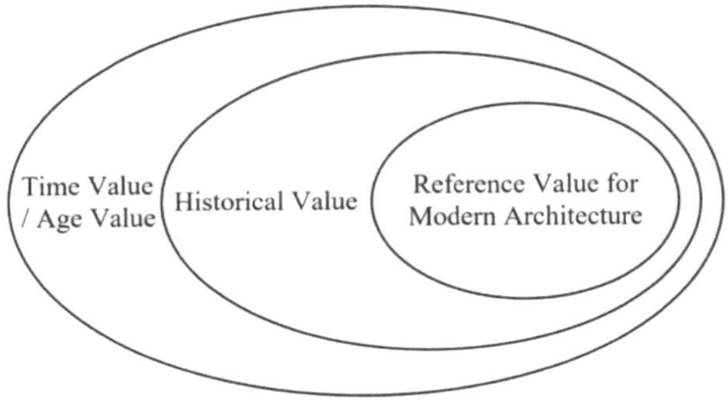

Fig. 1. Value exploration of old buildings

The old buildings and their decorative patterns in different historical periods show their "beauty" definitely, and also have a certain historical imprint [7–10]. Interpretation of architectural sites and decorative patterns revive the memories of that year and make them continue to reappear in contemporary scenes. Many architects in the world regard this as their responsibility. By excavating the historical traces of the site, Eisenman revived the historically existing ammunition depot in the design of the Vicksner Art Center of Ohio University, making it blend with the new building, so as to achieve symbiosis between the old memory and the new construction, and endow significance to the existence of the building in the specific environmental framework.

3 Teaching Practice of Old Building Reconstruction Design Based on Place Perception

3.1 Contents of On-site Teaching and Place Perception

On site teaching is an organizational form of teaching in natural and social activities, which is a necessary supplement to classroom teaching, and the continuation and development of classroom teaching either. It is a form of teaching organization associated with classroom teaching [11]. According to the arrangement of teaching tasks, organizing and arranging students to the construction site and personally completing teaching can significantly enhance students' intuitive feeling of theoretical knowledge and achieve twice the result with half the effort.

Architectural decoration pattern research is one of the important work in the early stage of old building reconstruction, and it is also the first practical link of old building reconstruction course. Through research activities and on-site teaching, it is available to obtain real place perception and provide basis and foundation for architectural design. In this sense, place perception refers to an experience. The place view of architectural psychology emphasizes people's perception of emotion and behavior in the place, and helps people obtain a sense of existence and belonging by establishing a place where people can connect with the world.

3.2 Organization of On-site Investigation Activities

The understanding and perception of the decoration design of old buildings are obtained through on-site investigation. According to the arrangement of the old building reconstruction course, a representative historical building was selected, and the students were divided into groups to carry out the research and practice of the old building reconstruction project design [10].

Teachers with rich practical teaching experience play a key role in the successful organization of research activities [12]. Before the research activity, the students were explained the relevant theories and methods, arranged to analyze the original drawings, understand the use function and moving line of the original space, and investigate the original use state of the building and the future use trend and possibility of the building. And the students were arranged to list research plans with project overview and existing problems, and conduct field research purposefully and directionally, so as to maximize efficiency. The research methods include on-site mapping, shooting, sketching, interview and dialogue with users and so on.

In the initial stage of the survey, teachers' work is mainly to guide students to carry out pattern mapping and comparative analysis of decorative connotation. On this basis, they can carry out graphic analysis, structural research, connotation study, deconstruction and reconstruction, which can deepen students' understanding of the function of architectural decorative components and the meaning of patterns.

General architectural space design is based on rational drawings to talk about design space and functional streamline. After integrating functional requirements, use streamline, material texture, spatial light and environmental factors through rational design process, plane layout and spatial representation diagram are formed to complete the design process [13]. The difference is that the research and teaching of the space transformation of the old building can not only be based on the original space given by the drawings, because the main body of the old building has been transformed for many times, the original space has been reconstructed and adjusted, and a large number of changes have not been marked on the original drawings, so the design guidance of the original space drawings has been greatly weakened. As the old building space has been set for a long time, the surrounding environment and internal structure have changed a lot, and the original design intention and logic may not be applicable. Simply repairing the old as the old does not necessarily reflect the original style of the building effectively. Based on this premise, the field investigation of the original space and the on-site survey of the original space community are more important in the process of the old building space transformation design course [14].

3.3 "Place Perception" On-site Teaching Method

The transformation of old buildings has outstanding comprehensive characteristics. According to the plane drawings, it is difficult to understand the complex current situation of old buildings, nor can we convey the real sense of space, scale and reality. This requires us to enter the real space, fully mobilize the five senses and perceive the original old buildings in multiple dimensions. Teaching method of "place perception" is to recognize space through perceptual walking and on-the-spot contact, and make spatial translation by combining rational analysis and evaluation of place, which is very effective in the teaching of space transformation design of old buildings.

In the on-site teaching of place perception, the teacher led the students to enter the Li-Fen folk house to collect pattern data, talk with the students according to the observation of the on-site environment, and guide them to pay attention to the elements that can reflect the unique style, culture and history of the old buildings. Most of these elements are decorative components and patterns [15].

The on-site presentation of architectural decoration patterns and construction is different from that expressed by two-dimensional paper media and screen media. The electronic information in drawings and three-dimensional models could reflect the proportion and color relationship between patterns and construction, but the perception of volume and texture can only have a comprehensive experience in the on-site light environment. In addition, the surrounding environment of the project, the user's use traces for many years, and the user's behavior will jointly form an overall design atmosphere. This ambient perception that requires perceptual and rational understanding needs to be understood by students.

The form of place perception on-site teaching makes the studying atmosphere relatively relaxed and natural. The feedback of students' on-site perception can be reflected in time through the communication between teachers and students. The conflict between the original design language and user behavior in old buildings is easy to be found by both teachers and students. Various details can become unique design entry points, resulting in different reconstruction and design ideas. This fully presents the advantages of on-site teaching.

3.4 Arrangement of Architectural Decoration Patterns

After the material information of architectural decoration patterns is collected, the sorting of data is a process of digestion and output of relevant knowledge.

3.4.1 Collection of Decorative Patterns

The collection of patterns is mainly to form the patterns into electronic files in the form of on-site photos, and then make them into vector CAD files and 3D files through vector transfer drawing. The texture and color information of materials used for the construction location and specific size of patterns are recorded, and the patterns are classified and compared according to the construction location, construction location, pattern content and material used. Thus, a set of system atlas is formed as a set of vertical or horizontal comparison. It also provides the basis and foundation for the later interpretation of the cultural implication of patterns and for the future deconstruction, reconstruction and re creation.

3.4.2 Combing of Decorative Patterns and Styles

The decorative pattern style of Wuhan Li-Fen residential buildings comes from the popular art deco in Europe and America in the early 20th century. This style emphasizes the industrial beauty of the machine age, geometric mechanical decorative lines and radial decorative patterns. It contains a strong sense of order and science and technology, and is the representative of rational graphics and patterns. Art decoration style is an art style with high integration and tolerance. The totem style of many ancient civilizations is the inspiration of this style. There are also decorative patterns of the combination of Eastern and Western cultures in Wuhan Li-Fen residential buildings. These decorative patterns and architectural structures are cleverly organized and dotted at the door head, corridor column and other positions of the building; All kinds of plant patterns with art nouveau style and water ripple lines imitating the river water used in the indoor ceiling of Wuhan Li-Fen residential buildings are flexible, implying strong natural vitality and perceptual beauty, showing the unique style of residential buildings in this period.

4 Comprehensive Evaluation of "Place Perception" On-site Teaching Method Based on Analytic Hierarchy Process

In order to make the "place perception" on-site teaching achieve the expected good results, combined with the on-site teaching in Li-Fen Kunhouli, a collection of ancient buildings in Wuhan, this paper makes a comprehensive analysis and evaluation of the survey data and teaching effect by using analytic hierarchy process. The field teaching process of "place perception" is a comprehensive and dynamic transmission process of architectural knowledge, and there are many factors affecting the teaching effect [16–19]. Combined with the actual situation, the paper establishes a comprehensive evaluation analytic hierarchy process model for the on-site teaching method of place perception, and calculates it, so as to find out the weak links in teaching and clarify the work focus in the next step.

4.1 Comprehensive Evaluation Index and Grade of "Place Perception" On-site Teaching Method

Integrated with the actual construction, main influencing factors and comprehensive evaluation characteristics of "place perception" on-site teaching method, the paper divides the basic indicators into three categories, including students' evaluation of curriculum, teachers' evaluation of curriculum and overall evaluation of teaching environment, with a total of 11 single indicators. The average scores of each part are calculated and summarized in the form of questionnaire survey. The main results are summarized in Table 1.

Table 1. Comprehensive evaluation indexes and relevant data of "place perception" field teaching method of analytic hierarchy process

Target *A*	Benchmark index *B*	Single index *C*	Average score *V*
Comprehensive evaluation of site perception teaching	Student evaluation a_1	Understanding of basic teaching methods a_{11}	82.1
		Achievement of learning objectives a_{12}	76.7
		Extracurricular content expansion effect a_{13}	84.3
		Foundation of professional knowledge a_{14}	85.5
	Teacher evaluation a_2	Degree of interaction a_{21}	87.7
		Teaching efficiency a_{22}	75.1
		Application proficiency a_{23}	76.2
		Teaching effect a_{24}	80.6
	Learning environment evaluation a_3	Security a_{31}	74.4
		Learning efficiency a_{32}	79.7
		Architectural characteristics of environment a_{33}	86.7

4.2 Solution and Consistency Test of Judgment Matrix

In order to make judgment quantitative, the scale method widely used in B and C layers is mainly used in this paper. In the scale method, the results of comparing the importance of any two indexes in the same layer to the former level indicators or targets are quantitatively described by specific values. The scale data in this paper is 1–9 [20].

According to the actual situation, one judgment matrix of layer A and three judgment matrices of layer B of "place perception" field teaching method based on AHP could be got, respectively in order:

$$J_1 = \begin{Bmatrix} 1 & 2 & 6 \\ 1/2 & 1 & 4 \\ 1/6 & 1/4 & 1 \end{Bmatrix}$$

$$J_1 = \begin{Bmatrix} 1 & 2 & 7 & 5 \\ 1/2 & 1 & 4 & 3 \\ 1/7 & 1/4 & 1 & 1/5 \\ 1/5 & 1/3 & 5 & 1 \end{Bmatrix}$$

$$J_3 = \begin{Bmatrix} 1 & 1/2 & 1/4 & 1/7 \\ 2 & 1 & 1/2 & 1/4 \\ 4 & 2 & 1 & 1/2 \\ 7 & 4 & 2 & 1 \end{Bmatrix}$$

$$J_4 = \begin{Bmatrix} 1 & 1/4 & 1/6 \\ 4 & 1 & 1/4 \\ 6 & 4 & 1 \end{Bmatrix}$$

The consistency of the four matrices is analyzed respectively, and the maximum eigenvalues of J_1, J_2, J_3, J_4 are obtained as follows: $\lambda_1 = 3.0092$, $\lambda_2 = 4.2206$, $\lambda_3 = 4.0022$, $\lambda_4 = 3.1078$.

The eigenvectors are:

$$u_1 = \{0.8685, 0.4779, 0.1315\};$$

$$u_2 = \{-0.8491, -0.4632, -0.0884, -0.2379\}$$

$$u_3 = \{-0.1149, -0.2222, -0.4444, -0.8602\}$$

$$u_4 = \{0.1128, 0.3255, 0.9388\}$$

The consistency test formula is shown in formula (1).

$$CR_i = \frac{CI_i}{RI} = \frac{\frac{\lambda_i - n}{n-1}}{RI} \tag{1}$$

The consistency value CR_i of judgment matrix i is considered as unacceptable when the consistency value is less than 0.1; the maximum eigenvalue λ_i of judgment matrix i is represented; n represents the dimension of judgment matrix i; RI can be obtained from the table [21–25]. The consistency values of the four judgment matrices are $CR_1 = 0.0079$, $CR_2 = 0.0817$, $CR_3 = 0.0008$, $CR_4 = 0.0929$; All of them are less than 0.1, and the degree of inconsistency can be considered acceptable.

Finally, the special vectors of each judgment matrix are standardized to form the weight matrix. The results are as follows.

$$w_1 = \{0.5849, 0.3255, 0.0896\}$$

$$w_2 = \{0.5182, 0.2827, 0.0539, 0.1452\}$$

$$w_3 = \{0.0700, 0.1353, 0.2707, 0.5240\}$$

$$w_4 = \{0.0819, 0.2364, 0.6817\}$$

4.3 Comprehensive Evaluation and Analysis of Evaluation Results

The comprehensive evaluation is mainly combined with the indicators, index average score and weight vector summarized in the previous analysis.

The comprehensive synthesis formula of comprehensive evaluation is presented as formula (2).

$$M = \sum_{i=1}^{n} w_i \times v_i \tag{2}$$

Where, M represents the composite score of the sub quota of index i according to its weight vector and the corresponding score matrix; wi represents the weight matrix of index i; vi refers to the score matrix of the index *i* [26]. The calculation results are shown in Table 2.

Table 2. Comprehensive evaluation score of "place perception" on-site teaching method

Index	Evaluation of students	Evaluation of Teachers	Environmental assessment	Comprehensive evaluation
Comprehensive score	81.1	79.2	84.0	80.8

By analyzing the comprehensive evaluation score table of "place perception" on-site teaching method, we can conclude that the comprehensive score of this teaching is higher than 80, and the comprehensive teaching effect is good. Generally speaking, teachers' maladjustment to the new teaching methods leads to the lowest score in the comprehensive evaluation. From the average scores of all sub indicators, it can also be seen that the lowest teaching evaluation lies in the consideration of the safety of the teaching environment, and this part can be optimized by improving the teaching route and corresponding measures of the "place perception" on-site teaching method after multiple teaching turns.

5 Conclusion

The course old building reconstruction design is a comprehensive course, which is of great significance for the restoration of cultural relics and ancient buildings, as well as the construction and reconstruction of urban buildings. In the course teaching, it is an important content to conduct comprehensive research on the original structural space form of the building and the external environment of the architectures. This is the on-site teaching based on place perception discussed in this paper, which guides students to complete on-site research before learning the reconstruction design course, systematically understand the form, material meaning and scale of architectural decoration patterns and construction, and analyze the environmental atmosphere elements of the site. Through the analysis of the data collected from the survey, the students' concerns are systematically directed to the key issues of transformation design, and the students' ability and enthusiasm to find and analyze problems independently are brought into full play in the process of on-site investigation, so as to provide new design ideas for the later transformation courses.

The purpose of using analytic hierarchy process to comprehensively analyze and evaluate the survey data and teaching effect is to detect the effect of "place perception" on-site teaching method and the problems existing in the teaching process. As a complex dynamic process, on-site teaching and its effect are affected by many factors. The establishment of analytic hierarchy process model for calculation can find out the crux of the teaching system and provide ideas and basis for improving the effect of on-site teaching.

Acknowledgment. This paper is supported by the 2020 fund project *Wuhan Li-Fen Architectural Decoration Structure and Pattern Research* of handicraft culture research center of Hubei Academy of fine arts, the key research base of Humanities and Social Sciences in Hubei Province (SGY-202004).

References

1. Ou, X., Wang, J., Liu, H., Wen, D., He, W., Pan, Y.: Research on the reform of practical teaching driven by practical training under constructivism. Int. J. Eng. Manuf. **1**(1), 77–83 (2011)
2. Ning, G.: Construction and implementation of innovation computer network practical teaching system. Int. J. Educ. Manag. Eng. **1**(2), 30–35 (2011)
3. Feng, W.: Walking perception and rational graphics – a new discussion on landscape design teaching. Decoration **4**, 74–76 (2014). (in Chinese)
4. Brenner, S.: Teaching cultural sensitivity at architecture schools for more sustainable buildings: lessons from reconstruction. In: Design for a Sustainable Culture, no. 5, pp. 197–213. Routledge, New York (2017)
5. Zhang, X., Chang, J., Li, Y.: Teaching reform of participating in interactive for principle of urban planning course in architecture specialty. J. Archit. Educ. Inst. High. Learn. **02**, 12–16 (2018)
6. Wu, H., Huang, X.: Research on the western pattern in Wuhan modern architectural decoration. J. Hubei Acad. Fine Arts **4**, 103–107 (2013). (in Chinese)
7. Liu, X.: Study on the Chu Style architectures and "New Chufeng" in the urban architectural design of Wuhan. Design Res. **3**, 83–88 (2014)
8. Zhang, N.: Research on image decoration of portal in Hankou Lifen building. Art Educ. **8**(2), 198–199 (2017). (in Chinese)
9. Lian, H., Li, G., Wang, X.: Steering toward history and novelty and searching for pulse – teaching reform of existing building renovation. China Archit. Educ. **1**, 84–89 (2018). (in Chinese)
10. Yang, W., Song, J.: Construction and research on the cognitive practice teaching mode of architecture specialty. Sci. Technol. Inf. **29**, 233–234 (2012)
11. Wall, A., Draper, C.: Taking it to the streets: teaching methods and curriculum courses on-site with partner schools. Middle Grades Rev. **3**(1), 1–8 (2017)
12. Shi, L., Yao, Y., Xia, X., Xie, X., Wu, L.: Practical teaching staff construction under new situation. Int. J. Educ. Manag. Eng. **1**(2), 57–61 (2011)

13. Ardana, I.M., Ariawan, I., Divayana, D.: Measuring the effectiveness of BLCS Model (Bruner, Local Culture, Scaffolding) in mathematics teaching by using expert system-based CSE-UCLA. Int. J. Educ. Manag. Eng. **7**(4), 1–12 (2017)
14. Yang, Y., Yu, K.: Construction of distance education classroom in architecture specialty based on internet of things technology. Int. J. Emerg. Technol. Learn. **11**(5), 147–148 (2016)
15. Yu, Q.: Reform and innovation of curriculum system of architecture specialty under the application-oriented training objectives. In: International Conference on Modern Educational Technology and Innovation and Entrepreneurship (ICMETIE 2020), pp. 231–236. Atlantis Press (2020)
16. Lu, L.: Discussion on the reform of practical teaching of building structure course. Mod. Econ. Inf. **24**, 394 (2016)
17. Han, Z., He, Y.: Discussion on teaching reform of landscape architecture design courses. In: 2nd International Seminar on Education Research and Social Science (ISERSS 2019), pp. 100–103. Atlantis Press (2019)
18. Bai, J., Qu, R., Li, H.: Evaluation of new engineering practice teaching effect based on fuzzy analytic hierarchy process. J. Higher Educ. **21**, 49–51 (2019). (in Chinese)
19. Quan, J., Quan, Y., Ma, L.: Fuzzy comprehensive evaluation of teachers' teaching quality based on analytic hierarchy process. J. Chongqing Technol. Bus. Univ. (Nat. Sci. Edn.) **3**, 34–38 (2017). (in Chinese)
20. Hu, Y.: Analysis on classification and evaluation of higher vocational teachers based on fuzzy AHP and teaching quality research method. Chinese J. Multim. Netw. Teach. (Medium Issue) **7**, 77–78 (2018). (in Chinese)
21. Singh, B.: Analytical hierarchical process (AHP) and fuzzy AHP applications-a review paper. Int. J. Pharm. Technol. **8**(4), 4925–4946 (2016)
22. Agarwal, P., Sahai, M., Mishra, V., et al.: Supplier selection in dynamic environment using analytic hierarchy process. Int. J. Inf. Eng. Electron. Bus. **6**(4), 20–26 (2014)
23. Liu, Y., Zhang, X.: Evaluating the undergraduate course based on a fuzzy AHP-FIS model. Int. J. Mod. Educ. Comput. Sci. **12**(6), 55–66 (2020)
24. Kurilovas, E., Vinogradova, I.: Improved fuzzy AHP methodology for evaluating quality of distance learning courses. Int. J. Eng. Educ. **32**(4), 1618–1624 (2016)
25. Mehregan M.R., Jamporazmey, M., Hosseinzadeh, M., et al.: Proposing an approach for evaluating e-learning by integrating critical success factor and fuzzy AHP. In: International Conference on Innovation, Management and Service, Singapore (2011)
26. Çebi, A., Karal, H.: An application of fuzzy analytic hierarchy process (FAHP) for evaluating students' project. Educ. Res. Rev. **12**(3), 120–132 (2017)

Incentive Mechanism and Teacher Growth in Colleges and Universities Based on Expectation Theory from Perspective of PCK

Ziye Wang[1(✉)], Hu Wang[2], and Changbin Jiang[2]

[1] College of Journalism and Communication, Wuhan Sports University, Wuhan, China
[2] School of Management, Wuhan University of Technology, Wuhan, China

Abstract. Focusing on the requirements for young teachers and their growth in the new era under the new historical environment, this paper analyzes the PCK (Pedagogical Content Knowledge) theory and discusses the basic abilities and qualities that teachers need to have; discusses the expectation theory, the relationship between work motivation, expectancy, valence and instrumentality, expounds the principles, contents and establishment methods of incentive mechanism, and discusses the growth of college teachers and the problems they face. Based on the analysis of PCK three-module, the four-module model is analyzed, and the five module model is deduced. The theme of the growth of teachers in colleges and universities is discussed from the perspective of PCK and its evolution model, the quality and ability of teachers under the guidance of PCK model, the enhancement of practical ability, the expectant formula of teacher's motivation, and the self-motivation of teachers, aiming at providing thinking and suggestions for talent training and teacher growth.

Keywords: PCK · Expectation theory · Incentive mechanism · Teachers' growth

1 Instruction

The development of higher education and social progress are a process of mutual promotion. The construction of colleges and universities needs to rely on the strength of teachers, especially young faculties. In this sense, the growth of young teachers determines the sustainable development of education [1]. Teachers need to build a perfect knowledge system themselves and have comprehensive quality. They need not only rich teaching knowledge and experience, but also the accumulation of scientific research and social practice. They also need to master certain teaching methods, understand students' pain points and requirements, and provide targeted help for students [2–4]. This is the requirement of society and times for teachers [5].

There is no doubt that the university is not a vacuum environment and will inevitably be affected by all kinds of social influences. The influence of social impetuosity on university teachers is inevitable. Therefore, to help young teachers progress and grow,

Z. Hu et al. (Eds.): AIMEE 2021, LNDECT 107, pp. 433–446, 2022.
https://doi.org/10.1007/978-3-030-92537-6_40

we must establish a scientific and perfect incentive mechanism on the basis of expectation theoretical analysis, form a joint force with positive influence, and create a good atmosphere, so as to make young teachers grow as educational backbone and serve the cultivation of excellent talents as quickly, steadily and reliably as possible [6–8].

2 Literature Review and Issue Analysis

The issues of this paper include PCK model, expectation theory, incentive mechanism and the growth of college teachers.

2.1 Concept and Model of PCK

PCK is the abbreviation of pedagogic content knowledge. It is a concept put forward by Schumann, a famous American educator, in the 1980s. It is defined as "the special integration of teachers' personal teaching experience, teachers' subject content knowledge and pedagogy". He believed that PCK is the product of the integration of subject knowledge, teaching knowledge and background knowledge; Teachers should have diversified and rich knowledge structure, correctly understand and internalize the subject knowledge into their own knowledge, and then transform the subject knowledge into teaching knowledge that can be accepted by students according to students' understanding [9].

Subsequently, many scholars continuously improved the meaning of PCK. Tamir believed that subject teaching knowledge should include "four parts: curriculum, students, teaching and evaluation knowledge" [10]. Marks pointed out that "teachers should master the knowledge about students, the knowledge about media teaching, have subject knowledge, be able to arrange students' activities in teaching and pay attention to teaching behavior" [11]. Cochran and others developed discipline teaching knowledge PCK into "pedagogical content knowing", referred to as PCKg for short [12].

Although different researchers give slightly different definitions, they agree that PCK is a dynamic system formed by the integration of multiple components "Teaching knowledge" should include knowledge factors about subject courses, students, teaching means, teaching methods, teaching activities and teaching situations. Its meaning mainly includes teachers' "what to teach" (understanding and grasping the content knowledge of the teaching subject), "how to teach" (teaching methods, teaching experience, etc.) and students' effective acceptance. The core is "effective teaching", i.e., the effectiveness of teaching. The "knowledge" and "cognition" here include not only the knowledge grasped by teachers themselves, but also the effective experience accumulated by predecessors, and even the formation of regular cognition. In order to make students accept effectively, it must include the understanding and grasp of students' reality and learning situation.

PCK is not only a theory and idea, but also a new teacher education framework, which integrates teachers' personal teaching experience, teachers' subject content knowledge and pedagogical knowledge into this framework. Through the integration of separate knowledge, better education and teaching results could be achieved [13, 14].

PCK is also an important factor to distinguish discipline experts, teaching experts, experienced teachers and new teachers. Because the subject experts who focus on building new knowledge in a subject field from the sight of research are different, subject teaching experts emphasize that they should start from the reality of teaching and help students understand and master these new knowledge. From this perspective, the knowledge covered by PCK is a necessary element for teachers' growth and development. The research on PCK is conducive to the improvement of teachers' teaching ability [15, 16].

2.2 Expectation Theory

Expectation theory was put forward by Victor Frum, a famous psychologist and behavioral scientist in North America in 1964, so it is also called Frum expectation theory, also known as potency means expectation theory. It is a theory combining management psychology and behavioral science, which has a good application effect in practice [17–19].

The expectation theory can be expressed as:

$$Working\,power = expected\,value \times titer \times Instrumental$$

Among them, expectation is people's judgment of the possibility of reaching a certain goal according to past experience, potency is the value of the return to meet personal needs, and instrumentality is people's belief that they can get a return once they complete the task. This is the so-called EVI model.

The formula of expectation theory can also be expressed as:

$$Incentive\,force = expected\,value \times titer$$

Compared with the previous formula, instrumentality is included in the potency.

The relationship between needs and objectives can be expressed in process mode, that is:

$$\begin{aligned} ''Personal\,effort- &\rightarrow personal\,achievement(performance)- \\ &\rightarrow organizational\,reward(reward)- \rightarrow personal\,needs'' \end{aligned}$$

The formula of this theory shows that the mobilization of human enthusiasm depends on the product of expectation and potency. In other words, the greater a person's grasp of the goal, the higher the probability of reaching the goal, the stronger the motivation and the greater the enthusiasm. In the management of various industries, including education, it is of certain significance to use the expectation theory to mobilize the enthusiasm of employees.

Basis of expectation theory: people can engage in a certain work and achieve organizational goals because these work and organizational goals will help them achieve their goals and meet their needs in some aspects [20, 21].

The premise of expectation theory: everyone has a preference for the expectation of results. When a person makes a choice among several feasible schemes with unpredictable results, his behavior is affected not only by his preference for the expected results, but also by the extent to which he thinks these results may be achieved.

Expectation theory reflects the relationship between needs and goals with three factors. To motivate employees, employees must be clear:

(1) Work can provide them with what they really need;
(2) What they want is linked to performance;
(3) As long as they work hard, they can improve their performance.

2.3 Incentive Mechanism

The connotation of incentive mechanism refers to the process of maximizing employees' commitment to organization and work through specific methods and management system. It is the sum of the structure, mode, relationship and evolution law of the incentive subject system interacting and restricting with the incentive object by using a variety of incentive means, standardizing and relatively fixing them in the organizational system [22]. It is also a connecting means for enterprises to transform lofty ideals into concrete facts.

Table 1. Elements and functions of incentive mechanism

Incentive mechanism	Constituent elements	Function
	Inducer set	Initiating behavior
	Behavior range system	Guide, standardize and restrict behavior
	Behavioral spatiotemporal system	
	Behavior domestication system	

As shown in Table 1, incentive mechanism includes five aspects: 1) set of inducing factors; 2) Behavior oriented system; 3) Behavior range system; 4) Behavior space-time system, and 5) behavior domestication system. It is the sum of five elements. Among them, inducing factors play the role of initiating behavior, and the latter four play the role of guiding, standardizing and restricting behavior.

According to whether the design and establishment of incentive mechanism meet the principles of scientificity and rationality, it determines if it has the role of promoting or weakening. The former is a positive impact, which can repeatedly strengthen and continuously enhance employees' behavior in line with the expectations of the organization, so as to make the organization grow and grow. The latter, however, shows that the organization does not show the expected behavior of employees, which inhibits and weakens the work enthusiasm of some employees.

The operation of incentive mechanism should adopt the whole process incentive operation mode based on two-way information exchange, which runs through the whole process of realizing organizational objectives.

There are three main principles of incentive mechanism: 1) Having both ability and political integrity; 2) Equal opportunity; 3) "ladder promotion" and "exceptional promotion" are combined.

2.4 Growth of Teachers in College and University

Colleges and universities are the base of talent training and undertake the great mission of exploring scientific mysteries and creating new knowledge for mankind. Young and middle-aged teachers are the future and hope of the school. To run a school well, it is necessary to have advanced school running ideas, sufficient hardware and software facilities, and a high-level and high-quality team of teachers. It can be said that teachers are the foundation of school construction and development [23].

The basic qualities that excellent teachers should have are: good professional ethics and noble ethics; Solid basic knowledge; Keen observation ability; Innovative thinking ability and habit; Strong teaching organization capability; Good teamwork spirit and healthy physical and psychological quality [24].

3 Growth of College Teachers and The Problems They Face

3.1 Growth and some Contradictions of Young Teachers

Nowadays, when colleges and universities are turning from elite education to popular education, the school has ushered in unprecedented favorable development opportunities, which enables the majority of young and middle-aged teachers to have the best stage to show their talents, but also faces great challenges and pressure. It is an urgent task for every young and middle-aged teacher to recognize the situation, change their ideas, and make psychological preparations and practical plans to meet the challenges.

The growth of teachers, especially young teachers, needs to solve several pairs of contradictions.

(1) Teaching and scientific research
From the school level, teaching and scientific research are the foundation of school development. However, it is also very difficult to grasp the balance of the relationship between the two and truly make teaching and scientific research go hand in hand.
(2) Career and family
The work of the school needs to be completed seriously, and the family also needs to operate carefully. This is a new normal that every young teacher must face.
(3) Individual and collective
Individual independence is what we pursue, and human natural group is also necessary. In teaching and scientific research, teachers need to cooperate with each other, learn from each other and overcome problems.

(4) Tutors and students
Teachers should take care of students, and students should be filial and respectful to teachers. They'd better be friends at the same time. Overcoming difficulties and sharing joy is a useful way to get along.
(5) Drip and accumulation
Every achievement we have made is not achieved overnight, but accumulated over time. For young teachers, we should not only have dreams, but also adhere to and accumulate. Jobs said: you can't connect the dots looking forward you can only connect them looking backwards. So you have to trust that the dots will somehow connecting your future“
(6) Opportunities and challenges
Behind the "scenery" on the ivory spire of young college teachers, there is actually a lot of hard work. There are challenges and rare development opportunities. The key is to accumulate at ordinary times and seize it when the opportunity comes. The so-called "coexistence of opportunities and challenges" is especially true for young teachers.

3.2 Difficulties Faced by Young Teachers

In recent years, great changes have taken place in the appearance of teachers. A large number of highly educated young teachers have stepped into the front line of teaching and scientific research and become an important force in teaching and scientific research of colleges and universities.

In depth investigation and research and obtaining first-hand information are effective ways to solve problems [25]. In order to understand the actual situation and problems of young teachers in colleges and universities, in line with the original intention of "discovering problems and exploring causes", we carried out a survey on young and middle-aged teachers. 92 questionnaires were sent out and 78 valid questionnaires were recovered. The number of questionnaires from male teachers and female teachers accounted for 64.1% and 35.9% respectively. At the same time, we also organized some interviews and symposiums to understand the real ideas of young teachers face to face.

The young teachers surveyed generally reflect that the first period of time after entering colleges and universities is the most stressful stage for young teachers and the period when they are most eager for special attention from the school. The main problems faced at this stage are: loneliness, housing conditions, teaching skills, introduction to scientific research and so on.

The survey shows that almost 85% of young teachers are married or ready to get married, and housing has become an important problem they need to solve urgently. For young teachers whose native families are not local, they are far away from their families and face a new environment alone. They especially need care and help.

As a newcomer in the workplace, young teachers face great pressure on teaching and scientific research. The accumulation of teaching experience needs a process. This is especially true for the application of scientific research projects. Because of the lack of basic achievements, it is difficult to apply for weighty scientific research projects; Because there is no scientific research project, it cannot support the development of research work. Because of insufficient accumulation and few achievements, it is difficult to promote professional titles; Because there is no senior professional title, it is more difficult to undertake scientific research projects. This is a dilemma called "Matthew effect".

The above problems can be summarized into three aspects. First, teaching crisis, lack of teaching experience, it is difficult to achieve good teaching results; Second, scientific research crisis, unclear scientific research direction and insufficient funds, making it difficult to obtain high-level scientific research projects; Third, physical crisis, heavy economic burden, great mental pressure and worrying health status.

4 Teachers' Quality and Growth Under Incentive Mechanism

4.1 Teachers' Quality and Ability Under the Guidance of PCK Model

According to the PCK theory, teachers should not only grasp the content knowledge of the subject they teach, but also understand the educational and teaching knowledge and teaching method skills, and pay attention to the learning and accumulation of teaching experience, so as to form their own teaching ability.

4.1.1 PCK Model and Its Evolution

In process of teaching, teachers should integrate subject content knowledge with teaching skills and educational knowledge according to their own understanding, follow the laws of education and teaching, and use certain methods to transform subject content knowledge into knowledge that students can accept and understand, so as to be effectively accepted by students. It includes not only teachers' deep understanding and comprehensive and correct grasp of subject content knowledge, but also teachers' problem of how to effectively impart knowledge to students. Figure 1 includes curriculum knowledge, teaching (method) knowledge and students' knowledge (understanding of students). The intersection of these three parts forms PCK.

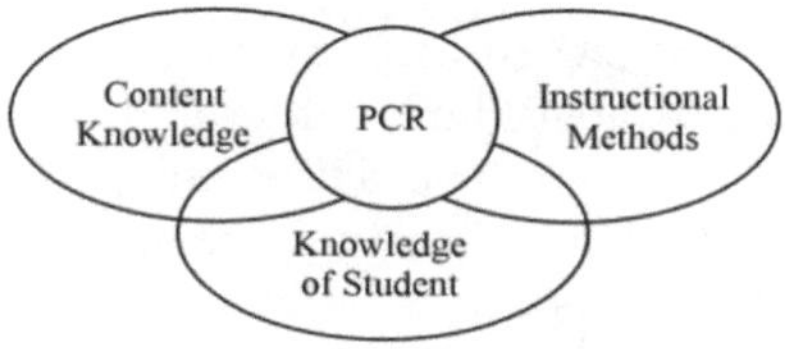

Fig. 1. Three module PCK model

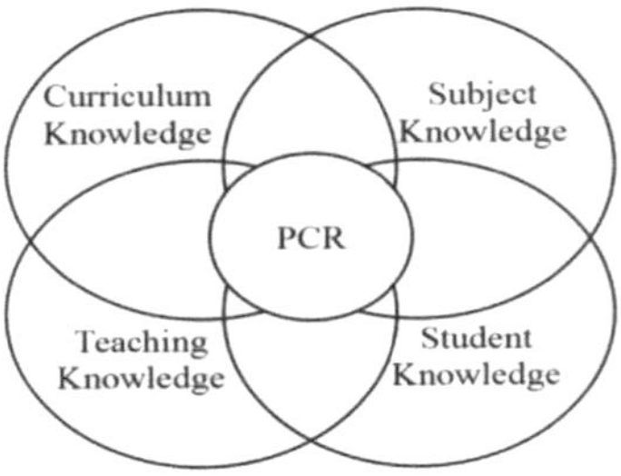

Fig. 2. Four module PCK model

To understand students, it is necessary not only to understand the number and basic level of students, but also need to understand their pain points in learning, understand the obstacles and difficulties they face in the process of growth, try to solve their difficulties and point out the direction of their efforts.

If it is not limited to a course itself, but to include the curriculum system and even the professional content, it becomes the four module PCK model shown in Fig. 2.

PCK theory emphasizes that teachers must have solid subject content knowledge and subject teaching knowledge. This requires teachers to have knowledge about students, pedagogy, psychology, curriculum materials, teaching methods, teaching management, teaching situations and other specific knowledge. They should also pay attention to the summary of teachers' personal teaching experience and integrate this knowledge. Teachers should attach great importance to solving the problem of "how to teach", that is, effective teaching, respect the law of education and teaching, and "be able to teach", so as to achieve the teaching goal and achieve the best teaching effect.

4.1.2 Practical Ability and PCK Model Improvement

In Aristotle's philosophy of practice, practice has a specific meaning. As a practical activity, the understanding of its practicality should surpass the simple judgment relative to theory. From the connotation of practice itself and its relationship with theory and production, the practicality of education can be understood as: education should take itself as the purpose; Education is not production, but it is inseparable from the skills of production; Education is not a theory, but theory is the highest educational practice.

Practice is a process of interaction between subject and object through a certain intermediary for the purpose of transforming the world. Whether for teachers or students, it is necessary to become talents in practice, grow up in service, effectively serve the society and reflect the University's own value and social value. Based on the experience of social practical activities, we should constantly expand the scope of practice and tap the potential of talent training, so as to realize the common growth of teachers and students.

This research holds that practical ability is also a very important item. Practicality is an important principle in education. Knowledge cannot be empty. Theory must be combined with practice in order to truly reflect value and play a role. In the process of creative thinking, people must participate in practice, promote the further development

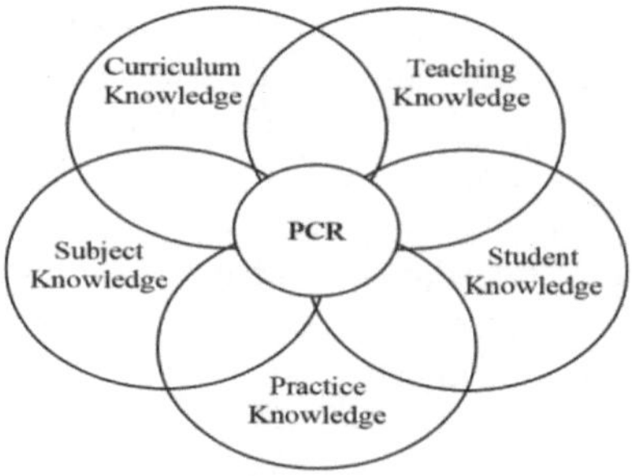

Fig. 3. Five module PCK model

of thinking ability and test the correctness of thinking results. Without practice, the development of thinking will lose power and there will be no creative thinking. The practical principle governs other teaching principles including independence, relevance and hierarchy. By adding "practical knowledge", a new PCK model with five modules in Fig. 3 is formed.

4.2 Incentive Mechanism Based on Expectation Theory

4.2.1 Significance and Role of Expectation Theory for College Teachers

According to the PCK theory, teachers should not only grasp the content knowledge of the subject they teach, but also understand the educational and teaching knowledge and teaching method skills, and pay attention to the learning and accumulation of teaching experience, so as to form their own teaching ability.

Expectation theory holds that when people expect a certain behavior to bring a specific result to the individual, and the result is attractive to the individual, the individual tends to take this behavior.

According to this theory, in order to motivate employees, managers must do and let employees know three things: work can provide what employees really need; As long as you work hard, you can improve performance; What employees want is related to performance. For college teachers, different from enterprise employees in the general sense, they have the following characteristics, such as strong autonomy, distinct creativity, desire for challenges, high labor complexity, strong willingness to flow, and dedication to personal development. At the same time, they also face complex and difficult development environment at any time in the process of professional development.

Therefore, how to establish a perfect incentive mechanism, coordinate and stimulate college teachers to work in a consistent way that can improve the common interests of students, teachers and schools is a more complex and worthy of serious research. Based on the expectation theory, according to the characteristics of college teachers, committed to the comprehensive and harmonious improvement of potency, expectation and instrumentality is one of the effective means to motivate college teachers, which should not only solve theoretical problems, but also solve practical problems. Expectation theory holds that when people expect a certain behavior to bring a specific result to the individual, and the result is attractive to the individual, the individual tends to take this behavior.

According to this theory, in order to motivate employees, managers must do and let employees know three things: work can provide what employees really need; As long as you work hard, you can improve performance; What employees want is related to performance. For college teachers, different from enterprise employees in the general sense, they have the following characteristics, such as strong autonomy, distinct creativity, desire for challenges, high labor complexity, strong willingness to flow, and dedication to personal development. At the same time, they also face complex and difficult development environment at any time in the process of professional development.

Therefore, how to establish a perfect incentive mechanism, coordinate and stimulate college teachers to work in a consistent way that can improve the common interests of students, teachers and schools is a more complex and worthy of serious research. Based on the expectation theory, according to the characteristics of college teachers, committed to the comprehensive and harmonious improvement of potency, expectation and instrumentality is one of the effective means to motivate college teachers, which should not only solve theoretical problems, but also solve practical problems.

4.2.2 Incentive Mechanism and Process for Teachers

Figure 4 shows the incentive process suitable for college teachers.

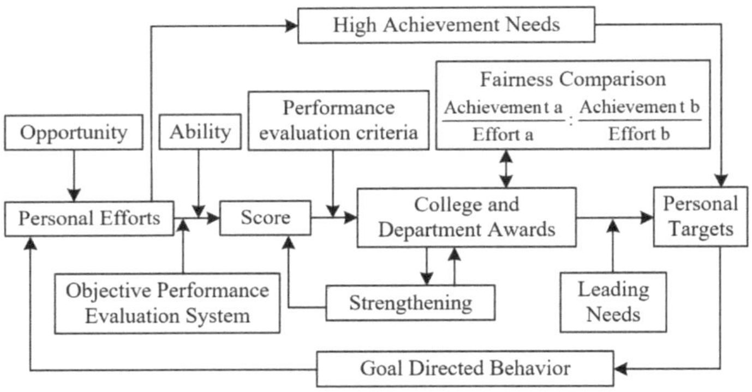

Fig. 4. Incentive process for college teachers

Special posts and sense of responsibility and mission provide opportunities and requirements for teachers' growth, and urge them to make efforts to continuously enhance their abilities. The objective performance evaluation system established by the university will evaluate the teachers' personal performance based on a complete set of performance evaluation standards. According to the completion of teachers, corresponding organizational rewards can be obtained. The fairness comparison between different teachers is essential. This comparison is the result of analysis according to everyone's output and input. After obtaining the organizational reward, it will strengthen individual performance to a certain extent. The need for individuals to complete tasks has grown, which forces teachers to constantly strive towards their personal goals, which will then guide teachers' personal efforts. Generally speaking, college teachers hope to achieve high achievements. Compared with people in other industries, this need for high achievements is more prominent.

4.2.3 Frum Expectation Formula in Teacher Motivation

The main content of expectation theory is Fromm's expectation formula and expectation model. According to Frum's expectation theory, employees' motivation power (*MP*, or *M*) comes from three factors: valence, or considered as value (*V*), expectation (*E*) and instrumentality (*I*), i.e.,

$$M = E \times V \times I,$$

The sum of the three constitutes the working motivation of teachers. Different from "sum of addition", the "product of multiplication" does not necessarily multiply much bigger. As long as one of them is zero, the result is zero. The relationship between the three is more subtle.

Expectation ("*E*" for short) is teachers' confidence that they can improve their work performance after paying their labor. In short, it is teachers' work confidence. If teachers' efforts can improve their work performance, their efforts will be willing, otherwise, they will not. Managers should establish confidence and conditions for teachers to create good work performance in order to improve their *E*-value.

Valence ("*V*" for short) refers to the value that teachers attach to the return they receive. It refers to the value of achieving the goal to meet their personal needs, that is, the return that teachers get must be what they need and value most. If the reward they get happens to be what they need most, the value of *V* will increase. What a teacher needs is actually his own psychological definition. For the same goal, due to the different environment and needs of each individual, the goal value of their needs is also different. The same goal may have "positive", "zero" and "negative" titers for everyone. If the individual likes the results available, it is a positive titer; Zero if the individual ignores its results; If you don't like the available results, it is negative titer. The higher the potency, the greater the incentive force should be. Therefore, college leaders find a balance point through negative communication, so that value-added plays a positive role in promoting work ability.

Instrumentality ("*I*" for short) is that teachers believe that their good work performance can get corresponding equivalent returns. The organization shall recognize and give certain forms of rewards (material or / and non-material). Therefore, the improvement of *I* value depends on the leaders' attention to performance.

When *EVI* reaches a higher value at the same time, MP must rise. It is more important to ensure that MP is at the high value for as long as possible. The charging and training of teachers and the reasonable incentive strategy of enterprises are effective measures.

If I is attributed to *V*, the formula is changed to a simplified form $M = V \times E$.

Different combinations of expectation and potency will produce different stimulating forces. There are several basic combinations.

We use "*H*" for high and "*L*" for low, then:

$V_H \times E_H = M_{H,}$ -a high bonus plan has a high incentive force;
$V_H \times E_L = M_{L,}$ -training, making heighten condition and ability of working;
$V_L \times E_H = M_{L,}$ -looking for other incentives;
$V_L \times E_L = M_{L,}$ -take new incentives and change working conditions.

This shows that in order to receive the expected incentive effect, organizational management must take the potency of incentive means (which can bring satisfaction to the incentive object) and the expectation of incentive to obtain this satisfaction as the premise.

4.2.4 Teachers' Self Motivation

As mentioned above, the so-called incentive includes both external incentive and internal one. The expectation theory can also be applied to the self-planning of young teachers. Take the planned goal as the direction of their efforts and constantly achieve higher results.

Facing many challenges and puzzles, from the perspective of young teachers themselves, it should be better to do the following: love education; Love students; Love their major and discipline; Stick to teachers' ethics; Improve professional quality; It should also be necessary to have tenacious perseverance and continue to practice; Cultivate a good character and ensure that every student is dealt kindly and fairly.

5 Conclusion

The quality of teachers determines the level of the school. The growth of young teachers determines the future development of colleges and universities. Establishing a perfect system, adopting various effective ways to help young teachers grow into the best ones, and building a high-level teaching staff are not only the needs of talent training and school development, but also the requirements of the society and the times.

Taking PCK as a guide is to help college teachers deeply understand the knowledge of courses they teach and curriculum system, comprehensively master teaching methods and means, accumulate rich practical knowledge and teaching experience, treat every student with love, know their difficulties and real inner needs, and provide them with the help they need.

Under the guidance of expectation theory, establishing a scientific and reasonable incentive mechanism has been proved to be an effective method. The combination of external incentive and internal incentive can create an atmosphere of fair competition on the one hand, so that all teachers have the motivation and opportunities for progress. On the other hand, it can establish the confidence of young teachers and make continuous progress and stable growth in their teaching and scientific research work, and make their career more colorful.

Acknowledgment. This paper is supported by: 1) Wuhan Sports University Young Teachers' Scientific Research Fund Project "The MECE study under the background of Sports Events media convergence"; 2) Teaching research project of Hubei Province "exploration and practice of integrating the technical courses of information management specialty into "Ideological and political education".

References

1. Qi-Ming, W., Xi-nan, Z., Li-li, Z.: A novel thought and method of university evaluation. Int. J. Educ. Manag. Eng. (IJEME) **3**(2), 59–65 (2013)
2. O'Hara, S., Bookmyer, J., et al.: Theory of action for resourcing professional growth. J. Prof. Capital Commun. **4**(01), 52–65 (2019)
3. Zhiqin, L., Jianguo, F., Fang, W., Xin, D.: Study on higher education service quality based on student perception. Int. J. Educ. Manag. Eng. (IJEME) **2**(4), 22–27 (2012)
4. Kaviyarasi, R., Balasubramanian, T.: Exploring the high potential factors that affects students' academic performance. Int. J. Educ. Manag. Eng. (IJEME) **8**(6), 15–23 (2018)
5. Shi, L.J., Yao, Y.J., Xia, X.Y., et al.: Practical teaching staff construction under new situation. Int. J. Educ. Manag. Eng. (IJEME) **1**(2), 57–61 (2011)
6. Restuccia, F., Das, S.K., Payton, J.: Incentive mechanisms for participatory sensing. ACM Trans. Sensor Netw. (TOSN) **12**(02), 1–40 (2016)
7. Yanjun, G.: The enlightenment of expectation theory on teachers' motivation in higher vocational colleges. J. Jiangsu Vocat. Tech. Coll. Econ. Trade **04**, 50–53 (2019). (in Chinese)
8. Chao, L., Ling, D.: Exploration on teaching incentive mechanism of college teachers based on expectation theory. J. Daqing Norm. Univ. **38**(03), 125–128 (2018). (in Chinese)
9. Börü, N.: The factors affecting teacher-motivation. Int. J. Instr. **11**(04), 761–776 (2018)
10. Roy, S., Bairagya, S.: Conceptualisation of Pedagogical Content Knowledge (PCK) of science from shulman's notion to refined consensus model (RCM): a journey. Educ. India Tourn. Q. Refer. J. Dial. Educ. **8**(2), 10–53 (2019)
11. Wei, Y., Yuying, G.: Influence and enlightenment of PCK on American science teacher education. Contem. Teach.Educ. **3**, 6–10 (2008). (in Chinese)
12. Feng, Z., Qu, T.: From PCK to pckg: a new turn in teachers' professional development. Foreign Educ. Res. **12**, 58–63 (2006). (in Chinese)
13. Liqin, Z., Dandan, W., Lehui, H.: The research on the program of educational technology ability building for vocational education teachers. Int. J. Educ. Manag. Eng. (IJEME) **2**(11), 46–51 (2012)
14. Marie, A.C., Robles, O.: The use of educational web tools: an innovative technique in teacher education courses. Int. J. Mod. Educ. Comput. Sci. (IJMECS) **5**(2), 34–40 (2013)
15. Xuyan, L.: Core characteristics of subject teaching knowledge (PCK) and its enlightenment to teacher education curriculum. Educ. Modern. **7**(18), 130–132 (2020). (in Chinese)
16. Yanfang, L.: Overview of research results of discipline teaching knowledge PCK. Curriculum Educ. Res. **37**, 229–230 (2016). (in Chinese)
17. Jie, T., Jianqi, Z.: Realistic dilemma of teacher development center in colleges and universities in China - based on expectation theory. J. Guangdong Vocat. Techn. Coll. Light Ind. **03**, 48–54 (2020). (in Chinese)
18. Gong, M.N.: An analysis of college Teachers' self-identity and its promotion strategies - an investigation based on Fromm's expectation theory. Educ. Theory Pract. **18**, 31–33 (2019). (in Chinese)

19. Xiangyu, Z., Jie, Z.: Research on incentive management of young college teachers from the perspective of expectation theory. Educ. Rev. **10**, 125–128 (2017). (in Chinese)
20. Xiao. H.Y: Management S.O. The incentive mechanism for promoting practice teaching ability of faculty in application-oriented university. Logist. Eng. Manag. **37**(11), 320–322 (2015)
21. Sun, R., Liang, S.: College young teachers' perception of the relationship between teaching and research and the incentive mechanism. Sci. Educ. Article Collect. **06**, 3–4 (2018)
22. Zhifang, X.: Teachers' discipline teaching behavior from the perspective of PCK. New Educ. **02**, 20–24 (2017). (in Chinese)
23. Hua, M., Min, Z., Liwei, H.: Innovation of management mechanism for faculty construction in colleges and universities. China Adult Educ. **22**, 51–53 (2018). (in Chinese)
24. Wang, X., Yingjie, W.: Exploration and practice for evaluation of teaching methods. Int. J. Educ. Manag. Eng. **2**(3), 39–45 (2012)
25. Meng, X.Y.: Survey of research quality and ability of vocational school teachers. Int. J. Educ. Manag. Eng. (IJEME) **10**(29), 9–16 (2012)

Dynamic Relationship Between Expenditure on Finance Education, Income Gap Between Urban and Rural Residents and Agricultural Economy Level

Lin Chen and Hongmei Gao(✉)

College of Economics and Management, Tianjin Agricultural University, Tianjin 300384, China

Abstract. In the implementation of the strategy of rural revitalization, it is of great significance to promote the fairness of urban and rural education and the rational distribution of educational resources to narrow the income gap between urban and rural residents and improve the level of agricultural economy. In order to find out the role and relationship between education, income and economy, by using cointegration test, impulse response function and variance decomposition method make an empirical analysis of the dynamic relationship between the three variables in VAR model. Firstly, examine the single integer order of the time series, all three variables are the first order single whole sequence. After that, the cointegration test shows a long-term equilibrium relationship between the three variables and the pulse response function is used to measure the impact of the random perturbation term on the current and future values of the variables. Finally, relatively important information about each random perturbation factor that affects the variables in the VAR model is given by variance decomposition. The results show that there is a long-term promotion between the three variables, but the short-term effect is not obvious.

Keywords: Expenditure on financial education · Income gap between urban and rural residents · Level of agricultural economy · VAR model

1 Introduction

Over the years, education expenditure has always been given priority as an important area of financial government expenditure to meet adequate financial support in the field of education. Individual' high level skills and the versatile knowledge acquired in educational context have become extremely important factors in evolving knowledge-based economies [1]. Cultivating high-quality farmers and providing talent support for agricultural rural modernization is the foundation of rural revitalization. Relevant research in Indonesia has found that through ICT education construction contributes to rural empowerment [2]. Higher education of students in rural and remote areas includes multiple factors. Improving educational participation rate is one of the important ways to evaluate the effectiveness of university training service-oriented talents [3, 4]. In general, increasing education expenditures will promote the reduction of the income

Z. Hu et al. (Eds.): AIMEE 2021, LNDECT 107, pp. 447–456, 2022.
https://doi.org/10.1007/978-3-030-92537-6_41

gap between urban and rural residents and increase the total output value of the agricultural economy. Many scholars have conducted extensive research on fiscal education, and some scholars have conducted empirical analysis on the relationship between fiscal education expenditures and narrowing the income gap between urban and rural residents [5–7].

However, few people have studied the dynamic relationship between fiscal education expenditure and the income gap between urban and rural residents and the level of agricultural economy [8, 9]. How does fiscal education expenditure affect the income gap between urban and rural residents? How to improve the level of agricultural economy? And how can the narrowing of the income gap and the improvement of the agricultural economic level counteract the fiscal education expenditure? Designing optimal policy and setting objectives have always been a concern area for the policy-makers [10, 11]. By establishing a VAR model of fiscal education expenditure, urban and rural residents' income gap and agricultural economic level, using Eviews software to analyze the relationship between the three variables, the countermeasures and suggestions are put forward to narrow the income gap between urban and rural residents and increase the agricultural economy and improve the utilization efficiency of financial education expenditure.

2 Theoretical Aspects of the Research

In this paper, the methods of unit root test, cointegration test, vector error correction (VEC) model, impulse response function and variance decomposition are mainly used to analyze the dynamic relationship between variables. All these methods are based on VAR models. Christopher Sims proposed VAR model in 1980. The VAR model does not have any prior constraints, but uses all current variables in the model to regress several lagged variables of all variables to estimate the dynamic relationship of the joint endogenous variables.

Before testing the cointegration or long-term equilibrium relationship of a set of time series, the single integer number of time series should be tested first. Cointegration is a statistical description of the long-term equilibrium relationship of non-stationary economic variables. A long-term equilibrium relationship between nonstationary economic variables is called cointegration. In the economic field,most macroeconomic variables are non-stationary, and it seems difficult to have a long-term equilibrium relationship between them,but in fact, the linear combination of certain economic variables may be stable. This means that there is a long-term stable equilibrium relationship between them. The cointegration test adopts the Johansen cointegration test method based on the VAR model. The Johansen cointegration test is a regression coefficient test method based on multiple equations, and there is no distinction between exogenous and endogenous. According to the Granger theorem, if several nonstationary variables have cointegration relations, then these variables have error correction (VEC) models [12].

The impulse response function is to analyze the dynamic influence of an error term on the system when a model is affected by a certain impact. It is used to measure the influence of a standard deviation shock from a random disturbance term on the current and future values of the internal variables over time. Christopher Sims proposed a variance decomposition method in 1980 based on the vector moving average model. Quantitatively and roughly analyze the contribution of each structural shock to the change of endogenous variables, evaluate the relative importance of different structural shocks and obtain relevant information about each random disturbance factor that affects the variables in the VAR model [13].

3 Practical Implementation

3.1 Data Source and Variable Selection

It has been proved to be an effective method to quantitatively analyze the research object by using mathematical methods and models [14]. The paper chooses the ratio of per capita disposable income of urban and rural households to reflect the income gap level, expressed by S, the ratio of gross output value of primary industry to gross domestic product to agricultural economic situation, expressed by G, the ratio of financial education expenditure to total financial expenditure reflects the level of financial education expenditure, expressed by E. The data of each variable are from the statistics of the National Bureau of Statistics for 1991–2020 and the Statistical Yearbook of China. For overcoming heteroscedasticity in time series, the natural logarithms of each variable are expressed as LNS, LNG, LNE respectively.

3.2 Unit Root Test

As shown in Fig. 1, the three variables LNE, LNG, LNS 1990–2020 shows obvious nonstationarity characteristics, which has a certain trend of change, while the first-order different sequence of the three variables ΔLNE, ΔLNG, ΔLNS show relatively stable characteristics. In order to further determine the stationarity and single-order number of each variable, it is necessary to carry out ADF unit root test for each variable.

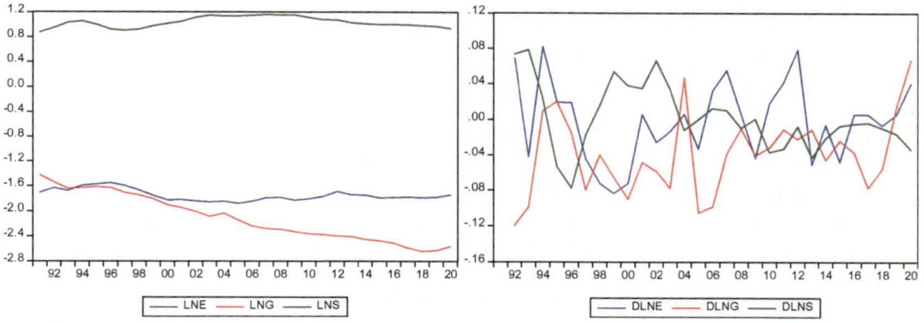

Fig. 1. Sequence diagram of LNE, LNG, LNS, ΔLNE, ΔLNG, ΔLNS

Because $\alpha = 0.05$, the ADF critical value of the regression formula of unit root test with time trend term under T = 30 condition is −3.57, the ADF values of variable LNS, LNG and LNE unit root test are −1.204, −1.296 and −1.303, are greater than −3.57, all three variables are non-stationary sequences. At this time, the nonstationarity of the first-order different series of three variables ΔLNS, ΔLNG and ΔLNE should be further checked. Because the unit root test of the original sequence already contains the time trend term, therefore the unit root test of the first order difference sequence does not need to contain the time trend term. The ADF threshold ($\alpha = 0.05$, T = 30) −2.97, the ADF values of ΔLNE, ΔLNG unit root test are −4.396 and −3.930 respectively, so they are all stationary sequences. And the ΔLNS unit root test is −2.717, close to −2.972, combined with its sequence diagram, ΔLNS can also be considered as stationary sequences. That is both LNS, LNG and LNE are first order monolithic sequences.

3.3 Cointegration Test

The number of lag periods of the VAR model should be determined before cointegration test k, because if the lag k is too small, the autocorrelation of the error term is sometimes very serious, which leads to the inconsistent estimation of the parameters. K is too large will lead to reduced degree of freedom and directly affects the effectiveness of the model parameter estimator. Usually, the lag period is judged according to the minimum principle of statistics such as AIC and SC of VAR models. The statistical values of the VAR model of lag 0–4 period established in this paper are shown in Table 1. Except for FPE and AIC, the other statistics are the minimum values of lag 1 period (with * in the table). The VAR model of lag 3 is sufficient to eliminate autocorrelation in random error term.

Table 1. Lagphase selection of VAR model on LNS, LNG and LNE

Lag	LogL	LR	FPE	AIC	SC	HQ
0	61.32407	NA	2.26E-06	−4.486467	−4.341302	−4.444665
1	172.31	187.8223*	8.92E-10	−12.33154	−11.75088*	−12.16433*
2	180.4117	11.84092	9.88E-10	−12.26244	−11.24628	−11.96982
3	192.2157	14.52805	8.66e-10*	−12.47813*	−11.02648	−12.06011
4	196.1762	3.960494	1.51E-09	−12.09048	−10.20333	−11.54705

* indicates lag order selected by the criterion
LR: sequential modified LR test statistic (each test at 5% level)
FPE: Final prediction error AIC: Akaike information criterion
SC: Schwarz information criterion HQ: Hannan-Quinn information criterion

The Johansen method is used to co-integration test the relationship between LNS, LNG and LNE. The test results are shown in Table 2. Since LR = 52.42299 > 35.19275, the zero hypothesis without co-integration equation is rejected at the 5%level, and at least one co-integration equation exists between the three variables. The zero assumption here is at most one co-integration equation is rejected at the 5% level, and there are at least two co-integration equations between the three variables.

Table 2. Results of co-integration test on LNS, LNG and LNE

Null hypothesis	Eigenvalue	Trace statistics	Critical value of 5% level	Prob.**
None *	0.680046	52.42299	35.19275	0.0003
At most 1 *	0.479019	21.65441	20.26184	0.032
At most 2	0.139269	4.049278	9.164546	0.4048

Trace test indicates 2 co-integrating eqn(s) at the 0.05 level
* denotes rejection of the hypothesis at the 0.05 level
**MacKinnon-Haug-Michelis (1999) p-values

Similarly, the existence of two co-integration relationships between the three should continue to be tested, since LR = 4.049278 < 9.164546 accept the zero assumption that there are at most two co-integration equations at the 5% level. Finally, it is concluded that there are two co-integration equations between the three variables standardized co-integration equation is as follows:

$$\begin{aligned} LNS &= -0.6769 - 0.7551LNE \\ &\qquad (1.06) \quad (2.07) \end{aligned} \tag{1}$$

$$\begin{aligned} LNG &= 2.3820 + 3.2003LNE \\ &\qquad (-1.13) \quad (-2.67) \\ Inside, e^1_{t-1} &= LNS_{t-1} - 0.7551LNE_{t-1} - 0.6759 \\ e^2_{t-1} &= LNS_{t-1} + 3.2003LNE_{t-1} - 2.3820 \end{aligned} \tag{2}$$

3.4 Analysis of Vector Error Correction Model

The cointegration test shows that there is a long-term equilibrium relationship between LNS and LNE, LNG and LNE. For each increase of financial education expenditure, the income gap between urban and rural residents decreases by 0.76%, and the primary industry economy increases by 3.20%. As a result, a VEC model of three variables can be established in the following matrix form:

$$\begin{aligned} \begin{bmatrix} \Delta LNS_t \\ \Delta LNG_t \\ \Delta LNE_t \end{bmatrix} = \begin{bmatrix} -0.2420 & 0.0763 \\ 0.1679 & 0.1151 \\ -0.0871 & 0.0008 \end{bmatrix} \times \begin{bmatrix} e1_{t-1} \\ e2_{t-1} \end{bmatrix} &+ \begin{bmatrix} 0.3657 & 0.0002 & 0.0448 \\ -0.1790 & -0.2657 & -0.0663 \\ -0.6727 & -0.5505 & 0.2965 \end{bmatrix} \times \begin{bmatrix} \Delta LNS_{t-1} \\ \Delta LNG_{t-1} \\ \Delta LNE_{t-1} \end{bmatrix} \\ &+ \begin{bmatrix} -0.5317 & -0.2425 & 0.0820 \\ 0.6038 & -0.2835 & -0.0486 \\ 0.7104 & -0.1934 & 0.2515 \end{bmatrix} \times \begin{bmatrix} \Delta LNS_{t-2} \\ \Delta LNG_{t-2} \\ \Delta LNE_{t-2} \end{bmatrix} \end{aligned} \tag{3}$$

Under the LNS error correction equation, in the short term, Coefficient for the lag periods 1 and 2 of the ΔLNG is 0.0002 and −0.2425, indicating that the narrowing of the income gap between urban and rural residents in the previous year has a positive effect on the improvement of agricultural economic level, while the widening of the income gap between urban and rural residents in the previous two years has a restraining effect on the improvement of agricultural economic level. The coefficients of the ΔLNE lag 1 and 2 periods are both positive, 0.0448 and 0.08, indicating that the increase of government financial education expenditure in the short term has an incentive effect on the narrowing of the income gap between urban and rural residents but not obvious.

In the error correction equation of LNG, from a short-term perspective, the coefficients of the lagging period 1 and 2 of ΔLNS is −0.179 and 0.6838, respectively, and the sum of the two is 0.5048, indicating that the income gap between urban and rural residents in the previous year has an impact on the agricultural economy of the current year. The increase in the level has a restraining effect, but it is not obvious, and the narrowing of the income gap between urban and rural residents in the previous two years has a strong positive incentive effect on the improvement of the agricultural economy; the coefficients of the lagging period 1 and 2 of ΔLNE are −0.0663 and −0.0486, indicating that the increase in government financial education expenditure in the short term has a restraining effect on the improvement of the agricultural economy. This is contrary to the long-term relationship between the two. This may be due to the long cycle of return on investment in education, which affects urban and rural residents. The facilitation of the narrowing of the income gap has been slow to respond, and for every 1% increase in government fiscal expenditures on education, the level of agricultural economy will increase by 3.2%.

In the error correction equation of LNE, the coefficients of the lagging period 1 and period 2 of ΔLNS are −0.6727 and 0.7104, respectively, and the sum of the two is 0.0377. It shows that narrowing the income gap between urban and rural residents in the short term will promote the growth of government fiscal education expenditure. The coefficients of the lag phase 1 and phase 2 of ΔLNG is 0.5505 and −0.1934, respectively. It shows that the improvement of the agricultural economy in the previous year had a great impact on the government's financial education expenditure, while the improvement of the agricultural economy in the previous two years had a small effect on the government's education expenditure. This is mainly because the government's education expenditure in the current year is mainly affected by the fiscal revenue of the previous year and has little relationship with the fiscal revenue of the previous two years. When the overall economic level grows rapidly, the agricultural economic level also shows a corresponding increase growth trend.

3.5 Pulse Response Analysis

The impulse response function is used to analyze the influence of a random disturbance of government expenditure on education, agricultural economic level and the income gap between urban and rural drama names on other endogenous variables. Only a stable VAR model can carry out impulse response analysis, and the sufficient and necessary condition of VAR model stability is that all eigenvalues of the parameter matrix of the model should be within the unit circle, or the modulus of the eigenvalue should be less than 1. As shown in Fig. 2, the eigenvalues of the VAR model are within the unit circle, which can be analyzed by impulse response. The values of impulse response function of three variables with 10-year lag are shown in Table 3.

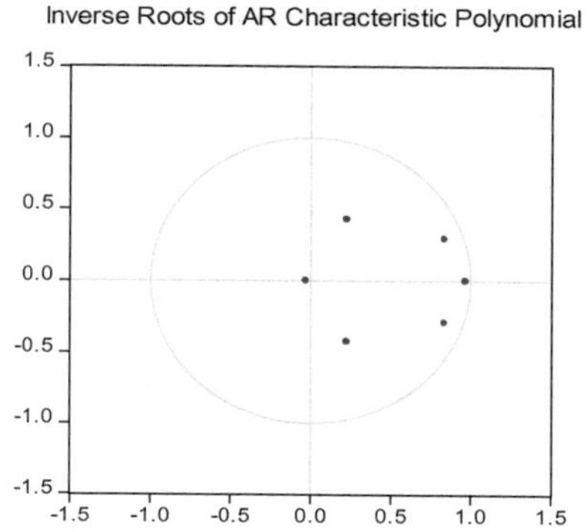

Fig. 2. Stationarity test of VAR model

Table 3. The impulse response value of 10-year lag period of LNS, LNG, and LNE

			1	2	3	4	5	6	7	8	9	10
①	LNS	LNS	0.0173	0.0157	0.0082	0.0025	−0.0006	−0.0025	−0.0038	−0.0045	−0.0046	−0.0040
		LNG	−0.0075	−0.0110	−0.0038	0.0059	0.0124	0.0153	0.0155	0.0141	0.0116	0.0084
		LNE	−0.0095	−0.0207	−0.0321	−0.0384	−0.0383	−0.0332	−0.0254	−0.0164	−0.0074	0.0005
②	LNG	LNS	0.0000	0.0121	0.0168	0.0165	0.0155	0.0153	0.0154	0.0153	0.0150	0.0144
		LNG	0.0365	0.0322	0.0214	0.0154	0.0136	0.0130	0.0125	0.0122	0.0121	0.0124
		LNE	0.0200	0.0278	0.0349	0.0369	0.0357	0.0333	0.0306	0.0276	0.0245	0.0216
③	LNE	LNS	0.0000	0.0031	0.0049	0.0063	0.0071	0.0073	0.0068	0.0058	0.0046	0.0034
		LNG	0.0000	−0.0080	−0.0115	−0.0122	−0.0113	−0.0093	−0.0064	−0.0031	0.0000	0.0027
		LNE	0.0436	0.0438	0.0395	0.0321	0.0233	0.0142	0.0058	−0.0011	−0.0060	−0.0088

(1) LNS response to random disturbances, as shown in part 1 of Table 3. LNS also reacts to a random disturbance of its own. The response to a random perturbation to the LNG equation is the strongest, but this response only began to appear in the second year and increased year by year. Its response value increased from 0.0075 in the first year to 0.0155 in the seventh year, and then gradually decreased small. The response to a random perturbation of the LNE equation changes from negative to positive and shows a trend of increasing year by year, reaching the maximum response value of 0.0005 in the tenth year.

(2) LNG response to random disturbances,as shown in part 2 of Table 3. LNG has the strongest response to random disturbances from the LNE equation.The response value in the first year is 0.02, and thereafter it increases to the maximum value of 0.0369 in the 4th year, and gradually decreases from the 5th year to 0.0219 in the 10th year. For LNS the random disturbance response of the equation is also large and stable. Its response value appears from the second year and remains between 0.0121 and 0.0168. The response to its own random disturbance shows a decreasing trend year by year, from the first year 0.0365 dropped to 0.0124.

(3) LNE reaction to each random perturbation as shown in part 3 of Table 3. LNE reacted greatly to each random disturbance, with its own random disturbance reduced from 0.0436 to −0.0088; the reaction to the LNS equation increased to 0.0034 in the following year; the reaction to the LNG equation started from the following year and was negative with the exception of the last two years.

3.6 Variance Decomposition

The variance decomposition results of three variables at the late 10 years are shown in Table 4.

Table 4. The variance decomposition of LNS, LNG, LNE in the 10-year lag period

		1	2	3	4	5	6	7	8	9	10
① LNS	S.E	0.0436	0.0624	0.0749	0.0826	0.0869	0.0888	0.0895	0.0897	0.0901	0.0906
	LNE	20.2122	41.8098	65.8368	78.1785	81.7937	81.8626	80.6692	79.1479	77.8298	77.0369
	LNG	12.5106	14.1533	8.0723	5.7924	6.9024	8.9627	11.0318	12.7937	14.0656	14.7537
	LNS	67.2772	44.0369	26.0909	16.0292	11.3040	9.1747	8.2991	8.0584	8.1046	8.2094
② LNG	S.E	0.0416	0.0608	0.0752	0.0867	0.0960	0.1036	0.1098	0.1149	0.1191	0.1225
	LNE	23.0106	31.7606	42.3142	49.8668	54.4838	57.1508	58.6106	59.2987	59.4704	59.2969
	LNG	76.9894	64.2600	50.0854	40.8125	35.3082	31.9094	29.6924	28.2386	27.3411	26.8613
	LNS	0.0000	3.9794	7.6004	9.3207	10.2079	10.9399	11.6970	12.4628	13.1885	13.8419
③ LNE	S.E	0.0211	0.0352	0.0485	0.0622	0.0741	0.0827	0.0880	0.0907	0.0918	0.0923
	LNE	100.0000	98.0919	95.9111	93.8808	92.0882	90.6657	89.7212	89.2336	89.0441	88.9428
	LNG	0.0000	1.6636	3.4977	5.0587	6.2773	7.0953	7.4972	7.5776	7.5240	7.5253
	LNS	0.0000	0.2445	0.5912	1.0605	1.6345	2.2391	2.7816	3.1887	3.4318	3.5320

(1) From the analysis of variance decomposition of LNS in Part ① of Table 4. LNE's contribution accounted for the vast majority, increasing from 20.21% in the first year to 77.04% in the tenth year. The second is the contribution of LNS to itself. The contribution in the first year reached 67.28% and gradually decreased with the extension of time. By the 10th year, the contribution ratio was only 8.21%. The contribution of LNG is relatively stable, with the lowest value and the highest value appearing in the 5th and 10th years, which is 5.79% and 14.75 respectively.
(2) From the analysis of variance decomposition of LNG in Part ② of Table 4. Its own contribution accounts for the vast majority, but the overall trend has been declining over time, from 76.99% of the maximum contribution value in the first year to 26.86%. Secondly, the contribution of LNE gradually increased, rising from 23.01% to 59.30%. The contribution of LNS was very small, increasing from 3.98% from the second year to 13.84% in the tenth year.
(3) Analyze the results of variance decomposition of LNE from part ③ of Table 4. Its own contribution accounted for the vast majority in the short term. The contribution in the first year reached 100%, and with the extension of the time interval, its contribution gradually decreased to 88.94% in the 10th year. Correspondingly, the contribution of LNG has increased year by year, from 1.66% in the second year to 7.53% in the tenth year. Compared with the former two, the contribution of LNS accounted for a small share, but the overall trend is still rising, from 0.24% in the second year to 3.53% in the tenth year.

4 Discussion

First of all, from the results of the model, narrowing the income gap between urban and rural residents has a certain promotion effect on fiscal education expenditures, and the improvement of the agricultural economy has a more obvious promotion effect on the increase of fiscal education expenditures. In fact, this promotion effect may not be direct, because financial education expenditures are mainly affected by the macroeconomic situation, government revenue and national fiscal policies. When the total economic volume grows rapidly, economic variables such as the level of agricultural economy and the income gap between urban and rural residents will generally change rapidly. Therefore, the key to increasing fiscal expenditures for education is to promote the steady and rapid growth of the economic aggregate.

Secondly, because the selection of variables will be affected by many factors such as time and data availability, the application of the model has certain deficiencies. Simple equations are far from determining the relationship between the three variables, and the model has yet to be improved. Expanding the proportion of fiscal education expenditure in GDP can effectively narrow the income gap between urban and rural residents, but relying only on education to narrow the gap between the rich and the poor is often limited. The cross-regional flow of resources hindered by the urban and rural dual structure, and capital tends to invest in urban areas with high return to maximize the benefits. To fundamentally change the excessive income gap between urban and rural residents, flexible and diversified economy policies measures should be taken to increase the income of rural residents.

5 Conclusion

In the long run, government expenditure on education plays a strong role in reducing the income gap between urban and rural residents and raising the level of agricultural economy, but in the short run, this role is not obvious. It shows that long-term financial investment in education is an inevitable choice to increase farmers' income and knowledge. Due to the unreasonable investment and expenditure structure of financial education expenditure, the Chinese government's financial education expenditure can't give full play to its promoting role. Therefore, it is not only necessary to increase fiscal expenditures for education, but also to adjust the structure of fiscal education expenditures, so as to improve the efficiency of fiscal education expenditures. To promote farmers' income, governments should increase the direct education subsidy to farmers and increase the education and training subsidy to green agricultural production mode.

The level of agricultural economy has little effect on reducing the income gap between urban and rural residents. It shows that the effect of promoting farmers' income increase simply by increasing agricultural output value is very limited. It is necessary to broaden and narrow the income gap between urban and rural residents and increase non-agricultural employment channels, such as adjusting the industrial structure as soon as possible and vigorously developing non-agricultural industries.

Reducing the income gap between urban and rural residents has a great effect on the promotion of agricultural output value. This shows that farmers are the main investment in agricultural production the income gap between urban and rural residents is narrowed which can increase the investment in agriculture and promote the increase of agricultural output value.

Acknowledgment. This research is supported by Wuqing District Science and Technology Development Plan Project of Tianjin Municipality, (No.WQKJ201803).

References

1. Kiss, J.T.: Examination of fiscal expenditures on education and research as human capital investments. MATEC Web Conf. **184**, 1–4 (2018)
2. Yasya, W.: Rural empowerment through education: case study of a learning community telecentre in Indonesia. Int. J. Mod. Educ. Comp. Sci. (IJMECS) **12**(4), 12–26 (2020)
3. Hossain, D., Gorman, D., Lawrence, J., Burton, L.: Factors that impact on rural and remote students' participation in higher education. Int. J. Mod. Edu. Comput. Sci. (IJMECS) **4**(7), 1–8 (2012)
4. Anand, R., Saxena, S., Saxena, S.: E-Learning and its impact on rural areas. Int. J. Mod. Educ. Comput. Sci. (IJMECS) **4**(5), 46–52 (2012)
5. Li, W., Cai, Z., Cao, S.: What has caused regional income inequality in China? effects of 10 socioeconomic factors on per capita income. Environ. Dev. Sustain. **23**, 13403–13417 (2021)
6. Luh, Y.: The impact of education on agricultural productivity: evidence from East Asian economies. Int. J. Food Agric. Econ. **5**(4), 11–24 (2017)
7. Yifan, W.: An empirical analysis of the factors affecting the income gap between urban and rural residents in Shanxi Province. Int. J. Appl. Math. Soft Comput. **6**(1), 30–35 (2020)
8. Huiyan, Z.: The verse effect of urban and rural income gap on agricultural economic development. Econ. Res. Guid. **18**, 17–18 (2017)
9. Liu, H., He, Q.: The effect of basic public service on urban-rural income inequality: a sys-GMM approach. Econ. Res. Ekonomska Istraživanja **32**(1), 3211–3229 (2019)
10. Singh, R.: Doubling farmer's income: the case of India. World Food Policy **5**(1), 24–34 (2019)
11. Niya, P.: An empirical study on the impact of education funding on farmers' income in poor areas. Stat. Dec. Making **37**(03), 67–70 (2021). (in Chinese)
12. Xiaotong, Z.: Eviews Use Guidelines and Case Studies. Machinery Industry Press, Beijing (2008).(in Chinese)
13. Sun, D.: Application of co-integration theory and VAR model: based on the perspective of farmer income increase. J. North China Univ. Technol. **17**(01), 28-32+42 (2017). (in Chinese)
14. Aliyev, A.G., Shahverdiyeva, R.O.: Application of mathematical methods and models in product - service manufacturing processes in scientific innovative technoparks. Int. J. Math. Sci. Comput. **4**(3), 1–12 (2018)

Evaluation Standards, Methods and System Guarantees for Selecting High-Quality Training Evaluation Organizations

Su Lifang(✉)

Wuhan Railway Vocational College of Technology, Wuhan 430025, China

Abstract. It is an important innovation in vocational education to combine academic certificate with vocational certificate skill grade certificate and explore the implementation of 1+X certificate system. China has now recruited nearly 400 training and evaluation organizations. It is an important task to select high quality training evaluation organizations from many training evaluation organizations. Based on the investigation of 176 training evaluation organizations, through exploratory factor analysis and confirmatory factor analysis, seven common factors of training evaluation organization were found. Through the analysis, multiple evaluation methods and dynamic supervision system of high-quality training evaluation organizations are proposed. And then play the leading and demonstration role of quality training evaluation organization.

Keywords: Training evaluation organization · Evaluation standard · Evaluation method · Evaluation system

1 Instruction

In 2019, the "National Vocational Education Reform Implementation Plan" lays out a top-level plan for the success of vocational education in the new era, and puts forward innovative solutions to the key and core problems that have been plaguing the development of vocational education. Among them, in order to alleviate the structural employment contradiction, improve the vocational education and training system; mobilize the enthusiasm of social forces to participate in vocational education, put forward to "start the 1+X certificate system pilot works" [1]. Training evaluation organization is an important subject in the 1+X certification systems, involving the development of vocational skill level certificates and standards, vocational training, and the transformation of talent training methods in vocational colleges. With the expansion of the implementation scope of 1+X certificate system, training evaluation organizations will have an impact on the whole vocational education. How to select high-quality training and evaluation organizations from numerous training and evaluation organizations? How to build a scientific evaluation standard system, evaluation methods and complete supervision system? It is a major subject of the 1+X certificate systems at the present stage.

Z. Hu et al. (Eds.): AIMEE 2021, LNDECT 107, pp. 457–468, 2022.
https://doi.org/10.1007/978-3-030-92537-6_42

2 The Birth and Connotation of Training Evaluation Organization

2.1 The Background of 1+X Certificate

In 2019, according to the National Vocational Education Reform Implementation Plan, the Ministry of Education launched the "1+X" pilot project focusing on serving national needs, market demand, and improving employability. "1+X" certificate refers to educational certificate + several vocational skill level certificates. "1" stands for academic certificate [2]. For students in vocational colleges, academic education helps them systematically learn professional knowledge and lay a solid professional foundation for their future career development. "X" represents a number of vocational skill level certificates, which can reflect the vocational skill level of learners. X" is an extension of "1" and is rooted in "1" [3]. The training objects of "1+X" certificate are divided into two kinds. One is the education that is carried out within the university for the students in school; the other type of education is outside school for the community. Of course, the "1+X" certificate system is not applicable to all majors. X certificate is mainly aimed at: First, it covers the fields of new technology, new technology and new standards related to agriculture, modern manufacturing industry and modern service industry [4]; Second, areas related to new infrastructure, new medical care, ensuring and improving people's livelihood; Third, areas that are urgently needed for national high-quality development, or areas that can meet the needs of people's livelihood [5].

2.2 The Birth of Training Evaluation Organization

In the process of implementing the modernization of national governance in China, the government has been changing its role to a service-oriented government through "government functions". In this 1+X certificate system pilot, the Ministry of Education no longer directly conducts X certificate evaluation and issuance, but builds a platform for colleges and training evaluation organizations to encourage and support school-enterprise cooperation in vocational skill level certificates. To play the role of a service-oriented government, on the one hand, it must meet the social and economic needs, guide the development direction of vocational education, let the colleges and universities meet the market, and give full play to the governments funding, services, policy guidance, and cooperation in promoting school-enterprise cooperation [6]. The role of publicity and guidance is to promote the entry of high-quality training evaluation organizations into colleges and universities, so that good vocational certificates and good schools can obtain more social resources and continuously optimize the allocation of social resources in colleges and universities. On the other hand, the non-profit nature of education also needs the government to supervise and manage colleges and training evaluation organizations. In view of this, the 1+X certificate system recruits the society in an open, fair and just way, and the Ministry of Education takes the lead in the recruitment and selection of education training evaluation organizations. Up to now, 301 enterprises have obtained the qualification of training and evaluation organization through four batches of recruitment for 1+X certificate, and the recruitment will continue in the future.

2.3 The Connotation of Training Evaluation Organization

In the qualitative aspect of the training evaluation organization, the specific analysis is mainly based on the provisions of the training evaluation organization in the "Vocational Education and Training Evaluation Organization Selection and Supervision and Management Measures (Trial)". Details as follows:

First of all, qualitatively speaking, the qualification of a legal person should first have the attributes of a legal person. Including enterprise legal person (such as company), association legal person (trade association), collective enterprise, in the number of training evaluation organizations, but also has the training evaluation organization selection qualification; Public institutions, which belong to non-profit legal persons according to the classification of legal persons, should also be training evaluation organizations.

Secondly, the training evaluation organization has legitimacy. The training evaluation organization has certain admittance threshold, needs the Ministry of Education promulgates according to the selection method carries on the recruitment selection. The meaning of open recruitment and selection on the basis of merit includes two layers: first, to become a training evaluation organization must meet the recruitment conditions; the second meaning, recruitment is a legal procedure, according to the requirements of the selection method for selection. The legal conditions and procedures for recruitment are based on the documents of the Ministry of Education, which is an administrative agency. The documents issued by the Ministry of Education are departmental regulations and laws in the broad sense. Only in accordance with the legal conditions can an organization become a training evaluation organization, so the training evaluation organization must have legitimacy.

Finally, Tasks of training evaluation organization. Responsibilities include the development and construction of vocational skill level certificates and standards, the implementation of vocational skill level assessment, certification and other related work, and the overall responsibility for the quality of certificates, reputation of the socialized evaluation organization. The vocational skill level is generally divided into primary, intermediate and advanced, and the corresponding vocational skill standard is formulated by the training evaluation organization for each level. In addition to the assessment of vocational skill levels, the training evaluation organization has an important responsibility to carry out training for teachers and students of colleges and people in need of the society, and in-depth school-enterprise cooperation.

The "1+X" certificate is a major innovative measure to promote high-quality professional development and enhance national competitiveness. It is an important mechanism for linking technological development trends and market demands, and it is also an important institutional guarantee for deepening the integration of industry and education and school-enterprise cooperation. The training evaluation organization is a new thing in vocational education. It is the main body of vocational education grade standard development, vocational skill evaluation and assessment, vocational education training implementer, vocational education resource aggregator and vocational education governance system optimizer, and it plays an important role.

3 Evaluation Criteria for Selecting High-Quality Training Evaluation Organizations

After recruiting 1+X certificate and evaluation organizations in four batches for two years, 301 training evaluation organizations have joined the 1+X certificate pilot works in China. However, the existing training evaluation organizations differ in terms of training experience, school-enterprise cooperation, and their own capabilities. The country vigorously promotes the 1+X certificate systems, and selects high-quality training evaluation organizations to demonstrate the role of pilot implementation of 1 +X certificate.

3.1 Selection of High-Quality Training Evaluation Organization Multi-dimensional Considerations

Randomly select 176 training evaluation organizations from the existing training evaluation organizations. After questionnaire surveys, interviews, etc., select 1+X training evaluation organizations in the questionnaire and perform exploratory factor analysis and confirmatory factor analysis to obtain seven Dimensions. The KMO value of the exploratory factor analysis is 0.799, indicating good sampling adequacy, and the Bartlett sphere test significance $p = 0.00$, indicating that there are common factors among the items, suitable for factor analysis, explained according to the total variance, based on values and extraction stone diagrams, Extract 7 principal components, and finally get 1–7 common factors.

Table 1. The training evaluation organization explores the factor analysis result of each dimension

Serial number	Common factor	Cumulative variance contribution rate
1	School-enterprise cooperation	26.068
2	Training evaluation organization's own foundation	37.349
3	Certificate implementation	44.072
4	Follow-up investment development	50.245
5	Certificate implementation conditions	55.396
6	Certificate social recognition	60.404
7	Training evaluation organization's previous training experience	65.294

The main method used by exploratory factors is principal component analysis, and the rotation method is the Caesar normalized most in-variance method. The cumulative explained variance of the seven common factors is 65.294%, which means that the total variance of the original variables is 65.294% Explanatory power. Both exploratory factor analysis and confirmatory factor analysis passed the statistical test and met the requirements of fitting. The results of exploratory factor analysis are shown in Table 1.

Principal component analysis (PCA) was used as the extraction method, and the Kaiser normalized maximum variance method was used to perform rotation. After 21 iterations, the rotation matrix was obtained. Based on the empirical research data, it puts forward three suggestions on the selection criteria, evaluation methods and supervision system of high-quality training evaluation organizations.

3.2 Selection Criteria for High-Quality Training Evaluation Organizations

In the 176 training evaluation organizations interviewed, including private enterprises, state-owned enterprises, public institutions, industry associations, collective enterprises, private non-enterprise units six types. The proportion of various training evaluation organizations is shown in (Fig. 1).

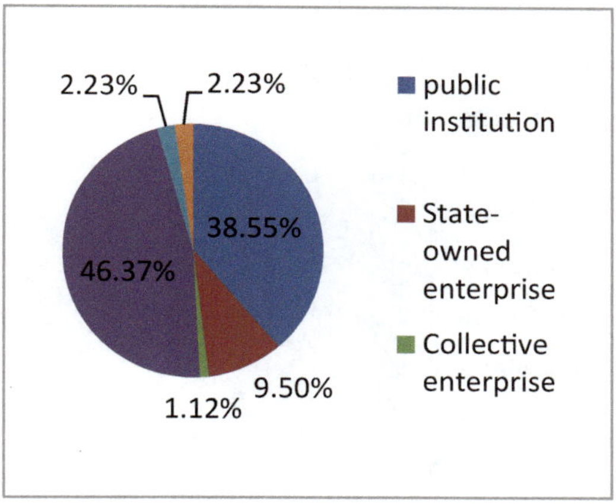

Fig. 1. Types of organizations involved in training evaluation

According to the above six training evaluation organizations, through the extracted 7 common factors, it is found that there are obvious differences in the three aspects of school-enterprise cooperation, 1+X certificate implementation, and certificate recognition, such as full-time certificate training teachers The number scale (Fig. 2), the training evaluation organization's certification scale in the past five years (Fig. 3), and the number of training students (see Fig. 4).

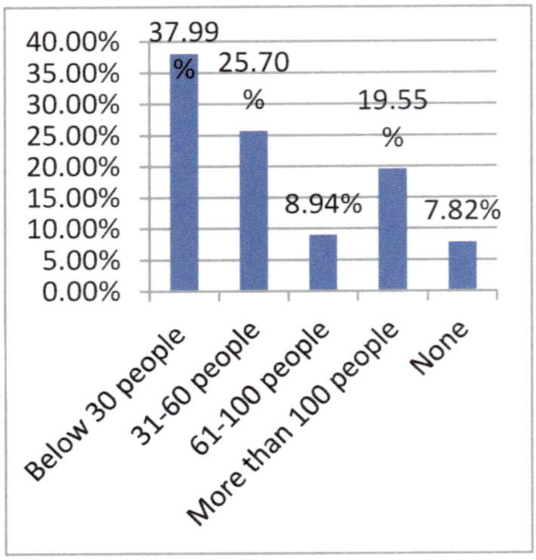

Fig. 2. Number of full-time training teachers

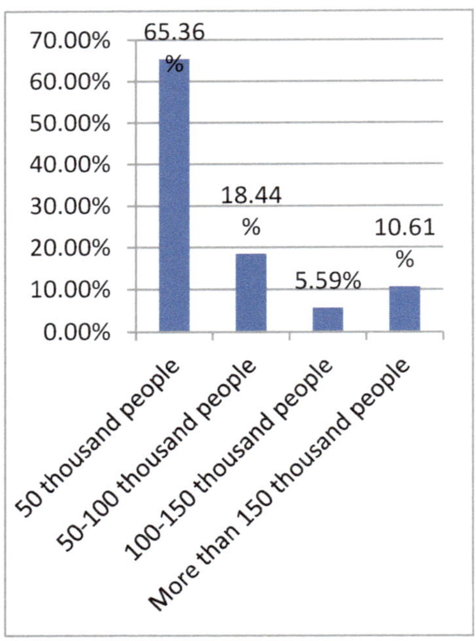

Fig. 3. The scale of the training evaluation organization in the past five years

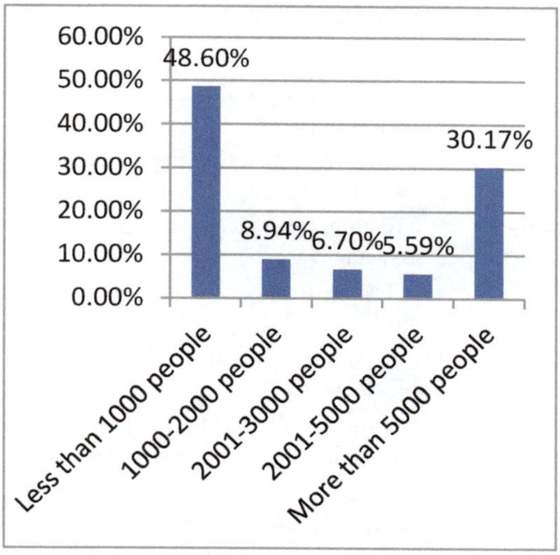

Fig. 4. Number of students in school

The data shows that there are differences in various evaluation organizations, and different assignments are given in the scoring system for selecting high-quality training evaluation organizations [7]. Based on the offline interviews of training evaluation organizations, it is concluded that a high-quality training evaluation organization needs to be measured from the following five aspects: influence in the industry, authority in standards, feelings in vocational education, strength in implementation, and market in certificates. In view of these five aspects, according to the seven common factors extracted, the sub-indexes are extended and designed respectively, and the index system of quality training evaluation organization selection is established (see Table 2).

In these five aspects, the certificate market is a key element of the premise that the 1 +X certificate can be implemented. Therefore, the weight given in the selection criteria is 35%.The quality of the actual implementation of the 1+X certificate is also an important measurement indicator for the training evaluation organization, and the weight given in the selection criteria is 25%. The industry influence of the training evaluation organization and the authority of the certificate standards will directly affect the social value of the certificate, so the weight given in the selection criteria each for 15%. At present, the main target of the implementation of 1+X certificate is school students. The training evaluation organization needs to go deep into the colleges, cooperate with the school to train the school students, and carry out 1+X teacher training for school teachers. In view of this, vocational education has feelings the selection criteria give a weight of 10%. The total of these five selection items is 100 points.

Table 2. Selection index system for high-quality training evaluation organizations

Items, weights	Indicators	Score
Influential in the industry M1 (15%)	1. The industry status of the training evaluation organization	52
	2. Training evaluation organization in the registered capital	48
Standards have authority M2 (15%)	1. Training evaluation organization's own advantages	60
	2. The basic situation of implementing the certificate	40
Vocational education has feelings M3 (10%)	1. Cooperative training with colleges in the past five years	60
	2. Provide 1+X certificate learning resources	40
Strong implementation M4 (25%)	1. Completeness of training rules and regulations	27
	2. Faculty team building	36
	3. Training site	15
	4. Implementation of certificate assessment	16
There is a market for certificates M5 (35%)	1. Certificate-oriented industry evaluation	30
	2. Applicable colleges	40
	3. Reflect new craftsmanship, new technology and new norms	30

Under the five items, two indicators are set and different points are given. The total score of each item is 100 points and the weight is multiplied to obtain each score.

The final score of the evaluated training evaluation organization is

$$\sum = M1 * 15\% + M2 * 15\% + M3 * 10\% + M4 * 25\% + M5 * 35\%$$

A score of 80 or more is classified as a high-quality training evaluation organization, and those with a score of less than 80 are classified as a training evaluation organization that requires three years of cultivation before evaluation, that is, a growing training evaluation organization.

4 Multi-subject Comprehensive Selection Mechanism

In order to ensure the fairness, openness and impartiality of the selection of training evaluation organizations, not only a set of scientific selection criteria is required, but also the participation of multiple selection subjects. Because colleges and universities and training evaluation organizations have different understandings of the orientation, connotation and characteristics of 1+X certificate in national education level due to the deviation of their own starting point and perspective, each subject has different understandings of high-quality training evaluation organizations. Therefore, in the selection of high-quality training evaluation organizations, education supervision institutions, colleges and training evaluation organizations should be included in the evaluation subject for multiple considerations. The comprehensive selection mechanism of multiple subjects shall be established, which shall be carried out according to the training evaluation organization's self-evaluation link, the expert evaluation link and the education supervision institution's evaluation link successively. The self-evaluation score of the training evaluation organization is the reference index for the selection of the high-quality training evaluation organization, the expert rating is the important basis for the selection of the high-quality training evaluation organization, and the evaluation of the education supervision organization is the key element for the selection of the high-quality training evaluation organization.

4.1 Self-evaluation of the Training Evaluation Organization

In the self-evaluation link of the training evaluation organization, the training evaluation organization scores itself one by one according to the selection indicators, and provides corresponding certification materials according to the evaluation method of the selection criteria. The certification materials submitted by the training evaluation organization shall be responsible for their authenticity. Once it is found that the training evaluation organization has submitted false certification materials, its qualification for selection will be cancelled.

4.2 Expert Evaluation

After entering the expert review process, the following issues need to be paid attention to in the review process: First, review experts can attract well-known experts and scholars who have long been engaged in vocational education and leading experts in the X-certificate industry under review. Establish a database of review experts, and review experts will be randomly selected from the expert database. Second, there should be no less than 5 evaluation experts for an X certificate, and there should be no avoidance issues such as interest relationship between the evaluation experts and the training evaluation organization to be reviewed. Thirdly, the expert review score is adopted as the average value. If an expert score is abnormally high or deformed, it is necessary to review, review or replace the expert review for re-evaluation.

4.3 Education Regulatory Agency Audit

In the review of education regulatory agencies, the Ministry of Education is the regulatory agency for the implementation of the 1+X certificate. However, due to the wide range and fast implementation of the 1+X certificate, there have been four batches of 301 training evaluation organizations so far, and the number will continue to increase in the future. The supervision of such a large training evaluation organization by the Ministry of Education will lead to oversight. Therefore, it is suggested that the 1+X certificate training evaluation organization carried out locally should be supervised by local education authorities familiar with local colleges. One of the main contents of supervision is to review and evaluate the training evaluation organization in accordance with the indicators of the selection criteria. The second is to review the certification materials submitted by the training evaluation organization. Once false materials are found, their certification qualifications will be cancelled and reported to the Ministry of Education for the record [8]. Third, to achieve the basic information of quality training evaluation organizations to be announced to the society, accept social supervision feedback, there are objections to the training evaluation organizations should be carried out in time, and the final opinions reported to the Ministry of Education and the society. Finally, after reviewing the evaluation scores of the training evaluation organizations, the education supervision institutions make the final conclusion by referring to the self-evaluation scores of the education and training evaluation organizations and the expert evaluation scores, and give the reasons for the selection and the improvement suggestions for the failure of the selection [9].

5 Supervision Mechanism of High-Quality Training Evaluation Organization

In order to ensure that high-quality training evaluation organizations provide long-term and sustainable high-quality education and training services, high-quality training evaluation organizations should implement timely supervision and dynamic adjustment mechanisms. The dynamic evaluation of the training evaluation organization suggests that the regulator should conduct an evaluation every three years and adjust the training evaluation organization according to the evaluation results [10]. At the same time, during the training and evaluation period, regulators can use big data to evaluate the implementation of the certificate on the one hand [11]. For example, the number of applicants for the 1+X certificate, the passing rate, and the employment situation of the personnel who have obtained the certificate, and the cooperation of the training evaluation organization and the school on the X certificate. Such as the number of school-enterprise cooperation with the X certificate, etc. Through big data detection, once it is found that the relevant data of the high-quality training evaluation organization is lower than the evaluation standard, the early warning mechanism can be activated.

On the other hand, in the evaluation at the end of the third year, the evaluation organizations that do not meet the high-quality training evaluation will be cleaned up, and the evaluation organizations that have been retired can be included in the growth training evaluation organization for training. If the training evaluation organization has honesty, fraud and other behaviors, one vote shall be rejected, shall not be included in the quality training evaluation organization, and shall be ordered to correct. For those who have not entered high-quality training evaluation organizations, but after three years of cultivation, can enter the high-quality training evaluation organization after self-evaluation, expert review and review by the competent authority.

Finally, in the supervision of training evaluation organizations, it is necessary to fully absorb the opinions of stakeholders such as students, colleges and universities, and corporate society [12]. The supervisory authority can supervise the training evaluation organization through online complaints and suggestions, online evaluation and other means [13]. Through the feedback of stakeholders, one can improve the efficiency of supervision, and the other can improve the overall planning in the implementation of the 1+X certificate system in the future for common problems.

6 Conclusion

The 1+X certificate systems is an innovative system proposed to alleviate the dilemma of decoupling my country's vocational education and market demand [14]. However, because of the public welfare of education, the 1+X certificate cannot be completely evaluated by the market.

In view of this, quality training evaluation organizations need to be evaluated from five aspects: internal influence, authoritative standards, feelings of vocational education, strength of implementation, and market of certificates. Use multiple evaluation methods of self-evaluation, expert review and regulatory review to select quality training evaluation organizations [15]. At the same time, a dynamic supervision mechanism should be established to ensure the healthy and sustainable development of quality training evaluation organizations.

References

1. Nanzhong, W., Haiying, X.: The basic process and quality assurance mechanism of 1+X vocational skill grade certificate development. Educ. Vocat. **24**, 26–32 (2019). (in Chinese)
2. Wang, Y., Lu, D.: German vocational qualification certificate system and its enlightenment to 1+X certificate system. Contemp. Vocat. Educ. **05**, 105–112 (2020). (in Chinese)
3. Tanaka, Y.: The economics of Cooperative Education: A Practitioner's Guide to the Theoretical Framework and Empirical Assessment of Cooperative Education. Routledge, London and New York (2014)
4. Kuhlee, D., Laczik, A.: Vocational Education and Training in Germany, p. 33. Research in Comparative & International Education, Oxford (2015)
5. Bundesinstitut für Berufsbidung: Evaluation des berufsbildungsgesetzes (BBiG) evaluierungsbericht. Bonn: Bundesinstitut für Berufsbildung, pp. 9–10 (2016)
6. Yang, L., Shuping, G.: The value implication, theoretical basis and realization conditions of 1+X cerificate system in the new era. Educ. Vocat. **23**, 12–17 (2019). (in Chinese)
7. William, W.B., Michael, L.W.: Shareholder primacy's corporatist origins: adolf berle and 'the modern corporation. J. Corp. Law, 34–99 (2008)
8. Kraakman et al.: The anatomy of Corporate Law: A Comparative and Functional Approach. Oxford University Press (2017)
9. Blasi, J., Kruse, D., Freeman, R.B.: Broad-based employee stock ownership and profit sharing: history, evidence, and policy implications. J. Participation Emp Owner. **1**(1), 38–60 (2018)
10. Bebchuk, L.A., Tallarita, R.: The illusory promise of stakeholder governance. Cornell Law Rev. **106**, 91–178 (2020)
11. Junhai, L.: On the rule of law guarantee system of factor market allocation. J. Lanzhou Univ. (Soc. Sci.) **48**(5), 28–38 (2020). (in Chinese)
12. Larry, F.: A sense of purpose. Harvard Law School Forum Corp Govern 17 (2018)
13. Friedman, M.: The social responsibility of business is to increase its profits. In: Zimmerli, W. C., Holzinger, M., Richter, K. (eds.) Corporate Ethics and Corporate Governance, pp. 173–178. Springer Berlin Heidelberg, Berlin, Heidelberg (2007). https://doi.org/10.1007/978-3-540-70818-6_14

14. Haisheng, P., Shangyun, Y.: Strategic choice of vocational education under the background of new economy and new technology: from the perspective of comparison. High. Eng. Educ. Res. **06**, 117–128 (2018). (in Chinese)
15. Sheather, G., Martin, R., Harris, D.: Partners in excellence: a new model for cooperative education. Innov. Educ. Teach. Int. **30**(1), 5–31 (1993)

Theme-Based Instruction Applied in Teaching of Non-english Majors in China and an Empirical Analysis

Yangqu Wang(✉) and Jinling Xia

School of Economics and Business Foreign Language, Wuhan Technology and Business University, Wuhan, China

Abstract. The traditional way of English teaching in China has been questioned because of the increasing concern about students' dumb English and high scores with poor English proficiency. Theme-Based Instruction (TBI) is to organize language teaching by themes or topics and attaches great importance to content. The TBI is expected to improve students' language proficiency and cognitive development. The effectiveness and feasibility of TBI on improving non-English majors' English proficiency in China by conducting a one-semester empirical project in a teaching university in China is still unknown. In this study, the class experiment was conducted at a teaching university in Wuhan. Two parallel College English classes were selected, one as a control group and the other as an experimental group. The control group was taught in the traditional way whereas the experimental group was taught with TBI. The result shows that the experimental group made significant progress in their overall English proficiency and listening comprehension after 12 weeks of TBI compared with the control group. The study indicates that TBI is feasible in College English teaching in China. It can be effective in improving non-English majors' English proficiency if it is applied consistently and longitudinally with well-designed activities and tasks.

Keywords: Time-series data · Correlation mining · WAMS/SCADA data fusion

1 Introduction

College English, a compulsory course for non-English majors in their first two years in mainland China, aims at preparing them in their English proficiency for their future jobs or academic fields. However, the outcomes of this course have not been satisfactory, especially in teaching universities [1]. Today, in the global context, the increasingly frequent interactions across countries set even higher standards for the talents in their English proficiency [2]. As a result, the gap between the undergraduates' English proficiency and their future careers' requirements has even grown.

Since 1980s, Chinese educators have been exploring the way of reforming College English teaching to narrow this gap [1]. To facilitate and guide the reform, the Ministry of Education of China has revised the syllabus for College English several times and launched College English Curriculum Requirements (2017), which requires that

Z. Hu et al. (Eds.): AIMEE 2021, LNDECT 107, pp. 469–480, 2022.
https://doi.org/10.1007/978-3-030-92537-6_43

English should be taught as a tool for communication rather than as isolated knowledge [3]. Under such a backdrop, numerous advanced English as a Foreign Language (EFL) teaching approaches of well-known scholars all over the world have been introduced to mainland China and a large number of college English teachers have made attempts to apply these approaches to their classroom to improve their teaching effectiveness [4].

Among them, Content-Based Instruction (CBI), which integrates content learning with language learning and thus involves students in a real language environment, aligns highly with the requirements as stated in the College English Curriculum Requirements (2007) for non-English majors [5, 6]. Theme-Based Instruction (TBI), one of the models of CBI, has drawn increasing attention of educators and EFL teachers in China since it was introduced and applied to English teaching in China in 1980s [1, 7].

The basic principle of TBI is to organize language teaching by themes or topics and attaches great importance to content so that students' language acquisition occurs as the "side products" of their comprehension of the content [7]. The meaningful content can arouse students' interest and curiosity in the outside world beyond a target language. Consequently, their language proficiency can be improved, and their cognitive development can be enhanced [8].

The traditional way of English teaching, which focuses on vocabulary, grammar and language points, has been dominant in China due to the exam-orientated English teaching in China [1]. Unfortunately, its effectiveness has been questioned on account of the increasing concern about students' dumb English and high scores with poor English proficiency. Great efforts and attempts have been made to search for more effective ways for English teaching at all school levels. In this study, the research was conducted to explore the effectiveness and feasibility of theme-based instruction in improving the English language proficiency of non-English majors in China. It also aims to probe into the effects of TBI on improving non-English majors' English proficiency in China by conducting a one-semester empirical project in a teaching university in China.

2 Methodology

2.1 Research Design and Data Collection

This research project was conducted in a teaching university in Wuhan, a central city of China. This teaching university aims at cultivating students' practical skills, among which the command of English ranks high. Students are encouraged to spend more time and energy learning English, but they are much lower in English proficiency and motivation for English learning than those in research universities. The research was carried out during the fall 2016 semester based on the two questions:

1) Can Theme-Based Instruction be more effective than the traditional teaching model of college English in improving non-English majors' English proficiency?
2) In what way does Theme-Based Instruction affect non-English majors' English proficiency?

2.2 Participants

The teaching university placed all non-English major freshmen into three level English classes – level A, level B and level C (level A being the highest), according to their scores on the National Matriculation English Test (NMET) 2016. Two classes at level B (the intermediate level) were chosen for this research project. One class served as a control group; 37 out of 53 students volunteered to participate in the project. The other class served as an experimental group; 36 out of 47 volunteered to participate in the project. The participants, all from the school of administration, are of similar age (18 to 19) and of similar education background (about 9 to 10 years' English learning experience).

2.3 Research Procedure

This research project was conducted from October 2016 to January 2017. The researcher taught both the control group and the experimental group with the same textbook New Horizon College English I but different teaching models. The researcher taught the control group in the traditional way, giving priority to the English language, including grammar and vocabulary. With regard to the experimental group, the researcher adopted TBI. Two exams were used respectively as the pretest and the posttest. The scores of the two exams were collected and compared to find out if TBI improved the English language proficiency of the participants from the experimental group.

2.3.1 Pretest

Since China's NMET is one of the most important national exams in China with high reliability and validity, it was adopted in this study as the pretest. All the participants' scores in NMET 2016 were collected. As mentioned previously, the English proficiency of the two classes were similar at the beginning of this study according to their scores on NMET 2016.

2.3.2 Teaching Design

College English is a four-credit course for all non-English majors in their first two academic years in China. The College English class for the participants met twice per week (90 min per meeting) for a total of 12 weeks. Textbook units. The textbook New Horizon College English (reading and writing) I, is composed of ten units, each focusing on a specific theme. All first-year College English instructors worked together and selected five units from the textbook (see Table 1).

In the control group, the researcher adopted the traditional way of English teaching featured by teacher-centeredness. The course procedures are presented as follows:

During the first phase, the researcher explained all new words and expressions to the students with many examples to show the students how to use them. To help the students consolidate their linguistic knowledge, the researcher asked them to do a lot of vocabulary exercises, such as filling in the blanks and making sentences.

After sweeping away vocabulary obstacles, the researcher explained the text in detail and when necessary, grammatical structures and points to help the students understand the text. The students were provided with abundant examples and supposed to finish exercises by mimicking these examples. Although the students were assigned to complete the exercises during class, the researcher's lecture dominated the whole class.

Table 1. The five units selected for fall 2016 first-year college english

No	Theme	Texts
Unit 1	Learning a foreign language	Section A: Learning a foreign language Section B: Keys to successful online Learning
Unit 2	Generation gap	Section A: Deep concern Section B: Is there a generation gap?
Unit 3	Human and love	Section A: Good heart to lean on Section B: The right son at the right time
Unit 4	First Impression	Section A: How wo make a good impression Section B: Body language
Unit 7	Gun	Section A: Face to face with guns Section B: Should I have a gun?

After class, the students were assigned to finish exercises, which revolved around the new words, expressions and grammatical structures. All the exercises were checked and explained if necessary, in the following class to ensure that the students had a good command of the linguistic knowledge.

Here is a sample lesson from the experimental group. In the experimental group, the researcher adopted TBI featured by theme-based activities such as group discussion, pair work, role play and presentations. A variety of listening and reading materials were chosen from newspapers, the internet, magazines, specialized books to supplement the textbook in the hope of providing authentic language environment and stimulating the students' interest.

The lesson on foreign language learning experience from Unit 1 Learning a Foreign Language is described below from Phase 1 to 6 in detail to demonstrate how the researcher designed the TBI class.

Phase 1: Warm-up (10 min)
The students were asked to watch a video of a piece of English news with both English and Chinese subtitles from www.CCTV.com about English learning in China. The news gave the students an updated overview of the current situation of English learning in China and provoked the students to think about the reasons behind it. The students were then asked to brainstorm the reasons why English learning was so popular in China. After the students shared their opinions, the instructor wrote all the possible reasons down on the blackboard, which prepared the students for vocabulary and sentence patterns needed in the following activities about their own foreign language learning experiences.

Phase 2: Interview and Report (20 min)
The students were divided into a few groups with 5 or 6 students in each. One student of each group acted as an interviewer and the others as interviewees. The interviewers were supposed to ask interviewees questions about their foreign language learning experiences, take notes and then report what they learned from the interview. They were provided with a handout including a list of questions (see below).

1) What foreign languages do you learn and when did you start to learn?
2) Could you describe your learning experience with one or two adjectives and why?
3) What impressed you most in your foreign language learning?
4) Do you have any problems while learning the foreign language and how do you deal with these problems?

After the group interview, the interviewers were invited to report their interviewees' foreign language learning experiences; meanwhile, the researcher took notes from these reports and listed student language problems on the board.

Phase 3: Further Discussion and Listening to a Lecture (10 min)
The researcher asked the students to brainstorm their solutions to the identified language problems on the board. All the vocabulary, expressions and their thinking involved in the brainstorming prepared the students for the following English lecture about how to learn English effectively. The lecture, given by a British teacher who taught English in China and targeted at EFL learners, provided students with authentic and meaningful input. While listening, the students were asked to complete the outline of the lecture by filling in the missing words. This task helped the students follow the lecture and focus on what they were listening to.

Phase 4: Skimming (20 min)
The students shared their foreign language learning experiences and identified common and specific problems they had in learning a foreign language. This activity prepared the students for the follow-up reading activity in that it helped the students predict the content of the reading. The students were then given 15 min to read the text *Learning a Foreign Language* and match 10 statements with their corresponding paragraphs. Each statement is about the main idea of a corresponding paragraph. The matching activity helped the students get the main idea and draw up the outline of the text.

Phase 5: Scanning and Group Discussion (20 min)
The students worked in groups of 5 or 6 members. They were allowed 10 min to scan the text. They were supposed to discuss with their group members and share what interested them or to associate the writer's learning experience with their own. Based on the students' discussion and information sharing, the researcher analyzed the writing techniques of the writer as well as explained new words or expressions when necessary.

Phase 6: Review and Assignment (5 min)
One student in each group was asked to retell one phase of the writer's learning experience, which reinforced the students' understanding of both the text and the way the paragraphs were developed. The students were assigned to write an essay about

their own foreign language learning experiences. They were supposed to apply the words, expressions and the writing techniques covered in this lesson to their essay.

2.3.3 Posttest

The final exam of College English in the 2016 fall semester was taken as the posttest to find out whether there were any differences in the two groups' English proficiency.

The final exam happened to have similar components as NMET, including listening comprehension, reading comprehension, cloze, translation and writing. Two English teachers with about 10 years' teaching experience from the English Department of the university were invited to grade the posttest using the same criteria. Each student's paper was graded by both teachers, but neither of them knew if this student was from the control group or the experimental group. The mean of the student's scores by both teachers was calculated for the student's final score.

2.4 Data Analysis

Data collection was completed around middle January 2017 after the posttest was held. The raw data from both the pretest and the posttest were recorded with Excel 2007. The researcher then utilized the software IBM SPSS 22.0 to process the data for statistical analysis. To find out whether there are any differences between the experimental group and the control group, the researcher calculated means and standard deviations and conducted *t*-Tests.

3 Results

3.1 The Pretest Results

The result of NMET was adopted as the pretest in this study. The pretest (100 points total) included four sections, 20 points for listening comprehension, 45 points for reading comprehension, 25 points for writing and 10 points for translation sections, but the researcher has no access to the score for each section. Therefore, only the participants' total scores were analyzed (see Table 2 for the results of the pretest).

Table 2. The participants' performance on the pretest based on the NMET

	Group	N	Mean	Standard deviation
NMET	Control	36	67.50	10.88
	Experimental	37	67.56	10.49

As Table 2 shows, the mean score of each group is about 67.50. The standard deviation of the control group is 10.88, a little bit higher than that of the experimental group (10.49), which suggests that the distribution of the scores in both groups is similar. Both the mean scores and standard deviations indicate that the two groups were roughly at the same English proficiency level at the beginning of the fall 2016 semester.

3.2 The Posttest Results

The posttest is the final exam of College English for non-English major freshmen, including same sections as the pretest – listening, reading, writing and translation. The total score for the posttest was also 100 points, 30 points for listening comprehension, 40 for reading comprehension, and 15 for writing and translation respectively.

Table 3 shows the comparison between the control group and the experimental group in terms of the total score as well as the score of each section in the posttest. As far as the total score is concerned, it is obvious that the mean score of the experimental group (i.e., 56.91) exceeds that of the control group (i.e., 50.17) by about 6.8 points; overall, the experimental group significantly outperformed the control group. There is also a significant difference between the standard deviation of the experimental group (i.e., 8.39) and that of the control group (i.e., 12.91). The higher standard deviation of the control group means that the total scores of the control group are more widely distributed than those of the experimental group.

As for listening comprehension, the mean scores of the two groups were greatly different from each other, 13.65 for the control and 17.22 for the experimental. With regard to the, standard deviation, the value for the experimental group was much smaller than that for the control one, 3.62 versus 5.66, which indicates that the scores of the experimental group varied less than those of the control group.

In the case of the other sections, reading comprehension, writing and translation, no significant differences between the two groups were found in terms of the mean. But as the mean score of each section indicates, the experimental group did surpass the control group in each section of the posttest. Specifically, the mean score of reading comprehension in the experimental group is about 1.6 point higher than that in the control group. The same difference was found in the mean score of translation between the two groups. The two groups were quite similar in terms of the mean score of the writing section, with only about 0.5-point difference.

Table 3. The participants' performance on the posttest

	Group	N	Mean	Standard deviation
Total	Control	36	50.17	12.91
	Experimental	37	56.91	8.39
Listening comprehension	Control	36	13.65	5.66
	Experimental	37	17.22	3.62
Reading comprehension	Control	36	23.06	5.88
	Experimental	37	24.66	3.40
Writing	Control	36	7.59	2.47
	Experimental	37	8.08	2.52
Translation	Control	36	5.86	3.27
	Experimental	37	6.95	2.91

In order to test whether these differences are significant, independent *t*-test was conducted, and results are shown in Table 4. Table 4 indicates that significant differences between the experimental group and the control group can be found in both total scores ($t = 2.616$, $p < 0.05$) and listening comprehension scores ($t = 3.414$, $p < 0.05$). The p value for all the other sections is not lower than 0.05, which suggests that there are no significant differences between the two groups in reading comprehension, writing or translation.

Table 4. Independent sample test of the participants' performance on the posttest

Levene's test for equality of variances	T-test for equality of means	
	t-value	p-value
Total	−2.616	0.011*
Listening comprehension	−3.414	0.001*
Reading comprehension	−1.413	0.162
Writing	−0.821	0.414
Translation	−1.464	0.148

*p < 0.05

4 Discussion

Based on the research result, remarkable differences between the two groups were found in both total score and listening comprehension score, which proved the effectiveness of TBI in improving the students' overall English proficiency and listening comprehension. Nevertheless, no differences were identified in reading comprehension, writing or translation.

4.1 Significantly Improved Overall English Proficiency

This section discusses the factors that may account for the experimental group's overall English proficiency.

According to the research result, the TBI significantly improved the overall English proficiency and the students' English study can be promoted from four aspects discussed as follows.

4.1.1 Motivation

The improved students' motivation may have contributed to the experimental group's overall English proficiency. Muoz et al. stated that language acquisition is greatly influenced by learners' affective conditions and motivation [9]. Motivation, to some degree, can determine the learners' success in language learning [10, 11]. No test of the students' motivation was carried out in this study. But based on the researcher's observation, the students in the TBI class were more active and fully involved in classroom activities. More students in the TBI class volunteered to answer questions, and they acted more positively when faced with difficult or multiple tasks. On the contrary, more silence and negative affection were observed in the control group. The students in the control group were more likely to be distracted and bored during the class. The different focuses of the two teaching methods may account for the students' opposite attitudes towards English learning. In the TBI class, language learning was contextualized, meaningful, and motivating to the students. In other words, the meaningful content aroused the students' interest more than dull grammatical points. For instance, the authentic texts were much more informative, revealing what was happening out of the classroom, which broadened the students' horizons and satisfied their curiosity about the outside world. In addition, involving the students in the authentic language environment and offering them more opportunities to use the language in real interactions may have greatly motivated them. With their improved motivation, they were willing to invest more time and effort in their language learning.

4.1.2 Comprehensible Input

Abundant comprehensible input offered in the TBI class may account for the significant difference in the two groups' overall performance on the posttest. Yan regarded the comprehensible input in the target language as an essential element in the learners' second language acquisition [12]. In this study, the students in the experimental group were more exposed to English from various sources than those in the control group. Besides the textbook, abundant theme-related materials were provided in the classroom. With plentiful videos, audios and articles from real-life settings in English-speaking countries, the researcher attempted to immerse the students in an authentic language environment. A lot of classroom activities and the researcher's lectures assisted the students in understanding course materials. The authentic and comprehensible materials enabled the students to acquire and use language for real purposes rather than memorize linguistic forms.

4.1.3 Pushed Output

Well-designed tasks and assignments may have helped the experimental group make greater progress. Braidi argued that output is an indispensable element in language learning and that output allows learners to receive feedback and develop linguistic awareness [13]. If learners are not pushed to produce output, their immersion in a target language may not necessarily lead to a successful language acquisition. In the TBI class, both input and output are of equal importance. Provided with sufficient comprehensible input, the students were supposed to produce output. Equipped with the information they gained in the class on a certain theme, the students were expected to form their own ideas and express them orally or in a written way. During discussion or other collaborative tasks, the students did negotiate for both meanings and language forms, which aroused the students' attention to linguistic forms and helped them correct their mistakes.

4.1.4 Learning Abilities

The students' autonomous learning ability and collaborative learning ability may also account for their significant improvement [14]. The two abilities were not evaluated in this study, but they were reflected by the students' assignments. The two groups were assigned different homework. The control group was usually required to work individually mainly on vocabulary and sentence translation, while the experimental group were required to complete a variety of tasks, including presentation, survey, report, role-play, and drama. Obviously, the homework for the experimental group required more time, creativity, group collaboration and commitment. Therefore, there were more opportunities for the students in the experimental group to develop their autonomous and collaborative learning abilities. The experimental group usually completed their assignments with more efforts and higher quality.

4.2 Significantly Improved Listening Comprehension

Listening comprehension being taught differently to the two groups may account for the difference in their performance on the listening comprehension section in the posttest. Firstly, the students in the experimental group received more oral input from the researcher (the instructor), which may have enhanced their listening comprehension.

Classes for the experimental group were mainly taught in English, attaching more importance to the students' understanding of the content. In contrast, Chinese was often used to help the students in the control group understand vocabulary and grammatical rules. Secondly, the students in the experimental group were involved in the real and meaningful interactions to practice listening. In the TBI class, listening skill was an integral component of every class, combined with the other language skills, especially speaking skill. Listening practice served as a way to gain information for real communication rather than for listening comprehension questions [15, 16]. In a regular class for the control group, listening comprehension was dealt with as an isolated skill. The students were asked to listen to some listening materials followed by comprehension questions. This traditional way of teaching listening comprehension was not as effective as TBI.

4.3 Non-significant Differences in Reading Comprehension, Writing or Translation

No significant difference was found in reading comprehension, writing or translation between the two groups in this study. As one of the receptive skills, reading comprehension was expected to be improved as greatly as listening comprehension. One possible reason for the minimal difference in reading comprehension is that the topics of the reading comprehension section in the posttest were not covered by any course unit. In other words, the experimental group did not have an obvious advantage over the control group in terms of vocabulary and background preparation for the test [17]. With regard to writing and translation, both groups performed poorly. Writing, a productive skill, is one of the greatest challenges for language learners and cannot be improved tremendously in a short period time [18]. It is not surprising that neither group improved their writing significantly after the first semester College English. Speaking was not tested in either the pretest or the posttest, so there is no evidence to show whether the students in the experimental group have improved their oral proficiency. However, based on the researcher's observation, the students in the experimental group were more confident and fluent while speaking English. This might be because the experimental group were offered more opportunities to develop their speaking skill. They were given much more oral tasks for real communication [19]. In other words, they practiced more in a natural way to develop their speaking skill. During communication, they probably negotiated for both meanings and forms and received feedback. Moreover, once they successfully accomplished speaking tasks, they gained a sense of achievement and confidence, which might have motivated them to speak more. If speaking were included in both the pretest and the posttest in this study, it is predictable that the experimental group would outperform the control group in the posttest.

5 Conclusions

This study aims to find out whether TBI is effective in improving non-English majors' English proficiency in mainland China. As the reviewed literature suggests, TBI, with a focus on content, provides an authentic language environment for learners and immerse them to develop their language skills.

In this study, the researcher conducted a one-semester experiment at a teaching university during the fall 2016 semester in Wuhan, China. Two College English classes were selected, one as the control group, and the other as the experimental group. The researcher applied the traditional teaching method in the control group, focusing on linguistic knowledge. While in the experimental group, the researcher adopted TBI, giving priority to content.

The data analysis shows that the experimental group made significant progress in their overall English proficiency and listening comprehension. Unfortunately, no significant improvement was found in the experimental group's reading comprehension or writing proficiency. In addition, since speaking was not tested in either the pretest or the posttest, no evidence was available regarding the effects of TBI on improving the learners' speaking ability.

In conclusion, the results from this study indicate that TBI is feasible in College English teaching. It can be effective in improving non-English majors' English proficiency if it is applied consistently with well-designed activities and tasks. In a TBI class, the emphasis on both content and target language can greatly arouse students' interest and thus motivate them to invest more time and energy in English study. In addition, abundant meaningful and comprehensible input along with various classroom activities or tasks provides the students with an authentic language environment.

Acknowledgment. This research is financially supported by 2020 Wuhan Technology and Business University Special Fund of Scientific Planning and Research of Private Higher Education Research Project "Research on Teacher Team Construction of Private Colleges from the Perspective of "Ideological and Political Theories Teaching in All Course" (GB2020002).

References

1. Jinfei, X., Yumei, F.: The evolution of the college English curriculum in China (1985–2015): changes, trends and conflicts. Lang. Policy **16**, 267–289 (2017)
2. Kormos, J., Kontra, E.H., Csölle, A.: Language wants of English majors in a non-native context. System **30**(4), 517–542 (2002)
3. Lin, L., Reinders, H.: Students' and teachers' readiness for autonomy: beliefs and practices in developing autonomy in the Chinese context. Asia Pac. Educ. Rev. **20**, 69–89 (2018)
4. He, D., Lindsay, M.: English teacher preference: the case of China's non-English-major students. World Englishes **30**(3), 428–443 (2011)
5. Wenmin, L.: College English teachers' "Content-Based" professional development in China. J. Appl. Res. Intellect. Disabil. **3**(5), 319–323 (2015)
6. Suwannoppharat, K., Kaewsa-Ard, A.: Utilization of content-theme-based instruction: an overhaul of english language learning for non-native English learners. Int. J. English Lang. Educ. **3**(1), 115–126 (2014)
7. Parkinson, J.: Acquiring scientific literacy through content and genre: a theme-based language course for science students. Engl. Specif. Purp. **19**(4), 369–387 (2000)
8. Herron, S.S., Parr, J., Davis, B., Nelson, P.: Theme-based instruction: making conceptual ties with the sickle cell story. Am. Biol. Teach. **72**(7), 422–426 (2010). https://doi.org/10.1525/abt.2010.72.7.6
9. Muoz, C., Tragant, E.: Second language acquisition and language teaching. Int. J. English Stud. **4**(1), 197–219 (2004)
10. Saleh, M., Lazonder, A.W., Jong, T.D.: Structuring collaboration in mixed-ability groups to promote verbal interaction, learning, and motivation of average-ability students. Contemp. Educ. Psychol. **32**(3), 314–331 (2007)
11. Ling, J., Kai, S.: Application of task-based approach in college English teaching based on internet-assisted multimedia. Int. J. Educ. Manag. Eng. **2**(8), 58–64 (2012)
12. Xiao, Y.: Second language acquisition: an introductory course. System **42**(3), 479–480 (2014)
13. Braidi, S.M.: Reconsidering the role of interaction and input in second language acquisition. Lang. Learn. **45**(1), 141–175 (1995)
14. Chang, W., Wang, T., Li, M.: Learning ability clustering in collaborative learning. J. Softw. **5**(12), 1363–1370 (2010)
15. Hale, G.A., Courtney, R.: The effects of note-taking on listening comprehension in the test of English as a foreign language. Lang. Test. **11**(1), 29–47 (1994)
16. Heping, W., Jinchuan, Z., Rongping, F., et al.: Teaching non-English-majored college students listening and speaking through multimedia and network. Int. J. Educ. Manag. Eng. **3**(1), 44–48 (2013)
17. Xiuli Tong, S., Deacon, H., Kirby, J.R., Cain, K., Parrila, R.: Morphological awareness: a key to understanding poor reading comprehension in English. J. Educ. Psychol. **103**(3), 523–534 (2011)
18. Webb, S.A.: The effects of pre-learning vocabulary on reading comprehension and writing. Canadian Mod. Lang. Rev. **65**(3), 441–470 (2009)
19. Jiafen, W.: Impact of foreign language proficiency and English uses on intercultural sensitivity. Int. J. Mod. Educ. Comp. Sci. **8**, 28–35 (2016)

Research Status of Online Course Teaching in Application-Oriented Universities in China

Li Bingbing(✉), Zhou Xiaofen, and Liu Zili

School of Logistics, Wuhan Technology and Business University, Wuhan 430065, China

Abstract. Online course teaching is very important in the context of the epidemic situation, but online course is relatively simple as so far, students generally feedback that learning enthusiasm is not good. Some students reported that the course content was difficult to digest. Online teaching methods have attracted the attention of many scholars. By searching the database of CNKI.net, we got 59 pieces of literatures. We start to analyze from these aspects: conceptual understanding, design model and implementation measures. Finally, countermeasures are put forward. Academia pays more attention to these: teach with the help of online communication tools, make use of online course platform, make use of the ideological and political education with the help of curriculum, and adjust different teaching methods to meet students' learning needs with the help of big data. By the bibliometric statistics, this paper explores the research status of teaching design of applied online courses in China, in order to provide some theoretical basis for the teaching research of online courses.

Keywords: Application-oriented universities · Online course · Teaching model

1 Instruction

Education should adapt to change of university outside, teachers must be able to innovate and use modern methods and tools to make course interesting and effective, for example online courses [1]. Online courses have mushroomed around the world, especially massive open online courses (MOOCs), making online education another hot topic in higher education. Advanced network information technologies such as social networks, cloud computing, artificial intelligence and big data provide a solid platform and broad space for the development of online education. Through online learning, all learners can share global excellent education resources through the Internet, breaking the limitation of learning space and time, realizing large-scale learning and personalized learning.

Recently, there are some literatures appearing for online course teaching research for improving class efficiency. Luo et al. uses "Xuexitong" mobile network platform to work out before, during, and after class, and realize integration of course teaching online and offline [2]. Yang refers to online course information acquisition by virtual simulation technology [3]. Zheng et al. provides a logistics model to offer optimization strategies for online courses during public safety incident [4]. Although the pilot

Z. Hu et al. (Eds.): AIMEE 2021, LNDECT 107, pp. 481–490, 2022.
https://doi.org/10.1007/978-3-030-92537-6_44

universities have also invested a lot of energy in doing a good job of network teaching content, and achieved some results. However, there are still few high-quality resources for online education suitable for application-oriented universities students, and the form of online courses is too single. Most of them convert traditional text teaching materials into digital network teaching materials and transmit video teaching videos to the Internet. They lack teaching design for learning needs and learning objects, and the effect of online learning is poor.

This paper investigated and analyzed the current situation of online courses development and instructional design in China, and explored the deficiencies of online curriculum instructional design in China from the perspective of learners.

2 Design

2.1 Data Source

The text data related to this study were collected by using the academic journal database of CNKI.net. The retrieval setting adopts article's title = "application-oriented university" and "online course" and "teaching design", literature retrieval type = "Journal", and the retrieval time span is from 2011 to 2020. The data are used to explore the development process and context of subject research.

2.2 Research Methods

Text analysis is to analyze the collected text, graphics, and other contents, which belongs to qualitative analysis. Visual analysis can be used to explore the emerging trends of research topics, and quantitative analysis can make up for the shortcomings of traditional content analysis methods, to explore the structure and evolution path of scientific paradigm. This study intends to mine teaching design of the application-oriented online course, and carry out text mining analysis, and explore its development process and model.

To summarize the trend and experience of online course teaching, we have searched literature from CNKI.net. And then, by description and analysis we get hot teaching model. Through reading higher cited literatures, we have summarized hot teaching methods and measures. Finally, we conclude our findings to improving teaching efficiency of online course teaching. Research path is shown as Fig. 1.

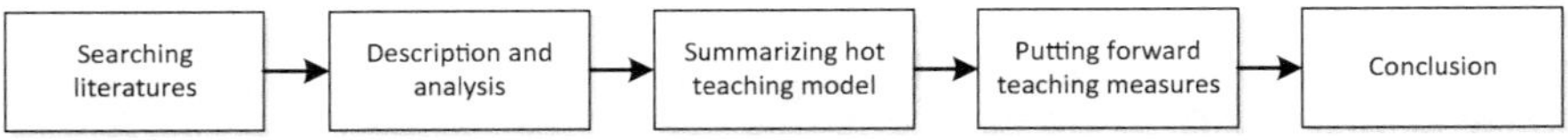

Fig. 1. Research path

3 Description and Analysis

3.1 Frequency Analysis

We get a total of 59 Chinese literatures. It is not difficult to see that 2013–2017 is the initial stage of Chinese literature publication, and the publication of papers also highlights that the research topic may not have attracted the attention of scholars, it is shown as Fig. 2. After 2018, the publication of Chinese literature showed a linear upward trend, and the instructional design of applied online courses began to attract more scholars' attention. Therefore, the current research on the instructional design of applied online courses has mainly experienced two stages: introduction period and development period.

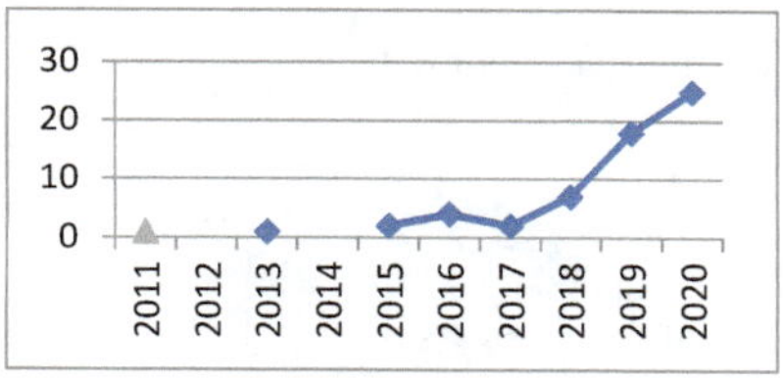

Fig. 2. Distribution of literature publication

3.2 Mutual Citation Analysis

There are multiple nodes from Fig. 3. The larger the dark blue node, the more frequently it is cited. Citation frequency reflects the importance of the literature in this field, and its content has great reference value for this subject research.

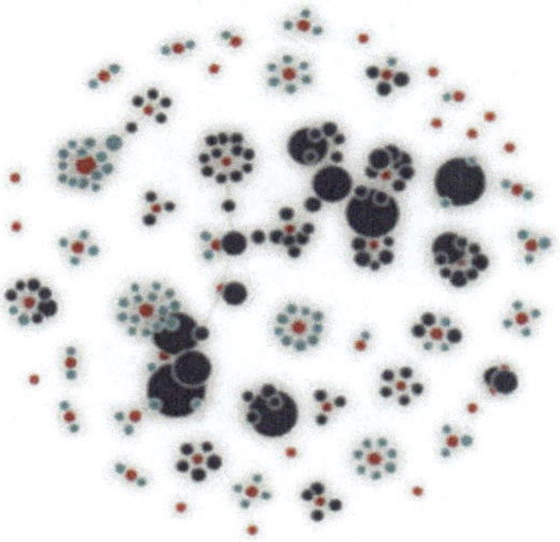

Fig. 3. Analysis of literature mutual citation network

The citation frequency of the top five documents is obtained as shown in Table 1. The mutual citation literature on the teaching design of applied online courses mainly focus on the mixed learning model, learning output based teaching model and collaborative learning model.

Table 1. Cited frequency of top five literatures

Author, year	Title	Frequency
Li, 2004	Principle and application model of blended learning	1855
Wu, 2107	Accelerating development and construction of emerging engineering, taking initiative to adapt to and lead the new economy	1089
Gu, 2014	OBE engineering education model in Shantou university	1085
Huang, 2006	Theory and practice of blended learning	946
Zhao, 2000	Collaborative learning and its model	886

3.3 Co-occurrence Analysis

According to the keyword co-occurrence network, it can be found that multiple keywords are highlighted. The larger the green node is, the more scholars pay attention to and recognize the keyword. Keyword co-occurrence network can reflect the research hotspot and development trend of the research topic from the side. There are six main keywords from Fig. 4.

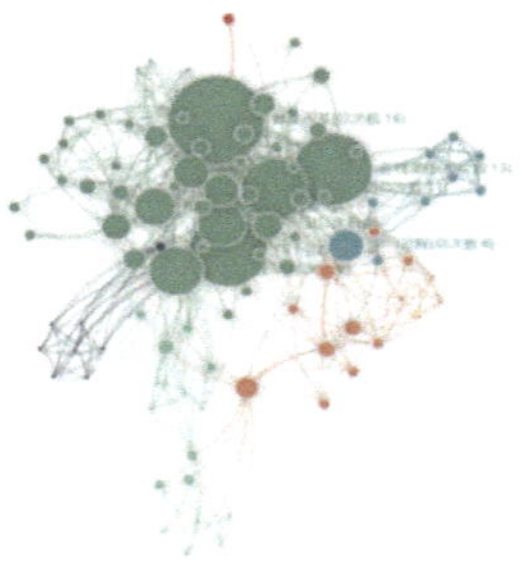

Fig. 4. Keyword co-occurrence network

Relevant data are obtained through keyword co-occurrence network. Shang's paper "The application of blended learning in the teaching of computer application foundation in universities" [5] had been downloaded 233 times, this paper was cited 14. From this paper we find that, hybrid teaching is the main teaching method of teaching reform. Better teaching effect can be achieved by integrating online courses and offline courses and adding autonomous learning. Through intensive reading of high cited literatures and literatures in recent three years, this paper helps us get concept understanding, design model, implementation measures and implementation effect of online course teaching in application-oriented universities.

3.3.1 Conceptual Understanding of Online Course Teaching

The initial application of online courses is not smooth, and the implementers have some misunderstandings about the concept. For example, the online course open platforms equivalent to the use of micro course platform [6], or upload the teaching materials to the network platform to facilitate students to learn online courses during the epidemic, such as superstar learning link [7], online course is to show students videos in class, etc. They are an innovative learning way. Compared with traditional courses, they can allow students to control learning time independently, place and progress [8]. Online courses take students as the main learning body and use a variety of resources to help themselves to learn independently in spare time. Teaching online is only the surface of online courses, not the core of implementation.

3.3.2 Design Model of Online Course Teaching

Online course teaching is divided into pure online course teaching, online and offline mixed teaching. Through intensive reading of the selected Chinese literatures, 30 references refer to mixed teaching, accounting for 50.85%, which is the main way of teaching reform in many universities. Hybrid teaching has become one of the hottest topics in teaching reform, also known as offline to online. It is characterized by the combination of the advantages of offline course teaching and online course learning, the transformation from a teacher-centered class to a student-centered class, the integration of "autonomous learning before class, in-depth learning during class, modification and feedback after class" and other defect detection and filling methods, and the formation of "clear teaching task, accurate design content, teaching resource sharing, learning task driven. The teaching process of "students' autonomous learning and free discussion" could promote students' learning outputs. Many teachers like to carry out the teaching design of online open courses, and determine the teaching structure of online courses by means of knowledge point review, new course introduction, case display, lectures, homework test, achievement display, homework expansion, etc. [8]. It is not difficult to find that although the theme is online courses, it still needs the cooperation of offline courses.

In the design process, hybrid teaching mainly includes two ways: teaching activity mixing and teaching material mixing. Hybrid teaching is a good teaching model that closely integrates the online course open platform and offline course teaching [6], and establishes a learning path of objectives, plans, autonomous learning, display and reflection. This teaching model can be comprehensively applied to basic courses with strong theory and skill courses with strong specialty, such as English and C programming class, architectural structure class, food biotechnology class, introduction to civil engineering [9], etc.

In addition, hybrid teaching can achieve classified teaching objectives [8]. By establishing the learning process of "memory, understanding, application, analysis, synthesis and evaluation", it can cultivate students' low-level learning ability from the level of "memory and understanding" and students' high-level thinking ability from the level of "application, analysis, synthesis and evaluation". Before class, we will focus on "memory and understanding", list the learning task and push it to students. The class focuses on "analysis and synthesis" to guide students' learning and simple application. After class, we will focus on "knowledge point evaluation", push some difficult training according to the current learning progress, improve the target threshold of knowledge point evaluation, reconstruct the knowledge system, and constantly achieve the highest classification goal of students. Hybrid teaching can be summarized as building a learning environment integrating scene, interaction, experience, and reflection for students before, during and after class, to promote students' learning enthusiasm and improve learning effect.

In addition to mixed teaching, some scholars have proposed other teaching models [8, 10]. Including project-based teaching and quaternion teaching, it has also been studied and supported by many scholars for open platform of online course. Project-based teaching is a kind of teaching closely related to mixed-teaching. It is a problem-oriented learning method. Based on real cases, students can complete the collection, design, implementation, and evaluation of information independently or cooperatively. Its remarkable feature is that it takes students as the main body, teachers as the guide and projects as the main line to cultivate students' ability to combine theory with practice. The four element teaching, namely 4C/ID teaching, takes the complete learning task as the core and integrates the basic elements such as learning task, relevant knowledge and ability, support procedure and special operation into the teaching. Learning task is the cornerstone, which is application-oriented and helps students build a cognitive system through inductive learning. Relevant knowledge and ability are the bridge to promote students to master the creative learning of a specific task. The program supported is an auxiliary program to help students play the regenerative function of learning tasks. Special operation is practice, which helps students improve their proficiency and regenerative skills.

No matter what kind of teaching model reform, it needs the support of different teaching methods. Many scholars believe that in the process of hybrid teaching, the certain teaching design thinking should be integrated to support the implementation of online course open platform, rather than just stay in online teaching. Through literature summary, the following teaching methods are mainly presented.

Firstly, "Helping-student-style App" is popular as a teaching tool. Student-centered internet teaching has been widely used during the epidemic period. For example, courseware, case videos, reference materials and course exercises are uploaded to the app by using apps such as learning link, cloud class, smart tree, and rain class, and learning points are set to allow students to study freely. Teachers could make usage of video production software such as micro class to customize digital learning materials suitable for students' independent collaborative inquiry learning.

Secondly, "production-research project" is often used in course teaching. Universities connect with enterprises and integrate enterprise projects into class teaching. Classes are guided by "problems-producing", bring the actual cases from companies, heuristic teaching is used to promote students to solve social practical problems with professional theoretical knowledge. Students could solve professional problems independently or cooperatively by observing phenomena, reviewing professional knowledge points, collecting various materials, discussing problems together and other methods.

Thirdly, "phased and comprehensive" is usually put into course teaching. The positive results of students' output learning are the goal of the online course open platform. Phased evaluation teaching can supervise and motivate the assessment of students' progress, so that teachers and students can master the actual learning results. It is recommended to regularly disclose the phased evaluation teaching results. Considering the assessment of the training effect of students' various abilities, it is necessary to adopt comprehensive evaluation teaching, including punching in, discussion, test, sharing, etc.

Fourthly, the teaching method of "combined" is paid attention by teachers. It is better to take students' output as the core, combine with the existing teaching types and curriculum characteristics for forming a combined teaching method. For example, "MOOC + SPOC + flipped classroom" combined teaching could build SPOC database, prepares teaching materials, guide students to learn and feedback teaching results. The combined teaching not only makes full use of the advantages of MOOC resource sharing, but also promotes the effective learning of offline courses.

3.3.3 Implementation Measures of Online Course Teaching Design

We have summarized some implementation measures to enhance teaching efficiency as-follows.

MOOC mixed teaching, namely massive open online course, is a large-scale online open course [11], which helps students use fragmented time to systematically learn courses, mainly focusing on the establishment of courses on the MOOC platform. Supported by micro courses, it can truly simulate online and offline teaching, cooperate with flipped classroom, and make full use of students' autonomous learning before class, classroom interactive communication and cooperative learning after class to improve teaching quality.

SPOC (mixed teaching) is small private online course, a small-scale restricted online course [12]. SPOC teaching model is developed through MOOC, which provides conditions for students' in-depth learning of courses, improves high-level thinking ability, and promotes the cultivation of professional talents. SPOC teaching model has the characteristics of small number of people. Small class teaching can effectively give students teaching guidance time, and the effect of solving doubts and answering questions is higher.

Flipped classroom is a course practice model to further implement blended learning [13]. Flipped classroom gives students a lot of practice for autonomous learning, so that students have greater mastery and control over learning, and promote students' active learning. Flipped classroom includes not only the process of knowledge transfer after class, but also the process of knowledge absorption. In the specific implementation process, it mainly includes learning situation analysis, and learning guidance arrangement. Generally, teachers and students ask each other, group discussion, knowledge combing and induction and other ways to turn the classroom.

OBE concept teaching, namely outcomes based education, is a teaching model guided by students' learning achievements [14], advocates "student-centered, output oriented and continuous improvement", mainly focuses on student orientation and students' self-education, pays attention to students' professional practice ability, promotes the organic combination of teaching activities with more flexible curriculum teaching, and helps students better absorb teaching contents.

CDIO teaching model, namely coherent design implementation operate teaching, is a teaching model advocating "learning by doing" and "project-based education and learning' [15]. With students as the main body, it is suggested to realize "school enterprise cooperation, mixed teaching and design-competition", and form a talent training model integrating "class, enterprise and competition". Integrating professional courses and basic courses, forming a professional course group with relevant majors as the core, and cultivating students' comprehensive practical ability through "conception, design, realization and operation" will be good.

BOPPPS (mixed teaching) is a closed-loop curriculum teaching model designed based on constructivism and communicative approach, which includes six teaching links: bridge in, objective, pre-test, participation, post-test, and summary [16]. Elicit knowledge points out practical cases and news reports, and throw out thinking questions and stimulate students' interest. Set verifiable and evaluable teaching objectives, and find the contact and melting point related to the specialty from moral education and emotion. Understand students' learning difficulties by questionnaire and adjust teaching ideas by test results. Carry out situational, research-based, heuristic and inquiry teaching to promote the communication between teachers, students, and students. Make use of the online network platform to check the deficiencies and make up the omissions of knowledge points with test question detection and mutual evaluation detection. Carry out "secondary thinking" on the learned knowledge points. Under the guidance of teachers, evaluate and feedback students' learning results from the levels of knowledge, ability, moral education, and ideological and political education.

Although scholars have put forward many teachings model, it can be obviously seen that teaching thinking of blended teaching course. A single teaching model may not be able to achieve the teaching objectives well. And it is important to develop and explore the teaching tools and ways to make full use of these teaching models.

4 Findings

In China's application-oriented universities there are both theoretical courses of management and practical courses of technology, which have different learning difficulties. Through literature review, it can be found that four ways to adopt online teaching for such courses mainly include the following.

Firstly, teaching with the help of online communication tools is popular used in class. Online communication tools mainly include WeChat, QQ, zoom, nailing, learning link, etc., and carry out live teaching. In order to guide students to concentrate on learning, it is better to set knowledge points as multiple small knowledge points, and make 10–15-min micro class videos.

Secondly, making use of online course platform is a good idea to improve class efficiency. First, hardware, mainly including cameras and recording pens, to record relevant videos and upload them to video broadcasting websites, such as BiliBili website, Youku video, Xuexitong, smart tree, excellent course alliance, school online, rain classroom, etc. to carry out course recording teaching. In the process of implementation, it is supported by the flipped classroom of micro class and micro video, highlights the key and difficult points of teaching, and can also integrate the characteristics of project teaching model, task driven model, analysis, and evaluation model.

Thirdly, making use of the ideological and political education with the help of curriculum can promote students' correct-values formation. Pay attention to the excavation of Ideological and political elements in the process of online course construction, and imperceptibly integrate the ideological and political content into the course teaching content. Fully excavate the ideological and political elements of each knowledge module and each course chapter, and instill the students with basic ethical awareness. Ideological and political education is urgent for the cultivation of applied university students. Through curriculum teaching, we can help students understand the professional connotation, stimulate students' patriotic enthusiasm and national pride, and Cultivate Four Haves youth with ideals, morality, culture, and discipline for the country.

Fourthly, adjusting combination of different teaching methods with the help of big data is also a good way to teach. Through the online course open platform, statistics on students' learning related data, including case length, video duration, times of discussions, etc., timely adjust the proportion of each teaching module in class according to big data, and adjust the key and difficult points of the course according to the syllabus. Treating students as learning centers, investigating students' autonomous learning ability and comprehensively are the useful ways to improve the quality of curriculum teaching.

5 Conclusion

Through the systematic analysis, we know the development trend of online courses in China's application-oriented universities. This paper summarizes the introduction and implementation of online course teaching mode. The scope of existing teaching and research is discussed, including not only management courses, but also engineering courses. Several teaching methods include MOOC, SPOC, flipped classroom, OBE, CDIO and BOPPPS are popular. In fact, to have a good course, a single teaching method is far from enough. Therefore, the combination of various teaching modes is worth recommending, which combination of ways, how much time allocation, depending on the nature of the course to decide. What's more, teachers should pay attention to online communication tools, online course platform and making full usage of big data into course teaching. However, literature cited network is used for literature analysis, empirical analysis has not yet been carried out. Empirical analysis about these summaries of teaching methods and models needs to do further study.

References

1. Bognár, L., Fauszt, T., Váraljai, M.: The impact of online quizzes on student success. Int. J. Emerg. Technol. Learn. **16**(11), 225–244 (2021)
2. Yueji, L., Yan, L., Meiling, L., et al.: Reform and application of biochemistry teaching based on formative evaluation of mobile network platform. J. Phys: Conf. Ser. **1533**(01), 1–6 (2020)
3. Lanzotti, A., Calise, M., Molaro, M., et al.: 'Federica's MOOC' (massive open online course): a blended course in engineering drawing at Federico II. Int. J. Interact. Des. Manuf. **13**, 1115–1128 (2019)
4. Haodi, Z., Fei, L., Liangquan, G., et al.: Optimization strategies for online courses in Chinese universities during COVID-19 pandemic. Int. J. Math. Sci. Comput. **1774**, 1–7 (2021)
5. Juanye, S.: Research on the application of blended learning in the teaching of computer application foundation. Educ. Modern. **5**(36), 289–293 (2018). (in Chinese)
6. Xiaoyue, W., Han, S., Lixin, L.: An innovative preschool education method based on computer multimedia technology. Int. J. Intell. Syst. Appl. **14**(14), 57–68 (2019)
7. Xiao, C., Hongmei, Z., Donghai, Z., et al.: A model-driven architecture approach for developing e-learning platform. Int. J. Intell. Syst. Appl. **6249**, 1–7 (2010)
8. Wei, W.: Application data mining technology in construction of college English blended teaching mode. Int. J. Intell. Syst. Appl. **1982**(01), 1–7 (2019)
9. Chen, T., Li, J.: Teaching reform of introduction to civil engineering under the background of online course. J. Yellow River Inst. Sci. Technol. **22**(05), 85–87 (2020). (in Chinese)
10. Shao, B., Li, Q., Sun, X., Yu, S.: Research on online curriculum design of vocational education based on 4C/ID. J. Xingtai Vocat. Techn. Coll. **37**(03), 5–8 (2020). (in Chinese)
11. Dongsong, C., Guangping, Y.: MPOC experiential learning model of engineering education based on distributed flip in local engineering colleges and universities in "Post MOOC" era. Int. J. Math. Sci. Comput. **20**(06), 1375–1379 (2017)
12. Yanhua, L., Ye, Z.: Design of implementing SPOC flipped classroom mixed mode in "post MOOC era": taking intermediate financial accounting course as an example. Account. Study **13**, 219–220 (2018). (in Chinese)
13. Qin, G.: Strategie on acclimatization of flipping classroom in China. Int. J. Soc. Sci. Educ. Res. **4**(09), 292–297 (2021)
14. Chen, J.: Strategies for cultivating autonomous learning ability in college English teaching based on OBE concept[J]. Int. J. Soc. Sci. Educ. Res. **4**(09), 390–396 (2021)
15. Lin, C.: Exploration and practice of industrial design specialty reform in local applied undergraduate colleges based on CDIO mode: taking Hunan institute of technology as an example. Educ. Teach. Forum **50**, 173–174 (2019). (in Chinese)
16. Xiaofu, L., Ying, G., Zhongqi, H., et al.: Practice of BOPPPS mode in the standardized trainging of non-radiology residents. China Contin. Med. Educ. **13**(24), 68–71 (2021)

Adaptive Adjustment and Circular Improvement of Exhibition Course for Visual Communication Specialty in the Age of Information

Xianhui Zeng[1] and Simeng Fan[2](✉)

[1] College of Art, Hubei University of Education, Wuhan, China
[2] Department of Environmental Art and Design, Hubei Institute of Fine Arts, Wuhan, China

Abstract. Exhibition design course is one of the professional compulsory courses of visual communication specialty, and the only comprehensive design course that introduces the concept of three-dimensional space and studies the relationship between environment and people. Based on expounding the characteristics of commercial display design in the information age, this paper analyzes the nature, requirements, and existing problems of the display design course of visual communication specialty, discusses the adaptive adjustment and cyclic optimization of the display design course, and uses the fuzzy analytic hierarchy process to analyze the links and effects of course theory teaching and practical teaching. The purpose is to find the weak links in teaching and make cyclic improvement, so as to play a better role in the cultivation of professional talents.

Keywords: Exhibition design · Informationize · Visual communication · Teaching research

1 Introduction

Display design is a compulsory course for visual communication major which takes the commodity as the main body, uses the artistic design language to carefully create the space and plane within the established time and space, so as to produce a unique spatial effect, and realize the perfect communication with the audience while explaining the theme of the exhibits.

The brand image is not only limited to the two-dimensional visual system design, but also should pay attention to the commercial display design to form an interactive space closest to consumers, so as to ensure the perfect presentation of products, promote sales and enhance the image and influence of the brand [1]. In the era of rapid development of information technology, commercial display design is particularly important for brand shaping and communication [2]. It is very necessary to carry out in-depth research on curriculum and teaching, and realize the adaptive adjustment and cyclic improvement of display design curriculum, so as to improve students' design ability. Using scientific evaluation methods, a comprehensive evaluation of the

Z. Hu et al. (Eds.): AIMEE 2021, LNDECT 107, pp. 491–502, 2022.
https://doi.org/10.1007/978-3-030-92537-6_45

adjustment and improvement of the curriculum will help to better the effect of curriculum teaching and play a more prominent role in the training of visual communication professionals [3–5].

2 New Features of Display Design in the Information Age

Different from the traditional business display, the commercial display design in the information age has the following four characteristics.

2.1 Focus on Brand Image

The commercial exhibition design idea has changed from the traditional product as the core to the core of brand [6, 7]. What people feel in the commercial space is not only the product, but also the brand image and its unique atmosphere through the display language in the space. This atmosphere forms an invisible field, which can make people form a good impression of the brand and achieve the coordination and unity of people, things, and fields [8, 9].

2.2 Topicality

The information brought to people in the Internet age is usually fragmented and one-sided, so it is difficult to correctly form the three-dimensional image of the brand. Businesses need to find ways to catch people's eyes quickly - set clear and interesting themes for their brand store space. The display design of the space is refreshing, so that the public can associate and resonate with the brand. The visual language of display design is the main means to express the theme [10, 11].

2.3 Interactivity

In order to improve the efficiency of display information transmission, commercial space display design pays increasingly more attention to the interaction with consumers, including situational interaction, multimedia device interaction, game interaction and other commercial activity design, so that the audience entering the commercial space can have the awareness of independent choice and active participation, enhance the sense of experience, and raise the popularity of the brand.

2.4 Entertainment

In the context of informatization, mass culture and consumption pursuit impact the development of commercial culture, showing a diversified development trend. Entertainment has become the "rigid demand" or "inelastic requirement" of mass consumption. The combination of business and entertainment is a new trend. The entertaining commercial space display is not only reflected in the theme design and visual symbols, but also in the store mode [12]. Many flash stores (such as Fig. 1 - L'Oreal New Year Express Flash Store in Paris) and internet sensation stores are produced to meet this entertaining demand. These new stores are sought after by consumers through the theme of full entertainment, fashionable and eye-catching visual design, and diversified interaction between offline and online, which greatly improves the communication speed of commercial brands.

Fig. 1. L'Oreal new year train flash store in Paris

3 The Nature and Requirements of Display Design Course for Visual Communication Specialty

3.1 Nature and Purpose of Exhibition Design Course

Exhibition design is a comprehensive, interdisciplinary, and practical course for art majors, involving multi-dimensional modeling design such as plane and three-dimensional, material and lighting, and emphasizing the reality and sense of form of spatial structure [13, 14]. In terms of function, it should be dealt with the relationship between exhibits, exhibition racks and the audience, to meet the needs of the audience through the application of ergonomics and design psychology, and enable people to obtain display information more efficiently [15]. Information transmission is the core purpose of display design, and the organization, guidance and planning of information is the key point in the process of display design.

3.2 Requirements for Exhibition Design Course

The exhibition design course is implemented based on brand thinking [16]. For the purpose of improving brand image and brand communication, carry out background and market analysis around the brand, summarize display needs, design display scheme, and launch project training. The teaching focuses on the display of commercial space, mainly including window display and store display design.

The development of informatization has had a significant impact on the social economy and even people's lifestyle [17]. The market demand determines the development direction of professional courses. It is the responsibility of visual communication major in colleges and universities to keep up with the development of the times and market changes and cultivate competitive design talents who adapt to social development [18, 19].

3.3 Problems in Exhibition Design Course

The problems existing in the display design course of visual communication specialty are mainly reflected in three aspects. The first trouble is the lack of differences in professional positioning. At present, most visual communication majors directly copy the exhibition course of environmental art specialty in the teaching of exhibition design course, ignoring the differences of students' knowledge system and training objectives, resulting in poor learning effect. Second, too much attention is paid to the operation of three-dimensional soft-ware, and the teaching of design thinking and spatial form is superficial, which leads to students only paying attention to the production of effect drawings, overlooking design ideas, and then the overall design level is low. Thirdly, students' participation in challenging practical projects is overemphasized, and the time limit of these projects is difficult to synchronize with the curriculum teaching, which increases the unpredictability of the curriculum and looks down upon the learning of the display design thinking system.

These problems are inconsistent with the talent requirements of the information age. In this sense, the innovation and adjustment of exhibition design course is imperative.

4 Adaptive Adjustment and Cyclic Optimization of Exhibition Design Course

In order to solve the problems existing in the current display course teaching of visual communication specialty, it is necessary to carry out differentiated teaching positioning according to different professional directions, systematically construct the theoretical and practical teaching system of the course, run the theoretical knowledge through the whole course, and comprehensively improve the teaching level and quality of the course [20]. The adaptability adjustment and cycle optimization of the course can be carried out from the following four aspects.

4.1 Adjustment of Course Theoretical Content

Online shopping born in the information age brings convenience to people, but after the initial novelty, people are still willing to return to offline stores and feel a richer shopping experience. Combining online stores with offline physical stores for management is an important link to make the brand image accepted and loved by the public.

Different from the traditional offline retail stores, today's commercial store design should pay more attention to the brand promotion and consumers' shopping experience. The display design of store space should firstly meet the functions, and consider consumers' sensory experience and spiritual needs at the same time, so as to create an attractive brand atmosphere. A newly born idea of "New retail" is launched in this context. It is a new-type retail model that enterprises rely on the Internet to upgrade and transform the production, circulation, and sales process of goods by using advanced technical means such as big data and artificial intelligence, so as to reshape the business structure and ecosystem, and deeply integrate online services, offline experience and modern logistics [21]. Under the influence of the new retail concept, the design of commercial stores needs to focus on the online and offline brand building and consumer experience. Representative flash store of new retail is a new type of brand retail store (usually short-term rental), which is generally selected in the commercial location with large flow of people. The goods operated have the internet sensation attribute, create topics in a short time, quickly attract people's attention, and cooperate with marketing means to promote brands and products.

In view of this phenomenon, the flash store design theory and case analysis led by brand activity planning are added to the content of the exhibition design course, so that students can understand more comprehensive exhibition design forms combined with the Internet and broaden their horizons and thinking.

4.2 Adjustment of Course Practice

According to the key adjustment of theoretical teaching, the display design course has made a series of adjustments in practice, and carried out a complete brand display design project in groups. The practice link is divided into four stages.

Stage I: arrange students to conduct research and analysis on theme (such as tea brand, trendy brand, comprehensive brand of life concept, etc.) in groups, collect relevant brand background information and store design cases, and summarize the design characteristics of theme brand industry, store advantages, space theme planning ideas, store space theme design ideas, etc.; collect the information of online and offline stores of similar brands horizontally to determine the general style and direction of commercial space design. This stage focuses on training students' theoretical knowledge, brand research ability, analysis ability and planning ability.

Stage II: each group displays the research results in class. After getting teachers' comments, classmates' mutual comments and modifications, students complete the plane design drawing, three-dimensional effect drawing and props and materials design drawing of the window. Window is not only an important carrier to express brands and commodities in commercial display design, but an effective way to attract consumers to enter the store also. This stage is to let students quickly enter the state of spatial visual design, deal with the relationship between plane and three-dimensional elements, and organize design formal languages such as color, 3D modeling and lighting.

Stage III: the main job is to design brand concept store or flash store according to the brand and theme determined by the team. This is the first time for students to design a comprehensive space display design. Teachers need to guide students to promote the design process with clear design ideas.

(1) According to the theme of the store, plan the display content and experience activities of the store, and draw a standard floor plan;
(2) Determine the visual elements, draw the visual graphics or symbols that can represent the key words, and realize the visualization of typical graphics;
(3) Design a distinctive sample of display design style to facilitate the cooperation and promotion of store space design in the later stage of the group.

(4) The team members divide the work to complete the store door design and the space design of each internal functional zone, organically combine the thematic visual elements with the space modeling, and cooperate to complete the preliminary design of the store display.

Stage IV: make entity and 3D model, and complete the design of mobile phone H5 page for online publicity. Shoot a 50 s video to show the space and details of the model, with music and subtitles. H5 page design style should highlight the characteristics of the store and have a clear display of activity content to enhance the publicity of stores and brands and meet the needs of information development. The workload at this stage is large and requires the close cooperation of team members (Fig. 2).

Fig. 2. Model making of solid l and 3D digital ones

4.3 Achievement Display and Performance Evaluation

At the end of the course, students will collect the project results of the group into a complete project book, show the design ideas and design process of each stage in detail, and mark the division of labor of the group members. The project book is used as an important basis for course evaluation. Whether the design idea is clear, the brand characteristics and theme are distinct, the display design language is rich and can maintain the unity of style, and the display space can achieve the ingenious combination of vision and function, clear primary and secondary division of space, reduction degree and details of model making are all the key points of the assessment.

The cultivation of students' comprehensive ability is the focus of this course. The group has two proposals in practice. At the end of the course, students are arranged to organize and plan the course exhibition, complete the design of exhibition visual posters and exhibition hall layout. The practice project is an opportunity for students to choose their interested flash shop as the design theme, complete the project proposal and model production, and finally show the results of the course studying. This is a diversified form of practice, which has achieved good results and has been proved to be a successful attempt. Every link in this process has well exercised students' oral expression ability, exhibition planning ability, communication ability and comprehensive quality.

4.4 Cyclic Improvement of Curriculum Adjustment

The process of continuous adjustment and improvement is the basis and guarantee for the success of a project [22, 23]. In order to ensure that the teaching of the display design course of visual communication specialty in the information age can enable students to have a deeper understanding of the cutting-edge information and theory of display design and better grasp the professional knowledge, the course has been continuously adjusted and circularly improved [24–26]. This paper establishes the main idea and mathematical model of fuzzy cycle analysis and adjustment of curriculum by using fuzzy analytic hierarchy process. Relevant model parameters and their meanings are shown in Table 1. Relevant model parameters and their meanings are shown in Table 1, which mainly includes the following five steps.

Table 1. Improvement of mathematical model parameters

Parameter	Description
G_A^s	Scores collection of students' theoretical courses before adjustment, $G_A^s = \{G_A^1, G_A^2, G_A^3, \ldots G_A^s\}$
G_B^s	Scores collection of students' practical courses before adjustment, $G_B^s = \{G_B^1, G_B^2, G_B^3, \ldots G_B^s\}$
G_C^s	Scores collection of students' theoretical courses after adjustment, $G_C^s = \{G_C^1, G_C^2, G_C^3, \ldots G_C^s\}$
G_D^s	Scores collection of students' practical courses after adjustment, $G_D^s = \{G_D^1, G_D^2, G_D^3, \ldots G_D^s\}$
S	The scores of students participating in the course, $S = \{1, 2, 3, \ldots s\}$
M_T	The eigenvalues of the scores of each part after processing, $M_T = \{M_a, M_b, M_c, \ldots M_t\}$ Among them, $T = \{a, b, c, \ldots t\}$
W	Fuzzy evaluation index vector of evaluation object
X_i	Fuzzy evaluation weight vector of index i
R_i	Fuzzy evaluation matrix of index i

The first step is to select the evaluation object.

After calculating, analyzing, and ranking the indexes at all levels, select the component with the lowest score for fuzzy analysis, and select the worst index for improvement.

There are many ways to express the performance level of students in each part, including taking the average, mode, median, weighted average, etc. This paper mainly uses the average with the maximum value removed as the characteristic value of the learning level of students in each part.

The average of the scores of the students participating in the general theory course is treated by removing the maximum value, which is expressed as the following formula, where M_A represents the characteristic value of the scores of this part, $\max G_A^s$ and $\min G_A^s$ represents the maximum and minimum values of the scores.

$$M_a = \frac{\sum_{s=1}^{S} G_A^s - \max G_A^s - \min G_A^s}{s-2}$$

Similarly, the remaining three parts are analyzed in turn:

$$M_b = \frac{\sum_{s=1}^{S} G_B^s - \max G_B^s - \min G_B^s}{s-2}$$

$$M_c = \frac{\sum_{s=1}^{S} G_C^s - \max G_C^s - \min G_C^s}{s-2}$$

$$M_d = \frac{\sum_{s=1}^{S} G_D^s - \max G_D^s - \min G_D^s}{s-2}$$

A	86	84	74	80	70	66	75	77
	81	87	89	87	76	85	67	75
B	88	81	74	85	77	80	82	66
	85	89	74	56	70	61	74	81
C	71	60	52	71	64	98	89	72
	79	75	70	61	76	72	66	94
D	89	80	65	44	54	95	69	84
	89	45	78	94	61	69	80	65

$$M_a = 78.8, M_b = 76.9, M_c = 72.7, M_d = 73$$

Because the third of the four values is the smallest, the corresponding adjustment part of the display design theory course of visual communication specialty in the information age is selected as the object of further optimization.

The second step is to establish an evaluation team.

In this paper, the implementation process of fuzzy evaluation mainly adopts the expert scoring method. Establish an expert group, mainly composed of education experts, exhibition design experts, senior practitioners in the industry and students. The composition of personnel shall be determined according to the actual situation.

The third step is to collect evaluation data.

In combination with the teaching situation of the course, the expert group is invited to conduct a comprehensive evaluation in combination with the students' achievements of the course.

The fourth step is to implement evaluation.

(1) Definding the fuzzy comprehensive evaluation method
The formula of fuzzy comprehensive evaluation method is: $W_i = X_i \cdot R_i$, where W_i represents the evaluation index vector of the evaluation object, X_i refers to the weight vector of the jth index, and R_i means the fuzzy evaluation matrix of the jth index.
(2) Determining the evaluation index set and comment set
The primary, secondary and tertiary layers of indicators in fuzzy evaluation are decided, as shown in Table 2.

Table 2. Three layer of indexes in fuzzy comprehensive evaluation

Index of layer 1	Index of layer 2	Index of layer 3
A- Theoretical courses before adjustment	A_1 teaching content	A_{11} teaching materials
		A_{12} Lecture notes and syllabus
	A_2 lectures	A_{21} teacher's language
		A_{22} blackboard writing and PPT
	A_3 classroom atmosphere	A_{31} student interest
		A_{32} classroom management
B- Practical courses before adjustment	B_1 course design	B_{11} course nature and function
		B_{12} course concept
		B_{13} course design
	B_2 course content	B_{21} course field
		B_{22} learning area
		B_{23} learning situation
	B_3 course implementation	B_{31} teaching method
		B_{32} teaching technology
C- Theoretical courses after adjustment	C_1 industry frontier trends	C_{11} brand design status
		C_{12} Design status of flash store
	C_2 brand activity planning	C_{21} flash store design theory
		C_{22} flash store design case
D- Practical courses after adjustment	D_1 frontier research and analysis	D_{11} characteristic analysis
		D_{12} case of store design
		D_{13} feature analysis
		D_{14} exhibition of research results
	D_2 window design	D_{21} window design theme
		D_{22} window design style
		D_{23} drawing of design and effect
		D_{24} props and materials
	D_3 store design	D_{31} store theme
		D_{32} visual representation
		D_{33} design idea
		D_{34} design schedule

Taking part B as an example, the fuzzy indexes of first, second and third levels are shown in the Table 2. The indicator sets are: Level 1 indicator $F = \{B\}$, level 2 indicator $B = \{B_1, B_2\}$, level 3 indicator $B_1 = \{B_{11}, B_{12}\}$, $B_2 = \{B_{21}, B_{22}\}$, and determine the comment indicator set $V = \{Good, Medium, Poor\} = \{8, 6, 0\}$.

(3) Determining of the weight vector of the evaluation index
In the adjustment of display design theory course of visual communication specialty in the information age, it mainly focuses on the introduction of cutting-edge trends in the industry, combined with the explanation of theoretical knowledge to make students better understand. The analysis of typical cases is mainly interspersed in the introduction process, so the weight vector of sub index b is: $F = \{1\}$, $B = \{0.75, 0.25\}$, $B_1 = \{0.64, 0.36\}$ and $B_2 = \{0.58, 0.42\}$.

(4) Setting fuzzy evaluation matrix
The elements of fuzzy evaluation matrix are mainly obtained by expert scoring method. After taking the average value, the corresponding fuzzy evaluation matrix is:

$$R_1 = \begin{Bmatrix} 0.70, 0.30, 0 \\ 0.89, 0.11, 0 \end{Bmatrix}, R_2 = \begin{Bmatrix} 0.68, 0.32, 0 \\ 0.85, 0.15, 0 \end{Bmatrix} \text{ and } R_3 = \begin{Bmatrix} 0.79, 0.21, 0 \\ 0.66, 0.34, 0 \end{Bmatrix}.$$

(5) Implementing of evaluation analysis

The formula can be used to make the calculation, and got the results as follows.

$$W_1 = B \cdot R_1 = \{0.75, 0.25\} \cdot \begin{Bmatrix} 0.70, 0.30, 0 \\ 0.89, 0.11, 0 \end{Bmatrix} = \{0.7475, 0.2525, 0\}$$

$$W_2 = B_1 \cdot R_2 = \{0.64, 0.36\} \cdot \begin{Bmatrix} 0.68, 0.32, 0 \\ 0.85, 0.15, 0 \end{Bmatrix} = \{0.7412, 0.2588, 0\}$$

$$W_3 = B_2 \cdot R_3 = \{0.58, 0.42\} \cdot \begin{Bmatrix} 0.79, 0.21, 0 \\ 0.66, 0.34, 0 \end{Bmatrix} = \{0.7354, 0.2646, 0\}$$

Through calculation, it can be concluded that the comprehensive evaluation of experts in part B is general, in which both evaluation verdicts of B1 and B_2 are general but B_2 is relatively worse a little. Therefore, considering the conclusion of this teaching evaluation, the overall improvement of exhibition design course should be attached importance to Part B, especially the optimization of part B_2.

The fifth step is the improvement and cycle optimization of the curriculum.

After optimizing of part B_2, based on the teaching practice evaluation process and relevant models of visual communication professional display design course in the information age mentioned above, repeat the five-step analysis in the new round of teaching activities of the course, and timely enhance the weak links and improve and optimize the course inch by inch, so as to effectively ensure the effect of course teaching and learning, and cultivate more useful talents.

5 Conclusion

Exhibition design is not simply a design course related to art, but a comprehensive course permeated with science, technology, and engineering. Especially in the information age, it provides diversified forms for the establishment of brand image and injects vitality into the communication of various brands with distinctive themes.

The course teaching should comply with the market development and public demand, pay attention to the consumers' experience in the display space, and organically adjust the course content and teaching methods in combination with the characteristics of the major.

Aiming at cultivating of application-oriented design talents with high artistic design literacy, systematic mastery of professional knowledge and skills of display design, accurate understanding of customer needs and grasp the trend of modern display design, and innovative consciousness and creative ability, it is necessary to use scientific methods to improve and consummate teaching contents and methods. It is undoubtedly very important to build a reasonable and advanced teaching model.

The teaching reform of display design course is mainly carried out from several aspects: the adjustment of course theoretical content, the trimming of course practical content, the mending of examination methods, and the cyclic improvement of the course. Using fuzzy analytic hierarchy process to analyze the links and effects of curriculum theory teaching and practical training is to find the weak links in teaching and make key improvements, so as to make the teaching of display design course play a better role in the growth and progress of students.

Acknowledgment. This paper is supported by the 2020 fund project *Wuhan Li-Fen Architectural Decoration Structure and Pattern Research* of handicraft culture research center of Hubei Academy of fine arts, the key research base of Humanities and Social Sciences in Hubei Province (SGY-202004).

References

1. Yanzhou, L.: Explore the role that chinese exhibition designers should play in exhibition design. Int. J. Soc. Sci. Educ. Res. **4**(3), 257–260 (2021)
2. Lustono, H.S., Wening, N.: Strategies to increase consumer purchase of ethnic restaurants through perception of brand loyalty authenticity, brand awareness and brand image. J. Glob. Econ. Bus. Finan. **3**(4), 384–392 (2021)
3. Liu, Y., Zhang, X.: Evaluating the undergraduate course based on a fuzzy AHP-FIS model. Int. J. Mod. Educ. Comput. Sci. (IJMECS) **12**(6), 55–66 (2020)
4. Liu, S., Chen, P.: Research on fuzzy comprehensive evaluation in practice teaching assessment of computer majors. Int. J. Mod. Educ. Comput. Sci. (IJMECS) **7**(11), 12–19 (2015)
5. Dayang, P., Petsou, C.S., Sambo, D.W.: Combining fuzzy logic and k-Nearest neighbor algorithm for recommendation systems. Int. J. Inf. Technol. Comput. Sci. **13**(4), 1–16 (2021)
6. Jiang, Q., He, R.: Research on exhibition design in the information age. J. Hum. Univ. (Social Science Edition), **S1**, 142–143+146 (2001). (in Chinese)

7. Peng, Z.: Theoretical research on exhibition architecture and exhibition space design in the information age. Tianjin University, Tianjin (2004).(in Chinese)
8. Xu N.: On the development trend of display design in the information age. Jinan, Shandong Institute of Light Industry (2010). (in Chinese)
9. Feng, W.: Walking perception and rational graphics-a new discussion on landscape design teaching. Decoration **4**, 74–76 (2014). (in Chinese)
10. Andrée, B., Charlotte, B., Martha, F.: Embodied ephemeralities: methodologies and historiographies for investigating the display and spatialization of science and technology in the twentieth century. Hist. Sci. **59**(2), 197–219 (2021)
11. Teresa Escano, M., Sumaylo, D.J.: Meaning-making exercise using images and soundscape: the case of San Pedro street, Davao city, Philippines. SAGE Open **7**(4), 215824401773635 (2017). https://doi.org/10.1177/2158244017736355
12. Bidjovski, G.: Displaying images and their characteristics from websites on users computers. Int. J. Image Graph. Sig. Proc. **6**(5), 36–44 (2014)
13. Yijin, Z., Juan, L.: Research on exhibition design in the information age. Beauty Times **12**, 97–98 (2018)
14. Yuxuan, H.: Teaching reform of exhibition design in the information age. Liter. Life Wenhai Art Garden **010**, 229–231 (2017). (in Chinese)
15. Lin, C.: Exploring the information transmission function of visual elements in exhibition space. Drama House **11**, 116–117 (2020). (in Chinese)
16. Zian, L.: Thinking evolution of spatial display design in the information age. Literary Life (Art China) **01**, 132–133 (2017). (in Chinese)
17. Zong, Y.: Discussion on the teaching method of studio system for art design major in higher vocational colleges. Adv. High. Educ. **4**(12), 230–232 (2020). (in Chinese)
18. Wang, X.: Research on the optimal path of practice teaching in teacher education. Educ. Res. Rev. **2**(4), 10–14 (2020). (in Chinese)
19. Nathaniel, L.G.: A conception of teaching. Springer, Boston, MA (2009). https://doi.org/10.1007/978-0-387-09446-5
20. Collazo, A.A.: A theory-based model for understanding faculty intention to use students ratings to improve teaching in a health sciences institution in Puerto Rico. J. Hisp. High. Educ. **17**(3), 229–248 (2018)
21. Hashimova, K.K.: The role of big data in internet advertising problem solution. Int. J. Educ. Manag. Eng. (IJEME) **6**(4), 10–19 (2016)
22. Wang, R.: Improvement of "flipped classroom" teaching model based on fuzzy comprehensive evaluation model. In: Proceedings 2016 Eighth International Conference on Measuring Technology and Mechatronics Automation ICMTMA, 335–338 (2016)
23. Wenting, M.: Research on fuzzy mathematics model of teaching quality evaluation. In: 2017 International Conference on Innovations in Economic Management and Social Science (IEMSS) (2017)
24. Zhou, L., Li, H., Sun, K.: Teaching performance evaluation by means of a hierarchical multifactorial evaluation model based on type-2 fuzzy sets. Appl. Intell. **46**(1), 34–44 (2017)
25. Zhang, L., Yangmei, G.: Research on fuzzy comprehensive evaluation of professional teaching standard setting effect in civil aviation higher vocational colleges based on analytic hierarchy process. IOP Conf. Ser. Earth Environ. Sci. **692**, 032–037 (2021)
26. Çebi, A., Karal, H.: An application of fuzzy analytic hierarchy process (FAHP) for evaluating students' project. Educ. Res. Rev. **12**(3), 120–132 (2017)

Research on Chinese Education for Preparatory Students Based on Diversified Development

Haifeng Yang[1], Junhua Chen[2], and Xinyi Zhan[3](✉)

[1] Center for Teaching Chinese as a Foreign Language, Wuhan University of Technology, Wuhan 430070, China

[2] School of Foreign Languages, Wuhan University of Technology, Wuhan 430070, China

[3] University of Chinese-Spanish Translation in Buenos Aires, 1031 Buenos Aires, Argentina

Abstract. With the accelerating internationalization of China's higher education, the international students scale in China has become larger and larger since 2000, and the education in China has become advancing. China has become the second largest destination country for overseas students in the world, and the education of overseas students is at a critical stage of conserved development from "scale" to "quality and efficiency improvement". The educational quality of international students is one of the important symbols of the internationalization of a country's higher education, and also the basis and key of the sustainable development of international students' education. The results of the development of academic education in China have promoted the return and rapid development of preparatory education in China. Through the analysis of Status quo of Chinese teaching for China preparatory students, this paper proposes we should improve professional quality, drop "more practice less theory" idea, emphasize theory; reform traditional "teacher-centered" classroom, absorb updated ideas of foreign functional and task-based teaching method, advocate "student-centered"; emphasize the integration of teaching resources, strengthen teaching management, implement partner teaching, conduct colorful extracurricular activities, and adopt the teaching mode of separate classroom.

Keywords: Diversified development · Preparatory students in China · Chinese teaching · Task-based method

1 Introduction

In recent years, with the ever accelerating globalization of China's higher education, China is gradually transforming from a huge country exporting students to a huge one importing students [1]. At a press conference in the news center after 19th People's Congress, Chen Bao-sheng, minister of the Ministry of Education, says, now over 180 countries and regions have established educational cooperation with China; 47 countries and regions have signed a degree mutual recognition agreement, that is to say the degrees between the two sides are mutual accepted, an internationalization. China has

Z. Hu et al. (Eds.): AIMEE 2021, LNDECT 107, pp. 503–515, 2022.
https://doi.org/10.1007/978-3-030-92537-6_46

become the third largest in the world and the largest in Asia overseas study destination. Since 2004, Teaching Chinese as a Foreign Language has developed prosperously, with the number of Chinese learners in mainland China increasing from 60,000 a year to over 100,000 a year. In 2018, 492,200 students from 196 countries and regions came to China, of which a total of 260,600 students from 64 countries along the Belt and Road route, accounting for 52.95% of the total number.

In 2019, China Education Online released the 2019 Higher Recruitment Survey Report showing that in recent years, more and more international students have come to China to pursue degrees, and the proportion of academic students and graduates has achieved double growth. In 2018, the total number of academic students adds up to 258,100, accounting for 52.44% of the total overseas students, increased by 6.86% year on year; the number of graduate students reached 85,000, up by 12.28% from 2017. At the same time, the academic structure of international students is becoming increasingly diversified, with more and more students studying engineering, management, science, arts and agriculture, with a year-on-year increase of more than 20%. (see Fig. 1).

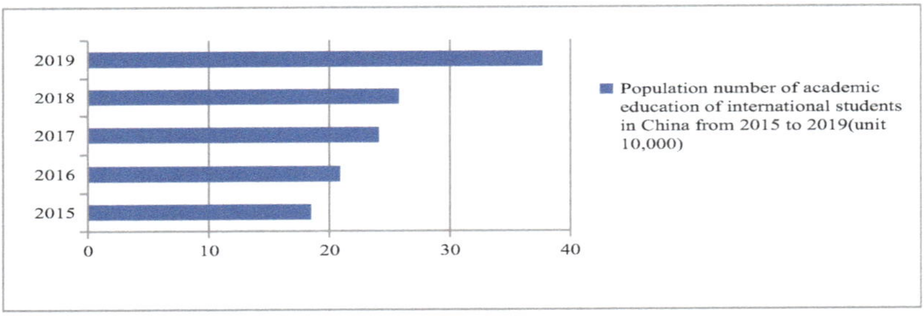

Fig. 1. Population number of academic education of international students in China from 2015 to 2019 (unit 10,000)

In this case, there is bound to be some students who have not met the requirements of professional enrollment and have to choose the preparatory study first. The preparatory education of international students in China is a preparation and transition stage of foreign students before receiving professional education, which is playing a crucial role in the academic education of institutions of higher learning [2]. On December 15, 2019, according to the Changjiang Daily report, Wuhan University has expelled 92 overseas students because of failing in their studies and violating the school regulations. Hubei University and Wuhan University of Technology also withdraw 1–2 international students every year. Statistically, most of them are preparatory students, that is, China preparatory students have more learning difficulties and obstacles. So the research of preparatory education in China is urgent and pressing, but our preparatory education in China began late and is still in the pilot exploration stage. How to adopt a scientific and effective teaching mode and quickly improve the quality of preparatory Chinese teaching is indeed a problem to be resolved urgently.

2 Chinese Teaching Status Quo for China Preparatory Students

China preparatory students usually fall into two categories: one is undergraduate students who apply for Chinese government scholarship, usually known as government-funded preparatory students; the other is at their own expense, known as self-funded preparatory students.

In 2005, the Ministry of Education and CSC (China Scholarship Council) implemented the pilot work for China preparatory students, and issued *the Notice of Extra teaching of Mathematics, Physics and Chemistry for the Chinese Government-funded Undergraduates from Some countries since 2005*, when the pilot universities were Tianjin University, Nanjing Normal University and Shandong University. In March 2009, the Notice of the Ministry of Education on *Preparatory Education for Undergraduate Students funded by China government scholarship (Teaching Foreigners [2009] No.20)* stipulates that all overseas students who apply for undergraduate majors in China universities must study Chinese for one year to pass Chinese HSK-5 for liberal arts students and HSK-4 for science students.

In terms of self-funded preparatory education, many colleges and universities have self-funded preparatory students. In 2016, entrusted by the International Department of the Ministry of Education, the preparatory education team of Tianjin University conducted face-to-face or email research on 47 universities such as Tsinghua University, Beijing University of Science and Technology, Southwest Jiaotong University, Southwest University, South China University of Technology and Shenzhen University with a research conclusion that 23 universities carried out self-funded preparatory teaching, but these universities arranged their teaching in accordance with the standards of their own, and there was no unified syllabus and textbooks [3]. Hu Hongjie put forward: "because the Ministry of Education has not made any clear requirements as to the goals, content, management and evaluation, so the enrollment colleges and universities have a lot of freedom and independence. From recent years of statistics analysis, the number of self-funded students is far greater than that of Chinese government scholarship students with wider professional selection, more countries, meanwhile due to low enrollment threshold, various students levels, and random selection, little regulation force, therefore, the institutionalization of undergraduate education problem needs to be solved [4]."

Due to the lack of teaching understanding of preparatory classes in China, ignoring the inherent particularity of preparatory students in China teaching, many problems arise, the teaching situation provokes concerns and worries. So where exactly is the question arising from?

2.1 The Interview

The interviewees are 40 students from Wuhan University of Technology, Hubei University and Central China Normal University who have finished preparatory Chinese learning, coupled with 10 HSK-4 tutors. The interview was conducted in two parts, each about 20 min, mainly focusing on "What is the main problem affecting

Chinese learning?" with a question and answer, and the whole interview was recorded. After the interview, all the whole collected content is transcribed into text in the first time to ensure that the content is accurate and convenient for research and analysis. After transcription, the text is analyzed and classified as: (1) It is not taught according to the actual level of Chinese language. For example," 2019–2020 preparatory class, some schools have 30–40 people per class and their Chinese are not in the same level.“ (2) not confident and shy to speak. For example, "a Chinese-Indonesian has repeatedly stressed that he did not dare to answer questions in class"; (3) insufficient learning time, such as "one year can only learn Chinese tutorial volume 3 (1st part)"; (4) problems with teaching methods, such as "whether in class and after class teachers use English to communicate with them, ignoring the real purpose of training their Chinese".

2.2 Questionnaire Survey

2.2.1 Preparation and Distribution of Questionnaire

The questionnaire is designed mainly in refer to Deterrents to Participation Scale-General, DPS-G) by Darkenwald & Valentine [5]. This table is mainly to measure adult learning disabilities with 34 projects, 6 factors: lack of confidence, lack of relevance, time limit, lack of interest, education costs and individual problems [6]. On the basis of interview, through induction, merge, delete, sorting into 42 projects, using Likert scoring method (Likert) five, score is proportional to the learning difficulties, In order to enable investigators to better understand the specific meaning of the questionnaire project, questionnaires have also been translated into English, French, Spanish, Korean, Vietnamese, etc. in addition to Chinese version.

The questionnaire is mainly distributed on the network platform. The specific feedback statistical results can be obtained directly from the network platform, which is convenient and fast.

2.2.2 Questionnaire Subject

In Wuhan University of Technology, Central China Normal University and Wuhan University, 200 questionnaires involving 42 projects were distributed, 196 questionnaires recovered with 190 valid, with an efficiency of 95%. The test subjects were 108 males, 82 females; 23 under 18, 80 aged 19–20, 58 aged 21–22, for ages 19 and above; 20 Chinese, non-Chinese 157, 3 unidentified; 68 from Asia, 90 from Africa, 12 from Europe, 5 from North America, 3 from South America, 2 of unknown nationality; 162 can speak in English, 18 can not. There are 60 in liberal arts, 98 in science and technology and 22 in economics and trade.

2.2.3 Questionnaire Method

The questionnaire subjects completed the test paper in 20 min, and the test paper was completed by voluntary gift. Data recruitment, project analysis and cause exploration are all using SPSS online version: SPSS analysis software SPSSAU.

2.2.4 Questionnaire Statistics Results

The project statistics mainly adopts the following methods: (a) with benchmarks 80, the questionnaire is divided into two groups: high and low scores, and respectively tests their distinguishing and identification power and eliminates the items that do not meet the standards; (b) Count the coefficient of items and total score one by one, and clear the items with the coefficient below 0.4. After continuous comparison and analysis, 34 items were finally retained to conduct factor exploration and analysis. Factor analysis of 34 items: KMO test value is 0.92, Bartletts' Test value is 10180.01, the correlation coefficient matrix p less than 0.001, which prove that the factor analysis can be conducted. Referring to Fan Wei and Li Xianzhuo, the main component analysis method and factor rotation method are adopted to abstract the common factors, the eigenvalues of the first five factors are greater than 1, the total explicable variance 51.3% [6].

Table 1. Items with factor loads statistics

Dimension	Second/Foreign language		Teacher		Time		Assessment	
	Items	Loads	Items	Loads	Items	Loads	Items	Loads
Items with factor loads	34	0.93	24	0.73	4	0.81	3	0.89
	30	0.88	26	0.75	8	0.72	5	0.87
	33	0.85	29	0.77	12	0.70	6	0.78
	32	0.83	23	0.73	18	0.68		
	31	0.81	22	0.71				
	29	0.91	30	0.72				
Total contribution	38.4		52.8		58.3		70.1	

According to the above verification results, 5 factors were selected for main component analysis. Main component analysis and factor rotation method were analyzed one by one again. Items with the factor load was less than 0.3, which is not easy to explain high correlation should be removed, and finally 24 items were retained. The main component analysis of 24 retained items is adopted again, indicating that according to the above statistics, the preparatory students now mainly have problems in second /foreign languages, teachers, time, assessment/evaluation and confidence. The total variance is 70.9%, and the interpretation variance of a single factor is: 35.9%, 18.6%, 8.9% and 6.8% respectively. As shown in Table 1.

Statistics result in Table 1 shows that the cronbach α value is 0.91, 0.90, 0.86, 0.85, 0.73 respectively. The total amount table of α value is 0.89. The original problems arising from Chinese students in China include five aspects: second /foreign languages, teachers, time, assessment/evaluation and confidence. The cronbach α value is greater than 0.73, the interpretation power of each component table is greater than 6%, the overall interpretation variance is greater than 60%, and the differentiation degree of each project reaches a significant level. All these explain that the items are well distinguished. Compare the five questions of the project with the 4 questions of the interview and we find some overlap. For example, "no confidence, dare not speak" can be classified as the confidence problem; "short time" can be classified as the time problem; "teaching method problems" can be classified as the teacher problem. The questionnaire covered the interview information, and all the measurement indicators were good.

2.2.5 Questionnaire Analysis

2.2.5.1. Reliability and Validity of the Questionnaire

A total of 24 questions for preparatory students, covering four elements: second/foreign languages, teachers, time, assessment/evaluation. The questionnaire was conducted among 200 students in Wuhan universities with internal consistency coefficient of 0.89 and good internal consistency coordination of the scale. The survey also noted the stability of time, after three weeks another round questionnaire for 30 students in each university, 90 students, with a confidence of 0.90. The reliability of the questionnaire is reliable.

2.2.5.2. Status Quo of Preparatory Students' Chinese Learning

According to the above statistical analysis, the second/foreign languages problem constitutes the first primary problem of Chinese learning for Chinese preparatory students, which can be interpreted as a 38.4% variation. The main task of the preparatory stage is to learn Chinese paving the way for academic learning, which shows the importance of Chinese learning. From the source of students, the three university students are mainly from Africa, followed by Asia, Europe; from the native language, most are non-Chinese-Tibetan language speakers, remarkably distant and different from Chinese in tone, word forms, meaning, especially in terms of Chinese characters [7]. In the survey, 95.1% of students thought that Chinese characters were too difficult, difficult to write, recognize, and remember and distinguish, such as "我"and"找". In the study of Chinese book 1 (1st part), 98% of the students felt confused with the two characters. 85% of students are affected by negative transfer of mother tongue and is difficult to rectify. Students from countries such as Ethiopia, Algeria, Liberia and Eritrea always read "我(wǒ)"as "wō"; students from Uzbekistan, Morocco, Sierra Leone and Mali read "睡觉(shuìjiào)" as "suíjiáo". The polysemic nature of Chinese characters also makes them more confusing. Bai Hassan believes that the idea of words first rooted in Chinese educators is the most abnormal phenomenon in the history of Chinese education, and even seriously affects the development of Chinese education. Chinese is not difficult, but it is we, educators that make it difficult to learn [7].

The second problem is the teacher qualification. In the questionnaire statistics, 93% of teachers have not clearly explained the vocabulary grammar in class, 10% teach in English, and selective use of English accounted for only 30%. 10% of teachers teach target language through target language by means of direct translation from Chinese to English of vice versa (see Fig. 2). The teachers of preparatory classes in China are basically part-time teachers of other majors, without corresponding professional training or professional teaching experience. Take the preparatory class teachers of XX University as an example. In the first semester of 2019–2020, three preparatory classes had seven teachers, six from Chinese Linguistics department, among them three from literature, three from linguistics; one being a sociology teacher. Seven teachers, without any training background of Chinese international education, are directly put on duty [8].

Teaching Chinese to preparatory students is quite different from teaching students pursuing master degrees or doctoral degrees, it is rapid and intensive [9]. Different from native Chinese speakers, preparatory students are zero basic Chinese learners. However, teachers and tutors untrained or non-professional are not clear about the language points of HSK, fail to grasp the key and difficult points of teaching, no or little idea of the

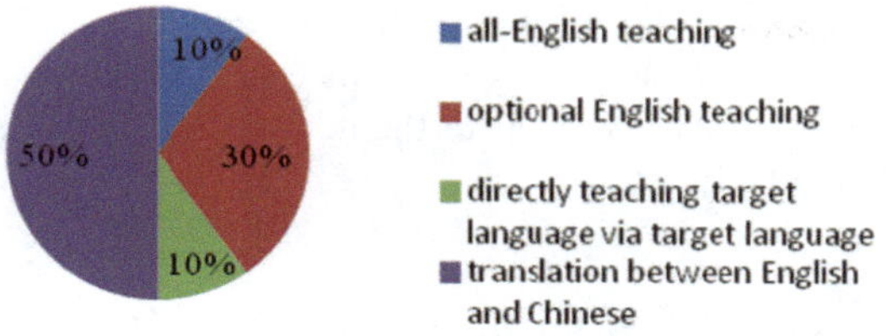

Fig. 2. Chinese preparory teachers language in Class

proportion of the class training or drilling. Some teachers good at English always say a lot of translation, without reducing translation with the increase of Chinese knowledge; the classroom language is very casual and random, not inputable language, not in accordance to students' actual level, violating the law of Chinese international teaching. In fact, although the Chinese international education has nearly 70 years, not everyone understands and recognizes it, and some people even hold the "discipline negation theory", and even think that "any Chinese can teach Chinese" without any more knowledge [9]. It is not easy to teach adults to master a second language more skillfully in a short period of time. Because of this, teachers as teaching Chinese for foreigners have special requirements to meet in their professional quality and teaching ability.

The third problem is about time. Pre-preparatory students unlike the general students learning a language, have their own particularity. They have to complete all the learning content within nine months or so which other ordinary Chinese students will finish in a year and even a year and a half, and in this period they must quickly improve their Chinese to HSK-4 to 5, which is undoubtedly a huge challenge for the preparatory students in China [10]. Every foreign student will gradually undergo a cross-cultural psychological adaptation process of U-shaped distribution. First, the honeymoon stage, This stimulus process often lasts for days or months, 6 months at most; Then it falls into the bottom of the "cultural imbalance", the Cultural Shock stage with language barriers and differences in their living environment and culture causing frustration, serious anxiety, helplessness and depression etc., This process tends to occur in the third and eighth months after reaching the destination, lasting time varying from person to person; Last is Adjustment stage, Their mood has gradually recovered, accepting a new culture, adapting to living in China. According to statistics, the highest rate of removing is preparatory students every year, mainly due to the failure to pass the "frustration period", resulting in weariness of learning. The author came into contact with a Libyan student. After he came to China, he soon became weary due to language barrister and other problems. The attendance rate of a semester was less than 30%. He usually killed the time by sleeping in the dormitory, playing games, chatting and other ways, and was finally dismissed. Therefore, how to face the challenge of the process of cross-cultural mental adaptation and successfully complete the preparatory learning task in a limited time is an urgent and pressing problem at present.

The fourth is the assessment or evaluation problem. At present, the Chinese level test for pre-preparatory students is mainly new HSK and "Chinese comprehensive unified examination". Among them, the scholarship students take the "Chinese Comprehensive Unified Examination", which mainly tests international students from three aspects of listening, reading and writing. Although the new HSK includes oral and written tests, preparatory students mainly participate in the written test. In fact, the oral ability of preparatory students is crucial to them, and directly involves their intercultural psychological adaptability.

3 More Understanding of Chinese Teaching for Preparatory Students

In view of the above problems existing in Chinese teaching for Chinese preparatory students, I mainly have the following shallow and humble understanding after years of teaching experience:

First, advocate the concept of vocabulary first, teaching beginning with words. Mr. Pan Weiguo believes that there are actually two languages in Chinese: spoken and written language, namely "oral Chinese" and "written Chinese". "Chinese international education" should focus on teaching "written Chinese". In Britain primary school students will first learn spelling, written forms of language. Since the creation of words, words are far more important than spoken pronunciation, especially for Chinese language. In the early 1950s, in the "Chinese language Class" of Tsinghua University equal importance is attached to bath Chinese language and literature as well. Later, because of the name rectification as "Chinese teaching" and English with double meaning as English language and English literature or written form, only English teaching theory was introduced to construct the theory of "Chinese teaching", which gradually led to the long-term atmosphere of emphasizing oral language over writing, listening and speaking over reading and writing in China. Chinese teaching should not only cultivate students 'language ability, but also cultivate students' reading and writing ability [11].

Second, improve teachers' professional quality. Mr. Zhu Dexi a distinguished linguist in China, put forward that Chinese teaching for foreigners "In fact, is a knowledge, has become a knowledge abroad, needing to be studied". In 1989, Mr. Zhu Dexi reiterated, "We claim that teaching Chinese for foreigners is an independent subject [12]. It is obviously wrong to think that teaching Chinese for foreigners is not a knowledge, and that all Chinese people can teach it". Teaching Chinese for foreigners is a language teaching, which necessarily involves a certain cultural content, and must also include the teaching of certain cultural factors, but the teaching of cultural factors must serve the language teaching. Language teaching is also different from linguistics teaching. Language is a communication tool, teaching language is to let learners master this communication tool, to cultivate their ability to use language to conduct communication. "Learning language is not about learning a set of knowledge, but about acquiring a skill [13]". Therefore, the language class is first a skill class, tool class. Language teaching must contain a certain language knowledge and language rule teaching, but teaching knowledge and teaching rule is to cultivate the ability to use language for communication purpose [9]. Some part-time Chinese teachers do not understand the nature and characteristics of teaching Chinese for foreigners, and so teaching a lot of knowledge of Chinese grammar or the various meanings and usage of vocabulary in class, which completely violates the law of language teaching. Therefore, it is very necessary to improve the professional quality of the preparatory student teachers in China.

Third. Implement the tutor system. According to the survey, foreign students in China, when finishing a year of pre-preparatory language study and entering into major studies still fail to understand professional or academic courses. The level of their Chinese and the understanding of Chinese culture, their daily life in China are also quite affected. They urgently long for further learning and improving the level of Chinese, deepening the understanding of Chinese culture, but are unable to get opportunities for further learning and cultural understanding. Establishing a tutorial system to help them remove barriers to language communication, to conduct daily communication, to understand Chinese culture and to enjoy Chinese life is an effective and efficient way.

Fourth, strengthen the oral language assessment. Wang Jimin believes that as a major country of overseas education, the construction and improvement of preparatory education examination system in China is an important part of preparatory education in China [14]. In many developed countries as in Germany DSH, PSP contain spoken English, while in Britain and Australia the spoken English scores are used as the basis for preparatory student enrollment and scholarships. Therefore, the setting of preparatory oral Chinese examination is the objective requirement to improve the evaluation of preparatory students in China [15–18].

4 The Diversified Development Measures for Preparatory Students in China

Pre-preparatory education is a relatively special language education, which requires in just 10 months, through strong, intensive and professional language training, so that students can use Chinese more accurately and fluently, and lay a foundation for their undergraduate major study in each school after passing the unified examination. This requirement is challenging for colleges, teachers, and preparatory students. With the development of language policy in some developed countries into internationalization and diversification, the preparatory language education in China can take the following measures:

First, strengthen and implement the educational concept of moral education, and formulate the path of Chinese curriculum moral education (Fig. 3). Educators should first comprehend the teaching objectives that the textbook compilers want to achieve through the selection of texts and the design of exercises. Good compilers will integrate core language knowledge and skills with values, emotional character, and self-management. More precisely, the cultivation of values, emotional character and self-management will be embedded in the teaching of language knowledge and skills. When teaching, teachers need to prepare two sets of goals: one is the teacher-student shared teaching goals, namely students' language goals and communicative goals; The other set is the teachers' special teaching goal, namely ideological and moral education goal, which need not be directly told to students and can be achieved unconsciously by teachers in the process of learning language knowledge and skills through concrete teaching materials and activities [19].

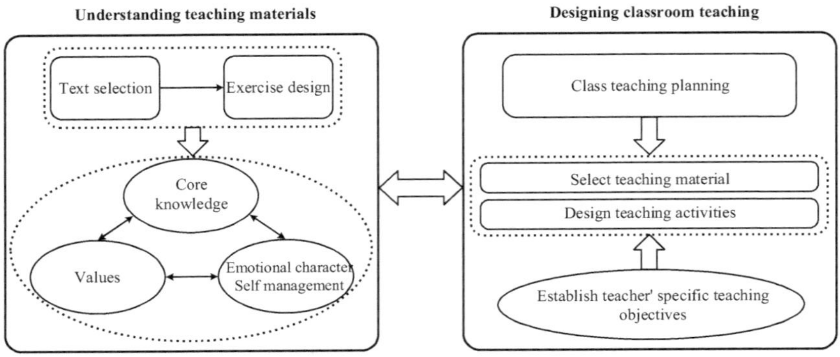

Fig. 3. Path of ideological and moral education for international students in China

Second, Integrate diversified teaching resources. We should attach great importance to the preparatory education work, give full play to their own characteristics and advantages, integrate college and school resources, take the preparatory education department in international school as the main force, take comprehensive Chinese, professional Chinese, Chinese listening, oral Chinese and other courses teaching and learning as major courses, integrate the Chinese history and culture, national conditions education and language major courses, combine the regular education and teaching with targeted intensive training, make great efforts to promote the comprehensive improvement of education and teaching quality.

Third, build a diversified teaching platform. By means of technology, Internet media, online education and others, and combined with Chinese ontology research, we should develop quality online open courses, quality video open courses, quality resource-sharing courses, establish curriculum resource database, case teaching database, international student associations, establish Moodle platform, etc. The constructed Moodle platform provides an important teaching resource platform for Chinese courses for preparatory students. Moodle platform has rich teaching resources for each chapter of *Chinese Coursebook*, including text teaching plans, PPT, teaching videos, reference materials, teaching cases and related exercises and tests. It is with the support of the Moodle platform that the various ability drilling-oriented language training can be carried out in the comprehensive Chinese teaching classroom, and it can achieve expected teaching results. At the same time, in the process of preparatory classroom teaching practice, it is also gradually found that instant communication tools such as WeChat groups and discussion groups, school journal network and Internet resources also have an important auxiliary role in preparatory teaching [20].

Fourth, carry out diversified extracurricular activities. By holding various forms, rich content of lectures, cultural display and perception of China activities, we can enable students not only through the classroom, but also through social practice, timely and accurately to understand Chinese law, Chinese history and humanities and contemporary Chinese national conditions as well as developmental achievements, to exchange culture, to strive to become hard working, law-abiding, knowledgeable and courteous and outstanding students and international youth of knowing China, China-friendly, and China-loving. At the same time, establish language companion learning with international students, combine the teaching of international students with the undergraduate teaching of Chinese international education by way of one-to-one guidance, tourism, joint interviews, rehearsal of drama, etc.

Fifth, adopt pair-classroom teaching method we should introduce pair-classroom teaching method to address the common phenomenon of low attendance, students playing mobile phones, watching computers, and inattentiveness to class. The core idea of pair-classroom teaching method is to allocate half classroom time to teachers and the other half to students to conduct interactive learning in the form of discussion. Similar to the traditional classroom, the emphasis is put on teaching before learning, teacher instructing first, students learning later. Similar to discussion classes, in pair-classroom teaching method the interaction between students and between teachers and students is emphasized to encourage independent learning. In pair-classroom teaching method teaching is depicted as three processes of distinct separation in time, namely Presentation, Assimilation, and Discussion.

5 Conclusion

In this paper, interviews, questionnaire analysis and content analysis are used to analyze the research status quo of preparatory education for international students through a combination of quantitative and qualitative research methods. In view of the above problems in Chinese teaching for preparatory students in China, strategic measures for diversified development are proposed based on my years of teaching experience. Specific conclusions are as follows:

1) At present, China has become the world's second largest destination country, with degree pursuing as the main body, and education informationization integrating into education. In this phase, information-based education means such as Moocs and online courses, multimedia technology are thriving, deeper penetrating into the physical education, especially when the Covid-19 occurs, so that many years of "hot for the theory of online education" spread in an unprecedented fashion, become the lighthouse of education history. The impact of online teaching on traditional offline teaching and the change of teaching mode brought by online teaching have brought the impetus of innovative ideas and teaching transformation to teachers.
2) Through interviews and questionnaires, we know that besides learning Chinese language, students want to know more about China and get comprehensive professional and vocational education in China. In this case, it is necessary to timely adjust the focus of Chinese education for preparatory courses in China, and realize transformation in teaching concepts, teaching modes, teaching contents and methods, so as to meet the needs of diversified development in the future.

3) On the basis of years of teaching experience and investigation, it is proposed that Chinese teaching for preparatory students in China must improve teachers' professional quality, change the teaching concept of "more practice less theory", and give more priority to the importance of theory; Change the "teacher-centered theory" of traditional classroom, absorb the new ideas of foreign functional method and task-based teaching method, advocate "student-centered", highlight the characteristics of Chinese language and Chinese characters to implement teaching and research; Emphasize the integration of teaching resources, strengthen teaching management, implement language partner teaching, carry out rich and colorful extracurricular activities, and adopt the teaching mode of dividing classes.

Acknowledgment. The program is supported by "The Fundamental Research Funds for the Central Universities" (WUT: 2021VI033). The Ministry of Education's Higher Education Department Industry-Academy-Education Project "Based on the Multi-terminal-based "Introduction to Teaching Chinese as a Foreign Language" Premium Online Course Construction" (201801049043), Wuhan The teaching and research project of the University of Science and Technology is based on the research results of the construction and practical application of the "integrated" ideological and political education system for Chinese International Education (w2021079), and the research results of the National Social Science General Project "The Semantic Evolution of Chinese Willing Verbs" (21BYY138).

References

1. Shengsan, C., Tao, L., Ying, W., Meili, X.: Chinese language teaching and research. Lang. Teach. Res. **6**(02), 15–25 (2021). (in Chinese)
2. Hongjie, H., Youqiang, L.: Review and reflection on preparatory education of international students in colleges and universities. Heilongjiang High. Educ. Res. **3**(04), 136–138 (2013). (in Chinese)
3. Yaqi, W., Jing, L.: Status quo and prospects of preparatory education in China. Overseas Chin. Educ. **23**(06), 109–113 (2018). (in Chinese)
4. Hongjie, H.: Operating within class, redefining extracurriculum – perspective on classroom teaching Chinese as a foreign language. Curriculum Educ. Res. **33**(06), 44–49 (2013). (in Chinese)
5. Xiaohui, Q., Tianshu, B., Qixun, Y.: A new method for hybrid nonlinear state estimation with PMU. Autom. Power Syst. **31**(4), 28–32 (2017). (in Chinese)
6. Fan, W., Li, X.: An empirical study on influencing factors of Chinese learning disabilities for college preparatory students in China. Lang. Appl. (4), 134–145 (2018). (in Chinese)
7. Bellassen, J.: Chinese teaching method with words and characters. Chin. Literat. (First Half) **35**(02), 43–50 (2019)
8. Huang, X., Hu, F.: A research review on preparatory education for international students (2005–2018). Int. Chinese Lang. Educ. (Chinese-English), **31**(02), 79–90 (2020). (in Chinese)
9. Xun, L.: Research on Chinese language teaching in international education. Int. J. Chin. Lang. Teach. **34**(01), 4–9 (2020). (in Chinese)
10. Chen, C.: A survey of Chinese Character Learning Strategies for Preparatory Students in China. Beijing Foreign Studies University, pp. 1–283 (2019). (in Chinese)

11. Wenguo, P.: Multidimensional thinking on the construction of foreign language discipline in the new era. Chin. Foreign Lang. **34**(04), 1–15 (2019). (in Chinese)
12. Zhao, J.: Introduction to Teaching Chinese as a Foreign Language. The Commercial Press, Beijing, pp. 1–560 (2004). (in Chinese)
13. Dexi, Z.: Speech at the opening ceremony of the third international seminar on Chinese language teaching. Langu. Teach. Res. **35**(04), 9–10 (1990). (in Chinese)
14. Jimin, W.: Chinese competence standard of preparatory education in China. Lang. Teach. Res. **31**(02), 26–34 (2021). (in Chinese)
15. Yun, D.: Study on the accelerated teaching mode of foreign students' foundation Chinese. J. Heilongjiang High. Educ. Res. **35**(04), 150–152 (2010). (in Chinese)
16. Quan, L.: Review on research on teaching Chinese as a foreign language in the past 20 years. J. Lang. Writing **03**, 100–106 (2002). (in Chinese)
17. De Matas, S.S., Keegan, B.P.: A case study on adult and workplace learning. Mod. Educ. Comput. Sci. **10**(01), 11–19 (2020)
18. Ebenezer, O.: Influencing children: limitations of the computer-human-interactive persuasive systems in developing societies. Int. J. Modern Educ. Comput. Sci. **12**(5), 1–15 (2020)
19. Qiufang, W.: Construction and practical application of "Integrated" ideological and political education system – a case study of cultivating "English and French bilingual + professional" international governance talents. Lang. world **4**(02), 2–9 (2021)
20. Dada, E.G., Alkali, A.H., Oyewola, D.O.: An investigation into the effectiveness of asynchronous and synchronous e-learning mode on students' academic performance in national open university (NOUN), maiduguri centre. Int. J. Mod. Educ. Comput. Sci. **11**(5), 54–64 (2019)

Information Technologies in Teaching the Discipline Project Management

Chang Shu[1], Nataliya Yur'ievna Mutovkina[2](✉), and Alexey N. Borodulin[2]

[1] Engineering and Technology College of Hubei University of Technology, Wuhan, China
[2] Tver State Technical University, Tver 170012, Russia

Abstract. Without introducing modern information technologies, communicational tools, the interaction of educational organizations with students and business representatives in the educational process, adequate training of future professionals who can successfully solve the tasks and develop the regional economy is impossible. The article discusses innovations in the organization and implementation of the educational process at Tver State Technical University, making the educational process as effective as possible for both teachers and students. In particular, the features and significance of the Project Management course included in the bachelor's degree program in the direction of 38.03.01 Economics, considering the competencies specified in the new educational standard are considered. A brief overview of the stages of training in this course, the methodological foundations of its teaching to future economists are presented. Furthermore, the leading software for students to perform practical work on project management is given. Finally, examples of projects carried out by students of economics in the framework of the training course "Project Management" and its development in competencies are given.

Keywords: Project management · Information technologies · Organization of the educational process · Software · Teaching methods

1 Introduction

One of the mandatory conditions for successful socio-economic development is to ensure adequate educational activities in educational organizations. New federal academic standards of higher education are applied to achieve the actual effectiveness of training. These standards contain a list of competencies, the formation of which will allow future graduates to become in demand in the labor market. Educational organizations (graduating departments) independently develop educational programs based on Federal-State Standards of Higher Education, including a list of academic disciplines, considering the content of competencies from the standard. Thus, when creating an educational program in the direction of 38.03.01 Economics by the Federal Standard of the Bachelor's Degree level, approved by Order of the Ministry of Higher Education and Science of Russia No. 954 of 12.08.2020, the competencies specified in the new standard were considered: eleven universal and five general professional competencies

Z. Hu et al. (Eds.): AIMEE 2021, LNDECT 107, pp. 516–525, 2022.
https://doi.org/10.1007/978-3-030-92537-6_47

[1]. As a result of the analysis of the competence set from the new normal, it was found that significant changes in the educational program are not advisable. Still, some training courses were replaced, considering the updated list of competencies in the standard. One of the academic disciplines in the new educational program in training 38.03.01 Economics is "Project Management." It significantly correlates with such competencies as:

- the ability to determine the range of tasks within the framework of the set goal and choose the best ways to solve them, based on existing legal norms, available resources, and limitations;
- the ability to offer economically and financially sound organizational and managerial decisions in professional activities.

The content of these competencies implies the analysis and evaluation of projects, plans of an economic entity for making informed decisions that optimize the company's financial condition and future activities. Therefore, the students' solution to choosing the optimal project from various possible alternatives seems to be relevant.

"Project Management" is an academic discipline that is included in the mandatory part of the list of fields formed in the new bachelor's degree program in the direction of training 38.03.01 Economics (Department of Accounting and Finance). The course "Project Management" is quite multifaceted and diverse. Traditionally, it is part of the curricula for training students in the areas of "Applied Informatics," "Information Systems and Technologies." The content of the course depends on the direction of students' training. So, for future economists, the most relevant forms of project presentation can be a business plan, a project for financing and introducing innovations, an investment project, a production program. In the conditions of digitalization of the economy, projects related to the search, development, and implementation of new information technologies, innovative software solutions, and the justification of their economic efficiency are of undoubted importance [2].

In connection with the highlighted features of the course and its purpose, the authors of the article see the purpose of the research in developing a scheme for using modern information technologies for conducting classes with students on the course "Project Management," choosing and adapting the most suitable software.

Appropriate information technologies should be used both in lectures and in practical classes.

2 Literature Review

2.1 The Essence and Content of the Course "Project Management"

A project is understood as a type of activity to achieve a specific goal with limited labor, material, and other resources.

The project approach has successfully proven itself to achieve goals and objectives in politics, economics, engineering, design activities, and the social sphere. Any project should have an actual problem for which it is being developed, clear goals and objectives, clear deadlines for the start and end of work, an action plan, a team of performers, specially allocated resources, quantitative and qualitative results that are predictable and determined before the start of the project.

The effectiveness of project management should be ensured through a high level of communication and interaction of all project participants, the precise distribution of responsibility among all performers with the allocation of a unique position – the project manager. At the same time, the main factors determining the project's success are quality, cost, and time. Therefore, all these aspects should be studied by students in the course "Project Management."

Within the course "Project Management," it is essential not only the organizational and methodological support of students by a leading teacher but also the practical implementation of projects. It consists of several stages: preparation of the project, discussion of the project, adjustment based on the meeting results, completion of the project, evaluation of the effectiveness of the project, and its presentation to the customer. Within the course's framework, the customer can be a teacher or a student subgroup. The game form of classes within the framework of studying the discipline "Project Management" in general seems to be quite effective [3]. It allows you to distribute the roles in the student group (the head, performers, responsibility centers, experts) and form such qualities as purposefulness, responsibility, the objectivity of assessment, the ability to think logically and express, defend your position.

2.2 The Place of Information Technologies in the Course "Project Management"

Information technologies used in the course "Project Management" should consider the features of this course. For example, many projects have a diverse composition of participants: performers, experts, partners, sponsors, customers, project managers. Therefore, all participants need an effective communication system that ensures close cooperation between them, exchanging information and ideas, and discussing the results obtained [4].

The model of any project consists of four components:

1) The goal, which is the achievement of the planned result.
2) Criteria and restrictions, among which it is possible to distinguish compliance with the requirements for the planned result, compliance with deadlines, and the cost limit.
3) Project management areas, namely: content management (definition of goals, results, and evaluation criteria); time management (dividing the project into separate works, determining their sequence and duration); cost management (determining the types, quantity, and cost of resources and jobs); quality management (defining quality standards, ways of their implementation); personnel management (distribution of roles, responsibilities, and coordination of teamwork); communication management (dissemination of information inside and outside the project); deviation management (identification and timely response to risks, problems, undesirable changes); contract management (identification of required goods and services, interaction with suppliers).

4) Organizational and technical solutions, for example, organizational support, system infrastructure, system integration, information support, packages of application programs for project management, personnel, documents, information, statistical data, etc. [5, 6].

Information technologies should support the study of the project life cycle, namely, provide opportunities for designing the pre-investment, investment, and operational stages. Various functions and tasks are performed at each location, corresponding to a particular set of documents that provide legal, organizational, financial, social, personnel, and other information and documentation support for the activities performed. The probability of successful implementation of the project directly depends on the correctness of the project documents. Therefore, the course "Project Management" should provide for the use of electronic document management systems.

3 Using Specialized Software in the Course "Project Management"

3.1 Overview of Software Products

Depending on the complexity, the project can contain hundreds of operations that require different time and resource costs. Some of these operations must follow in strict order, while others can be performed simultaneously (in parallel). A delay in the start of execution or an increase in the duration of some works may lead to an increase in the project's duration as a whole; for other jobs, these delays do not affect the time of the project. Therefore, for complex projects, planning, control, and management procedures may be complex without unique methods and tools for quantitative analysis.

After the approval of the project idea (the project is approved considering all the justifications of its significance, applicability in natural conditions, consistency with the current legislation), students (individually or in groups) perform all stages of design and prepare the final presentation of the project. When designing, it is advisable to use various online applications and specialized software, for example, Asana, Basecamp, GanttPRO, Genius Project, Jira, Microsoft Project, Todoist, Trello, Wrike and others. Detailed information about these software products is provided on the official websites of the developers [7–15]. In addition, there is also a description of the different versions of the programs and the conditions for their installation on the computer and further use. Some of these programs are free so that all students can take advantage of them [9, 14, 15].

The presentation of the project can be made using Microsoft PowerPoint, with which all students are already perfectly familiar by the third year of study [16].

Of course, the best software products for use in the educational process, in our opinion, are Microsoft Project and Primavera Project Planner Professional. However, it is not always possible to install these programs in computer classrooms for conducting classes. The way out of this situation may be modeling project management using Excel add-ins. This approach allows you to get good results when studying network planning and optimization methods.

3.2 MS Excel Features for Project Management

In MS Excel and specialized programs, it is possible to create screen forms, develop a database, calendar planning, develop mathematical models, and perform calculations that can be used in maintaining project documentation. In MS Excel, it is possible to build a network schedule of the work project under consideration to consider the technological sequence of work. You can also calculate the central time characteristics of the project, namely the early and late dates of the occurrence of events (outcomes). The time attributes can be specified on the network schedule of the project. MS Excel allows you to quickly determine the critical deadline for completing the project, crucial work for a specific state of the network schedule (calculating the time reserves for possibly necessary work). On the MS Excel worksheet, you first need to formulate a linear programming problem equivalent to a network graph, taking the late dates of network graph events as variables. The goal will be to minimize the late deadline of the final event. To solve this optimization problem, the MS Excel add-in "Solver" is used. MS Excel allows you to output a stability report obtained by the "Solver" add-in [17].

MS Excel has built-in functions that allow you to perform the necessary calculations, including tasks from the "Date and Time" category. For example, by calculating the start date of a project and setting the start time, you can calculate each work's start and end times [18].

MS Excel allows you to create a project model using tables, forms, and macros that consider the specifics of the project and the features of the economic entity for which the project is being implemented.

4 Example of a Task for the Course "Project Management" and Its Implementation in Microsoft Excel

As an example of a task for the course "Project Management," offered to students for their implementation in practical classes, using the MS Excel add-in "Solver" for optimizing the cost of the project is considered.

4.1 Mathematical Model of the Project

First, students must build a mathematical model that will allow them to find the optimal calendar plan of the minimum cost, implemented in the interval between the points of typical and maximum intensive modes of project work. Then, for the formulation of the linear programming problem, the following notation is introduced:

W is a lot of works of the project.
ET_i is the beginning of the i-th work.
$x_{(i,j)}$ is the amount of time reduction for work (i,j).
n is the number of works that need to be performed on the project.
T_0 is the desired project execution time (the project's directive deadline).
$c_{(i,j)}$ is the money saved from reducing the time of execution of work on the project.

$T_{(i,j)}$ is the allowed time for the work to be completed. During this time, you can implement all the work on the project at a relatively low cost and decent quality. $T^*_{(i,j)}$ is the desired time for completing the work. This is the shortest possible time for the execution of work on the project. To complete all the work in the specified time with a quality that is not worse than in the time $T_{(i,j)}$, it is necessary to increase the cost of the work. Reducing the execution time while maintaining high quality, as a rule, always leads to an increase in the project's cost.

With the entered designations, the target function of the project will take the form:

$$\min Z(\vec{X}) = \sum_{(i,j)}^{n} c_{ij} \cdot x_{ij} \tag{1}$$

The system of model constraints is defined as follows:

$$\begin{cases} ET_j - ET_i + x_{(i,j)} \geq T_{(i,j)} \\ x_{(i,j)} \leq T_{(i,j)} - T^*_{(i,j)} \\ ET_n \leq T_0 \\ ET_i \geq 0, x_{(i,j)} \geq 0, (i,j) \in W \end{cases} \tag{2}$$

If $\left\{ET_i^*, x^*_{(i,j)}\right\}$ is the optimal plan, then $x^*_{(i,j)}$ is when the duration of work (i,j) should be reduced.

4.2 Generating Source Data for a Task

The initial data for compiling the project model is presented in Table 1. This data is conditional. The time to complete work on the project can be expressed in any period, for example, in days, weeks, months.

All work can be performed independently of each other. The condition for the connectivity of the project works is shown in the second column of Table 1. The work specified in the first column cannot be performed until the previous work (defined in the second column) is completed. It follows from the table that the permissible project execution time is $T(P) = 14$, the possible project execution time T_0 is 10, and the total project costs $C(P)$ under the acceptable mode will amount to 8 290 monetary units. Therefore, it is necessary to determine the time by which it is possible to reduce the execution of individual works so that the project's cost with increasing costs due to urgency is minimal.

Each student of the group receives such a task and performs it in Microsoft Excel. It is advisable to make a unique version for each student. The options may differ in numerical values of the time for completing each work on the project, the cost of performing the work, the number of jobs on the project, different combinations of preliminary work.

Table 1. Initial data for the project

Current work	Previous work	Acceptable execution time	Possible execution time	Costs under the acceptable mode, den. Units	Costs under the possible mode, den. Units	Money saved, den. Units
A	–	3	2	800	1600	800
B	–	2	1	1200	2000	800
C	A	5	3	1650	2950	1300
D	B	5	3	1740	2480	740
E	C, D	6	4	1800	2700	900
F	C, D	2	1	600	900	300
G	F	2	1	500	1000	500

4.3 Implementation of the Project Model in Microsoft Excel

Following their version, each student enters data into an Excel spreadsheet and composes a linear programming problem. An approximate version of the data representation is shown in Fig. 1.

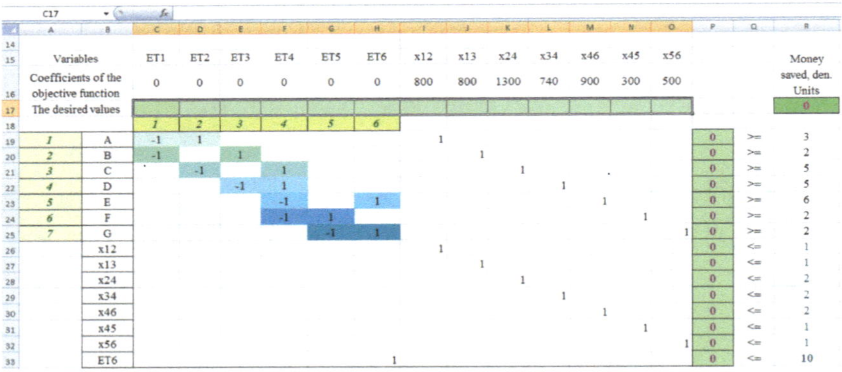

Fig. 1. The original project model

Next, students call the "Solver" add-in and fill in the necessary fields: set the target cell, specify the range of cells to be changed, add restrictions according to the statement (1)–(2).

The result of solving the problem is shown in Fig. 2.

Then the student must make a conclusion based on the results of the decision.

It is advisable to evaluate student work according to three criteria:

1) The level of understanding of the optimization problem in terms of the quality of compiling a table with initial data and the task (1) - (2).
2) The quality of creating an optimization model in MS Excel and the possession of the "Solver" add-in.
3) The correctness of the interpretation of the results obtained.

	A	B	C	D	E	F	G	H	I	J	K	L	M	N	O	P	Q	R
14																		
15	Variables		ET1	ET2	ET3	ET4	ET5	ET6	x12	x13	x24	x34	x46	x45	x56			Money saved, den. Units
16	Coefficients of the objective function		0	0	0	0	0	0	800	800	1300	740	900	300	500			
17	The desired values		0	2	2	6	8	10	1	0	1	1	2	0	0			4640
18			1	2	3	4	5	6										
19	1	A	-1	1					1							3	>=	3
20	2	B	-1		1					1						2	>=	2
21	3	C		-1		1					1					5	>=	5
22	4	D			-1	1						1				5	>=	5
23	5	E				-1		1					1			6	>=	6
24	6	F				-1	1							1		2	>=	2
25	7	G					-1	1							1	2	>=	2
26		x12							1							1	<=	1
27		x13								1						0	<=	1
28		x24									1					1	<=	2
29		x34										1				1	<=	2
30		x46											1			2	<=	2
31		x45												1		0	<=	1
32		x56													1	0	<=	1
33		ET6						1								10	<=	10

Fig. 2. The result of solving the optimization problem

When defending works on the course "Project Management," students' answers are evaluated according to the following criteria:

- Professional, intelligent solution of the problem
- Novelty and originality of the resolution of the issue (task)
- Brevity and clarity of the theoretical part of the solution of the problem
- Ethics of discussion
- The activity of all members of the micro-group

The best projects can be presented at student project competitions, registered on the grant application sites. Therefore, in teaching the course "Project Management," it is essential to consider students' characteristics, organize project work, and evaluate.

5 Conclusion

The course "Project Management" plays a vital role in the training of undergraduate students in the field of Economics since it provides students with the most critical competencies that are so necessary for their future professional activities. The most important thing here is that students learn the methodology of project management, namely:

1) The project lifecycle (both cascade and iterative models are studied). Within the life cycle study, students get acquainted with various models of behavior of the customer, manager, and project team.
2) The optimal cost of the project.
3) The number of project participants and the types of communication between them.
4) The criticality of the project product being developed about the consequences of the production of a low-quality, defective product (service), as well as approaches to anticipating risks and responding to them in case of occurrence.

Using MS Excel to solve problems from the Project Management course allows you to speed up the process of finding the optimal solution and consider projects with many operations in practice. In addition, modeling optimization processes will enable students to focus not only on obtaining a solution but also on analyzing the results.

It also seems appropriate to use several software products to solve problems in project management; students work with several software applications. A successful example of such work is shown in [18]. As a result, students will learn to work in different software environments and determine in which cases it is necessary to use a particular program, information technology.

Thus, the use of information technologies and specialized software in the course "Project Management" allows you to solve complex optimization problems and ensures the formation of the student's competencies that will be useful to him in his future professional activity.

In the future, the authors of this article plan to develop their methodology for teaching the course "Project Management" with an emphasis on the active use of information technologies and possible software products in solving project tasks. It is also planned to evaluate the results of training through a survey of students, analysis of the current and final certification results.

Acknowledgment. This scientific work was partly supported by RAMECS.

References

1. Order of the ministry of education and science of the Russian Federation No. 954 of 12.08.2020. On the approval of the Federal State Educational Standard in the field of training 38.03.01 Economics [in Rus]. ConsultantPlus Legal Reference System. URL: http://www.consultant.ru/document/cons_doc_LAW_361147/ Accessed 25 May 2021
2. Stone, J.A.: A sustainability theme for introductory programming courses. Int. J. Mod. Educ. Comput. Sci. (IJMECS) **11**(2), 1–8 (2019)
3. Alvarez-Dionisi, L.E., Mittra, M., Balza, R.: Teaching artificial intelligence and robotics to undergraduate systems engineering students. Int. J. Mod. Educ. Comput. Sci. (IJMECS) **11**(7), 54–63 (2019)
4. Adekunle, S.E., Adewale, O.S., Boyinbode Olutayo, K.: Appraisal on perceived multimedia technologies as modern pedagogical tools for strategic improvement on teaching and learning. Int. J. Mod. Educ. Comput. Sci. **11**(8), 15–26 (2019)
5. Newton, R.: Project Management. Step by Step. How to Plan & Manage A Highly Successful Project. Prentice Hall, Pearson, p. 168 (2007)
6. Tovb, A.S., Tsipes, G.L.: Project Management: Standards, Methods, Experience, p. 240. Olymp-Business, Moscow (2005)
7. Work on big ideas, without the busywork. https://asana.com/. Accessed 26 May 2021
8. The all-in-one toolkit for working remotely. https://basecamp.com/. Accessed 26 May 2021
9. Online Gantt chart for project management. https://ganttpro.com/ru/?redirectByBrowserDetectedLocale. Accessed 27 May 2021
10. Project management software. https://www.geniusproject.com/. Accessed 26 May 2021
11. Atlassian. Jira Software. https://www.atlassian.com/software/jira. Accessed 27 May 2021
12. Microsoft Project. https://www.microsoft.com/ru-ru/microsoft-365/project/project-management-software?market=ru. Accessed 27 May 2021

13. Get everything in order with Todoist. https://todoist.com/ru. Accessed 27 May 2021
14. Trello helps teams effectively solve work tasks: the official website of the company. https://trello.com/. Accessed 26 May 2021
15. Wrike. Online project management software: official website. https://try.wrike.com/rus-project-management/. Accessed 26 May 2021
16. Fetaji, B., Fetaji, M., Ebibi, M., Kera, S.: Analyses of impacting factors of ICT in education management: case study. Int. J. Mod. Educ. Comput. Sci. (IJMECS) **10**(2), 26–34 (2018)
17. Popovich A.T., Savinykh V.N.: Operational analysis of complex work management in the Excel environment. In the collection: Scientific search in the modern world. In: Collection of materials of the Fifth International Scientific and Practical Conference, pp. 97–100 (2014)
18. Supchinsky, O.P., Kapustyan, M.F.: An integrated approach to organizing and planning production processes based on network planning and cloud technologies. Omsk Sci. Bull. **3** (133), 138–142 (2014)

Feedback Controlled English-Medium Engineering Education in the Materials Forming Principles Course

Qilai Zhou(✉), Qingying Zhang, Wei Feng, and Shangyu Huang

Wuhan University of Technology, Wuhan 430070, China
zhou.qilai@whut.edu.cn

Abstract. The English-medium instruction (EMI) in engineering courses at the university level is expanded globally. The most critical issue for the EMI is how to implement it in an academic course. In this study, feedback controlled EMI model for engineering education is proposed based on the data accumulated in a typical engineering course. The reflective information is accumulated in terms of observations, interactive activities, case-based student presentations as well as questionnaires. The instruction strategy can be adjusted timely based on the student feedback. The results show that specific data are acquisitive in the EMI class to track the student's learning state. Most importantly, it is confirmed based on the reflective practice that the students have a positive attitude towards the EMI class at the current stage. The study concludes that a more programmatic EMI strategy should be developed based on the feedback information to meet the internationalization needs of the students in a global context.

Keywords: English-medium instruction · Engineering education · Feedback

1 Instruction

Over the last five years, the comprehensive education reform for the college engineering course is pushed forward in China. New concepts, proposed domestically by the Chinese Education Department, e.g., Emerging Engineering Education [1], and that introduced abroad, e.g., Outcomes-Based Education (OBE) [2, 3], are required to integrate into college education. Besides, China signs the Washington Accord in 2016, which is an international agreement between bodies responsible for accrediting engineering degree programs [4]. As a result, the implementation of the advanced education concepts for engineering students is an argent task. For the improvement of the education level in a short time, it needs to understand the international teaching concepts. One efficient route is to learn the documents directly and absorb teaching experience directly from the original copy in the foreign language from the developed countries, e.g., the United States, Britain, Japan, and also including South Africa. Similarly, this learning method is appropriate for university students to learn the latest specialized engineering knowledge. However, the reality for engineering education in most universities is that classroom teaching is still based on Chinese publications rather than the original copy. In addition, for the scientific and engineering fields, most of the Chinese

Z. Hu et al. (Eds.): AIMEE 2021, LNDECT 107, pp. 526–537, 2022.
https://doi.org/10.1007/978-3-030-92537-6_48

academic expression is not directly translated from English. It was created by the Japanese in the recent 100 years, which was then adopted by Chinese scientists and engineers. Certainly, most Japanese Kanji (Chinese characters) interprets the original meaning of the phrase from Europeans accurately. However, for the university student, some translated Japanese phrases are difficult to comprehend by the literal meaning although they have been widely used in the Engineering fields. Further, for the Japanese students themselves, Mankul reports that the post-graduate student in Hokkaido University does not have an ideal environment for the improvement of their English [5]. Therefore, it is important to offer an international environment for college students to catch on to the originality of each point in their specialized course and enhance their internationalization consciousness. English-medium instruction (EMI) is the direct way for the increment of the internationalization of college education in a non-English speaking country [6]. However, the use of the English language to teach an academic subject is changeable. It was pointed by Lasagabaster et al. that EMI courses are mostly pushed by policymakers in a top-down manner, but the teachers do not receive specific implementation instruction [7]. Hence, the development of a specific EMI strategy for teaching academic subjects is an urgent task. First of all, we should be clear about the obstacles in EMI courses:

1) English level for each student is varied, and it is difficult to promise that every student can understand the content in classroom teaching.
2) For most of the present specialized courses, the teaching material or the teaching textbook is insufficient.
3) The enthusiasm of the student to take part in the English-medium teaching is inexplicit.

The above are the main issues that obstruct the implementation of EMI at various universities, which are derived from the Ref. [8–11]. Although there exist some difficulties, as educators, we should find ways to solve them and develop corresponding classroom instructions.

To understand the effectiveness in an EMI classroom, real-time feedback from the students is significant. Farrell has discussed various reflective tools that EMI teachers can use to facilitate the EMI program, e.g., dialogue, classroom observations, narratives, and team-teaching etc. [6]. The classroom instruction should be adjusted based on the analysis of the feedback or reflective data. Also, the OBE system includes the concept that the ability to solve practical problems should be endowed to each student. If some students have a relatively low language level, additional time should be paid to promote his or her ability to adapt to the teaching pace. Therefore, a feedback model is necessary for the modification of the English-medium classroom instruction.

In this study, the EMI is adopted in a typical engineering course named “Materials forming principles” at Wuhan University of Technology. This course is the most basic academic subject for the students who majored in “Materials forming and processing”. The main content in this course involves the basic industrial procedures including welding, forming and casting, which are the fundamental techniques that support the manufacturing industry. This investigation of implementing EMI is the first attempt, therefore only two chapters, i.e., solidification for the welding poor and heat-affected zone (HAZ) are chosen for the teaching in English. Note that the above two sections are the core issues in welding technology, which involves the fundamental issues in physics, chemistry, and metallurgy. Hence, if the student could understand these issues in the English-medium teaching, the classroom instruction in the materials forming principles course could be applied to most of the engineering courses.

2 Classroom Instruction

Implementing EMI for the materials forming principles course consists of three aspects:

1) Teaching in English by the teachers.
   To realize the EMI for the engineering course, the teachers must have the ability to speak fluent English and express their professional knowledge clearly.
2) Interaction with the student
   It can be anticipated that a considerable portion of the students would feel fresh and could not follow the teacher's pace at the beginning. Hence, it is of great importance to observe the student's state in the classroom based on interactive activities.
3) Student presentation
   The student presentation can further enhance their analysis, summary, and expression ability. Also, the teachers can evaluate the teaching effectiveness by the student performance in presentation.

2.1 Teaching Strategy

First of all, for the students who have the opportunity to receive higher education at Wuhan University of Technology, the educators should believe or assume that the students have enough language basis to attend EMI courses. Then, the corresponding teaching strategy should be made on this presupposition. However, as the experience for the EMI class for the materials forming principles is not available at Wuhan University of Technology, a corresponding teaching strategy should be designed and the reflection or feedback information should be accumulated while implementing EMI. Then prompt adjustment on the instruction based on the student feedback should be made to integrate content and language. Especially, in the first lesson, the academic words and phrases may seem unacquainted to most students. If we simply introduce them without any techniques, the teaching effectiveness would be even lower than using the native language. Detailed instruction should be introduced to bridge the gap between English and specialized English.

Figure 1 gives a specific example for the interpretation of a new academic word in the EMI class of the materials forming principles course. In this course, there is a very significant academic phrase called "martensite". As shown in the figure, the etymology and history of the important academic word should be mentioned when it is taught for the first time. Most youngsters are energetic, and they are interested in listening to scientific history. The non-native speaking students start learning a foreign language at elementary school by reading stories in a foreign language. Therefore, if the teachers could initiate their lecture in the form of scientific stories, the students would not feel strange and can be immersed in the lecture. For the students, the academic word, martensite, is a new concept with complicated pronunciation. Note that this is the first lesson in the EMI class for materials forming principles course. If the students found that the key academic knowledge point is difficult to understand, they would lose interest to continue, or they would constantly think about what the word "martensite" means. In this case, the student's focus shifts to the language issue that deviates from the academic point. What's worse, if the professor was unconscious of the student's reaction, which could be read from the students' face and eyes, and he/her repeatedly explained the meaning of martensite from the academic view, the students would feel even boring and this kind of EMI course would fail.

At this point, the teacher should spend effort on the student state via interaction with the student including observation and questions. Base on the reflective information, the teacher should adjust the instruction strategy immediately if the negative feedback information from the student has been noticed. For the case of interpretation of the word martensite, the following issues should be addressed in terms of etymology and history:

1) The reason for the naming of martensite.
2) The related scientist, Adolf Martens, and his achievements.
3) The feature of the martensite structure.

The teaching strategy based on these 3 bullets is effective to make the student immersed in the class. The correlation between martensite and hardness can be interpreted to the student in EMI class. After the class, even the student read the phase "Martensite transformation" for the first time, they can get the point without translation of it in his/her native language, which was confirmed by the after-class questionnaire. Furthermore, some instructions can be added after class, e.g., reading the history of other academic words, to enhance the EMI mode and increase the student's ability. In the following lesson, the teacher can ask the student to give a presentation on other academic words. Based on this strategy, not only does the interactive activity enhance, but also more reflective data can be accumulated.

Another point that makes it more difficult for EMI in China needs to mention. The simplification of the Chinese translation of the academic phrase is quite different from most non-English-speaking countries. In most cases, the transliteration in Chinese only includes the first syllable by one Chinese character for the simple pronunciation, especially in the fields of chemistry and metallurgy. For instance, the translation of "Martensite" in Chinese only use the first syllable, i.e., "Ma", as the name for this kind of material structure. As a result, even for the students who have already know the martensite in the prerequisite course taught by native language, they would be puzzled when having a first glance at the English word "Martensite". Note that this situation is more common for the translation of chemical elements. Thus, more efforts should be paid for the interpretation of the academic words by the classroom instruction, especially at the beginning of English-medium education.

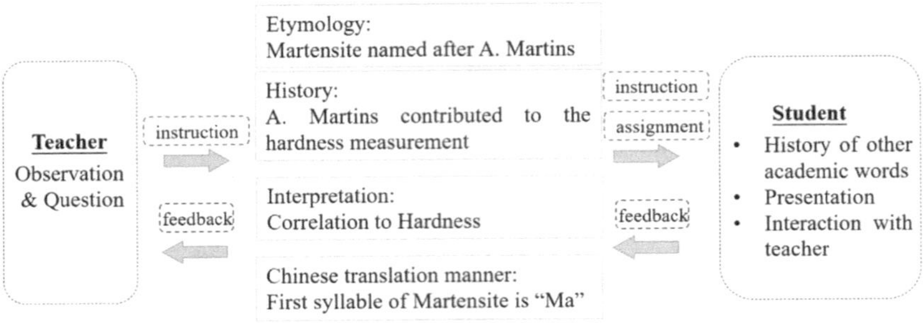

Fig. 1. Schematic for the interpretation of academic word "martensite" in the EMI class.

2.2 Interaction in the Classroom

For the confirmation of the effectiveness in EMI class, the teachers should know the student learning state by classroom observations. Further, the teachers are encouraged to allocate a certain time in the class for the interaction with students to achieve the following aims:

1) Awareness of the state of the mind than the attendee is.
2) Question and answer, including academic knowledge as well as other aspects.

At this stage, it is assumed that all the students have emerged in the classroom teaching and the thinking flowchart can be guided by the lecturer through the aforementioned teaching strategy. In this situation, the language is no longer the obstacle in the EMI class. For the students, one fundamental difference in the EMI lecture is that they may constantly translate the previously learned knowledge points into English or vice versa if all the prerequisite courses were taught in their native language. Therefore, the teachers should lead the students thinking in English, and guide them to comprehend the academic knowledge solely in English.

Specifically, there are two interactive ways to know the learning state of the students. First, the teachers should ask a question and discuss it with the student to confirm whether they are concentrating on the class. Some questions can be related to the course content, and the rest can be free, which is used to make the student more concentrated on thinking in English. Second, the teachers can use one-quarter of the class time as the group discussion section. To avoid an awkward pause in the section, some topics should be prepared in advance. After the discussion, each group is required to assign one group leader, who gives the conclusion to the teacher and other groups.

2.3 Case-Based Student Presentation

Case-based presentation is one of the most important steps in EMI class in an academic subject. The teacher sets specific topics in advance to guide the student in preparing their presentation. Based on our reflective practice, most of the students tend to make the English presentation, which is further confirmed by the questionnaire. Thus, the teacher should try the utmost to offer the presentation environment. To an extent, the environment of the presentation should be made as an international academic conference. Besides, the student presentation also offers a good opportunity for the teachers to evaluate the teaching effectiveness and accumulate reflective data. However, the reality for the EMI engineering education practice is that the student does not have the chance to discuss a case in the form of an academic presentation. One main reason is that most teachers do not believe the students can make an academic presentation on their own after the finish of the class. The EMI education at Wuhan University of Technology is also confronted with the same situation. The available information on the student academic presentation in the class is scarce.

For the materials forming principles course, six topics were set to guide the student for the preparation of the presentation, which are listed in Table 1. One can notice that the specific content in the guiding topics is not mentioned. Instead of direct instruction on the specific issue, the students are required to find the practical engineering case by themselves. Their ability to collect information and solve practical problems can be improved in this process. In addition, if they entered a company after graduation, the presentation ability is also expected by the employers. In each presentation, the group leader should introduce the issue encountered in the Engineering production, e.g., cracking, hardening, and segregation etc., and propose the appropriate solutions. For the student, the ability to collecting information, analysis, application of specialized knowledge, etc., can be effectively enhanced. This instruction is consistent with the OBE concept that the teaching at university should be focused on the cultivation of students' practical engineering ability [12, 13].

Table 1. Guiding topics in the student presentation

Number	Topic
1	Application of the welding materials
2	Effect of alloying on the structure and property for the welding seam
3	Structure of the welding heat affected zone
4	Examples of the welding crack
5	Welding for the nonferrous metals/different materials
6	Advanced/special welding method

3 Feedback System

Before the implementation of the EMI in the materials forming principles course at Wuhan University of Technology, there is no existent feedback controlled model which can be referred to. Even for the most of courses that are taught in the native language, there is also a lack of an effective feedback system. The awareness of the learner-centered approach concept that comes into the university also starts to be enhanced since 2016 when China signs the Washington Accord. In simple words, the teachers should place the students into the center position, and imagine that if the teacher himself were the student, could he or she understand the teaching content. The feedback from the students does not only has great importance in the English-medium course but is also essential in the common courses which are taught in the native language. In the current attempt, the reflective data were drawn from the students, and the feedback controlled English-medium engineering education (FCEE) approach is proposed based on the data accumulated in this study. The schematic of the FCEE model is shown in Fig. 2. The ultimate aim of this model is the fusion of the OBE concept into classroom teaching, the effectiveness of which was evaluated by the objective achievement scale [14]. It must be clear that EMI is a feasible tool for the accomplishment of this purpose. The evaluation of the teaching effectiveness can be performed in four aspects: student state, interaction frequency, academic presentation in the classroom, and examination score, as shown on the right side in Fig. 2. The most direct and convincible way to collect reflective information is a questionnaire. To further increase the efficiency of the information collection, the teachers should know the most important issue in their course, and make the questionnaire in the form of multiple choice. The feedback can be done in the classroom or online after the class. Therefore, it is possible to conduct the feedback investigation easily through the operation on electronic devices including tablet PC and smartphones, and the student's feedback can be passed to the model timely.

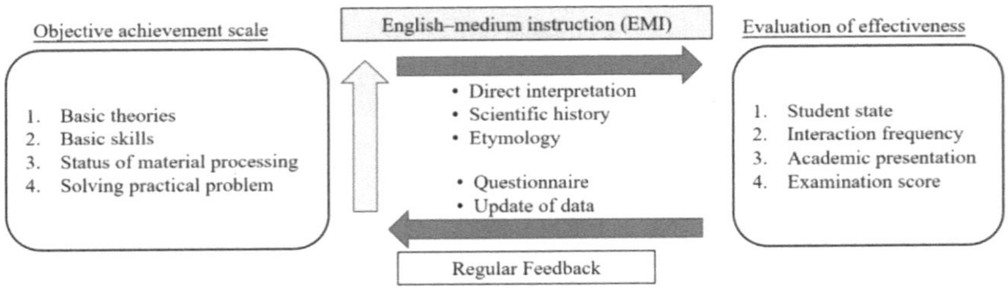

Fig. 2. Schematic for the feedback controlled English-medium engineering education (FCEE) model developed for the materials forming principles course.

Table 2 shows the multiple-choice question used in the questionnaire for this course, and the corresponding reflective results are shown in Fig. 3. It is interesting to find that 94.9% of students agree that EMI class helps them understand the origin of academic words and phrases, and special attention should be paid to this phenomenon. It has been mentioned that certain amounts of the common academic words and phrases in modern Chinese were imported from Japan. The original meaning of the English academic words can be interpreted precisely in Japanese for most cases, but some adopted Japanese Kanji (Chinese Character) is difficult to understand. For instance, the phrases, ageing treatment, plasticity, tension, and stress, are difficult to comprehend for university students who major in Engineering. On the contrary, if these words were explained to the student directly in English, the students would find it easier to understand. As a result, the student will know the precise meaning of the academic words vividly than the translation of it into their native language.

Table 2. Questions that used for the collection of students' feedbacks

Number	Question
1	Does the implementation of English-medium teach help to comprehend the meaning of the academic words and phrases?
2	Have your thinking ability and self-learning ability been improved by the English-medium teaching strategy?
3	Do you think that your ability or awareness of searching information from an academic press, e.g., Elsevier and Springer, has been improved?
4	Do you fully understand the basic knowledge of the solidification of the molten pool and the heat-affected zone of welding after classroom teaching?
5	Can you use the fundamental knowledge to solve practical engineering problems?
6	Do you think that the implementation of English-medium, teaching in specialized courses improves the language ability, or improves the comprehension of the knowledge points?
7	For the understanding of the knowledge points in English, e.g., molten pool solidification and welding heat affected zone is it as difficult as you think
8	Through the study of this course, have you understood what inquiry-based research is?
9	Does the implementation of the English-medium teaching mode improve your knowledge, ability, and international perspective?

It is noteworthy that most of the students consider that their ability to analyze problems is improved after EMI class. Meanwhile, the students begin to be aware that academic knowledge can be searched in English, which is more convenient and efficient. They find that their horizons have been broadened. Besides, the students agree that the difficulty in the EMI class is not increased compared to the native language-medium learning, which is confirmed by bullet 6 in Fig. 3. On the other hand, for the bullet 4 and 5, it can be noticed that over 70% student does not fully understand the specific knowledge points. This phenomenon is not surprising because the knowledge itself is difficult. Thus, in the following lesson, the explanation for the difficult knowledge should be taught again and acquire the feedback till most of the students master it. The accumulation of more detailed feedback in the reflective practice within a short interval is scheduled in the following semester. The FCEE system will be modified and programmed in future research.

Further, the suggestions and comments are encouraged in the feedback model, but it is not mandatory for the student. As is shown in Fig. 4, half of the students do not have a suggestion.

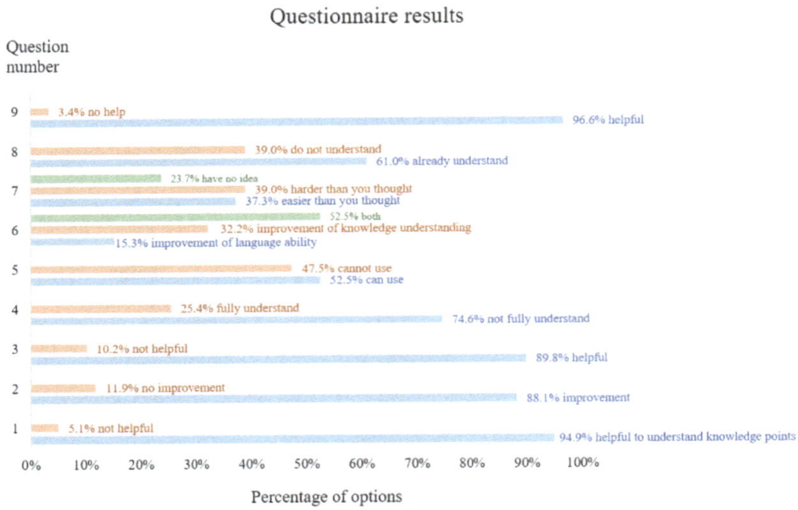

Fig. 3. Results of the questionnaire after the English-medium teaching.

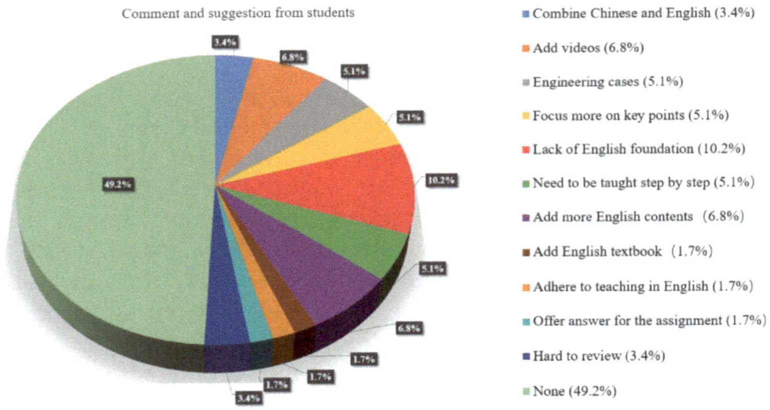

Fig. 4. Comment and suggestions offered by the student after the teaching.

Therefore, it can be seen the EMI in the materials forming principles major can be smoothly pushed forwards, which is not as difficult as it seems. Also, it is noticed that 10% of students think that their English foundation is not good, but they dose not negate themselves. On one hand, this student crowd should increase their English level by themselves if they do not understand the scientific story in the EMI class. On the other hand, the EMI class itself offers an ideal environment for the increase of English language level. In any case, the reflective information should be continued to monitor the state for the student crowd who feels difficult in language.

4 Discussion

The EMI in engineering education can effectively arise the interest of the student as it is a very new and different experience for the university student. Although the Chinese Education Department propel the education reform for the university, the traditional idea and perspective hinder the implementation of the new concept [15]. From the feedback of the students who have attended the EMI class, it can be seen that the students are willing to accept new concepts and new teaching instructions. Hence, the main obstacles are not the student themselves, but the EMI teachers or the educators who are afraid to practice new teaching strategies in class. The reflective information indicates that this timidity or worry is unnecessary.

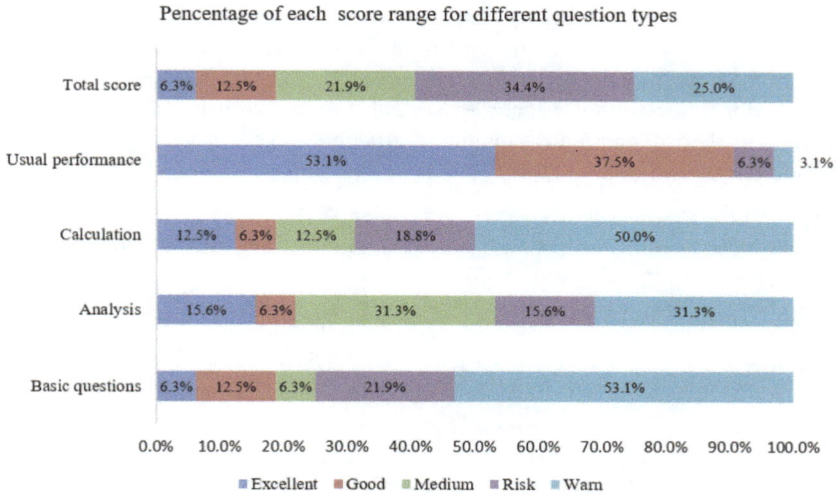

Fig. 5. Analysis of the exam score of materials forming principles course that two primary chapters were taught in English

Figure 5 shows the examination scores and objective achievement scale based on the OBE concept for the materials forming principles course where the primary sections were taught in English. It can be seen that the EMI class does not affect the examination scores that most teachers concern about. Besides, the objective achievement scale is normal. Therefore, more efforts should be paid for the establishment of effective teaching instructions and the accumulation of the classroom teaching data, which can facilitate the discovery of the laws in the EMI course. In addition, the feedback from the teachers and other engineering courses is also essential for the improvement of the model.

5 Conclusions

The EMI in the Engineering course at materials processing and forming major was attempted at Wuhan University of Technology. The teaching effectiveness shows that EMI class can replace the native language-medium teaching mode. Most of the college students have sufficient language foundation to adapt to EMI class rapidly. The student is not the obstacle in the implementation of the new teaching strategy. For the increment of the teaching efficiency, more efforts should be paid in the teaching strategy and accumulation of the student reflective information for the improvement of the FCEE model. In the current stage, the essential preparation work in an EMI engineering class is listed as follow:

1) In the first lesson, the student should be immersed in the English environment by the scientific history and etymology of academic phrases.
2) The student presentation at the end of the course offers an opportunity for the student to show their ability in English.
3) The feedback from the student is critical for the adjustment of the teaching instruction due to minimum amount of students have a relatively lower foreign language level.

Acknowledgement. This work was supported by the Teaching Reform Project of Wuhan University of Technology (No. w2020062).

References

1. Case, J.M., Light, G.: Emerging research methodologies in engineering education research. J. Eng. Educ. **100**(1), 186–210 (2011)
2. Macayan, J.V.: Implementing outcome-based education (OBE) framework: implications for assessment of students' performance. Educ. Measur. Eval. Rev. **8**(1), 1–10 (2017)
3. Gouws, F.E.: Teaching and learning through multiple intelligences in the outcomes-based education classroom. Afr. Educ. Rev. **4**(2), 60–74 (2007)
4. Hanrahan, H.: The Washington accord: history, development, status and trajectory. In: 7th ASEE Global Colloquium on Engineering Education, pp. 19–23 (2008)
5. Manakul, W.: English in engineering education for Japanese graduate students. Australas. J. Eng. Educ. **13**(2), 53–63 (2007)
6. Farrell, T.S.C.: Professional development through reflective practice for English-medium instruction (EMI) teachers. Int. J. Biling. Educ. Biling. **23**(3), 277–286 (2020)
7. Lasagabaster, D.: Fostering team teaching: mapping out a research agenda for English-medium instruction at university level. Lang. Teach. **51**(3), 400–416 (2018)
8. Wu, H., Zhao, J., Fan, R., Su, Z.: Teaching Non-English-Majored College Students Listening and speaking through multimedia and network. Int. J. Educ. Manag. Eng. **3**(1), 44–48 (2013)
9. Jiang, L., Sun, K.: Task-based approach in college English teaching of digital-based context. Int. J. Mod. Educ. Comput. Sci. **2**(1), 32 (2010)

10. Macaro, E., Akincioglu, M., Han, S.: English medium instruction in higher education: teacher perspectives on professional development and certification. Int. J. Appl. Linguist. **30** (1), 144–157 (2020)
11. Ishikura, Y.: International learning experiences at home in Japan: The challenges and benefits of taking English-medium courses for Japanese students. Learn. Teach. **9**(3), 42–61 (2016)
12. Shao, L., Wang, J., Cheng, M., et al.: Reform and practice of modern electrical control technology based on OBE[C]. In: 2021 2nd International Conference on Computers, Information Processing and Advanced Education, pp. 336–340 (2021)
13. Rajak, A., Shrivastava, A.K., Bhardwaj, S., et al.: Assessment and attainment of program educational objectives for post graduate courses. Int. J. Mod. Educ. Comput. Sci. **11**(2), 26–32 (2019)
14. He, G.: Blended teaching mode of art course based on objective achievement scale. Int. J. Emerg. Technol. Learn. **15**(13), 289–302 (2020)
15. Macaro, E., Han, S.: English medium instruction in China's higher education: teachers' perspectives of competencies, certification and professional development. J. Multiling. Multicult. Dev. **41**(3), 219–231 (2020)

Features of the Architecture and Models of Decision Support Systems in Individual Sports

Egor A. Timme(✉)

Russian Association of Computer Science in Sport, Moscow, Russia

Abstract. This article is devoted to methodological issues of decision-making in certain sports in the process of sports training of athletes. The interrelations between various components of athletes' sports training are revealed and presented in the form of cognitive schemes. Approaches to the description of the process of sports training are formed, summarizing the experience of coaches and sports scientists. On this basis, conceptual models are built that include factors that influence decision-making, such as training conditions and effects. The definition of training effects observed in sports, such as delayed, residual, saturation, plateau, super-compensation, rebound, tolerance, overtraining, is given. A brief overview of mathematical models describing these effects and how they are implemented in the architecture of decision support systems (DSS) is given. The definition of the so-called "paradox of variability of physical activity" is given, when a more variable and diverse training program leads to better results than a monotonous one. The algorithm of decision-making by trainers and the features of using DSS in practice are described in detail.

Keywords: Decision support systems · Sport training · Sport performance · Individual sports · Training program

1 Introduction

The high-tech sports market is developing dynamically and will reach $42 billion by 2026. Leading sports clubs are implementing digital transformation strategies, sports federations are cooperating with scientific centers in the field of sports medicine and neurotechnology development. The coaching staffs have adopted decision-making support systems based on artificial intelligence (AI) technologies, which allow analyzing numerous performances of teams and individual athletes' indicators, developing optimal training strategies and adjusting tactical actions. One of the directions in sports medicine has been formed, where sports doctors analyze the level of training of athletes on the basis of data arrays obtained from wearable devices and "smart" equipment. According to the study, in which text arrays of data from various sources were analyzed using the iFORA big data mining system, including scientific publications, patents, regulatory framework, market analytics, industry media, materials of international organizations, vacancies and other types of sources, the first place in the ranking of top technologies in the sports industry is occupied by decision support systems (DSS) in

Z. Hu et al. (Eds.): AIMEE 2021, LNDECT 107, pp. 538–545, 2022.
https://doi.org/10.1007/978-3-030-92537-6_49

sports [1, 2]. One of the main problems of modern sports, the solution of which is focused on applied research, which is essential for the management of sports training, is the problem of optimal construction of the training process and the generation of individual training plans.

2 Decision-Making Process in Sports Training

The system-structural approach in the training of athletes allowed to establish a relationship between the elements of the system, to determine its hierarchy, and on this basis it was proposed to divide the functions of management at different levels [3]. Considering the management process as a complex hierarchical system, three main levels were identified, which allow to present the entire system in full and to determine the relationship of impacts and controlled elements. The theory of three-level system of management of preparation of athletes was developed: 1st level – indicators of competitive activity; 2nd level – indicators of various parties of skill of athletes and 3rd level – indicators and estimates of a condition of the leading systems of an organism of athlete.

At the same time, the following subjects are distinguished in the training process management system that influence decision-making: the athlete himself, the coach (or the coaching staff, which includes: the head coach, the physical training coach, the technique coach, etc.), the sports doctor, the sports manager (team manager). There are other influencing factors, such as family, friends, rivals, team members, but we do not take them into account in this work. At the same time, the following subjects are distinguished in the training process management system that influence decision-making: the athlete himself, the coach (or the coaching staff, which includes: the head coach, the physical training coach, the technique coach, etc.), the sports doctor, the sports manager (team manager). There are other influencing factors, such as family, friends, rivals, team members, but we do not take them into account in this work.

For each sport, decision-making systems have both common features and differences specific to these sports. For individual sports, the decision-making process is as follows (Fig. 1). In this case, the coach is the decision-maker, and he must take into account a number of factors, both internal and external, plan a training program and bring the athlete to the peak of athletic form, taking into account training effects, such as acute, urgent, cumulative, set aside, residual effects.

The training process is organized in accordance with certain objectives, which are expressed in the amount of growth of sports results and determine the necessary for their implementation training program. The value of the increase in sports results is a criterion for the effectiveness of training. Streamlining external interactions and increasing the proportion of those forces that contribute to the successful achievement of goals is an essential condition for the growth of sports skills (Fig. 2).

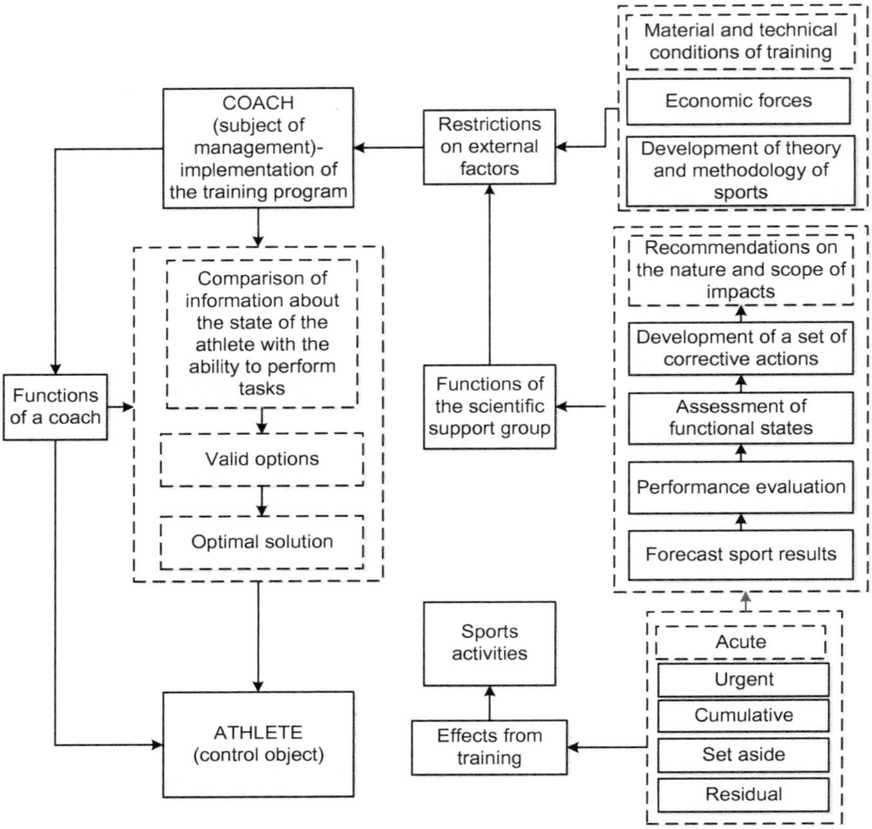

Fig. 1. The scheme of decision-making by the coach in the management of sports training

Decision support systems for solving selection problems in conditions of uncertainty and large error of the initial data, using fuzzy logic have a great prospect for the development of DSS in sports [4].

The method of multi-criteria optimization with the determination of the resulting objective function, which allows us to simultaneously consider various aspects of choice problems, was developed in and is successfully used to optimize training plans [5–7].

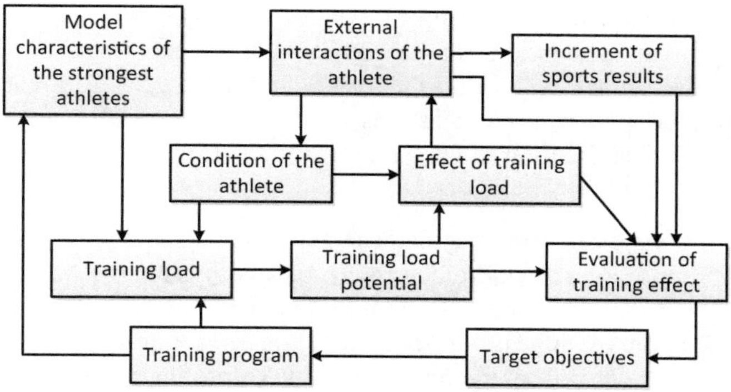

Fig. 2. Relationships in the training process

3 The Sequence of Making Coaching Decisions

Decisions in the course of the coach's activity consist of a cyclic sequence of steps.

Step 1. Determining the initial level of the athlete. The initial level means the vector of the functional state (values of the level of development of physiological systems that ensure the athlete's performance: maximum oxygen consumption of the aerobic and lactate thresholds, efficiency, kinetics of oxygen consumption, etc.), the level of sports results, technical level, finishing abilities (buffer capacity, strength indicators, kinetics of oxygen consumption, etc.), trainability (ability to respond to the load), psychological characteristics, etc. This vector can be constructed in different ways. It can be used in a variety of models.

Step 2. Setting the main goal for the season (macrocycle) – the desired sports result is decomposed into target indicators (components of the final state vector), expressed in scalar, interval, fuzzy values.

The timeline from the current moment to the moment of the main competitions.

Setting the main goal for the season is possible in 4 ways:

1) Coaching intuition.
2) Coaching experience.
   Coaching experience and intuition have limitations, and a period is needed to clarify the goal (for example, additional testing, studying the athlete's reactions to the load, calculating the adaptive profile).
3) The study of the model characteristics of an athlete from the point of view of comparison with the model characteristics of elite athletes (taking into account anthropometric, physiological, biomechanical indicators).
4) The use of the adaptation profile of previous years, or the selection of the initial adaptation profile from athletes of a similar level.

The final goal is specified in the process of calibrating the model, clarifying the characteristics of the sport, implementing a specific amount of training load, etc.

Step 3. Develop a training strategy. Periodization is determined (rough division into macrocycles, determination of the list of competitions and training camps).

Approximate determination of the total level of training load and its approximate distribution by orientation (within the blocks), as well as the formulation of intermediate goals for each block (and informing/discussing these goals to the athlete).

The preparation strategy is the least susceptible to changes. However, unforeseen circumstances (reduced funding of fees, illness, injury, etc.) may force a change in the training strategy.

Step 4. Planning the current mesocycle. The subsequent mesocycle is rescheduled in real time, based on the results of forecasting progress towards the target indicators in the current mesocycle. The principle of a sliding window is used when there is a smooth recalculation of the volume and direction of the training load. The dose-response model [9], as well as the adjusted adaptive profile, provide information for adjustment. At the planning stage of the subsequent mesocycle, the "alphabet" of training loads is used. Each training load is an independent object, characterized by orientation, performance conditions (running uphill, on sand, smoothly), intensity, duration, structure (interval, continuous training, etc.). The expert system accumulates the experience of experts (the best coaches) from the positions of training compatibility, conflict adaptation mechanisms, etc., i.e., there is a formalization in the form of restrictions on the properties of loads in the form of objects ("you can not give a high-intensity load after a day of rest"). The system generates a set of possible combinations, a sequence of training exercises according to the optimality criterion. The optimality criterion can be different – the lowest level of training impulse to achieve the tasks set, or to maximize the results at a given level of load. Since the vector characterizing the athlete's condition is multicomponent, a particular indicator (a set of indicators, for example, strength training) can act as a particular criterion of optimality.

Step 5. The implementation of the current mesocycle is carried out with parallel monitoring of the athlete's reactions, fatigue markers (biochemical analyzes, self-assessment of muscle pain, etc.), questionnaires of psychological and/or social stress, etc. While several options for taking into account the current state of the athlete for the implementation of the training process are possible. Firstly, a flexible training program based on an algorithm is possible, when a particular decision is made depending on the current functional state of the athlete. At the same time, the targets for training load (general or specific, for example, strength) and intensity must be met. This approach is called an adaptive training plan. Secondly, monitoring provides information about anxiety states that require prompt decisions on the training load and its type (for example, the cancellation of high-intensity training, reducing the duration and/or intensity of training), the introduction of a recovery microcycle or the beginning of a program of recovery measures. Thus, the cancellation or postponement of any training session will lead to an automated recalculation of subsequent training sessions of the current mesocycle, based on certain criteria (for example, minimizing the load, maximizing the result, etc.) in the current mesocycle. The relationship of different markers, their combination and dynamics make it possible to make decisions (a syndromic approach). Or we use alarm thresholds based on the available range of accumulated data (for a particular population of athletes). Other approaches to the formation of a decisive rule are also possible. The accumulation of individual data on markers and indicators of fatigue makes it possible to formulate refined "decisive rules" (thresholds in production models, fuzzy rules, neural networks, etc.). A training load is an object characterized by the following properties: orientation, duration, intensity, structure, compatibility with other types of load, relations with the previous and subsequent load. Actions can be performed on objects - they can be combined into chains taking into account restrictions on their properties, their properties can be added and changed, new load variants can be generated.

4 Quantification of the Training Load and Mathematical Models Describing Training Effects

Training load is a combination of intensity, duration and frequency of training. In the existing models of the development of training effects, there are three groups of key variables: quantified training load; quantified sports results; variables describing the state and dynamic characteristics of fatigue and fitness. One of the widely used methods of the so-called training impulses (TRIMPS) was developed by a group of sports physiologists. This indicator is calculated using the formula [8]:

$$TRIMP = txy \tag{1}$$

where t is the training session time in minutes, x is the heart rate reserve, y is the correction factor calculated as:

$$y = ae^{bx} \tag{2}$$

where y is the calculated concentration of lactate in the blood, x is the heart rate reserve, a and b are the parameters selected by the least squares methods.

The heart rate reserve is calculated as:

$$x = \frac{HR_{train} - HR_{rest}}{HR_{max} - HR_{min}} \tag{3}$$

where HR_{train}, HR_{rest}, HR_{max}, HR_{min} – heart rate during exercise, at rest, maximum and minimum, respectively.

In addition to the traditional one, there are also other methods for calculating this indicator: based on the lactate profile, based on generalized training zones, based on an individual oxygen consumption profile. Each method has its advantages and disadvantages.

Currently, many approaches to the mathematical modeling of sports training have been developed, at the same time, almost all of them are a "black box" and do not describe the details of the physiological processes occurring in the body when exposed to various physical loads.

The practice of sports training allowed us to identify certain effects observed in almost all athletes:

- The effect of supercompensation – an increase in performance and sports result, after some time after a single training load;
- Plateau effect-stopping the growth of sports results at a certain level and direction of a multi-day training load;

- Adaptive limit or "saturation" - the effect expressed in the fact that with a sufficiently large load, the results stop growing;
- The effect of overtraining – exceeding the adaptation limit, which leads to negative consequences, expressed in a decrease in sports results, illness or sports injury;
- The "rebound" effect-the term came from facmacology and is expressed in a sharp increase in performance and results with a decrease in training load;
- The effect of tolerance-also came from pharmacology and is expressed in the habituation of the body to the same training stimuli;
- The effect of load variability is a paradoxical effect, which is expressed in the fact that more diverse training programs lead to better athletic results compared to more monotonous ones with the same training volume.

  All these effects should be described by mathematical models. Existing models describe these effects, but, unfortunately, to varying degrees.

The classical model is a discrete model proposed by the physiologist Eric Bannister and developed by Thierry Busso and his followers, which is the following expression [9–11]:

$$p(t) = p(0) + k_1 \sum\nolimits_{i=1}^{t-1} e^{-(t-i)/\tau_1} w(i) - \sum\nolimits_{i=1}^{t-1} k_2^i e^{-(t-i)/\tau_2} w(i) \tag{4}$$

$$k_2^i = k_3 \sum_{i=1}^{t-1} w^j e^{-(i-j)/\tau_3} w(i)$$

where p – performance, g – function of fitness, h – function of fatigue, k_1, k_2, τ_1, τ_2 τ_3 – coefficients, which are called the athlete's adaptive profile.

The phenomena of saturation and the adaptation limit can be described by the following nonlinear equations having 7 parameters: k_1, k_2, τ_1, τ_2, α, β, σ [12].

$$\dot{g} + \frac{1}{\tau_1} g^{\alpha} = k_1 \sigma \tag{5}$$

$$\dot{h} + \frac{1}{\tau_2} h^{\beta} = k_2 \sigma$$

$$p = p_0 + g - h$$

If $\alpha = \beta \to 1$ the equations turn into a continuous analogue of the Bannister linear model.

Mathematical models with distributed delays have great prospects for explaining training effects with delay, but they are not currently implemented [13].

5 Conclusions

The architecture of decision support systems in individual sports should reflect the processes occurring in the system "coach-athlete-training conditions – training loads". At the same time, they should be able to solve several types of tasks that use models of the training process. The first task is to identify the model, when the parameters of the model, called the adaptive profile of the athlete, are estimated using various methods based on training loads and test data. The second task is to produce simulation modeling with known model parameters, which aims to check what kind of sports result certain training plans can lead to. The third task is to optimize the training plan according to criteria such as the maximum sports result and the minimum training volume. And in general, the appointment of DSS in individual sports should help the coach and the athlete to set and achieve goals, optimize training loads, while avoiding diseases and injuries.

References

1. HSE University, Research Departments, Institute for Statistical Studies and Economics of Knowledge. https://issek.hse.ru/news/484743102.html?fbclid=IwAR2cc_n3G7BpukteWqNWC4XQk9XlUeP0YtNvLyQxRxi09k4kSJBhASHzanM. Accessed 27 July 2021
2. Calof, J., Meissner, D., Vishnevskiy, K.: Corporate foresight for strategic innovation management: the case of a Russian service company. Foresight **22**(1), 14–36 (2020)
3. Mouratidis, A., Vansteenkiste, M., Lens, W., Sideridis, G.: The motivating role of positive feedback in sport and physical education: evidence for a motivational model. J. Sport Exerc. Psychol. **30**(2), 240–268 (2008)
4. Van Dinh, N., Thao, N.X.: Some measures of picture fuzzy sets and their application in multi-attribute decision making. Int. J. Math. Sci. Comput. **4**(3), 23–41 (2018)
5. Barabash, O., Shevchenko, G., Dakhno, N., Neshcheret, O., Musienko, A.: Information technology of targeting: optimization of decision making process in a competitive environment. Int. J. Intell. Syst. Appl. **9**(12), 1–7 (2017)
6. Faradian, A., Manjafarashvili, T., Ivanauri, N.: Designing a decision making support information system for the operational control of industrial technological processes. Int. J. Inf. Technol. Comput. Sci. **7**(9), 1–7 (2015)
7. Timme, E.A., Dayal, A.A., Kukushkin, Y.A.: Training plans optimization using approximation and visualization of Pareto frontier. In: Lames, M., Danilov, A., Timme, E., Vassilevski, Y. (eds.) Proceedings of the 12th International Symposium on Computer Science in Sport (IACSS 2019), pp. 69–76. Springer International Publishing, Cham (2020). https://doi.org/10.1007/978-3-030-35048-2_9
8. Borresen, J., Lambert, M.I.: The quantification of training load, the training response and the effect on performance. Sports Med. **39**(9), 779–795 (2009)
9. Banister, E., et al.: A systems model of training for athletic performance. Aust. J. Sports Med. **7**(3), 57–61 (1975)
10. Busso, T.: Variable dose-response relationship between exercise training and performance. Med. Sci. Sports Exerc. **35**(7), 1188–1195 (2003)
11. Clarke, D.C., Skiba, P.F.: Rationale and resources for teaching the mathematical modeling of athletic training and performance. Adv. Physiol. Educ. **37**(2), 134–152 (2013)
12. Turner, J.D., et al.: A nonlinear model for the characterization and optimization of athletic training and performance. Biomed. Hum. Kinet. **9**(1), 82–93 (2017)
13. Karabutov, N.: System with distributed lag: adaptive identification and prediction. Int. J. Intell. Syst. Appl. **8**(3), 1–7 (2016)

Computer-Aided Instruction for Efficient Academic Writing

Ekaterina Isaeva(✉)

Perm State University, Perm, Russia

Abstract. The paper tries to solve the problem of poor academic writing due to insufficient authors' experience in written academic communication in a foreign language, their poor knowledge of international academic writing culture, and underdeveloped language skills. This issue is especially relevant to the universities situated in non-English speaking countries. To solve the problem, we equip our readers and potential authors with user-friendly instruction of aiding their writing with computer tools, which allow for the automation of particular writing tasks. These include paper planning, text structuring, guided writing, assessment, and proofreading. Automation is achieved through the usage of open-source online academic writing services, freely distributed desktop programs, applications, and add-ons for browsers. Among them are CohVis, De-Jargonizer, Gingko, Grammark, Marking Mate (standard version), Paper rater, Rationale, Research Writing Tutor, Write & Improve, and Writefull. The paper also provides some case studies to illustrate the efficiency of the tools in focus in teaching and implementing academic writing, which makes up the paper's pragmatic value. All these tools have been checked for applicability and efficiency in the academic writing process, as well as consistency and digital security.

Keywords: Computer-aided instruction · Academic writing · Natural language processing · Paper structuring · Proofreading · Mind map · Concept map · IMRAD

1 Introduction

In the field of international academic communication, we still do not have a universal mediation methodology to enhance knowledge transfer and acquisition.

The authors face the problem of effective communication of their ideas and finding in a foreign language in an academic paper. This results in rejection of papers, which do not meet journal requirements by the publishers, avoiding writing papers in a foreign language (mainly in English) by the authors, concentration of local science within one language group, unavailable to the international scientific community.

Meanwhile, the current state of computer technology and artificial intelligence provides a great opportunity to automate routine tasks in academic writing.

In the paper, we provide user-friendly instruction on embedding opensource instruments for writing automation in international academic communication practice.

Z. Hu et al. (Eds.): AIMEE 2021, LNDECT 107, pp. 546–555, 2022.
https://doi.org/10.1007/978-3-030-92537-6_50

For this reason, the paper will be of interest to a wide range of academics, young researchers, and academic writing instructors.

The first problem is concerned with a different approach to planning the paper. This problem is especially relevant to the authors in the field of humanities, where the authors might be accustomed to the principles of creative writing and writing a story "from a scratch". The flow of their thought may be good for story writing but seem rather ambiguous to a foreign reader.

Secondly, the culture of text structuring is different. As the Head of the department of English for Professional Communication at the university, I came across two sides of the problem. On the one hand, students of different departments whom we teach English for Professional Communication are unexperienced in the well-known IMRAD (Introduction, Methods, Results, Discussion and Conclusion) structure or are not skilled enough in the usage of transitional words and special phrases for guiding the reader along their writing. On the other hand, the Department professors feel inconvenience in restructuring the logic of their thinking and presenting their ideas in the way a foreign reader is accustomed to.

Thirdly, the problem relates to paper reviewing, assessment, and proofreading. The usage of spelling checkers embedded into traditional text editing programs is a good habit of today's authors. These programs are being developed and extend their functions and capabilities, yet they do not cover all the problems relevant to the routine of writing a paper, e.g., style, logic, and field specifics and others. Additionally, the service does not include all the media of academic writing, i.e. do not work equally efficiently in the browser and email systems.

2 Methods

To write the paper we have used the following methods: estimation of accessibility of tools for free distribution, consistency, and efficiency and case study on the tools for academic writing.

2.1 Estimation of Accessibility of Tools for Free Distribution, Consistency, and Efficiency

We have analyzed 50 instruments (47 available through the database compiled by Carola Strobl et al. and 3 recommended by various centres for academic writing support for international University students) [1].

Among 47 resources included in the database, only 10 are currently available open-source for free distribution. Among them are CohVis, De-Jargonizer, Gingko, Grammark, Marking Mate (standard version), Paper rater, Rationale, Research Writing Tutor, Write & Improve, and Writefull. We have tried to access, install (if required), and apply for writing an academic paper all of the above. We have also classified the tools according to the major tasks they solve, particularly paper planning, structuring, guided writing, paper assessment, and proofreading. Some of the tools are multitasking, i.e. solve several problems, e.g. guided writing and assessment are often accompanied by proofreading mechanisms. Additionally, we extend the database with the tools we use in our academic writing practice, such as Mirro for paper planning, as well as Grammarly and Writing Mentor for grammar correction.

We have tried these tools for academic writing automation while teaching Russian master students at the course of Written Academic Communication in a Foreign Language and for writing our scientific articles to be published in English in international journals and book series. The case study has revealed qualitative improvement in academic writing including logic, style, word usage, grammar, and readability, both in students and academics.

2.2 Mind Map for Paper Planning

The idea of drawing a mindmap to be used for logical paper laying out is a reasonable solution to this problem.

Rationale is an online tool for developing your ideas in a graphical form. This might be effective at various stages of academic presentation, but we find it particularly reasonable to use it before writing a paper. Rationale provides instruments for making argument maps to structure arguments, analyze reasoning, identify assumptions, and evaluate evidence. The interface is intuitively clear, and the toolbar reminds ones in standard programs like MS Word and Paint.

Miro is becoming popular nowadays due to its rich functional capabilities and editing options. This resource reminds Rationale and is also convenient for visualising your ideas. It provides options for setting whiteboards like real ones and drawing mindmaps on them with linked shapes (fields), sticky notes, and the possibility to upload various multimedia. The resource is good for interactive collaborative work in the agile style.

2.3 Guided Writing for Paper Structuring

An effective solution to the problem of mismatch in the concept of structuring the paper might be guided writing.

Gingko Writer is an easy to master online tool for guided writing and paper structuring. The resource provides interactive templates for different tasks, including academic papers. The template is done in a form of a tree as a tree-based word processor. Each node or card has tips for writing and is linked to one or more child nodes. The nodes can be moved and rearranged. Each node is branched to further detailed cards [2]. The users write their papers on cards with further explication. The cards connected to one "branch" are highlighted, which makes the ideas traceable and helps to preserve clear logic during the whole paper.

For our case study, we have experimented with writing the current paper using Gingko. It is worthwhile mentioning, that this is our first experience of creating a complete piece of writing using this tool. The program proved its resourcefulness for guiding the writer along the process of paper writing and gradual development of the ideas according to the instructions provided on the templates. The visual representation of the process is given in Fig. 1.

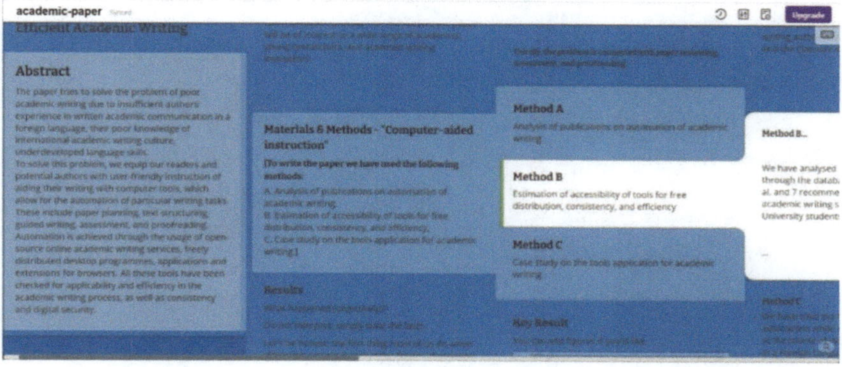

Fig. 1. Guided writing with Gingko

The resource provides a user-friendly interface for achieving cohesion in writing and improving your academic writing strategy.

Write & Improve is an online resource developed by the University of Cambridge. It helps every learner to improve their English writing [3]. This is a resource bank of interactive writing tasks sorted by levels. The paper is assessed automatically and feedback on where the writing may need improvement is provided. The changes are reassessed immediately. The service is valuable both for self-learning and blended teaching.

2.4 Proofreading Apps

There is a wide range of online services, desktop programs, applications, and add-ons for automated paper reviewing, assessment, and proofreading.

CohViz is a graphical feedback system for the automatic generation of concept maps as an external representation of the text to improve cohesion and self-regulation during writing [4, 5]. The software extracts the keywords from the text and uses automatic tagging to cluster the concepts. This helps to visualize the main topics of the text and the vocabulary used to elaborate the ideas.

We have tested the system using a short student's essay summarising NLP methods. We faced a technical issue when adding a multi-paragraph text. For this reason, all the paragraphs have been written as one. As a result, the following mindmap has been generated (see Fig. 2).

Fig. 2. CohViz concept map

In our sample text, we can see five nodes corresponding to the main areas discussed in the essay: article, analytic, method, and representation. Separately, are located the nodes of ranking, embeddings, and wordrank. We can infer, these three nodes need cohesive links, to express the main approaches to NLP addressed in the essay.

This tool is good for analyzing the visibility of the ideas and the adequacy of vocabulary for their representation.

De-Jargonizer is an online resource that helps to assess the lexical availability of the paper for a potential reader [6]. If the paper is written for an expert in NLP and data science, it is wise to include relevant terminology in these fields. Let us see analytics on the student's sample essay. To upload the text, one needs to choose a relevant period among 2012–2015, 2013–2016, 2014–2017, 2015–2019, 2016–2019. For our sample, we have chosen the period of 2016–2019. Then, the text is to be uploaded or pasted into a text box. The result is provided as a mark-up text with highlighted mid-frequency and jargon words. The results we got are as follows (Fig. 3).

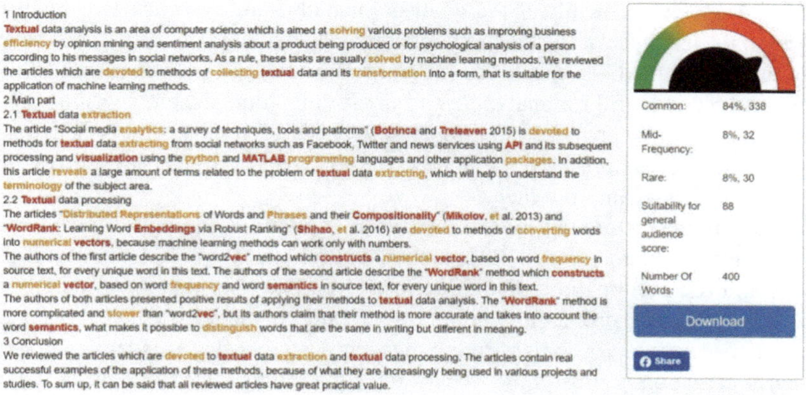

Fig. 3. De-Jargonizer text analysis

As is seen from Fig. 3, the text is suitable for the general audience due to its low terminological density. The developers believe, that "ideally readers, including second language readers, should be able to understand 98% of the words in a text… the writer should review the words in red and decide to either leave them, delete them or provide an explanation" [6].

Grammark is an online resource for checking the academic style, grammar, and readability of the document [7]. The instrument helps to identify potential errors connected with passive voice, wordiness, sentence length, transitions, academic style, nominalizations, spelling, and other grammar rules. The settings on the norms are customizable for humanities, science, and web publications, which is reasonable due to the difference in various genres' requirements. Let us check our sample (see Fig. 4).

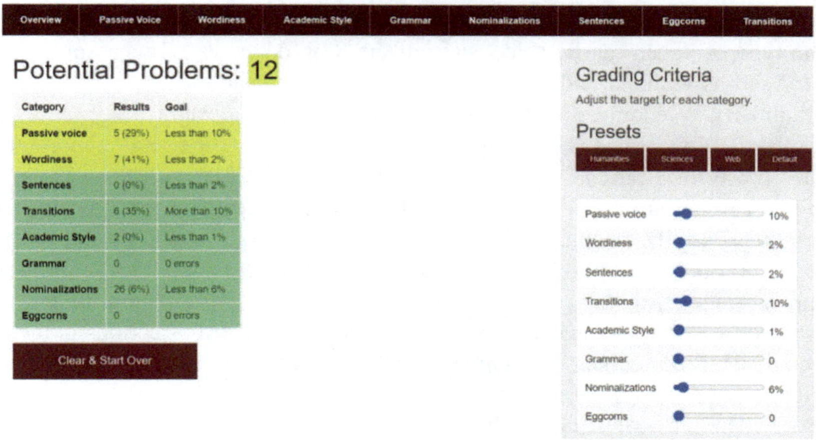

Fig. 4. Grammark proofreading.

Figure 4 shows the result of the analysis, namely, the overview tab. Each of the categories is clickable for further explication and available for the author's proofreading the highlighted issue, correction, and resubmission for further analysis.

The resource is useful for analyzing and improving your academic writing style and grammar skills, for it provides additional references to relevant tools and language norms. The developers preserve the open-code ethics and provide the source code for download and further improvement.

The next online instrument for text assessment and proofreading we will consider is Marking Mate: Standard Version. This is intended for checking academic writing (such as essays and reports) of more than 200 words [8]. The tool identifies potential style, spelling, and grammar problems. The reporting card contains such fields as paragraph rating, sentence rating, vocabulary rating, academic style rating, academic style rating, grammar/punctuation rating, general writing style rating. Marking Mate gives the overall score of the paper (maximum 12) and provides general (not personalized) tips on the problems found. For our sample example, the overall score is 9/12 with the problems connected with paragraph rating, academic style, grammar and punctuation, and general writing style. The recommendation to focus on paragraph and general writing style is given, the issues are highlighted, and specific tips for improving each case are available.

This tool seems particularly useful for academic writing teaching, developing self-assessment competence and academic stale and correct grammar awareness in students. The feedback is available for printing and could be sent to the teacher.

PaperRater is also available free of charge and does not require any downloads, signing up, or logging in. The resource employs artificial intelligence methods for grammar and syntax checks, writing suggestions, and plagiarism checking (https://www.paperrater.com/free_paper_grader). The feedback reports on bad phrases usage, transitional phrases score, the number of passive constructions, the simple sentence starts, usage of academic vocabulary, and contains analytics on the average length of sentences and the balance between simple and complex sentences. What we find especially convenient is personified approach to data analysis, i.e. the author's level and the type of paper can be selected from the drop-down list. This resource is also good for self-assessment and increasing style awareness and clear writing.

Grammarly is recommended by the Center for Communication Excellence, Graduate College of Iowa State University: "Grammarly is a digital writing-support tool that provides powerful grammar, spelling, and plagiarism checking based on natural language processing and machine-learning technologies" (https://cce.grad-college.iastate.edu/resources/writing-resources). The tool is available both as a browser and word-processor add-ons and desktop application. Installed as an app for a browser, it becomes a routine artificial proofreader of your online writing and contributes greatly to the author's literacy. A free version of Grammarly provides corrections in the issues connected with correctness (grammar, spelling, punctuation) and delivery (tone detection). The premium version contains much more useful features, such as clarity-focused sentence rewrites, compelling vocabulary, confident writing, formality level, inclusive language, plagiarism detection, and others.

Another tool recommended by Iowa State University is Writing Mentor [9]. This Google Docs add-on uses NLP methods to provide user-friendly chatbot support and intuitive feedback on the topic development, mood, coherence, editing, and convincing moves. All identified errors are highlighted, and instant feedback is available for every case. To analyse our sample essay, we need to choose between paragraph writing and extended writing and then among the essay, narrative, letter, and other. We choose an essay option. We receive the following basic feedback: well developed (the estimated topic is compositionality, which is supported with a list of keywords, including value, tools, numerical, distributed, etc.), the mood is positive (the positive terms are solving, improving, efficiency, successful, etc.; negative terms are problem and problems), coherent (the follow of ideas is analysed and marked through topic terms, transition words, the balance of long and short sentences, the usage of the title and section headers, and the use of pronouns), well-edited (grammar usage & mechanics, 15 potential problems identified; no problems with word choice or contractions). The advanced feedback states the text to be convincing (usage of hedges, boosters, standard vocabulary, claim verbs and synonyms). The author can apply changes according to the feedback and get an updated report.

An add-on for MS Word is designed for "academic and technical writing … Using language models trained on millions of journal articles, Writefull corrects grammar, wording, sentence structure, and more - aimed specifically at academic writing" [10]. To check if the add-on provides better features than those included in the text editor by default, we have applied it for the sample essay proofreading. Note, the text has been checked and edited by the author using the default MS Word corrector. As a result, we have 3 suggestions for the text improvement: deleting the comma before the "that"-clause, and 2 cases connected with the use of the so-called "Oxford" comma. The paragraph check has not returned any problems. Additionally, Writefull provides extra prompts with academic patterns sorted by the IMRAD structure and an external search for an academic vocabulary to be presented in a concordance form.

The proofreading features provided by Writefull are quite useful and are worth using in case the add-on does not slow down the work of the text editor. In our case, this has not influenced the processing power.

3 Results

As a result, we provide an updated feature specific list of open-source tools for computer-aided academic writing instruction. Computer-aided tools for organizing ideas and paper planning include Rationale and Miro. Online instruments for guided writing described in the paper are Gingko and Write & Improve. The list of proofreading options contains CohViz, De-Jargonizer, Grammark, Markin Mate: Standard Version, PaperRater, Grammarly, Writing Mentor, and Open Writefull.

Our case study has included writing the current paper with the usage of the tools described here, as well as assessing and proofreading a sample essay written by a non-English speaking Master student. As a result, we have determined that the tools described in the paper are applicable for the given tasks and provide diverse options for automation in academic writing.

4 Discussion

The paper has a pragmatic focus on using particular tools to solve the most urgent academic-writing tasks of a non-native language speaker. However, modern information technology provides a range of tools for improving particular language skills. For instance, there are systems for assessing English Speaking skills based on certain parameters, improving vocabulary, and suggesting relevant topics for a given text [11].

Other tasks, such as automatic summarization and referencing are also of great help for a digitally sophisticated writer. The former is implemented with modern software, which is smart enough to include important information of input documents, skip unnecessary and extra information, organize consecutive sentences coherently, and ensure their thematic relation [12]. The latter is effectively realized with such services as MS Word and its referencing function, EndNote by Clarivate, Mendeley by Elsevier, and others.

And finally, there are huge advances in the field of machine translation based on nonfinite machine learning from the user, who uses the opensource software online and provides his or her corrections to an automatically translated text.

5 Conclusion

Based on the experience received while working on the current paper, we see that automation of routine processes of academic writing is possible with the development of NLP and means of Artificial Intelligence. The results show that there are multiple tools for facilitating written communication in a foreign language. The paper provides only a few of them focusing on the problems of paper planning, structuring, and proofreading. The first problem can be solved with online tools for creating mind maps and concept maps. Using them will contribute to the development of the academic tradition of agile planning and conceiving paper writing as work on a project.

The problem connected with paper structuring is effectively solved through automated guided writing. The available online instruments provide clear instruction for paper development and clear structuring. Finally, the problem with poor style and literacy has received most of the developers' attention and can be solved with multiple tools for paper assessment and proofreading. We believe that mastering these tools will contribute to international academic communication and the exchange of scientific knowledge. More than that, the tools described can be used by academic instructors to teach students to develop and present their ideas clearly and logically, increase their style and literacy awareness.

References

1. Strobl, C., et al.: Digital support for academic writing: a review of technologies and pedagogies. Comput. Educ. **131**, 33–48 (2019). https://doi.org/10.1016/j.compedu.2018.12.005
2. Gingko Writer. https://app.gingkowriter.com/Hz1zA. Accessed 28 July 2021

3. Write and Improve. https://writeandimprove.com/. Accessed 28 July 2021
4. CohViz. http://132.230.54.216:8080/. Accessed 28 July 2021
5. Burkhart, C., Lachner, A., Nückles, M.: Assisting students' writing with computer-based concept map feedback: A validation study of the CohViz feedback system. PLoS ONE **5**(6), e0235209 (2020). https://doi.org/10.1371/journal.pone.0235209
6. De-Jargonizer. http://scienceandpublic.com/. Accessed 28 July 2021
7. Grammark. https://grammark.org/dist/#/. Accessed 28 July 2021
8. Marking Mate: Standard version. http://writingtools.xjtlu.edu.cn:8080/mm/markingmate.html#RESULTS. Accessed 28 July 2021
9. ETS's Writing Mentor. https://mentormywriting.org/. Accessed 28 July 2021
10. Writefull2021. https://www.writefull.com/writefullapp. Accessed 28 July 2021
11. Haji, E.E., Azmani, A.: Proposal of a digital ecosystem based on big data and artificial intelligence to support educational and vocational guidance. Int. J. Mod. Educ. Comput. Sci. **12**(4), 1–11 (2020). https://doi.org/10.5815/ijmecs.2020.04.01
12. Abdolahi, M., Zahedi, M.: Textual Coherence improvement of extractive document summarization using greedy approach and word vectors. Int. J. Mod. Educ. Comput. Sci. (IJMECS) **11**(4), 23–31 (2019). https://doi.org/10.5815/ijmecs.2019.04.03

Author Index

Z. Hu et al. (Eds.): AIMEE 2021, LNDECT 107, pp. 557–559, 2022.
https://doi.org/10.1007/978-3-030-92537-6